Contents

Preface vii

CHAPTER 1 **Data Analysis** 1

1.1   Some Thoughts about Models and Mathematics 2
1.2   A Modeling Activity 4
1.3   Using Scatter Plots to Analyze Data 10
1.4   Investigations Using Data Collection 20
1.5   The Median-Median Line 25
1.6   How Good is the Fit? 35
1.7   The Least Squares Line 40
1.8   Error Bounds and the Accuracy of a Prediction 49
Chapter 1 Review Exercises 53

CHAPTER 2 **Functions** 55

2.1   Functions as Mathematical Models 56
2.2   Characteristics of Functions 61
2.3   A Toolkit of Functions 64
2.4   Finding the Domain of a Function 73
2.5   Functions as Mathematical Models Revisited 79
2.6   Investigating a Conical Container 93
2.7   Investigating Graphs of Transformations of Functions 95
2.8   Basic Transformations of Functions 97
2.9   Combinations of Transformations 104
2.10 Composition of Functions 112
2.11 Investigating Graphs of Compositions of Functions 117
2.12 Composition as a Graphing Tool 119
2.13 Inverses 128
2.14 Using Inverses to Straighten Curves 138
2.15 Investigations Using Data Collection 146
2.16 Introduction to Parametric Equations 149
Chapter 2 Review Exercises 156

CHAPTER 3   **Exponential and Logarithmic Functions**   159

3.1   Functions Defined Recursively   160
3.2   Loans and the Binary Search Process   167
3.3   Geometric Growth Models   173
3.4   Investigating the Mantid Problem   180
3.5   Geometric Series: Summing Geometric Growth   182
3.6   Investigating Garbage Disposal   188
3.7   Compound Interest   190
3.8   Graphing Exponential Functions   197
3.9   Introduction to Logarithms   207
3.10 Graphing Logarithmic Functions   214
3.11 Solving Exponential and Logarithmic Equations   220
3.12 Logarithmic Scales   229
3.13 Data Analysis with Exponential and Power Equations   235
3.14 Error Bounds in Re-expressed Data   251
3.15 Investigating More Data Collection   255
3.16 Investigation: Assessing Your Model   262
Chapter 3 Review Exercises   264

CHAPTER 4   **Modeling**   267

4.1   The Tape Erasure Problem   268
4.2   Radioactive Chains   273
4.3   Free Throw Percentages   275
4.4   Choosing the Best Product   279
4.5   The Tape Counter Problem   288
4.6   Developing a Mathematical Model   295
4.7   Some Problems to Model   297

CHAPTER 5   **Circular Functions and Trigonometry**   301

5.1   The Curves of Trigonometry   302
5.2   Graphing Transformations of Trigonometric Functions   313
5.3   Investigating a Predator-Prey Relationship   321
5.4   Sine and Cosine on the Unit Circle   323
5.5   Getting to Know the Unit Circle   329
5.6   Angles and Radians   337
5.7   Solving Trigonometric Equations   350
5.8   Investigating Trigonometric Identities   358
5.9   Using Trigonometric Identities   360
5.10 Inverse Trigonometric Functions   366
5.11 The Double Ferris Wheel Investigation   373
5.12 Composition with Inverse Trigonometric Functions   376
5.13 Solving Triangles with Trigonometry   383
5.14 Investigating Hanging Pictures   396
Chapter 5 Review Exercises   398

CHAPTER 6   **Combinations of Functions**   401

6.1   Introduction to Combinations of Functions   402
6.2   Investigating Sums and Products of Functions   404
6.3   Sums and Products of Functions   408
6.4   Investigating CO_2 Concentration   421
6.5   Investigating Beats   424
6.6   Introduction to Polynomial Functions   426
6.7   Investigating Polynomial Functions   431
6.8   Polynomial Functions   433
6.9   Rational Functions   442
6.10   Application of Rational Functions   456
6.11   Investigating a Traffic Flow Model   461
Chapter 6 Review Exercises   464

CHAPTER 7   **Matrices**   467

7.1   Introduction to Matrices   468
7.2   Matrix Addition and Scalar Multiplication   470
7.3   A Common-Sense Approach to Matrix Multiplication   474
7.4   Computer Graphics   485
7.5   The Leontief Input-Output Model and the Inverse of a Matrix   495
7.6   Additional Applications of the Inverse of a Matrix   509
7.7   The Leslie Matrix Model   512
7.8   Markov Chains   523
Chapter 7 Review Exercises   541

Appendix A: Complex Numbers   543
Appendix B: Derivations of Linear Least Squares Parameters   545
Answers to Selected Exercises   547
Index   575

Dedication

To Henry Pollak for his inspiration and support and to the students of NCSSM who have explored mathematics with us.

Preface

Contemporary Precalculus through Applications provides students with an applications-oriented, investigative mathematics curriculum in which they analyze complex situations and use technology to solve problems and to enhance their understanding of mathematics. The topics presented lay a foundation to support future course work in mathematics including calculus, statistics, discrete mathematics, and finite mathematics. The topics also provide an introduction to the mathematics used in engineering, the physical and life sciences, business, finance, and computer science. Whenever possible, new material is presented in the context of real-world applications. Students are active learners who generate ideas for both the development of problem statements and for the solution of problems. They learn to use a variety of techniques to solve problems that are investigated from a number of perspectives as students proceed through the course. Since problems are presented in the context of real-world applications, the interpretation of solutions is given strong emphasis.

The goals of *Contemporary Precalculus through Applications* parallel those of the National Council of Teachers of Mathematics *Standards,* a document developed simultaneously with this textbook. A primary goal of the authors is to foster the development of mathematical power in students. Other goals for students addressed include: exposure to real-world applications of mathematics in a wide variety of disciplines so that students can learn to value mathematics; preparation for future course work in mathematics; development of the self-confidence necessary to undertake further study; development of collaboration skills, and opportunities to use modern technology to enhance understanding and to solve problems.

The fabric of *Contemporary Precalculus through Applications* is woven from five spiraling themes treated with increasing depth and breadth at each exposure as follows.

Mathematical Modeling

The use of mathematics to model a wide variety of phenomena is central to the course. The modeling approach is used in analyzing sets of data, introducing and applying the various elementary functions, and applying matrices to various problem-solving situations. Additionally, the modeling approach provides motivation for student study. Problem situations are often presented in such a way that the student must supply the mathematical framework required in the solution process.

Computers and Calculators — Technology as Tools

Technology has lessened the need for extensive computation using pencil and paper techniques. Now students can focus on mathematical concepts and structure while calculators or computers carry out the computations. In this text, calculators are used in evaluating expressions, in applying numerical algorithms, in matrix calculations, in presenting a graph or table for gaining insight or solving problems, and as an important aid in making conjectures.

The graphing calculator and the computer are used for calculations related to functions and matrices and also for quick and accurate graphing. Empirical models are central to the course, and the graphing calculator and the computer are used to develop such models from actual data (which students sometimes gather themselves) through re-expression and curve fitting. Throughout, the design of the material is based on the assumption that appropriate technology is available to all students. This might include a single microcomputer available for demonstrations by the teacher, graphing calculators for individuals, or a computer lab for class and individual use.

Applications of Functions

The overall goal of the study of elementary functions is to illustrate how functions serve as bridges between mathematics and the situations they model. The study of specific functions is motivated by the need for tools to build empirical models and to approximate trends in data. The importance of compositions and inverses of functions is heightened by the frequency of their use in building mathematical models. The focus on understanding the behavior of functions leads to an emphasis on graphing, using hand-drawn sketches and graphs created by technology computer or graphing calculator. The geometry of the functions is often used to enhance understanding of the algebra.

Data Analysis

The principal goal of data analysis is to deduce information from data. A conceptual treatment of resistant and least squares techniques of curve fitting is given here. Students are reminded that observation and measurement in the real world result in values for variables, rather than formulas for functions. Techniques of data analysis allow students to uncover what, if any, functional relationship exists between the relevant variables. Re-expression of data by means of elementary functions is helpful in extracting information and gaining insight from data. Through discussions of residuals and causation versus association, students learn that predictions based on techniques of data analysis have limits to their accuracy and reliability.

Discrete Phenomena

Because discrete techniques are required for the mathematical analysis of many phenomena, discrete mathematics topics are included in this text. Some of these topics include recursion, finance, population growth, economic production, and Markov chains.

Contemporary Precalculus through Applications is designed to encourage students to approach mathematics in new and innovative ways. A conscious effort is made to combine several of the major themes in examples and problems, to ask familiar questions in new contexts, and to apply new concepts to familiar questions. Each theme that spirals through the course has an effect on the instructional approach, but none more than the use of the computer and graphing calculator as tools. Regularly having access to technology during discussions — using a computer in front of the classroom or regularly using a graphing calculator — enables students to ask and answer "What if …?" questions, to make conjectures, to check their guesses and analyses, and to work with real data. All these activities are invaluable contributions to the learning process.

The models that are developed in the course come from many diverse areas. Naturally, models from the sciences are included, but in addition, models from banking and finance, anthropology, economics, sociology, sports, and environmental issues also appear. The applications vary in complexity and depth and are often revisited as new techniques are learned. For example, characteristics of the simple exponential function are investigated and applied. Later, data analysis is used to model the phenomenon of population growth, with the computer or calculator assisting in the consideration of long-run expectation.

The course is not designed for a head-down march through the syllabus. The authors expect and have worked to create opportunities for discussions in class, additional questions, and reflection. For some problems there is no single "best" answer. The course has been constructed with the philosophy that the quality of learning is more important than the completion of a syllabus.

The Second Edition

The second edition of *Contemporary Precalculus through Applications* is the result of rethinking the curriculum based on students having daily access to technology as a tool in learning mathematics and a greater commitment to put mathematics in context. The problem solving and technology capabilities of our students inspired the development of new activities related to the curriculum contained in the first edition. These capabilities also inspired the addition of new topics that were not included in the first edition. As calculators have become more readily available and more powerful and as students have expanded their power as mathematicians, precalculus students have been enabled to do far more than we first envisioned. Consequently, the list of topics for study and activities for investigation and study has been expanded. The following are important new changes in this edition:

- Data analysis is integrated in all topics of study. The course begins with an introduction to data analysis using both the median-median line and the least squares line. The study of data analysis spirals throughout the course and is enhanced with the study of each new function.
- Investigations provide students with active opportunities for hands-on exploration, conjecture and discovery, and extended problem situations. These investigations provide valuable class activities that can be done in small groups, as projects, or as extended writing assignments.
- Recursive functions and the study of geometric series introduce students to the exponential function.
- Parametric equations are included in the study of functions, trigonometry, and combinations of functions.
- The study of combinations of functions encompasses all types of functions. This topic is not limited to traditional polynomial and rational functions. Combining all types of functions allows students to develop sophisticated functions for describing complex real-world situations.
- The variety of problems in the Modeling chapter provides an extensive resource of real-world applications that can be studied throughout the course.
- The strong emphasis on data analysis and modeling provides excellent preparation for students who take the Advanced Placement Statistics course following the precalculus course.

The text structure of the new edition enhances opportunities for group activities, writing, lively class discussions, and projects for students. Whether using problems and questions from class practice, exercise sets, or investigations, teachers and students will find many opportunities for a variety of classroom activities.

Acknowledgments

The first edition of this textbook was developed with funding from the Carnegie Corporation of New York and the National Science Foundation. New topics, investigations, and problems of this edition were developed for our students and were often shared with other teachers through our newsletter, Teaching Contemporary Secondary Mathematics. This newsletter was produced as part of the NCSSM Teacher-Leader program that was funded by the National Science Foundation. Any opinions, findings, and conclusions or recommendations expressed through this support are those of the authors and do not necessarily reflect the views of the Carnegie Corporation or the National Science Foundation.

Three chapters of the first edition were published previously by the National Council of Teachers of Mathematics. NCTM has given permission for those chapters to be included in this textbook.

A number of colleagues at NCSSM gave us support and help throughout the revision. Special thanks go to Julie Allen, Laurie Cavey, Peggy Craft, Robin Cunningham, Tracey Harting, Maria Hernandez, and Marilyn Schiermeier. Terry Brown tirelessly supported the production of the text for our students and for publication. The teacher-leaders of our NSF-funded institutes spent three summers at NCSSM and provided valuable feedback during the development of new ideas. The annual Teaching Contemporary Mathematics Conference gave the authors the additional opportunity to discuss and share new ideas with our long distance colleagues.

We thank our publisher, Everyday Learning Corporation, for its confidence in our work and its support in producing a significantly different new edition.

1 Data Analysis

1.1	Some Thoughts About Models and Mathematics	2
1.2	A Modeling Activity	4
1.3	Using Scatter Plots to Analyze Data	10
1.4	Investigations Using Data Collection	20
1.5	The Median-Median Line	25
1.6	How Good Is the Fit?	35
1.7	The Least Squares Line	40
1.8	Error Bounds and the Accuracy of a Prediction	49
	Chapter 1 Review Exercises	53

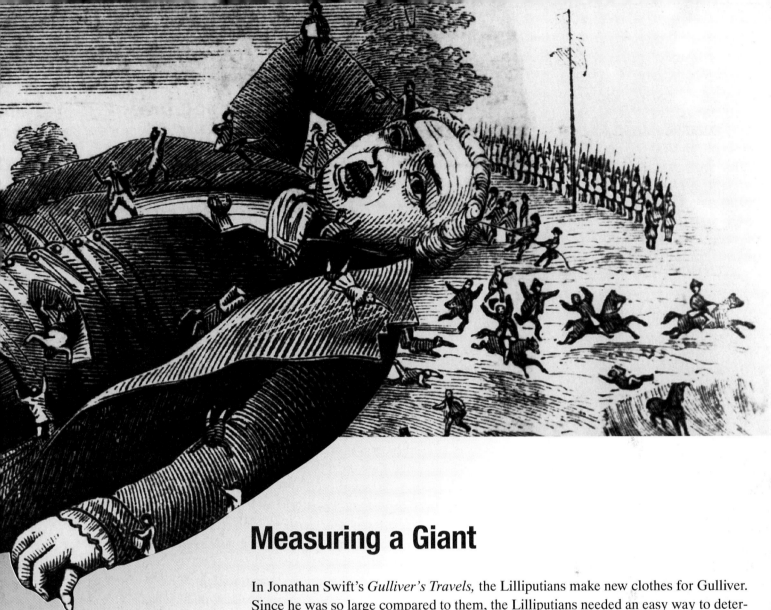

Measuring a Giant

In Jonathan Swift's *Gulliver's Travels,* the Lilliputians make new clothes for Gulliver. Since he was so large compared to them, the Lilliputians needed an easy way to determine his measurements. According to Gulliver, "Then they measured my right thumb, and desired no more; for by a mathematical computation, that twice round the thumb is one round the wrist, and so on to the neck and waist; and by the help of my old shirt, which I displayed on the ground before them for a pattern, they fitted me exactly."

Are the Lilliputians correct? Is it true for people in general that twice the circumference of the thumb is equal to the circumference of the wrist, and twice the circumference of the wrist is equal to the circumference of the neck, and twice the circumference of the neck is equal to the circumference of the waist?

Some Thoughts About Models and Mathematics

When children think about models, they generally consider some kind of a toy, perhaps an airplane or a dinosaur. When scientists and mathematicians think about models, they are generally considering a model as a tool, even though they may be thinking about the same airplane or dinosaur. Scientists and mathematicians use models to help them study and understand the physical world. People in all walks of life use models to help them solve problems; problems in this course will involve models used by bankers, anthropologists, geologists, and many, many others.

What is a model?

So, just what is a model? Models are representations of phenomena. To be useful, a model must imitate important characteristics of the phenomenon it represents, and it must also be simpler than what it represents. A model usually differs significantly from what it represents, but these differences are offset by the advantage that comes from simplifying the phenomenon. A good example is a road map, which models the streets and highways in a particular area. Clearly, a map has a lot in common with the actual streets and highways: It shows how roads are oriented and where they intersect. A road map simplifies the situation; it ignores stoplights, hills, and back alleys and instead focuses on major thoroughfares. Such a map is very useful for traveling from one city to another but is not very good for finding the quickest route to the shopping mall or the best street for skateboarding. Road maps, and most other models, are useful precisely because they ignore some information and thereby allow you to see other information more clearly.

Another fairly common model is an electrocardiogram (EKG), which models the electrical activity of the heart. The EKG is an excellent model when used to determine the heart rate or to find which regions of the heart may be damaged after a heart attack. It is not a useful model for determining the volume of blood flowing through the heart. Different models emphasize different aspects of a phenomenon; the choice of which model to use depends on which aspect is under investigation.

The ability to predict is the ultimate test for a model. A good model allows us to make accurate predictions about what will occur under certain conditions. If what actually occurs is consistently different from our predictions, then the model is of little use. Scientists and mathematicians often update or revise models as they learn more about the phenomenon under study. Sometimes a model needs to be completely discarded and replaced with a new one. For example, before Columbus sailed to the Americas, many people believed the world was flat, but that model was quickly abandoned in light of new information.

Even though Isaac Newton's models for the actions of a gravitational field have been replaced by Einstein's relativistic model, we still use Newtonian physics in many situations because it is easier and because it gives reasonably accurate results. The aspects of Einstein's mechanics that are ignored are largely irrelevant in most everyday applications, so the Newtonian model is still a good one.

As we proceed through this course, we will encounter phenomena that we will want to know more about. Our task will be to find a mathematical expression or a graph that mimics the phenomenon we are interested in. This model must accurately represent the aspects of the phenomenon that we care about, but it may be very different from the phenomenon in other ways. To find a representative model requires a large toolkit of mathematical information and techniques. The fundamental concepts of algebra and geometry are all a part of that toolkit. The graphing calculator and computer will serve as tools to construct and analyze models for the phenomena we study. Probably the most important tools necessary for making models are an inquisitive mind and a determined spirit.

Often we will not stop after one model has been developed but will form two or three models to get a better view of the subject. For example, suppose a rock is thrown into the air. How can its height be modeled? One way to model this phenomenon is with the graph in Figure 1.

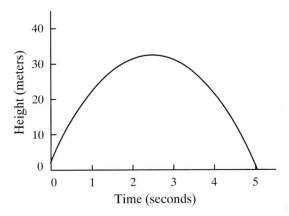

Figure 1   Height of a rock over time

Another way to describe the height of the rock with respect to time is with an equation. If the height is represented by h and the time is represented by t, then our model is

$$h = -5t^2 + 25t + 2.$$

Notice that these models do not give complete information about the problem. We cannot tell from the models what type of rock was thrown, who threw it, or why. These aspects of the problem are not relevant, since the main question concerns height and time. Both models provide this information. What distinguishes one model from another is what each model emphasizes and what each model ignores.

A Modeling Activity

The process of developing a mathematical model is often challenging, and it is almost never a one-step process. In modeling, it is important that you think about both the mathematics you know and the phenomenon you are trying to model. Moving back and forth between the two in a thoughtful, organized manner is essential. Precalculus is the study of the basic functions that are used to describe our world. Throughout this course, you will learn how to use your growing knowledge of functions to model real-world situations. In this section, you will work through a sample modeling problem to demonstrate some useful modeling techniques. You are not expected to be able to do this problem by yourself at this point. By the end of the course, however, you should be comfortable with the modeling process and confident in your ability to solve problems similar to this one.

EXAMPLE 1   **Determining the Best Ticket Price**

The senior class at the local high school wants to raise money to support the athletic program by selling tickets that will allow a family to attend all athletic events at the school. The class officers are trying to decide the price for a single ticket. Some students argue for setting the price low, believing that a low price would bring a large response. Others want to set a higher price, so that even if not many tickets were sold, the school would still make money. The students decide to ask the parents what they would be willing to pay for an all-sports ticket. Students assume parents want the sale to be a success and, therefore, parents will give them accurate information. A survey is sent to all 811 families with students in the school asking, "What is the most you would be willing to pay for an all-sports ticket good for this school year?" The results are given in Figure 2.

Figure 2   Results of the survey

Maximum price ($)	50.00	75.00	90.00	95.00	115.00	135.00	150.00	175.00
Expected ticket sales	145	80	45	85	120	80	60	150

Take a minute to consider the following questions that will help in organizing your thinking. What do you expect to be the relationship between the price the students set for the tickets and the response to the sale? How can the class officers use this information to determine the "best" price for a single ticket? Imagine that you are in charge of the sale and it is your responsibility to determine the price of the tickets. Where do you begin?

To determine the price that will bring in the most money to the school, you need to develop a mathematical model relating ticket price and total revenue, or amount of money brought in by the sale. To develop this model, you must understand the information provided by the data collected. What information about the parents and their

scatter plot

support for the sale is contained in the data? Do the data support the conjecture that the more the ticket costs, the fewer families would be interested?

Information that is presented as a list of numbers is often difficult to evaluate. One step in analyzing the relationship between two variables is to make a *scatter plot*. A scatter plot is simply a graph in a rectangular coordinate system of all ordered pairs of data. A scatter plot displays data so we can see the general relationship between two variables. The relationship (or lack thereof) should be more obvious if we plot the data. When making a scatter plot, either variable may be plotted on either axis. If we suspect that one variable depends on the other, however, we usually plot the dependent variable on the vertical axis and the independent variable on the horizontal axis. Figure 3 shows a scatter plot of the ordered pairs (*maximum price, expected ticket sales*).

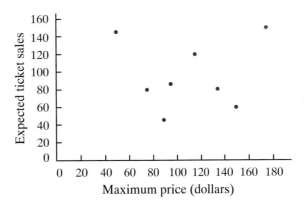

Figure 3 Scatter plot of the data in Figure 2

It is always a good idea to plot the data, although this particular plot does not seem to give us much useful information. There is no obvious pattern in the data. Perhaps we need to think harder about what the data are telling us.

Did all of the families respond to the survey question? If a family did not respond, what does this mean about their interest in the tickets? All models begin with some simplifying assumptions. One way to think about the families that didn't respond is to assume that they are not interested in supporting the athletic program and will not buy a ticket at any price. There are other assumptions we could make about those families that didn't respond to the survey, of course, and we will consider some of them in the homework exercises. For now, assume that only those families that responded will purchase a ticket. With this assumption, we can interpret the data.

According to the results of the survey, 150 families are willing to pay as much as $175 for each ticket. How many families will be willing to pay $150 per ticket? There are 60 families willing to pay up to $150, and families willing to pay $175 will certainly be willing to pay $150 for each ticket. Therefore, we would expect $150 + 60 = 210$ families to purchase tickets priced at $150.

Continuing to sum each successive number of ticket buyers, you can create a new table (Figure 4) showing how many families you can expect to purchase tickets at each price. If you knew how many families would buy a ticket at each price, then you can use that information to predict the price that will bring in the most money.

Figure 4   Price and cumulative ticket sales

Price ($)	50.00	75.00	90.00	95.00	115.00	135.00	150.00	175.00
Cumulative ticket sales	765	620	540	495	410	290	210	150

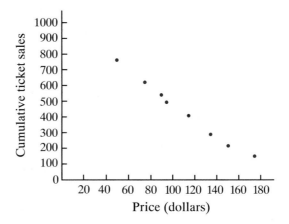

Figure 5   Scatter plot of the data in Figure 4

The scatter plot of the data, shown in Figure 5, provides useful information about the relationship between the ticket price and the number of tickets the students can expect to sell. This graph confirms that higher prices result in lower sales. If we can find a mathematical equation relating price and number of tickets students expect to sell, we can approximate the number of tickets that will be sold at prices that are not on the list.

The general pattern of the points in the scatter plot is linear. If you place a pencil over the graph of the data, the pencil does a good job of modeling the relationship between the two variables. (See Figure 6.)

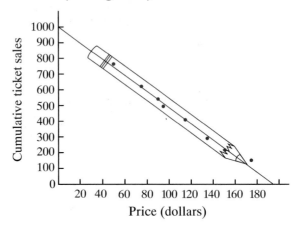

Figure 6   Pencil model

What is the equation of the line represented by the pencil? There are a number of ways to find an equation of a line that does a good job of modeling the data set. We will look at two standard techniques later in this chapter. For now, a quick estimate will do. It appears from the graph that the *price*-intercept (where the ticket sales are zero) is approximately $195. The *ticket*-intercept (where the price is zero) is approximately 1000 tickets. This means that the two points (195, 0) and (0, 1000) lie on our line. The equation of the line passing through these two points is

$$Tickets = -\frac{200}{39} \cdot Price + 1000.$$

Using function notation, we say that the number of tickets sold, T, is a function of price, P, and write

$$T(P) = -\frac{200}{39}P + 1000.$$

Using this function, we can predict how many tickets will be sold for any particular price. For example, if the price is $60 per ticket, the school expects to sell $T(60) \approx 692$ tickets. If the students charge $110 per ticket, the school can expect to sell only $T(110) \approx 436$ tickets.

When working on multistep problems, it is easy to lose your focus and forget how the process helps you reach the final goal. It is important to stop periodically to compare where you are in the process of solving the problem to the original goal. The students want to find a relationship between the price they charge and the revenue from the sale of the tickets. They want this model so they can determine the price to charge to make the most money. The function $T(P) = -\frac{200}{39}P + 1000$ doesn't answer this directly. It only indicates how many tickets the students can expect to sell for a specified price. It is important to note that this is not what they wanted to find, but it is what they could find from the data they gathered. You can now use this function to address the question whose answer the students really want to know; that is, what price will bring in the most money?

If the students charge $60 per ticket, they can expect to sell about 692 tickets, with a revenue of $41,520. If they sell the tickets at $110 each, they may sell only about 436 tickets, but that would bring in $47,960. So, a price of $110 is better than $60. The revenue expected from the sale of the tickets is the product of the number of tickets students expect to sell, given by $T(P)$, and the price, P. In function notation, the revenue, R, generated from the sale of $T(P)$ tickets is

$$R(P) = P \cdot T(P) = P\left(-\frac{200}{39}P + 1000\right).$$

In this case, revenue is a quadratic function of price. A view of the revenue function is given by looking at its graph (Figure 7) on the next page, which you will recognize as a parabola.

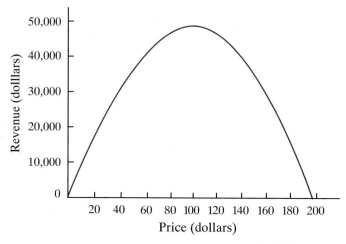

Figure 7 Graph of revenue function $R(P) = P\left(-\frac{200}{39}P + 1000\right)$

To find the price that will generate the maximum revenue, recall what you learned about quadratic functions in earlier courses. A parabola has a vertex, which in this problem represents the maximum revenue for the students' project. You could use your calculator to approximate the coordinates of the vertex, but it is quicker to recall that the vertex of a parabola is mid-way between the zeros. For this function, $R(P) = 0$ at $P = 0$ and $P = 195$, so the vertex is located at $P = 97.5$. Thus, the students should charge \$97.50 for the tickets. If they charge \$97.50, they can expect to sell about 500 tickets and receive \$48,750 in revenue. ▪

Now review the process that helped us arrive at a solution in this problem. We were faced with a question, "What price should the students charge to maximize the money brought in by the project?" The students had some ideas about the relationship between price and participation. They believed that the more a ticket costs, the fewer tickets they would sell. But, these ideas were not quantified. So, we needed to determine how much participation would drop with each dollar increase in price.

To quantify this relationship, we looked at the information from the parents' survey. However, the data generated from the survey did not directly lead to the desired relationship. To find the number of tickets students could expect to sell at each price, we had to create a new data set by accumulating the survey data. After creating a data set that represented this relationship, we looked at the graph of the scatter plot and observed a linear pattern. We then used our knowledge of lines to fit a linear model to approximate the number of tickets sold at prices that were not on the survey.

Once we had a linear model for the expected level of participation, we used this to generate a quadratic function that modeled the expected revenue. Last, knowledge about quadratic functions helped us find the optimum price and the maximum revenue.

Throughout the process, we had to stay focused on the question at hand. Sometimes we made progress by using mathematics, and at other times we used our understanding of the problem's context. In this particular problem, the mathematics of finding the equation of a line and finding the vertex of a parabola should be familiar.

However, the process of modeling and knowing when and how to use those mathematical techniques may be new. Don't be concerned if you could not have done this problem on your own. Learning precalculus mathematics and how to use that mathematics in problem settings such as this one is what this course is all about.

Exercise Set 1.2

1. In the linear equation for the number of tickets sold, $T(P) = -\frac{200}{39}P + 1000$, interpret the meaning of the slope, the P-intercept, and the T-intercept in the context of the model.

2. What question should the students ask if they want to generate the values in Figure 4 directly from their questionnaire?

3. Consider the responses received from the questionnaire as representative of all 811 families. Those who didn't return their surveys are still interested, but they either forgot or didn't have time to fill out the survey. In this case, assume that the number of interested families follows proportionally from the results of the survey. That is, since 80 of the 765 responses, or 10.46%, reported that the most they would pay would be $75, then 10.46% of the total population of 811 families would be willing to pay at most $75 for the tickets. Rework the problem based on this assumption and determine the "best" price.

SECTION 1.3   Using Scatter Plots to Analyze Data

Graphical models will serve as a first step in analyzing data. These models provide information so you can answer questions such as the following:

- How are heating bill costs related to average temperature?
- What will be the winning speed for the New York Marathon in the year 2010?
- Is there a relationship between the amount of time a student spends studying for a test and the grade he or she gets on the test?

Though the questions posed above are different in many respects, each requires the collection, organization, and interpretation of data. To answer each question, we need to analyze the relationship between variables. This course will concentrate on paired measurements, that is, the analysis of two variables. Sometimes one variable depends on the other. For example, we expect that blood pressure in adults of the same height in some way depends on their weight, and that crop yields depend on the amount of rainfall. At other times there is a relationship between the variables, but it is not one of cause and effect, or dependence. For example, we might show that there is a relationship between points scored and personal fouls committed by college basketball players, but one of these variables is not dependent on the other. Sometimes there is no relationship at all between the two variables, such as the distance a student lives from school and his or her height.

To determine whether there is a relationship between two variables, we must analyze data consisting of ordered pairs. Sometimes these data are gathered from a well-designed, carefully controlled scientific experiment, and at other times the data may exist in the world around us.

EXAMPLE 1   Average Test Scores

The data in Figure 8, provided by the College Board World Wide Web site, shows the average test scores for high school seniors graduating in 1996 and the percentage of graduates taking the test from each state.

Do you think there is a relationship between the average test score and the percentage of graduates taking the test? Study the data in Figure 8 to determine whether or not you think a relationship exists.

Figure 8 Average test scores by state and percentage of graduates taking the test in 1996

State	Verbal	Math	Total	%	State	Verbal	Math	Total	%
Alabama	565	558	1123	8	Montana	546	547	1093	21
Alaska	521	513	1034	47	Nebraska	567	568	1135	9
Arizona	525	521	1046	28	Nevada	508	507	1015	31
Arkansas	566	550	1116	6	New Hampshire	520	514	1034	70
California	495	511	1006	45	New Jersey	498	505	1003	69
Colorado	536	538	1074	30	New Mexico	554	548	1102	12
Connecticut	507	504	1011	79	New York	497	499	996	73
Delaware	508	495	1003	66	North Carolina	490	486	976	59
District of Columbia	489	473	962	50	North Dakota	596	599	1195	5
Florida	498	496	994	48	Ohio	536	535	1071	24
Georgia	484	477	961	63	Oklahoma	566	557	1123	8
Hawaii	485	510	995	54	Oregon	523	521	1044	50
Idaho	543	536	1079	15	Pennsylvania	498	492	990	71
Illinois	564	575	1139	14	Rhode Island	501	491	992	69
Indiana	494	494	988	57	South Carolina	480	474	954	57
Iowa	590	600	1190	5	South Dakota	574	566	1140	5
Kansas	579	571	1150	9	Tennessee	563	552	1115	14
Kentucky	549	544	1093	12	Texas	495	500	995	48
Louisiana	559	550	1109	9	Utah	583	575	1158	4
Maine	504	498	1002	68	Vermont	506	500	1006	70
Maryland	507	504	1011	64	Virginia	507	496	1003	68
Massachusetts	507	504	1011	80	Washington	519	519	1038	47
Michigan	557	565	1122	11	West Virginia	526	506	1032	17
Minnesota	582	593	1175	9	Wisconsin	577	586	1163	8
Mississippi	569	557	1126	4	Wyoming	544	544	1088	11
Missouri	570	569	1139	9					

How did you get information from the list of numbers in Figure 8? Did you actually read all of the data, or did you look for your state and then skip to this paragraph? Presented as just a table of numbers, the data are difficult to interpret. There should be some way of organizing and simplifying the data set so that its essential characteristics are apparent. Then you can decide whether or not average test scores are related to the percentage of students taking the test.

The scatter plot in Figure 9 on the next page shows average total test scores plotted on the vertical axis and the percentages of graduates taking the test from each state plotted on the horizontal axis. For example, the ordered pair representing North Carolina, (59, 976), indicates that 59 percent of the 1996 high school graduates living in North Carolina took the test and their average score was 976.

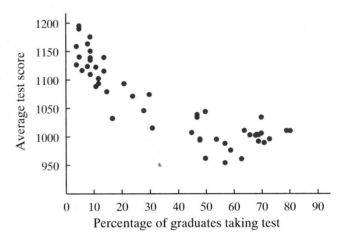

Figure 9   Average test scores and percentage of graduates taking test in 1996

negative association

The scatter plot in Figure 9 convinces us that there is a relationship between these variables. The points tend to slope downward to the right, so the states with a higher percentage of graduates taking the test generally have lower average test scores. When one variable decreases as the other increases, we say there is a *negative association* between the two variables. Notice also that the states with the lowest percentages of students taking the test have the highest average test scores. Can you give a plausible explanation for this? Which display do you find easier to interpret, the table or the graph?

A scatter plot is an effective tool for analyzing data. Special characteristics of data that may go unnoticed in a table are more obvious from a graph. If there is some relationship between the variables, a pattern or trend is usually apparent in the scatter plot. In this case, the scatter plot shows some spread in the data. Though the variables are related, we would have to consider the relationship somewhat loose, or weak, in the sense that knowing a value of one variable does not necessarily give us confidence in predicting a value for the other variable.

Notice that the data points appear to cluster in two groups with a gap between them. In about one-half of the states, less than 30% of the high school graduates took the test in 1996. In the other half of the states, more than 40% took the test. ■

EXAMPLE 2   The Leaning Tower of Pisa

The Leaning Tower of Pisa is a famous tower in Pisa, Italy. The tower was built on soft ground, and over the years it has achieved its characteristic lean as one end of its foundation sinks into the soil. The amount that the tower leans is measured by comparing a point on the tower to where it would be if the tower were straight. (See Figure 10.) We want to determine the relationship between the year and the amount the tower leans. Since the amount the tower leans increases every year, we might say the amount of lean is a variable that depends on a second variable, the year.

Figure 10 Measuring the amount of lean for the Leaning Tower of Pisa.

Data collected concerning the Leaning Tower of Pisa is shown in Figure 11.

Figure 11 **The amount of lean for the Leaning Tower of Pisa by year**

Year	1975	1976	1977	1978	1979	1980	1981	1982	1983	1984	1985	1986	1987
Lean (mm)	2964.2	2964.4	2965.6	2966.7	2967.3	2968.8	2969.6	2969.8	2971.3	2971.7	2972.5	2974.2	2975.7

Source: G. Geri and B. Palla, "Considerazioni sulle piu recenti osservasoioni ottiche alla Torre Pendente di Pisa," taken from THE BASIC PRACTICE OF STATISTICS by Moore ©1995 by W. H. Freeman and Company. Used with permission.

positive association

The scatter plot in Figure 12 shows a strong relationship between these variables. The points tend to slope upward to the right, so in this case there is a *positive association* between the variables. That is, both variables increase together. There does not appear to be any obvious curvature; rather, the points seem to be increasing steadily. Therefore, the shape is linear.

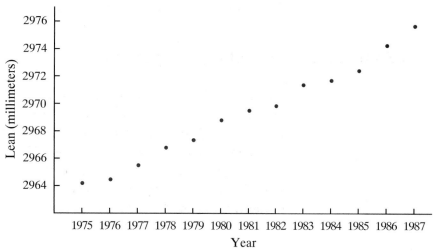

Figure 12 Amount of lean by year

Sometimes a scatter plot may have a point or points that appear to stand out from the rest. These points may follow the general pattern of the data but are far removed from other points. At other times, there are points that are inconsistent with the general trend. Such points may indicate errors in measurement or in plotting that need to be corrected, or they may indicate the presence of some factor that deserves special attention. Whatever the cause, you should look for, and attempt to explain, odd points, called *outliers,* that do not appear to fit the general pattern of the scatter plot.

outliers

CLASS PRACTICE **1.** Complete the following four tasks for each of the six scatter plots shown in Figures 13–18. The labels describe the variables on each axis.

- Determine the shape of the relationship (linear or curved).

- Determine whether the relationship is positive or negative.

- Determine whether there are gaps, clusters, or outliers apparent in the data.

- Determine whether you would feel confident making predictions from the scatter plot. Do you consider the association weak or strong?

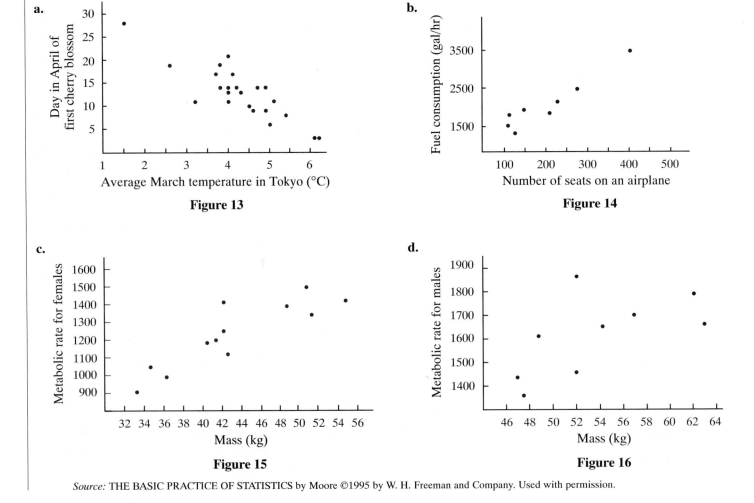

Figure 13

Figure 14

Figure 15

Figure 16

Source: THE BASIC PRACTICE OF STATISTICS by Moore ©1995 by W. H. Freeman and Company. Used with permission.

e.

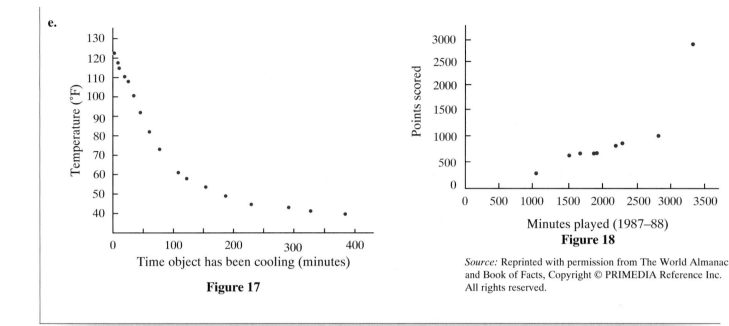

Figure 17

Figure 18

When you examined the scatter plots in the Class Practice, you should have noticed several graphs with outliers. In Figure 13, there is a point in the upper left corner that fits the general trend of the data but is noticeably removed from the other points. This point, which is an outlier, represents a year in which the average March temperature was unusually low, and the cherry blossoms did not appear until late in April. In Figure 14, the point in the upper right corner represents an airplane with an exceptionally high number of seats and exceptionally high fuel consumption. Cover the outlier with your finger and use the other points to predict fuel consumption for a plane with 400 seats. Do you consider this point to follow the general trend of the other data points? The final outlier is in Figure 18. The outlier represents the player with the most minutes played in the 1987–88 season. This player had the highest playing time, but his playing time is not inconsistent with that of the other players. This point is an outlier because of the very high number of points scored. This player does not have statistics that follow the general pattern of the other players.

When a relationship is suggested by a scatter plot, the next step is to describe it mathematically by finding an equation that summarizes the way the two variables are related. Such an equation is another example of a mathematical model. When we discussed mathematical models at the beginning of the chapter, we pointed out that a good model simplifies the phenomenon it represents and gives us the ability to predict. If we can find the equation of a curve that closely "fits" a scatter plot, we can focus on the important characteristics of the relationship between the variables without the clutter of a scatter plot. We can also use this equation to predict the values of one variable for specific values of the other variable. Sometimes we use the model to

interpolate

extrapolate

interpolate, or estimate new values among data values, and sometimes we use the model to *extrapolate,* or predict values outside the region of the data. To extrapolate, we must have good reason to believe that the pattern observed in the data continues.

What is the equation of the line that best fits the data?

To obtain an equation to model the Leaning Tower of Pisa data, you could sketch a line that passes through the data and follows the general path of the data. What is the equation of the line that best fits the data? The process of fitting a linear model to a set of data is an important aspect of data analysis. With the help of graphing calculators or computers, we can quickly fit a line to a given set of data. For the moment, we will just estimate the location of the linear model to demonstrate how you can use this line. Figure 19 shows a line through the data of our scatter plot.

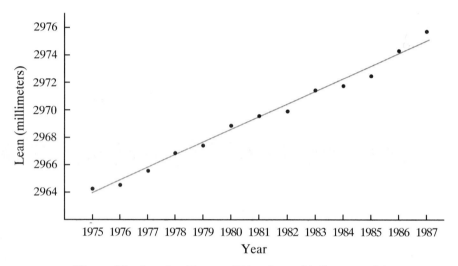

Figure 19   Leaning Tower of Pisa data with linear model

Notice how the line follows the pattern of the points in this scatter plot. Some of the points are above the line, some are below, but they are all close to the line. Since the data points are closely following the path of the line, you can conclude that the relationship between these variables is strong, and you can feel very confident that the line does a good job of describing this particular phenomenon. Expecting the same trend to continue into the near future, you could also feel confident in using the model to extrapolate or predict the value of the lean in future years.

How close to the line do the points have to be to consider the model good? Think back to Example 1, and re-examine the scatter plot in Figure 9 on page 12. Try to sketch a line that follows the path of the data. Figure 20 shows one possible line that could be used to model the test data, but many of the points are not close to the line. There is quite a spread in the data. Also, the line tends to overestimate the scores in the middle and underestimate the scores at the ends. In this case, a linear model does not do a good job of describing the phenomenon. If we tried to fit a curve to the data, we still would not be too confident in our model because the data are so scattered. The issue of closeness is relative and depends on the particular variables and the size of their values. For now, we'll say simply that the relationship between the percentage of students taking the test and the average test score is weak, which means the points are too spread out for us to feel very confident making predictions.

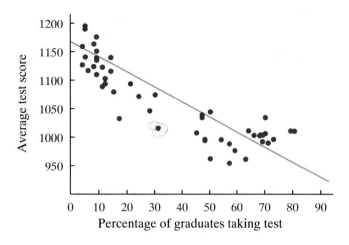

Figure 20 Scatter plot of test scores with a linear model

Exercise Set 1.3

1. In Iowa City, Iowa, a monthly utility bill provides the customer with information about the daily cost of gas and electricity as well as the average temperature during the month. The information in Figure 21 has been taken from a household in Iowa City for the months of August through March.

Figure 21 Heating bill data

Month	Aug.	Sept.	Oct.	Nov.	Dec.	Jan.	Feb.	Mar.
Average temperature (°F)	70	69	58	44	31	23	27	27
Average daily cost of natural gas ($)	0.35	0.38	0.78	1.41	1.86	1.94	1.97	1.76
Average daily cost of electricity ($)	0.98	0.78	0.82	0.77	0.86	0.65	0.80	0.73

Source: Adapted with permission from *Algebra in a Technological World, Curriculum and Evaluation Standards for School Mathematics,* copyright 1995 by the National Council of Teachers of Mathematics. All rights reserved.

    a. Make two scatter plots of the data. One scatter plot should show average cost of gas as the dependent variable and average temperature as the independent variable. The other scatter plot should show average cost of electricity as the dependent variable and average temperature as the independent variable.

    b. Describe the relationship between each pair of variables.

    c. Sketch a freehand line through each set of data, and find the equation of each line. Use the equations to estimate the gas bill and the electric bill if the average temperature for next February were 19°.

2. Complete the following four tasks for each of the three scatter plots shown in Figures 22, 23, and 24.

- Determine the shape of the relationship (linear or curved).
- Determine whether the relationship is positive or negative.
- Determine whether there are gaps, clusters, or outliers apparent in the data.
- Determine whether you would feel confident making predictions from the scatter plot. Do you consider the association weak or strong?

a. The scatter plot in Figure 22 shows the length (cm) of a pendulum and the period. Period is the time in seconds it takes to complete one oscillation, returning to the starting position. The ordered pairs are (*length, time*).

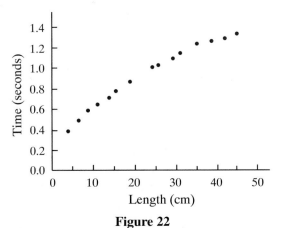

Figure 22

b. The scatter plot in Figure 23 shows the size of the engine (in liters) for a variety of different cars and the number of miles per gallon an owner might expect in city driving. The ordered pairs are (*engine size, mpg*).

Source: http://www.eren.doe.gov/feguide

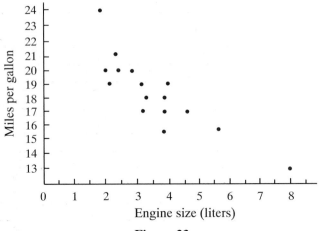

Figure 23

c. The scatter plot in Figure 24 shows the number of years played and the average number of yards gained per run for 20 veteran running backs in the National Football League at the start of the 1995 football season. The ordered pairs are (*years played, average number of yards*).

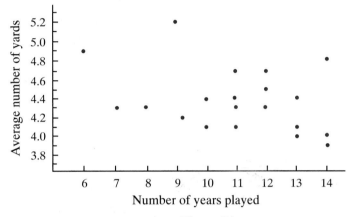

Figure 24

3. Characteristics of different aircraft flying in the United States are shown in Figure 25. From what you can see in the table, try to predict which pairs of variables have a linear relationship. Check two of your predictions by creating scatter plots of the data. Were your predictions correct?

Figure 25   Aircraft data

Airplane	Number of seats	Speed (mph)	Flight length (miles)	Fuel consumption (gallon/hour)	Operating cost ($/hour)
B747-100	405	519	3149	3529	6132
L-1011-100/200	296	498	1631	2215	3885
DC-10-10	288	484	1410	2174	4236
A300 B4	258	460	1221	1482	3526
A310-300	240	473	1512	1574	3484
B767-300	230	478	1668	1503	3334
B767-200	193	475	1736	1377	2887
B757-200	188	449	984	985	2301
B727-200	148	427	688	1249	2247
MD-80	142	416	667	882	1861
B737-300	131	413	605	732	1826
DC-9-50	122	378	685	848	1830
B727-100	115	422	626	1104	2031
B737-100/200	112	388	440	806	1772
F-100	103	360	384	631	1456
DC-9-30	102	377	421	804	1778
DC-9-10	78	376	394	764	1588

Investigations Using Data Collection

A large quantity of data is available from almanacs, newspapers, journals, and other written sources. But it is also important to collect your own data in an informal laboratory setting. The investigations that follow require minimal equipment and can be completed in a short amount of time. Ideally, you should work together with your classmates in groups of three or four. Since it is not practical for each student in a given class to explore all the investigations in this section, you should be prepared to share your results with the class.

Before taking measurements you should agree on the level of accuracy and the measuring techniques. When asked to describe a scatter plot, you should use vocabulary developed in the previous section and make a careful sketch of the scatter plot on paper. Measurement error involved with the data collection should be explained whenever possible.

INVESTIGATION 1   Body Proportions

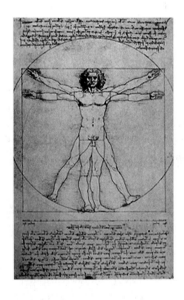

According to Leonardo da Vinci, all human bodies have predictable proportions. For example, he stated that arm span is equal to height, forearm length is one-tenth of height, length of the foot is one-sixth of height, and length of the hand is three times the width of the wrist. See if you and your classmates agree with da Vinci's claim by collecting the following data sets.

Arm span/Height

Question   Is a person's arm span equal to his or her height?

Data Collection   Measure the height and arm span (from fingertip to fingertip) of at least 10 people.

Analysis   Make a scatter plot of (*arm span, height*). Describe the relationship, and compare it to the idealized human form as expressed by da Vinci.

Forearm/Height

Question   Is a person's forearm length equal to one-tenth of his or her height?

Data Collection   Measure the height and forearm length of at least 10 people.

Analysis   Make a scatter plot of (*forearm length, height*). Describe the relationship, and compare it to the idealized human form as expressed by da Vinci.

Foot/Height

Question   Is a person's foot length equal to one-sixth of his or her height?

Data Collection   Measure the height and foot length of at least 10 people.

Analysis Make a scatter plot of (*foot length, height*). Describe the relationship, and compare it to the idealized human form as expressed by da Vinci.

Hand/Wrist

Question Is a person's hand length three times the width of his or her wrist?

Data Collection Measure the hand length and the wrist width of at least 10 people.

Analysis Make a scatter plot of (*wrist width, hand length*). Describe the relationship, and compare it to the idealized human form as expressed by Da Vinci. ■

INVESTIGATION 2   Overhead Projector

Sitting in class, you have noticed that the image projected onto a screen from an overhead projector gets larger as the overhead projector is moved farther away from the screen.

Question Is the relationship between the distance an overhead projector is from a screen and the height of the image projected on the screen linear or curved?

Equipment Overhead projector, transparency with an image to focus, meter stick or ruler to measure the height of the image, tape measures to measure distance from the projector to the screen.

Data Collection Place the overhead projector as close to the screen as possible with the image in focus. Measure the distance from the screen to a fixed point on the projector. Also measure the height of the image on the screen. Move the overhead projector slightly away from the screen, focus the image, and take both measurements again. Repeat this process to collect at least 10 data points.

Analysis Make a scatter plot of (*distance, image height*). Describe the relationship. ■

INVESTIGATION 3   Pennies

If you take a jar containing a collection of 100 pennies and empty it onto a table, how many pennies would you expect to land heads? If you remove the pennies that show heads, return the remaining pennies to the jar, shake it up and empty the jar again, how many do you expect to land heads? What happens in the long run?

Question What is the relationship between the number of times you have emptied the jar and the number of pennies that remain after you remove those that show heads?

Equipment One hundred pennies, jar.

Data Collection   Take a jar containing a collection of 100 pennies, shake the jar to mix the pennies, and empty it onto a table. Remove the pennies that are showing heads and record the number of pennies remaining. Return the remaining pennies to the jar, shake it well, empty the jar again, remove the pennies that are showing heads, and record the number of pennies remaining. Continue this process until no pennies remain.

Analysis   Make a scatter plot of (*number of times you empty jar, number of pennies remaining*). Describe the relationship.   ▪

INVESTIGATION 4   Bouncing Ball

If you drop a ball from the ceiling of your math classroom, it will bounce higher than if you drop it from desk level. What is the relationship between the height of the drop and the height of the bounce?

Question   How is the height from which a ball is dropped related to the height of its first bounce?

Equipment   Bouncing ball, tape measure, tape.

Data Collection   Tape or hang the tape measure on a wall. Measure the height from which you plan to drop the ball. Drop the ball and measure the height of the first bounce. Error can be minimized by having two or three students sight the rebound height and averaging their results. Repeat this process until you collect at least 10 data points.

Analysis   Make a scatter plot of (*drop height, rebound height*). Describe the relationship. How high will your ball bounce if it is dropped from a height of 3 meters?

Other Questions to Consider   Do all balls bounce in the same way? You can try this investigation with different types of balls and make a comparison.   ▪

INVESTIGATION 5   Circles

You have learned the relationship between the diameter of a circle and its circumference. Can you use data from circular objects to confirm this result?

Question   How is the circumference of a circle related to its diameter?

Equipment   Empty cans or jar lids, tape measure.

Data Collection   Measure the diameter and circumference of empty cans, jar lids, or other circular objects until you have collected at least 10 data points.

Analysis   Make a scatter plot of (*diameter, circumference*). Describe the relationship. Is this the relationship that you expected? Use your model to find the circumference of a can with a diameter of 3 centimeters, and compare it to the known result.   ▪

INVESTIGATION 6   Index Card

If you are sitting in the second row of a movie theater and someone sits directly in front of you, your view is probably not obstructed. However, if you are sitting towards the back and the same thing happens, it will be significantly harder to see the movie screen, especially if you are not very tall. The following experiment investigates this issue by using a tape measure in place of the movie screen and an index card in place of the head of the person who is blocking your view.

Question   How does the distance you are away from the wall affect the length of the tape measure that is obscured by an index card?

Equipment   Index card, tape measure, tape.

Data Collection   Attach a tape measure horizontally to a wall. Have a student close one eye, and hold an index card at arm's length. Record the student's distance from the wall and the length of the section of the tape measure that is obscured by the card. Have the student take one small step back (about 12 inches or 30 centimeters), close one eye, and again record the distance from the wall and the length of the tape measure that is obscured. Repeat this process until you collect at least 10 data points.

Analysis   Make a scatter plot of (*distance from wall, length of tape measure obscured*). Describe the relationship.

Other Questions to Consider   How does the size of the card affect the data and therefore the scatter plot? (Experiment by simply rotating the index card 90°. Compare results.) How does the length of the person's arm affect the data and therefore the scatter plot? (Experiment by having a different person hold the index card. Compare results.) ▨

INVESTIGATION 7   Pendulum

If you swing a long pendulum, it takes more time to complete one swing than if you swing a short pendulum. This experiment allows you to investigate the relationship between the length of a pendulum and its period.

Question   How does the period of a pendulum depend on its length?

Equipment   Pendulum (constructed by tying a small nut or several washers onto a string at least two meters long), meter stick, stopwatch.

Data Collection   Vary the length of the string by about 20 centimeters from one trial to the next, and measure how the period of the pendulum (time to complete one swing across and back) changes. To measure the period, students should hold one end of the string stable, pull the weight slightly (about 20°) to one side, let the weight make 10 complete swings, record the time, and then divide the time by 10. Collect at least 10 data points. Be sure to include some long lengths as well as short lengths.

Analysis   Make a scatter plot of (*length of string, period*). Describe the relationship. ▨

INVESTIGATION 8   Road Map

Road maps provide a legend for computing straight-line distances as well as mileage between points along roads shown on the map. How do these distances compare in your state?

Question   How does the straight-line distance between two cities relate to the shortest travel distance between the cities?

Equipment   State road map, ruler.

Data Collection   Use the ruler to measure the straight-line distance between two cities and convert this distance to miles using the legend on the map. Compute the travel distance by adding the distances along the shortest route between the two cities. Repeat this process to collect at least 10 data points. Be sure to include a variety of distances.

Analysis   Make a scatter plot of (*straight-line distance, travel distance*). Describe the relationship. How many miles would you expect to travel between two cities that are exactly six inches apart on the map?

Other Questions to Consider   How do you think the scatter plot for Nevada would compare to the scatter plot for West Virginia? Is the relationship observed in your scatter plot the same for all states?

The Median-Median Line

The investigations and exercises in the previous sections illustrate that freehand methods of curve fitting, though helpful, may vary from person to person. To some extent this allows personal bias to enter the data analysis process. Generally, mathematics is objective and exact, so it becomes a concern when different people obtain reasonable but different lines to model the same set of data. Which line is best? In this context, the concept of best has a mathematical definition, which will be discussed in Section 1.7. For now, however, we will be satisfied with a standard, repeatable procedure that produces a reasonable line, called the *median-median line*. This line is also referred to as a *median-fit line* or a *resistant line*.

median-median line

resistant line

As stated previously, the first step is to look at a scatter plot of the data to be sure that the relationship between the variables is linear. You do not want to blindly fit a line to a data set that is obviously nonlinear. A look at the scatter plot will also make you aware of outliers. Such points may indicate errors in measurement, errors in data entry when using technology, or errors in graphing when working by hand. Errors should be corrected before proceeding. If the points are correct, it is helpful, if possible, to gather additional data near each outlier to determine whether it is an isolated extreme point.

EXAMPLE 1   **Radioactive Contamination**

To explain the procedure for finding a median-median line, data has been taken from an article in *Journal of Environmental Health,* May–June 1965, Volume 27, Number 6, pages 883–897. Robert Fadeley, author of the article, explains that the Atomic Energy Plant in Hanford, Washington, has been a plutonium production facility since the Second World War. Some of the wastes have been stored underground in the same area. Radioactive waste has been seeping into the Columbia River, and eight Oregon counties and the city of Portland have been exposed to radioactive contamination. Figure 26 lists the number of cancer deaths per 100,000 residents for Portland and these counties. The table also includes an index of exposure that measures the proximity of the residents to the contamination. The index is based on the assumption that county or city exposure is directly proportional to river frontage and inversely proportional both to the distance from the Hanford, Washington, site and to the square of the county's (or city's) average distance from the river.

Figure 26   Radioactive contamination data

County/ City	Umatilla	Morrow	Gilliam	Sherman	Wasco	Hood River	Portland	Columbia	Clatsop
Index	2.5	2.6	3.4	1.3	1.6	3.8	11.6	6.4	8.3
Deaths	147	130	130	114	138	162	208	178	210

Is there a model that describes the relationship between the index of exposure and the rate of cancer deaths? The scatter plot provided in Figure 27 has the exposure index on the horizontal axis.

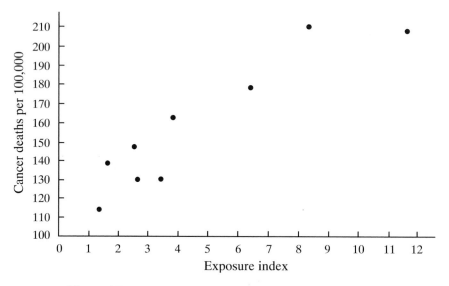

Figure 27 Scatter plot of radioactive contamination data

The scatter plot in Figure 27 indicates a positive linear relationship between these two variables. To fit a median-median line to these points, divide them into three groups. The grouping is based on the x-values of the points. To do this graphically, create a left-most group consisting of points with the smallest x-values, a middle group, and a right-most group. For this data set, the number of data points is divisible by three. If the number of points is not divisible by three, extra points should be assigned symmetrically. If there is only one extra point, it should be placed in the middle group; if there are two extra points, place one in each outer group. However, points with the same x-values must always be placed in the same group. In this example, the groups consist of the following points: the left-most group contains (1.3, 114), (1.6, 138), and (2.5, 147); the middle group contains (2.6, 130), (3.4, 130), and (3.8, 162); and the right-most group contains (6.4, 178), (8.3, 210), and (11.6, 208).

Now consider each group of observations separately and order the values of each variable. The way in which the x- and y-values are paired should be ignored at this stage. For example, in the right-most group, the ordered x-values are 6.4, 8.3, and 11.6; the ordered y-values are 178, 208, and 210. Now create a *summary point* for this portion of data by using the median x-value, 8.3, and the median y-value, 208, and combining them to create the ordered pair (8.3, 208). Repeat this process to obtain (1.6, 138) and (3.4, 130) as summary points for the other two groups. Notice that two of the summary points are actual data points, but one is not. These three summary points are marked on the scatter plot in Figure 28.

summary point

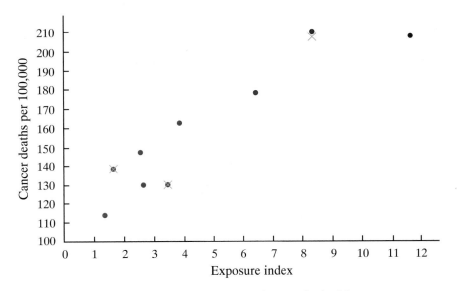

Figure 28 Data and summary points marked with an ×

You will now use the summary points to determine the median-median line. The summary points from the two outer groups will determine the slope, and all three summary points will determine the *y*-intercept. To understand the process graphically, place a ruler on the scatter plot to connect the two outer summary points. Now move the ruler one-third of the way toward the middle summary point, being sure to keep it parallel to its original position. If you trace the ruler, the line you draw is the median-median line. By moving the ruler one-third of the way toward the middle point, you give each summary point equal weight in determining the *y*-intercept. In Figure 29, the dotted line is the line through the outer two summary points, and the solid line is the median-median line.

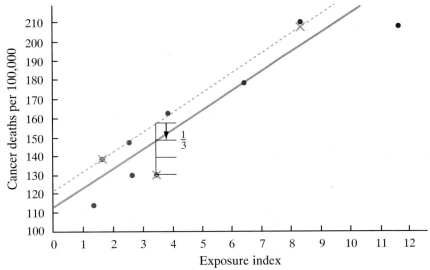

Figure 29 Sliding line toward the middle summary point

To find the equation of the median-median line, first compute the equation of the line passing through the two outer summary points. This line passes through points with coordinates (1.6, 138) and (8.3, 208), so the equation is given by

$$y - 138 = \frac{208 - 138}{8.3 - 1.6}(x - 1.6),$$

which simplifies to

$$y = 10.44776119x + 121.2835821. \tag{1}$$

This line is shown on the scatter plot in Figure 30. We will retain all the digits displayed by the calculator or computer as we work through the numerical calculations. We will round numbers only at the end of the process.

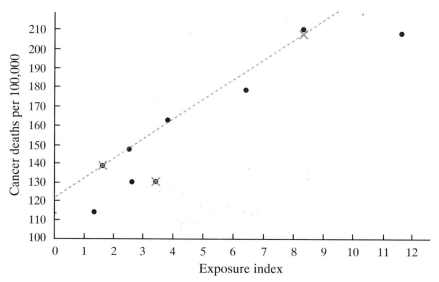

Figure 30   Line through two outer summary points

To incorporate the middle one-third of the data, slide this preliminary line one-third of the distance toward the middle summary point. To determine how far to slide the line, hold the x-value of the middle summary point constant and compare the y-values of the point on the line and the middle summary point. What value of y does equation (1) produce when $x = 3.4$? Since this is an intermediate step in the problem, we use all the displayed digits in the computation. Substituting $x = 3.4$ in equation (1), we obtain

$$y = 10.44776119(3.4) + 121.2835821 = 156.8059701.$$

The point (3.4, 156.8059701) lies on the line connecting the two outer summary points. To slide the line one-third of the way toward the summary point (3.4, 130), subtract the y-values, $130 - 156.8059701 = -26.8059701$ and take one-third of this value, which is -8.935323367. Since the directed distance is negative, move the line down approximately 8.94 units. To adjust equation (1), subtract 8.935323367 from the original y-intercept to find a new y-intercept of $121.2835821 - 8.935323367 = 112.3482587$. The equation we use as our model is

$$y = 10.44776119x + 112.3482587.$$

Now that all calculations have been completed, you can round the constants to display fewer decimal places. You should retain enough decimal places to be able to round predicted values to the same accuracy as the given data. The index values were provided to the nearest tenth and death rates were provided to the nearest integer. If you round the constants in the equation to the nearest hundredth, you will be able to make substitutions for either variable and round the final answer to the same decimal accuracy as the data. Equation (2) will serve as a model for the effects of radioactive contamination.

$$y = 10.45x + 112.35. \tag{2}$$

This line is drawn on the scatter plot in Figure 31.

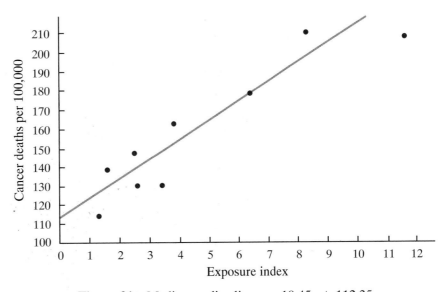

Figure 31   Median-median line $y = 10.45x + 112.35$

What information does the model provide?

Now we have a model that describes the relationship between the index of exposure to radioactive contamination and the rate of cancer deaths in the region near Hanford, Washington. What information does the model provide? First consider the y-intercept in the equation. Its value is 112.35. In terms of the model, it is the cancer death rate we predict when the index of exposure is zero, that is, when there is no radioactive contamination. Does it seem reasonable for there to be approximately 112 cancer deaths per 100,000 residents in an area without exposure to radioactive contamination? Now consider what values the variables can assume. Algebraically, we can extend the line over all real number values of both variables. Many real numbers, however, are not reasonable values for the exposure index defined for this analysis. There is no way to have a negative index, and on the basis of the indices in the data set, we would question how large an index could reasonably be. Finally, consider the slope. Its value is 10.45. In terms of the model, the slope indicates that for each unit increase in the exposure index, we expect the cancer deaths to increase by about 10.45 per 100,000 residents. ▧

Keep in mind that data sets contain measurement errors, so models based on data only approximate the relationship between variables. When you are able to find a good fit, you can feel confident that your model provides reliable information about this relationship. You cannot assume, however, that a change in one variable necessarily causes a change in the other. Nor can you assume that the model will be appropriate outside the domain of observed values. You must be careful, therefore, when using models to extrapolate.

Before proceeding, review the summary for finding the equation of the median-median line through a set of points.

Summary of the Procedure for Finding the Equation of the Median-Median Line

- Separate the data into three groups of equal size (or as close to equal as possible) according to the values of the independent variable.

- Within each group, find the median x-value and median y-value. The three points whose coordinates are (*median x, median y*) are called the summary points.

- Find the equation of the line through the summary points of the outer groups. Call this line L.

- Slide L one-third of the way toward the middle summary point (x_m, y_m).
  1. Find $L(x_m)$, the y-coordinate of the point on L at the x-value of the middle summary point.
  2. Find the vertical distance between (x_m, y_m) and $(x_m, L(x_m))$ by subtracting y_m and $L(x_m)$.
  3. Calculate one-third of the vertical distance, and adjust the y-intercept of L so that the vertical distance between L and (x_m, y_m) is reduced by one-third.

- The resulting line is known as the median-median line.

EXAMPLE 2   The First Blooms of Spring

Many people eagerly anticipate the first blooms of spring flowers. One of the most beautiful of the spring blossoms is that of the Japanese cherry tree. Experience demonstrates that if the spring has been a warm one, the trees will blossom early, but if the spring has been cool, the blossoms will arrive later. Mr. Yamada is a gardener who has been observing for the last 24 years the date in April when the first blossoms appear. His son, Hiro, went to the library and found the average temperatures for the month of March during those 24 years. The data are given in Figure 32.

Figure 32   **Average March temperature and days in April until first bloom**

Temperature (°C)	1.5	2.6	3.2	3.7	3.8	3.8	4.0	4.0	4.0
Days in April until first bloom	28	19	11	17	19	14	21	13	11
Temperature (°C)	4.0	4.1	4.2	4.3	4.5	4.6	4.7	4.9	4.9
Days in April until first bloom	14	17	14	13	10	9	14	9	14
Temperature (°C)	5.0	5.1	5.1	5.4	6.1	6.2	3.5		
Days in April until first bloom	6	11	11	8	3	3	?		

Source: Atarasi Sugaku, Tokyo Shoseki Textbook Company.

This year, Hiro took the daily temperature for each day in March. At the end of the month, he found the average temperature in March to be 3.5°C. The spring was a cool one. Hiro would like to predict the date in April when the first blooms will appear on the cherry trees.

Solution   Hiro first makes a scatter plot (Figure 33) to get a picture of the data. If you compare this scatter plot to the original data, you will notice that only 23 data points are shown in this graph, yet there are 24 data points in Figure 32. Two years had identical values for both variables and are both represented by the single point in the graph.

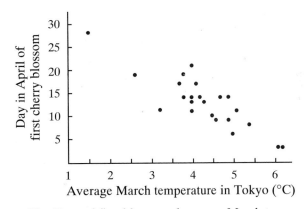

Figure 33   Date of first bloom and average March temperature

The scatter plot reveals a decreasing linear relationship, so Hiro decides to fit a median-median line to the data and use the line to predict the date for this year. When he attempts to separate the 24 data points into three groups, he finds a complication. Since 24 is divisible by three, he expected each group to contain eight points. The break between the left-most and middle thirds should come between the eighth and ninth points. Observe in Figure 34 on the next page that the seventh, eighth, ninth, and tenth data points have an x-coordinate of 4.0. When we divide the data into thirds, we are separating the points with two vertical lines. Points with the same x-coordinate cannot be separated by a vertical line, so we must adjust the break between the groups so that points with the same x-coordinates are grouped together.

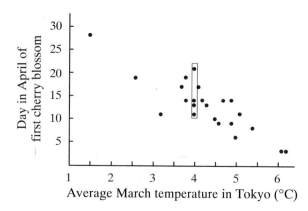

Figure 34   Points at $x = 4$ that cannot be in different groups

If the natural division points have the same x-coordinates as other points, they are generally not placed in the middle group, since it is the outer summary points that determine the slope of the median-median line. In Figure 35, the left-most group has 10 points, the middle group has 6 points, and the right-most group has 8 points. Once the divisions are determined, the remainder of the procedure remains the same.

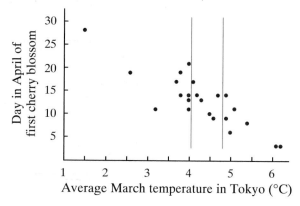

Figure 35   Partitions used for median-median line   ▦

1. Following the procedure outlined in this section, find the median-median line for Hiro's data. Use it to predict the date on which the first blossoms will appear.

2. Data for the Leaning Tower of Pisa were given in Figure 11 on page 13.

    a. Find the equation of the median-median line for the data. Keep all decimal places in your computations and in the equation as you answer Parts b, c, and d.

    b. Use a calculator or computer to examine this line superimposed on the scatter plot. Do you think it is a good model? Explain why or why not.

    c. If we assume the lean has increased in a manner consistent with the data and nothing has been done to halt the process, what would you predict the amount of lean to be in 1997?

    d. According to your model, in what year was the lean only 2.9 meters?

    e. Round the constants in the equation for the median-median line to two decimal places. Use a calculator or computer to examine this line super-imposed on the scatter plot. Do you think it is a good model? Explain why or why not. Use this model to predict the amount of lean in 1997. How does this prediction compare to the one from Part c?

    f. Modify the original data so that the independent variable is the number of years since 1900 (i.e., 1975 becomes 75, 1976 becomes 76). Recompute the equation of the median-median line. Use the equation, first with all deci-mals retained and then with constants rounded to two decimal places, to predict the amount of lean in 1997. Explain why rounding off does not cre-ate the same problems that you observed in Part e.

3. Fit a median-median line to the data sets about gas and electricity consump-tion found in Exercise 1 of Exercise Set 1.3. Use these equations to complete the following:

    a. If the average temperature for next February were 19°, how much would you expect the gas bill and the electric bill to be? how do these estimates compare to the ones you made in Exercise Set 1.3?

    b. Explain the meaning of the slope and intercepts of each line in the context of this problem.

4. The record times for the mile run are given in Figure 36 for the years 1880 to 1985. Fit a median-median line to the data. Use that line to predict the records in 1990 and 1995. Look up these records in an almanac and compare the actu-al records to the predictions.

Figure 36 Record times for running a mile

Year	Time (seconds)	Year	Time (seconds)
1880	263.2	1945	241.4
1882	261.4	1954	239.4
1882	259.4	1954	238.0
1884	258.4	1957	237.2
1894	258.2	1958	234.5
1895	257.0	1962	234.4
1895	255.6	1964	234.1
1911	255.4	1965	233.6
1913	254.6	1966	231.3
1915	252.6	1967	231.1
1923	250.4	1975	231.0
1931	249.2	1975	229.4
1933	247.6	1979	229.0
1934	246.8	1980	228.8
1937	246.4	1981	228.5
1942	246.2	1981	228.4
1942	244.6	1981	227.3
1943	242.6	1985	226.3
1944	241.6		

5. The metabolic rates of males and females with the associated lean body masses are given in Figure 37.

Figure 37 Metabolic rate data

Female mass (kg)	36.1	54.6	48.5	42.0	50.6	42.0	40.3	33.1	42.4	34.5	51.1	41.2
Metabolic rate	995	1425	1396	1418	1502	1256	1189	913	1124	1052	1347	1204
Male mass (kg)	62.0	62.9	47.4	48.7	51.9	51.9	46.9	54.1	56.8			
Metabolic rate	1792	1666	1362	1614	1460	1867	1439	1654	1702			

a. Make a scatter plot of the data using mass as the independent variable and metabolic rate for females as the dependent variable. Find the median-median line for this data set. Does the median-median line model the data well? Explain why or why not.

b. Repeat the procedure in Part a for males.

c. Repeat the procedure in Part a for males and females combined.

d. Would you have more confidence in your predictions using the lines calculated separately for males and females or using the line calculated for the combined data? Explain.

SECTION 1.6 | How Good Is the Fit?

residual

When we fit a curve or a line to a data set, we should examine that curve in relation to the scatter plot to determine how well it models the relationship between the variables. Most of the points on the scatter plot will not fall exactly on the curve. The question is, "How close to the curve are the data points?" Since the curve is used to predict *y*-values, it is important to look at the vertical distance between each observed point and the fitted curve. This difference in *y*-values, obtained by subtracting the *y*-value on the curve from the *y*-value of the data point, is called a *residual,* or a *deviation.* Every ordered pair in the data set has a residual value associated with it. These residuals measure the difference between the actual data values and the values predicted from the model.

The graph in Figure 38 shows the median-median line we previously computed in the radioactive contamination example on pages 25–29. Recall that the horizontal axis gives values of the exposure index and the vertical axis gives cancer deaths per 100,000 residents. The vertical distances marked show the residuals or vertical deviations of the data points from the fitted line. Positive residuals are associated with data points above the line, and negative residuals are associated with data points below the line.

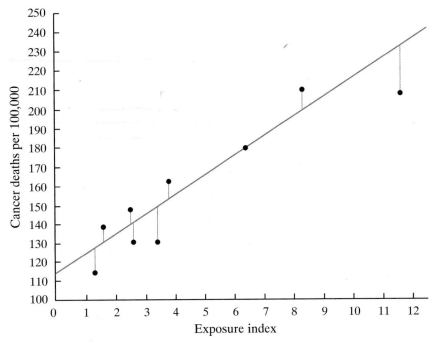

Figure 38   Median-median line with residuals indicated

We will use the notation (x_i, y_i) for the observed data points. This subscript notation means that the point $(1.3, 114)$ is (x_1, y_1), $(1.6, 138)$ is (x_2, y_2), $(2.5, 147)$ is (x_3, y_3), and so forth. Suppose the linear equation fit to the data is denoted $y = L(x)$, so the points on the line corresponding to data values of the independent variable are $(x_i, L(x_i))$. The residual associated with each point is $y_i - L(x_i)$.

Figure 39 Data with values from median-median line and residuals

x_i	y_i	$L(x_i)$	$y_i - L(x_i)$
1.3	114	126	−12
1.6	138	129	9
2.5	147	138	9
2.6	130	140	−10
3.4	130	148	−18
3.8	162	152	10
6.4	178	179	−1
8.3	210	199	11
11.6	208	234	−26

An examination of the residuals in Figure 39 will give us a better sense of how well the curve fits the data, as well as highlight characteristics of the data that had not been previously noticed. Check the residuals to see whether they follow any trend or pattern as the x-values vary. Are residuals all positive in the middle and all negative at both ends (or vice versa)? If some pattern exists, it might indicate that we chose the wrong curve to fit the data. In Figure 38, positive and negative residuals are scattered throughout the graph, so we do not have this problem.

You should also look at the relative size of the residuals. Residuals that are small relative to observed y-values provide evidence of a good fit. One or two relatively large residuals may draw attention to outliers that were not previously detected. You should attempt to determine why these points do not fit the general pattern of the data. If the outliers result from errors in measurement, the data points should be corrected or excluded from the analysis. If, on the other hand, the outliers are not the result of errors, a relatively large residual may provide interesting and useful information. If the nonconformity can be explained, we often gain additional information about the data we are trying to model. In Figure 38, it is clear that the line is closer to some of the points than to others, but the only residual that seems to stand out is the final one, which is associated with the point (11.6, 208). Look back at the scatter plot in Figure 27 on page 26. Do the points level off as the exposure index becomes large? We really cannot answer this question without more data. However, this observation would probably make us feel less confident when we use the model to extrapolate for large x-values.

It is easier to analyze the residuals when they are paired with corresponding x-values and studied as a new data set (x_i, r_i), where $r_i = y_i - L(x_i)$. A scatter plot of this data set, called a *residual plot*, allows us to check the size and pattern of the residuals. If the model provides a good fit for the original data, the residual plot should show points scattered randomly within a horizontal band about the horizontal axis with relatively small y-values. Figure 40 shows the ordered pairs (x_i, r_i) created from the radioactive contamination example. This plot shows that the residuals are relatively small in magnitude and do not follow a trend or pattern; therefore, this linear model is a good one.

residual plot

Residuals

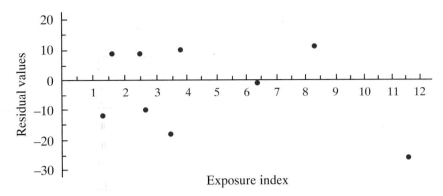

Figure 40 Residual plot for radioactive contamination data

After doing a median-median fit, always examine the residuals to determine how good the fit is. We can fit a median-median line to any data set; the technique will work regardless of whether or not using it is appropriate. That is why it is very important to look at the scatter plot of the data to decide whether a linear model is reasonable. If we fit a straight line to data when the true relationship is not linear, the residual plot will usually show a pattern as you will see in Exercise Set 1.6, Exercise 1. Because the scale on a residual plot is usually much smaller than on the original scatter plot, curvature is more easily noticed on the residual plot. Therefore, we should use both the scatter plot and the residual plot in the curve-fitting process to help ensure a good fit.

Exercise Set 1.6

1. Sketch the residual plots for the linear models in Figures 41, 42, and 43. In each case answer the question, "How good is the fit?" Your answer should include comments on the absence or presence of a pattern in the residuals and on the relative size of the residuals.

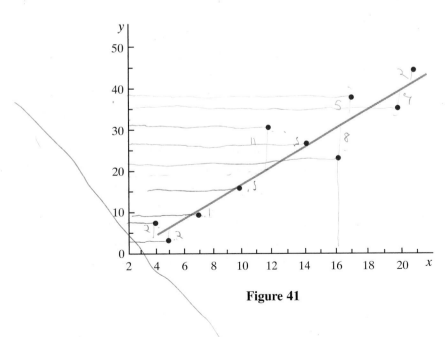

Figure 41

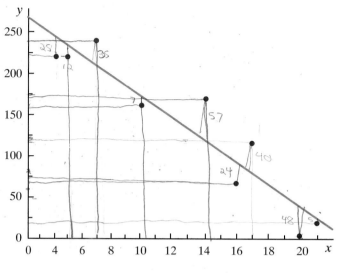

Figure 42

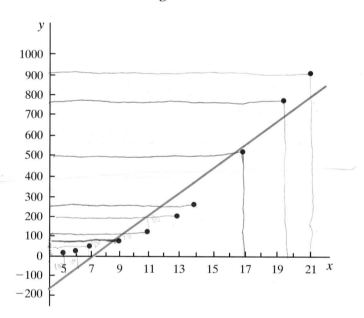

Figure 43

2. The data in Figure 44 represent the temperature of a cup of water as it cools and the associated time.

Figure 44 **Cooling data**

Time (minutes)	0	5	10	16	20	35	50	65	85
Temperature (°F)	124	118	114	109	106	97	89	82	74
Time (minutes)	128	144	178	208	244	299	331	391	
Temperature (°F)	64	62	59	55	51	50	49	47	

a. Find the equation of the median-median line for the data.

b. Use a calculator or computer to make a residual plot and then transfer the residual plot to paper.

c. Examine the residual plot and answer the following questions. Do you notice a pattern in the residuals? If so, what does the pattern tell you about the model? Is there a better model? Why or why not? If so, what does its graph look like?

3. The data in Figure 45 represent the distance an object has fallen and the associated time since it was dropped.

Figure 45 **Falling object data**

Time (seconds)	0.16	0.24	0.25	0.30	0.30	0.32	0.36	0.36	0.50	0.50
Distance (cm)	12.1	29.8	32.7	42.8	44.2	55.8	63.5	65.1	124.6	129.7
Time (seconds)	0.57	0.61	0.61	0.68	0.72	0.72	0.83	0.88	0.89	
Distance (cm)	150.2	182.2	189.4	220.4	254.0	261.0	334.6	375.5	399.1	

a. Find the equation of the median-median line for the data.

b. Use a calculator or computer to make a residual plot. Examine the residual plot and answer the following questions. Do you notice a pattern in the residuals? If so, what does the pattern tell you about the model? Is there a better model? Why or why not? If so, what does its graph look like?

SECTION 1.7 The Least Squares Line

Up to this point, the median-median line has served as the model for a linear data set. You may have wondered, however, if some other line might do a better job of fitting the data than the median-median line. If so, what line would be better? What line would be best? If this question were posed for class discussion, you would probably get many different opinions, some of which are listed below.

Student A: The best-fitting line passes exactly through the greatest number of data points.

Student B: The best-fitting line results in equal numbers of data points above the line and below the line; that is, the number of positive and negative residuals must be equal.

Student C: The best-fitting line is influenced most by data points near the middle of the data set and does not give much weight to extreme values.

Student D: The best-fitting line minimizes the sum of all the perpendicular distances between the data points and the line.

Student E: The best-fitting line minimizes the sum of all the vertical distances between the data points and the line.

A discussion about a line of best fit cannot be resolved without a definition of the word "best." Since mathematical models are often used for interpolating and extrapolating, mathematicians have defined the best line to be the line that minimizes errors in predictions. If you consider several lines through a data set, you would probably agree that small residuals are preferable to large ones.

Consider the data in Figure 46 from the radioactive contamination example.

Figure 46 **Radioactive contamination data**

County/City	Umatilla	Morrow	Gilliam	Sherman	Wasco	Hood River	Portland	Columbia	Clatsop
Index	2.5	2.6	3.4	1.3	1.6	3.8	11.6	6.4	8.3
Deaths	147	130	130	114	138	162	208	178	210

Recall the median-median line (to two decimal places) was $y = 10.45x + 112.35$. Figures 47 and 48 illustrate the fit of the median-median line and its associated residual plot.

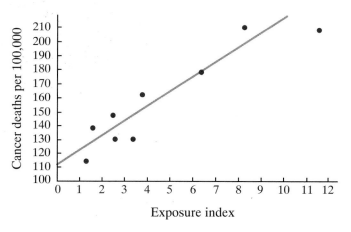

Figure 47   Data and median-median line

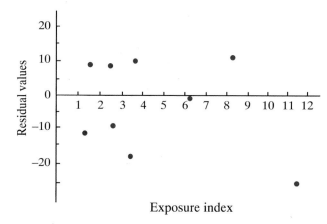

Figure 48   Residual plot

Figures 49 and 50 show a slightly different line, $y = 9.27x + 114.68$, with its residual plot.

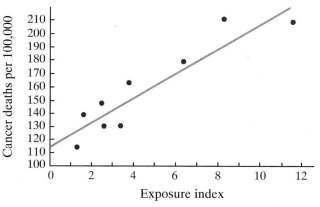

Figure 49   Data and linear model $y = 9.27x + 114.68$

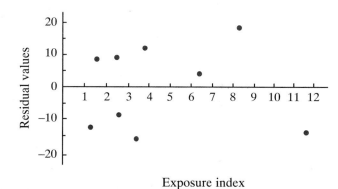

Figure 50   Residual plot

The Least Squares Principle

Finding the best-fitting line requires that the residuals be as small as possible, since they represent the errors in predictions. How can we accomplish this? Your first thought may be to minimize the sum of the residuals. However, since residuals can be either positive or negative, their sum could be close to zero even for a poorly-fitting line. Your next thought might be to minimize the sum of the absolute values of the residuals. However, historically mathematicians have found absolute value inconvenient to work with. Therefore, to accomplish the goal of keeping residuals as small as possible, we minimize the sum of squared residuals. This criterion, called the *least squares principle*, dates back to the nineteenth-century work of Adrien Legendre.

least squares principle

least squares line

regression line

According to the least squares principle, the best-fitting line is defined to be the one that minimizes the sum of the squares of the residuals. Such a line must have residuals close to zero so that their squares will be small, and therefore the sum of their squares will be as small as possible. This line is referred to as the *least squares line* or the *regression line*. The line through the data in Figure 49, $y = 9.27x + 114.68$, is the least squares line with two decimal place accuracy. You should notice that it is similar, but not identical, to the median-median line.

Techniques from calculus or algebra can be used to derive formulas for the slope and y-intercept of the least squares line. An algebraic derivation is given in the appendix. Graphing calculators can compute the slope and y-intercept of the least squares line. We will assume that you have the technology available to find the equation of the least squares line for a data set. Our attention will focus on analyzing rather than calculating this line.

As you proceed with the process of fitting lines and curves to data sets, remember to examine a scatter plot of each data set for linearity. This check is important, because it is possible, though not wise, to find the least squares line for data sets in which the relationship between the variables is not linear. Examination of a scatter plot will also allow you to observe outliers, if they exist.

Comparison of Median-Median and Least Squares Lines

Now that we have two methods of modeling a data set with a line, we will consider the advantages offered by each method. The least squares line is the more popular method; it is the one generally included in statistics textbooks. The advantage of the least squares method is that it offers a mathematical standard by which the line is developed. By minimizing the sum of the squared residuals, we minimize the total error in the line. However, the least squares line does not always capture the major thrust of the data. The formulas used to determine the slope and y-intercept of the least squares line involve every point. Therefore, if the data set has outliers, the least squares line may be "thrown off" a great deal. The median-median line, in contrast, is highly resistant to the effects of outliers.

The goal of the least squares method is to make the sum of the squared residuals as small as possible. According to the least squares criterion, we prefer to have 8 points off by one unit rather than a single point off by 3 units since $1^2 + 1^2 + 1^2 + 1^2 + 1^2 + 1^2 + 1^2 + 1^2 = 8$ and $3^2 = 9$. In contrast, each data point affects the median-median line only indirectly, via the three summary points. These ideas are explored further in the Class Practice on the next page and in Exercise Set 1.7.

1. a. Find the residuals for the median-median line for the radioactive contamination data. Make a list of the squared residuals and find their sum for the median-median line.

    b. Find the residuals for the least squares line for the radioactive contamination data. Make a list of the squared residuals and find their sum.

    c. How do the two sums of the squared residuals compare?

    d. Examine squared residuals for the median-median line and the least squares line. In each case, which residual has the largest effect on the sum?

Keep in mind that linear models involve two parameters, the slope and the y-intercept of the line. These two values are not of equal importance in the model. Of the two, the slope gives more information about the relationship between the two variables. A slope of -3 indicates that an increase in the independent variable results in a threefold decrease in the dependent variable. The intercept, on the other hand, simply gives an initial condition or starting point. Knowing that the intercept is 5 tells us very little about the relationship between the variables. As a result, errors in the slope of a linear model are much more costly in terms of understanding the relationship than errors in the y-intercept.

EXAMPLE 1

Chicago Bulls Basketball Data

Figure 51 shows the data and a scatter plot of the number of points scored and the number of minutes played during the 1987–88 basketball season by Chicago Bulls players who played 600 minutes or more. Figure 52 on the next page shows the same data, but for the 1993–94 basketball season. Compare the median-median and least squares models for both data sets.

Figure 51   Total points scored and total minutes played for the 1987–88 Bulls

Player	Minutes played	Points scored
Corzine	2328	804
Grant	1827	622
Jordan	3311	2868
Oakley	2816	1014
Paxon	1888	640
Pippen	1650	625
Sparrow	1044	260
Sellers	2212	777
Vincent	1501	573

Minutes played (1987–88)

Figure 52 Total points scored and total minutes played for the 1993–94 Bulls

Player	Minutes played	Points scored
Armstrong	2770	1212
Blount	690	198
Cartwright	780	235
Grant	2570	1057
Kerr	2036	709
Kukoc	1808	814
Longley	1502	528
Myers	2030	650
Pippen	2759	1587
Wennington	1371	542
Williams	638	289

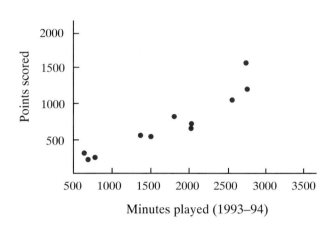

Solution By fitting median-median and least squares lines to both data sets, we can see how a single point affects the model. The equations are

1987–88 Season median-median line: $P = 0.3354M + 48.6946$

least squares line: $P = 0.9296M - 1009.6476$

and

1993–94 Season median-median line: $P = 0.4522M - 102.7593$

least squares line: $P = 0.5086M - 165.3197$,

where P is the number of points scored and M is the number of minutes played. Study the linear models shown with the data points in Figures 53 and 54. The solid line in each graph is the least squares line. Notice the effect of the outlier on the least squares line in Figure 53. If our goal was to use a line for the 1987–88 season to predict the number of points a player might score, we would choose the median-median line for our model. The slope of the least squares line reflects the outlier of Michael Jordan. For this problem, the points that represent the records of other Bulls players more accurately reflect the typical NBA player. Therefore, we would prefer a line that is a good model for all of the data except the outlier. The 1993–94 Bulls roster did not include Michael Jordan, and no other player shows up in the scatter plot as an outlier. Consequently, the median-median line and least squares line in Figure 54 are not very different from each other.

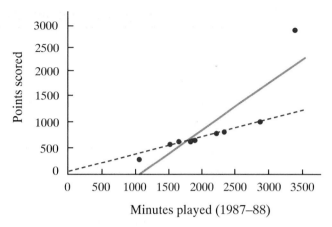

Figure 53   Two linear models (1987–88)

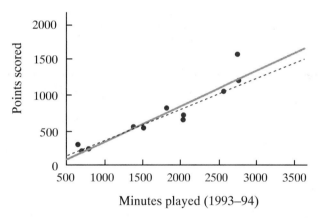

Figure 54   Two linear models (1993–94)

In the preceding example, notice that the least squares line is very sensitive to outliers, particularly at the extremes of a data set. As you complete the exercises at the end of this section, you should notice that outliers near the middle of a data set tend to affect the *y*-intercept of the least squares line, and outliers near the extremes of a data set tend to affect the slope. The farther an outlier is from the middle of a data set, the more influence it has on the slope. In contrast, the median-median line is relatively resistant to the effects of outliers. If you want extreme points to influence your linear model, the least squares line is preferred over the median-median line. If, on the other hand, you prefer to fit a line that is not influenced by extreme points, the median-median line is often a better choice.

In practice, both lines can be drawn quickly with a graphing calculator or computer, and the extent to which outliers affect each line can be readily determined. Unlike the median-median line, the least squares line carries with it a lot of statistical theory that is very useful for more in-depth statistical investigations. A complete description of this theory is beyond the scope of this course.

1. According to the least squares principle, can the sum of squared residuals be zero? If so, what would that mean about the fit?

2. The data in Figure 55 are points on the line $y = x$.

Figure 55 Data set for Exercise 2

x	1	2	3	4	5	6	7	8	9
y	1	2	3	4	5	6	7	8	9

Make a conjecture about how equations of the median-median and least squares lines will be altered if the data points are changed in the following ways. Use a calculator to check your predictions.

a. The data point (7, 7) is changed to (7, 3).

b. The data point (5, 5) is changed to (5, 1).

c. The data point (1, 1) is changed to (1, 2).

d. The data point (1, 1) is changed to (1, 20).

3. Most of the data points in Figure 56 lie on the line $y = x$. The point (8, 9) is above the line, and the point (9, 8) is below the line.

Figure 56 Data set for Exercise 3

x	1	2	3	4	5	6	7	8	9
y	1	2	3	4	5	6	7	9	8

a. You should expect one point to pull the least squares line up and another point to pull the line down. Predict whether or not these two effects will offset each other when fitting a least squares line to the data.

b. Use a computer or calculator to find the equation of the least squares line.

c. Have the effects on the least squares line of the two variant points offset each other? If not, which point wins the "tug of war"? Explain why.

d. What are the summary points for the median-median line?

e. What effect, if any, do the variant points have on the median-median line?

4. The symbol $\bar{x}$ represents the average, or mean, of the x-values, and $\bar{y}$ represents the mean of the y-values. Verify that the least squares line passes through the point $(\bar{x}, \bar{y})$ in the radioactive contamination data listed in Figure 46 on page 40.

5. A biology student noticed that crickets seemed to chirp faster in the summer than in the spring or fall. Her grandmother had told her that she could determine the temperature by listening to the crickets. Over the next season she counted the chirps per minute of a cricket and recorded the temperature. Her data are provided in Figure 57.

Figure 57 Cricket data

Chirps (per minute)	55	67	75	83	91	99
Temperature (°F)	50	54	55	58	58	60
Chirps (per minute)	119	134	140	149	164	178
Temperature (°F)	67	69	70	74	77	79

  a. Find a least squares line that the student can use to estimate the temperature by listening to the crickets. Interpret the slope and y-intercept in terms of the phenomenon.

  b. Explain how this model could be used to estimate the temperature very quickly by counting chirps for only 15 seconds.

  c. If you wanted to describe mathematically the relationship between temperature and cricket chirps, which variable is more appropriate to consider as the dependent variable? Is this the same variable that you treated as the dependent variable in Part a? If not, find a new model. Interpret the slope and y-intercept. Predict the number of chirps per minute given that the outside temperature is 65°.

6. Two students were trying to find a quick way to approximate the length of the hypotenuse of a right triangle if they know the lengths of both legs. They used a random number generator to create random numbers from 1 to 10 to use as the lengths of the two legs. They compared the sum of these two numbers to the length of the hypotenuse of a right triangle with those two sides. They decided that they could approximate the length of the hypotenuse by taking two-thirds of the sum and adding one.

  a. Using the random number generator on your calculator, generate 10 pairs of numbers representing the lengths of the legs of 10 right triangles.

  b. For each pair of legs, use the Pythagorean theorem to calculate the hypotenuse. Also use the rule "two-thirds of the sum and add one" to approximate the hypotenuse. Make ordered pairs of the type (*Pythagorean theorem hypotenuse, approximation hypotenuse*).

  c. If the approximation is a good one, the line $y = x$ should be a good model for the data. How confident are you in the approximation? Why?

  d. How accurate is this approximation if the lengths of the legs are greater than 10? (Generate more data and test the model.)

7. According to the *MINITAB Handbook* (Second Edition by Barbara F. Ryan, Brian L. Joiner, and Thomas A. Ryan, Jr., PWS-KENT Publishing Company, 1985, page 254.), a statistician named Frank Anscombe constructed the data sets listed in Figure 58.

Figure 58   Anscombe's data sets

A	10.00	8.00	13.00	9.00	11.00	14.00	6.00	4.00	12.00	7.00	5.00
B	8.04	6.95	7.58	8.81	8.33	9.96	7.24	4.26	10.84	4.82	5.68
C	9.14	8.14	8.74	8.77	9.26	8.10	6.13	3.10	9.13	7.26	4.74
D	7.46	6.77	12.74	7.11	7.81	8.84	6.08	5.39	8.15	6.42	5.73
E	8.00	8.00	8.00	8.00	8.00	8.00	8.00	19.00	8.00	8.00	8.00
F	6.58	5.76	7.71	8.84	8.47	7.04	5.25	12.50	5.56	7.91	6.89

a. Find the equation of the least squares line and median-median line for each of the four data sets composed of ordered pairs (A, B), (A, C), (A, D), and (E, F).

b. Compare the median-median and least squares lines you obtain for each data set.

c. Compare the least squares lines for the four data sets.

d. If you knew nothing but the equations of the least squares lines, how much information would you really have about the data sets?

e. Examine the scatter plots for the four data sets. For which data sets is a linear model appropriate? Discuss problems that might occur in fitting a line to each data set.

f. Anticipate what the residual plots will look like when the linear models are fit to each data set. Use a computer or calculator to check your predictions.

SECTION 1.8　Error Bounds and the Accuracy of a Prediction

Recall that Hiro wanted to predict the date on which the first cherry blossoms appear based on the average temperature in March. (See page 31 in Section 1.5.) The least squares line relating the average temperature in March and the number of days in April before the first cherry blossoms appear is

$$y = -4.69x + 33.12,$$

and its graph is shown in Figure 59. The average temperature in March was 3.5°C, so Hiro expects the blossoms to appear on a date close to April 17 if he uses the least squares line to make his prediction. Unlike models that we develop based on scientific theory, the least squares line is a model based on data. We do not assume that it is perfect. In fact, a different set of ordered pairs would result in a different model. When you use the least squares line to predict a single number for the date of the first blossoms, you should expect some error in this prediction. Information about the magnitude of the error is contained in the residuals. The residual plot in Figure 60 shows that all but two of the residuals are between -4 and 4. So, it is quite likely that the blossoms will occur on April 17 $\pm$ 4 days, or between April 13 and April 21. You have no guarantee that this interval contains the actual date of the first blossoms, but the information from previous years makes you feel fairly confident that the date of the first blossoms will be in this interval.

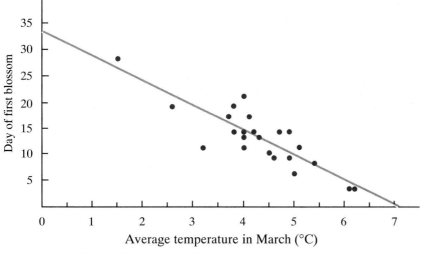

Figure 59　Scatter plot with the least squares line

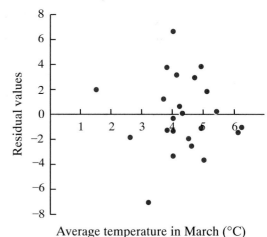

Figure 60　Residual plot

Is there an interval around April 17 that would make you even more confident that it includes the actual day of the first blossoms? Again, the residuals will help answer this question. Since the largest residual associated with the model is 7 days, you could report with very high confidence that the first blossoms will appear on April 17 ± 7 days or between April 10 and April 24. Of course there is no guarantee that this will happen, but based on 24 years of data, the model has never been off by more than 7 days.

The method used to create an interval estimate involved a number that was added to and subtracted from the single value predicted by the least squares line. There are different choices you could make about how much to add and subtract. As you might expect, statisticians have objective ways to produce these intervals. We will not consider those methods in this course, but we will suggest other ways that provide useful information. One way to create an interval is to add and subtract the residual that is largest in magnitude. However, in most cases an interval created in this way will be unnecessarily large and therefore not particularly useful to a person using the model to make decisions. Another way to create an interval is to add and subtract the average value of the residuals. But the average residual value will always be zero, since the positive residuals balance the negative ones. To avoid this cancellation, you can take the absolute value of each residual and then calculate the average. An interval can then be created with this average absolute residual by adding to and subtracting from the y-value produced by the least squares line.

standard deviation

Another option would be to use the standard deviation of the residuals as the number that is added to and subtracted from the y-value predicted by the least squares line. The *standard deviation* of the set of residuals is a number that measures the spread of the residuals about their average. When a least squares line is fit to a linear data set, approximately 95% of the residuals will fall within an interval of two standard deviations on either side of the average of the residuals, which is always zero. Thus, you will find that approximately 95% of the data will fall within an interval of two standard deviations on either side of the least squares line. In the blossom example, the standard deviation of the residuals is 2.955. Adding and subtracting 2.955 × 2 or 5.91 days from the predicted y-value should give you a reasonable interval to estimate the date of the first blossoms. Since the average temperature in March was 3.5°C, Hiro would expect the first blossoms to appear between April 11 (April 17 − 6) and April 23 (April 17 + 6).

error bounds

Once you choose a technique for calculating intervals, you can produce *error bounds* for your linear model. The model we developed for Hiro's data is $y = -4.69x + 33.12$. If you create an interval by adding and subtracting twice the standard deviation, 5.91 days, then you can be fairly certain that for any x-value the actual day of first blossoms falls between the two linear models $y = -4.69x + 33.12 - 5.91$ and $y = -4.69x + 33.12 + 5.91$. These equations simplify to $y = -4.69x + 27.21$ and $y = -4.69x + 39.03$. The graphs of these equations are error bounds and are shown with the least squares line and data in Figure 61. All but two data points are between these error bounds, so the error bounds appear to do a good job of capturing the variation in the original data.

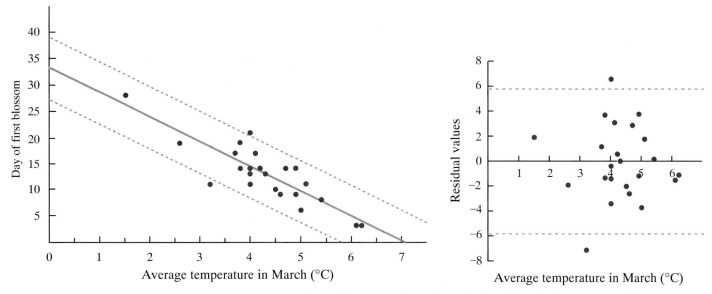

Figure 61 Error bounds created by adding and subtracting 5.91 days from the least squares line

Using the *y*-value predicted by the least squares line and two times the standard deviation of the residuals, Hiro can predict to within about 6 days the date on which the first blossoms will appear. His least squares line predicts that the first blossoms will appear on April 17. But he probably would be more comfortable predicting that the first blossoms will appear between April 11 and April 23. Graphing the error bounds together with the least squares line makes it easy to visualize their relationship.

Exercise Set 1.8

1. Calculate the error bounds for the radioactive contamination data (Figure 46 on page 40) and the Leaning Tower of Pisa data (Figure 11 on page 13). Explain how you determined your error bounds.

2. A lumberjack has been studying the trees on the mountain near his house. Last summer he measured the diameter of the trees at a point 1.2 meters above the ground. He then used a tool to measure the height of the trees. The data he collected are given below in Figure 62.

Figure 62 Tree data

Diameter (cm)	16	18	19	20	21	21	23	24	25
Height (m)	12.9	13.2	12.8	15.0	15.7	14.9	14.7	17.3	19.2
Diameter (cm)	25	26	28	30	31	32			
Height (m)	15.2	18.6	17.9	16.4	19.5	17.9			

a. Make a scatter plot of the data; then find an equation of a line to fit the data.

b. If a tree has a diameter of 36 centimeters, what height does your equation predict? Use error bounds to determine how close you expect your prediction to be to the actual height.

c. If you know the diameter of a tree, based on your model above, is it likely that your prediction of the height of the tree will be within 2 meters of the actual height?

d. Use the linear equation from Part a to determine a model for the relationship between circumference and height. Which do you think the lumberjack actually measured, the diameter or the circumference? Is one measurement better to use than the other? Explain your reasoning.

3. Manatees are large, gentle sea creatures that live in the shallow waters along the coast of Florida. Manatees are slow moving and can easily be injured by rapidly moving motorboats. Figure 63 shows the number of powerboat registrations in thousands in Florida from 1977 to 1990 along with the number of manatees killed in boating accidents.

Source: THE BASIC PRACTICE OF STATISTICS by Moore © 1995 by W.H. Freeman and Company. Used with permission.

Figure 63   Manatee data

Year	1977	1978	1979	1980	1981	1982	1983	1984	1985	1986	1987	1988	1989	1990
Powerboat registrations (1000s)	447	460	481	498	513	512	526	559	585	614	645	675	711	719
Manatees killed	13	21	24	16	24	20	15	34	33	33	39	43	50	47

a. Use your calculator to find a linear model relating the number of powerboat registrations to the number of manatees accidentally killed each year. Interpret the slope and *x*-intercept of this line. Look at a residual plot. Is your linear model appropriate?

b. Use your calculator to find a linear model relating the year since 1977 to the number of powerboat registrations. (The first ordered pair is (0, 447).) Interpret the slope and *y*-intercept of this line. Look at a residual plot. Is your linear model appropriate?

c. Use your calculator to find a linear model relating the year since 1977 to the number of manatees accidentally killed each year. Estimate the number of manatees that will be accidentally killed in the year 2000. Calculate error bounds for the model you found. Explain how you determined the error bounds, and use them to write an interval estimate for the number of manatees accidentally killed in the year 2000.

4. Fit a line to each data set you gathered in the investigations in Section 1.4. Describe each residual plot. For those data sets that appear to be linear, find the error bounds. How confident are you in each model?

Chapter 1 Review Exercises

1. The students in a precalculus class wonder how long it will take the student body, which consists of 300 students, to complete the wave if all 300 students are sitting in one continuous row. (The wave is formed by students standing up and then sitting down, one after the other.) To investigate this question, the class recorded the length of time it took student groups of various sizes to complete the wave. The data are shown below.

Figure 64 Wave data

Number of students	6	7	11	12	14	16	17	20	22	26
Time (seconds)	2	3	2.9	3.4	4.5	4.5	3.9	4.5	5.3	6.2

a. Make a scatter plot of the data. Do the data appear linear?

b. Find the equation of the least squares line for this data set.

c. Interpret the slope of the least squares line in the context of the problem setting.

d. Interpret the intercept on the vertical axis in the context of the problem setting.

e. What is the residual for the point (12, 3.4)?

f. According to your model, how long will it take the student body to complete the wave? How confident are you in your answer? Explain.

g. Find the equations for error bounds for your model.

h. If you were asked to compute the median-median line for these data, what are the coordinates of the three summary points?

2. Under what conditions will the median-median line be a better model for data than the least squares line? Explain.

2 Functions

2.1	Functions as Mathematical Models	56
2.2	Characteristics of Functions	61
2.3	A Toolkit of Functions	64
2.4	Finding the Domain of a Function	73
2.5	Functions as Mathematical Models Revisited	79
2.6	Investigating a Conical Container	93
2.7	Investigating Graphs of Transformations of Functions	95
2.8	Basic Transformations of Functions	97
2.9	Combinations of Transformations	104
2.10	Composition of Functions	112
2.11	Investigating Graphs of Compositions of Functions	117
2.12	Composition as a Graphing Tool	119
2.13	Inverses	128
2.14	Using Inverses to Straighten Curves	138
2.15	Investigations Using Data Collection	146
2.16	Introduction to Parametric Equations	149
	Chapter 2 Review Exercises	156

The Flying Hat

During the winter of 1993, a blizzard struck the East Coast. A man 6 feet 3 inches tall went outside during the storm. A gust of wind, reported to be as high as 40 miles per hour, blew his hat from his head. The hat flew over a fence 4 feet high that was 25 feet away. By how much did the hat clear the fence, and how long did it stay in the air before it landed?

Functions as Mathematical Models

Suppose that a pharmaceutical researcher has developed a new medicine for pain relief. He wants to determine the relationship between the level of medicine in a person's body and the time since the medicine was taken. Determining this relationship will help the researcher provide accurate information regarding dosage and frequency of medication for individuals taking this medicine.

This example, and others in the preceding chapter, illustrate the need to know about relationships in order to make accurate predictions. You may already be familiar with scientific formulas that express the relationship between variables; often these formulas are used to make predictions. But it is not only scientists who use knowledge about relationships this way. To make reasonable decisions, we must know certain relationships and use them to make informed choices. For instance, stopping distance is a function of driving speed, and we use this information to decide on a safe following distance as we drive.

In relationships like those described above it is usually assumed that one of the quantities determines the value of the other. When this is true, the quantity that depends on the other is called the *dependent variable;* the other quantity is called the *independent variable.* For example, a safe following distance depends on driving speed, so driving speed is the independent variable and following distance is the dependent variable. In the relationship between the level of medicine in the body and time since it was taken, the level of medicine depends on the time since it was administered. Thus, the amount of medicine is the dependent variable and the independent variable is the time since the medicine was taken.

dependent variable

independent variable

How can we display or describe the relationship between two quantities or variables?

How can we display or describe the relationship between two quantities or variables? One way is to list all the ordered pairs of values that occur, but it is very difficult to see the characteristics of the relationship in this form. For example, your doctor probably has a chart that shows ideal weights for various heights. Another way is to give a formula or equation that relates the two variables. For some relationships, formulas are already known that express how the variables are related. For other relationships, data can be gathered and a mathematical model determined using techniques of data analysis. A third way to show the relationship between two variables is to draw a graph in a rectangular coordinate system. Such a graph can reveal many of the properties of the relationship and help us to better understand it.

You will spend a fair amount of time in this course studying graphs of the relationship between two variables. You will learn to extract information from graphs that have been drawn for you and also to make graphs to display information.

EXAMPLE 1 A sales representative for a major food company receives a bonus each quarter that is 1% of his total sales for that period of time. In Figure 1, the graph on the left shows the relationship between amount of bonus and amount of sales.

Since the total sales amount determines the amount of the bonus, and not vice versa, the total sales is the independent variable and is graphed on the horizontal axis. It should be clear that a larger bonus is associated with larger total sales. The graph indicates an *increasing relationship*. That is, as the total sales for the quarter increases, the bonus also increases. The graph is also *continuous* because it does not have any breaks in it.

increasing relationship

continuous graph

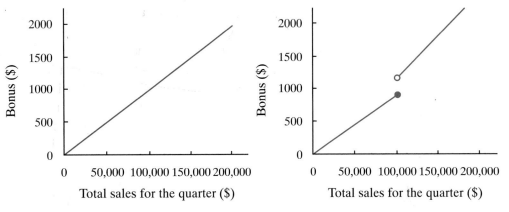

Figure 1   Bonus plans for two different companies

Another company that markets similar products has a different bonus plan. It offers a smaller percent of sales bonus (0.8%) for a sales volume of $100,000 or less. For sales volumes exceeding $100,000, this percent is increased to 1.2%. The graph showing the relationship between bonus amount and sales for this company is on the right in Figure 1. The break in the graph occurs because the percent used to calculate the bonus changes when the total sales amount exceeds $100,000. This graph is *discontinuous* at a point. This means the graph has a break that requires you to pick up your pencil from the paper as you draw the graph.

discontinuous graph

relation

In any situation in which the values of one variable are paired with the values of another variable, we have a *relation*. All the relationships between variables considered so far are examples of relations. For example, a driving speed is paired with a following distance, or an amount of sales is paired with a bonus amount.

Another relation is the price you pay for an airline ticket and the distance you fly. It would seem reasonable that the cost of an airplane ticket should depend on the distance you fly. However, it is entirely possible that two tickets for flights of 400 miles might have very different prices. The price of a ticket is not uniquely determined by the length of the flight; other factors, such as the size of the airports, the day of the week, and the date of purchase, all influence the ticket price.

The relation between price and distance differs from other relations we have considered. The difference lies in our ability to use values of the independent variable to predict values of the dependent variable. Given a value for the independent variable (*distance*), we cannot confidently determine the value of the dependent variable (*price*). In many relations, knowing the value of the independent variable guarantees that we can find a unique value for the dependent variable. Whenever this is true, the relation is called a *function*. In the airplane ticket relation we are not guaranteed a unique value of the dependent variable; this relation is not a function.

function

A *function* is often defined as a set of ordered pairs for which each first coordinate has one and only one second coordinate. It is also useful to think of a function as a process that maps, or sends, each permissible first coordinate to a unique second coordinate. When we think about a function as a mapping, the mapping emphasizes the dynamic process of pairing values. In this course it is important to think about functions as being dynamic, as doing something to an *x*-value to get a *y*-value.

A variety of notations is used to define particular functions. Functions are usually named with a letter, and *f* is a common choice. The familiar notation $y = f(x)$, which is read "*y* equals *f* of *x*," shows that *f* does something to an *x*-value to produce a *y*-value. We can also write $f : x \rightarrow y$ to show that the function *f* sends a value of *x* to a particular *y*. For example, the notations

$$f : x \rightarrow x^2 + 1$$

and

$$f(x) = x^2 + 1$$

and

$$y = x^2 + 1$$

input/output

all indicate that a given value of *x* is paired with a *y*-value that is obtained by squaring the *x*-value and then adding 1.

argument of the function

The function *f* makes ordered pairs of the form $(x, x^2 + 1)$. The *input* is *x* and the *output* is $x^2 + 1$. In this notation, *x* is often called the *argument of the function*—it is the variable that the function acts on to produce the second coordinate of each ordered pair. The output of the function is sometimes referred to as the *value of the function*.

value of the function

Exercise Set 2.1

1. For each relationship described below, identify the two quantities that vary and decide which should be represented by an independent variable and which by a dependent variable. Sketch a reasonable graph to show the relationship between the two variables. You should be concerned only with the basic shape of the graph, not with particular points. Write a sentence or two to justify the shape and behavior of your graph.

    a. The amount of money earned for a part-time job and the number of hours worked

    b. The number of people absent from school each day of the school year

    c. The temperature of an ice-cold drink left in a warm room for a period of time

    d. The amount of daylight each day of the year

    e. The water level on the supports of a pier at an ocean beach on a calm day

f. The population of the United States according to each census since 1790

g. The height of a pop fly in baseball after being hit into the air

h. The distance between the ceiling and the tip of the minute hand of a clock hung on the wall

i. The height of an individual as he or she grows

j. The size of a person's vocabulary from birth onwards

k. The number of bacteria in a culture over a period of time

2. Sketch a graph of the relationship presented in each table below. Identify dependent and independent variables. State any conclusions you can make about the relationship based on the shape of your graph.

a. Refer to the postal rates given in Figure 2.

Figure 2   Third class postage rates

Weight	Rate
not more than 1 oz	0.32
greater than 1 and not more than 2 oz	0.55
greater than 2 and not more than 3 oz	0.78
greater than 3 and not more than 4 oz	1.01
greater than 4 and not more than 5 oz	1.24
greater than 5 and not more than 6 oz	1.47
greater than 6 and not more than 7 oz	1.70
greater than 7 and not more than 8 oz	1.93
greater than 8 and not more than 9 oz	2.16
greater than 9 and not more than 10 oz	2.39
greater than 10 and not more than 11 oz	2.62
greater than 11 and not more than 13 oz	2.90
greater than 13 and not more than 16 oz	2.95

b. Refer to the tax rates given in Figure 3.

Figure 3   Income tax table

Taxable Income	Income Tax
over $0 but not over $23,350	15% of income
over $23,350 but not over $56,550	$3,502.50 + 28% of income over $23,350
over $56,550 but not over $117,950	$12,798.50 + 31% of income over $56,550
over $117,950 but not over $256,500	$31,832.50 + 36% of income over $117,950
over $256,500	$81,710.50 + 39.6% of income over $256,500

3. Let $f(x) = 3x - 1$.

    a. Find $f(4)$.

    b. Find x such that $f(x) = 4$.

    c. Write an expression for $f(2x)$, $2f(x)$, $f(x + 2)$, and $f(x) + 2$.

4. Let $g(x) = x^2 - x$.

    a. Write an expression for $g(a)$.

    b. Find x such that $g(x) = 6$.

    c. Write an expression for $g(\frac{1}{x})$, $\frac{1}{g(x)}$, $g(x - 1)$, and $g(x) - 1$.

5. If $f(x) = x^2 + x$, write and simplify an expression for each of the following.

    a. $f(x + 1)$ b. $2f(x) - 3$

    c. $f(0.5x)$ d. $f(\frac{1}{x})$

    e. $f(|x|)$ f. $f(x + h)$

6. Let $f(x) = x^2$ and let $g(x) = \frac{1}{x}$. Express each of the following functions as a transformation of f or g. For instance, $y = (x + 1)^2$ can be expressed as $y = f(x + 1)$.

    a. $y = (x - 4)^2$ b. $y = \frac{3}{x}$

    c. $y = x^2 + 5$ d. $y = \frac{1}{x - 1}$

    e. $y = 2 + 7x^2$ f. $y = \frac{1}{5x}$

SECTION 2.2 Characteristics of Functions

Communication is important in the study of mathematics. A specialized vocabulary has been developed to help provide accurate and succinct information about mathematical relationships. The following examples illustrate some of these mathematical terms and definitions.

EXAMPLE 1

After breakfast, a cup of hot chocolate is left sitting on the kitchen table. When you come home for lunch, you contemplate drinking the hot chocolate and wonder how its temperature has changed throughout the morning. Figure 4 shows a sketch that represents the relationship between the temperature of the hot chocolate and the length of time it has been sitting on the table.

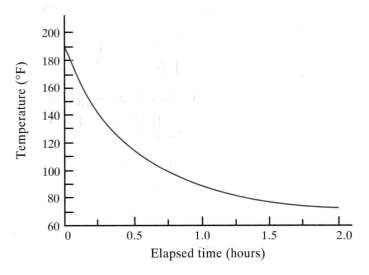

Figure 4 Temperature of hot chocolate versus elapsed time

Since the temperature depends on the time, the temperature of the hot chocolate is the dependent variable and the amount of time that has passed since the hot chocolate was left on the table is the independent variable. You can think about a function that has ordered pairs (*elapsed time, temperature*). The numbers that represent elapsed time make up the *domain* of the function: they are the first coordinates in the ordered pairs of the function. In this case you have a limited domain. Domain values extend from $x = 0$, which represents the time the hot chocolate was put on the table, until the time you arrived home. If, for example, you were gone two hours, the domain would be $0 \leq x \leq 2$. The numbers that represent the temperatures during this time make up the *range* of the function: they are the second coordinates in the ordered pairs of the function.

domain

range

Judging from the graph in Figure 4, we estimate that the range in this example is $72 \le y \le 190$. This graph doesn't have an *x*-intercept since there is no time at which the temperature is zero. It does, however, have a *y*-intercept. The *y*-intercept represents the temperature when the *x*-value equals zero. In this situation, the *y*-intercept represents the initial temperature of the drink. Another important feature of the graph is the upward curvature. In mathematics, this curvature is called *concave up*. In contrast, if the graph opened downward, the curvature would be described as *concave down*. If you examine the graph from left to right beginning at the *y*-intercept, you will notice that *y*-values continue to decrease as *x*-values increase indicating that the temperature continues to fall. Thus, the function that models the relationship between temperature and time is said to be *decreasing*. As the values of the independent variable, time, get larger and larger, the values of the dependent variable, temperature, get smaller and smaller. Because there are no breaks in the graph, this is a continuous function. ■

concave up

concave down

decreasing relationship

EXAMPLE 2   A spring is suspended from a doorway and a weight is attached to the spring. The weight is pushed up from its equilibrium position and then released. Assume that the spring and weight will continue to oscillate indefinitely. Figure 5 shows a sketch that represents the relationship between the elapsed time and the distance the spring moves from its equilibrium position. The distance is positive when the weight is above equilibrium position.

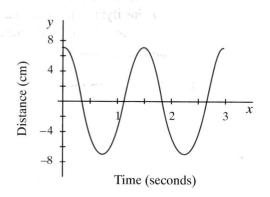

Figure 5   Distance from equilibrium versus elapsed time

A function that describes this situation has ordered pairs (*elapsed time, distance from equilibrium*). The domain is all real numbers greater than or equal to zero, written $x \ge 0$. Reading from the graph, it appears the range is $-7 \le y \le 7$. The range is limited because the spring and weight can only travel up and down a specified distance from equilibrium. The positive *y*-intercept indicates that the spring was pushed up from its equilibrium position before being released. After the spring is released, it travels back to its equilibrium position and then to its lowest position as illustrated by the decreasing portion of the graph. The spring then assumes an upward motion traveling from the lowest point back to its equilibrium position and then to its highest point. This is illustrated by the increasing portion of the graph. The displacement of the spring from the highest point to the lowest point and back to the highest point requires 1.5 seconds. The motion then repeats indefinitely, reaching its highest point at 3 seconds, 4.5 seconds, and every 1.5 seconds thereafter. Because the graph of this function repeats for each interval of 1.5 seconds, we say that the function is *periodic* with period 1.5 seconds. Each time the spring passes through its equilibrium position, the *y*-coordinate is zero. Thus the *x*-intercepts of the graph represent times associated with equilibrium position. These *x*-values that produce a *y*-value of zero are called *zeros* of the function. The low and high points can be described as *turning points*. A turning point is where a function changes from increasing to decreasing or vice versa. This function has infinitely many turning points. This function also has many changes in curvature. From the time the spring is released until it passes through its equilibrium position, it is concave down. Once it has passed equilibrium it is concave up until it comes back to equilibrium. ▩

periodic

zeros of the function

turning points

Exercise Set 2.2

1. In Example 1 we examined a graph showing the temperature of hot chocolate over a two-hour period. Sketch a graph for this relationship using the domain $0 \le x \le 4$. What is the range of the function you graphed?

2. In Exercise Set 2.1, you sketched graphs of relationships and wrote a sentence or two justifying the shape and behavior of each graph. Given the new vocabulary and different ways you now have of characterizing functions, return to Exercise 1 on page 58 and write a more complete description for each Part a through k.

3. The function f is graphed in Figure 6.

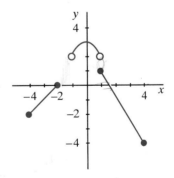

Figure 6   Graph of f for Exercise 3

a. What is the domain of f?

b. What is the range of f?

c. Where is f an increasing function? a decreasing function?

d. Where is f concave up? concave down?

A Toolkit of Functions

Graphing Functions—Developing a Toolkit

One of the goals of this course is to develop your ability to graph relationships between variables. To be proficient, you must be able to recognize which functions are best left to a computer or calculator to graph and which are more efficiently tackled with paper and pencil. In this chapter you will develop knowledge and techniques to facilitate paper-and-pencil graphing.

Most of the graphs you make by hand do not require pinpoint accuracy. Rather, your graphs should display basic characteristics of the function and of the relationship between the variables. To begin, you will be introduced to a collection of nine functions that make up your toolkit. These functions are important for their intrinsic mathematical content and as a set of tools for mathematicians, scientists, economists, and anyone else who creates mathematical models to solve problems. This toolkit is not sufficient to tackle every problem, but it provides a foundation that you will use as you expand your knowledge of functions. A thorough understanding of these toolkit functions is essential and will make the future study of functions and mathematical modeling more meaningful. As you read about each of the nine toolkit functions you should plot a few points to help you understand each graph.

Constant Function

constant function

One of the simplest functions is one in which all first coordinates are paired with the same second coordinate; such a function is called a *constant function*. For example, $f(x) = 4$ consists of ordered pairs all with second coordinate 4. The graph is the horizontal line $y = 4$. (See Figure 7.)

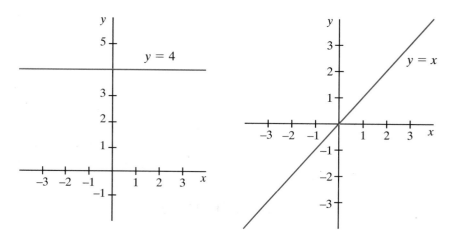

Figure 7 Constant and linear functions

Linear Function

linear function

identity function

general linear function

The simplest *linear function* is $f(x) = x$. It is often called the *identity function* because coordinates in each ordered pair are identical. The graph of this function is the diagonal line shown in Figure 7. Notice that this graph is everywhere increasing and continuous. The *general linear function* is $f(x) = mx + b$; its graph is a variation of the line $y = x$ that has slope m and y-intercept b. You should be familiar with these variations from study in previous mathematics courses and from your work in Chapter 1.

Quadratic Function

quadratic function

The toolkit *quadratic function* is $f(x) = x^2$. You can understand the shape of the graph of $f(x) = x^2$ by thinking about the ordered pairs (x, x^2) that belong to the function. Whenever x is greater than 1, $x^2 > x$, so the graph is above that of $y = x$. When x is between 0 and 1, $x^2 < x$, and the graph is below that of $y = x$. The graph of $f(x) = x^2$, shown on the left in Figure 8, is called a *parabola*. Notice that it is decreasing for negative x-values and increasing for positive x-values. The graph has a turning point at $(0, 0)$, which is called the *vertex* of the parabola. The curvature of this graph is concave up.

parabola

vertex

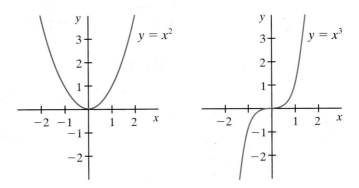

Figure 8   Quadratic and cubic functions

Cubic Function

cubic function

The graph of the toolkit *cubic function* $f(x) = x^3$ is shown on the right in Figure 8. This graph is increasing for all x-values, and it is steeper than the graph of $f(x) = x^2$ for $x > 1$. Notice that the graph lies entirely in the first and third quadrants; in the ordered pairs (x, x^3), either both coordinates are positive or both are negative.

Square Root Function

square root function

principal square root

The *square root function* is defined by the equation $f(x) = \sqrt{x}$. Since the symbol $\sqrt{x}$ represents only the *principal* (positive) *square root* of x, each x-value is paired with a unique y-value. (Think about what your calculator does when you compute the square root of 9; the calculator displays 3, not ± 3.) The graph of $f(x) = \sqrt{x}$ is shown on the left in Figure 9. Notice that it is increasing for all x-values, but the rate of increase is very slow. The curvature of this graph is concave down.

We can compare the steepness of $f(x) = \sqrt{x}$ with that of linear, quadratic, and cubic functions by determining for each function the x-value that is paired with a y-value of 64. For $f(x) = x^3$, the ordered pair is (4, 64). For $f(x) = x^2$, the ordered pair is (8, 64). For $f(x) = x$, the ordered pair is (64, 64). For $f(x) = \sqrt{x}$, the ordered pair is (4096, 64). Notice that the cubic function attains the value of 64 when $x = 4$, whereas the square root function doesn't attain this value until $x = 4096$.

It is also interesting to study the steepness of $f(x) = \sqrt{x}$ in another way. When $x > 1$, $\sqrt{x} < x$, so the graph is below the line $y = x$. In effect, taking the square root pulls y-values down if $x > 1$. On the other hand, when $0 < x < 1$, $\sqrt{x} > x$, and the graph is above the line $y = x$. For these x-values, taking the square root increases the y-values.

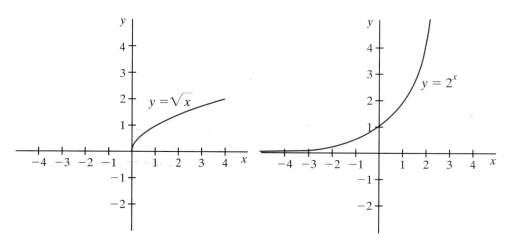

Figure 9   Square root and exponential functions

Exponential Function

exponential function

The toolkit *exponential function* is $f(x) = 2^x$. The graph, shown on the right in Figure 9, is increasing and concave up for all x-values. As the x-values get more and more negative, y-values get smaller and approach zero, so the graph gets closer and closer to the negative x-axis. We say that the graph is *asymptotic* to the x-axis. In general, any function of the form $f(x) = a \cdot b^x$, where a is any nonzero constant, $b > 0$, and $b \neq 1$, is called an exponential function.

asymptotic

Reciprocal Function

reciprocal function

The graph of the *reciprocal function* $f(x) = \frac{1}{x}$ is shown on the left in Figure 10. Notice that the graph is discontinuous at $x = 0$. The function f produces an output by taking the reciprocal of its input. Since taking the reciprocal does not cause a sign change, the graph lies entirely in the first and third quadrants. In the first quadrant where x-values are positive, the reciprocals of small numbers are big numbers and the reciprocals of big numbers are small numbers. As *x*-values get larger, *y*-values get smaller and approach zero, so the graph gets closer and closer to the *x*-axis which is an *asymptote* for the function. Because *x*-values close to zero are paired with large *y*-values, the graph also has the *y*-axis as an asymptote. Similar analysis for negative values of *x* explains the behavior of the graph in the third quadrant.

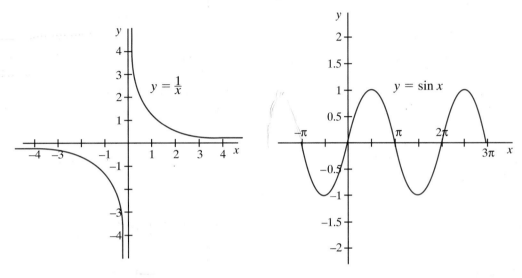

Figure 10   Reciprocal and sine functions

Sine Function

sine function

The toolkit *sine function* defined by the equation $f(x) = \sin x$ is graphed on the right in Figure 10. The graph of the sine function oscillates in a regular pattern between a maximum *y*-value of 1 and a minimum *y*-value of -1. The function has a value of 0 at $x = 0$. It increases to its maximum of 1 at $x = \frac{\pi}{2}$, returns to 0 at $x = \pi$, decreases to its minimum of -1 at $x = \frac{3\pi}{2}$ and returns to 0 at $x = 2\pi$. Then it repeats the pattern indefinitely. If you look at the section of the sine function between -2π and 0, or between 2π and 4π, you will notice that it has the same shape as the section between 0 and 2π. Thus the sine function is periodic. Since the portion being repeated is 2π units long, the *period* is said to be 2π. That is, the sine function completes one cycle every 2π units. The values $\frac{\pi}{2}$, 2π, and so on may seem unusual to you. Since π is related to circles, you may guess that the sine function is somehow related to circles. This is, in fact, true, but we will wait until a later chapter for a thorough discussion of the relationship. For now, it is important to know the basic shape of the sine function and coordinates of intercepts and turning points. You can verify that $\sin \frac{\pi}{2} = 1$, $\sin \pi = 0$, and so forth by using your calculator in radian mode. Radian mode is what your calculator uses to evaluate the sine of a real number.

period

Absolute Value Function

absolute value function

The graph in Figure 11, which resembles the shape of the letter "V," is the toolkit *absolute value function*, $f(x) = |x|$. Recall that the symbol $|x|$ is defined as follows:

$$|x| = x \text{ if } x \geq 0 \quad \text{and} \quad |x| = -x \text{ if } x < 0.$$

piecewise-defined function

For positive x-values and zero, $|x|$ is equal to x. For negative x-values, $|x|$ is equal to the opposite of x. The absolute value function is actually an example of a *piecewise-defined function*. This means that the function is defined differently for different x-values. The function $f(x) = |x|$ is the same as the piecewise-defined function

$$f(x) = \begin{cases} x & \text{if } x \geq 0 \\ -x & \text{if } x < 0 \end{cases}.$$

Notice that f pairs positive x-values with themselves and negative x-values with their opposites. For positive x-values the graph of f is identical to the line $y = x$; for negative x-values the graph is identical to the line $y = -x$.

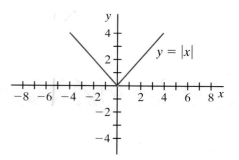

Figure 11   Absolute value function

EXAMPLE 1   Sketch a graph of the following piecewise-defined function.

$$f(x) = \begin{cases} x^2 & \text{if } x \leq -1 \\ \dfrac{1}{x} & \text{if } -1 < x < 0 \\ \sin x & \text{if } x \geq 0 \end{cases}$$

Solution   The graph is shown in Figure 12. For x-values less than or equal to -1, the function is defined by the equation $y = x^2$, and the graph consists of a section of the toolkit parabola. For x-values between -1 and 0, the function is defined by the equation

$y = \frac{1}{x}$, and the graph consists of a section of the toolkit reciprocal function. Note the open circle at $(-1, -1)$; this indicates that the point is not on the graph of the function. For x-values greater than or equal to 0, the function is defined by the equation $y = \sin x$.

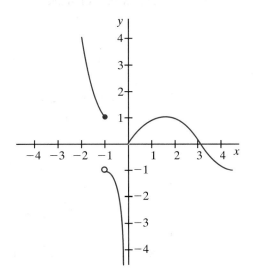

Figure 12 Graph of piecewise-defined function

Exercise Set 2.3

1. Sketch graphs of $y = x^4$, $y = x^5$, $y = x^6$, and $y = x^7$. Make a generalization about the graph of $y = x^n$ for n an even integer and for n an odd integer.

2. Identify the domain and the range of each toolkit function.

3. Functions of the form $f(x) = a \cdot x^n$ where $a \neq 0$ are called *power functions.* Which of the toolkit functions are examples of power functions?

power functions

4. Let $p(x) = \sqrt{x}$, $q(x) = \frac{1}{x}$, and $r(x) = x^3$. Express each of the following as a transformation of p, q, or r. For instance, $y = \sqrt{x + 1} + 2$ can be expressed as $y = p(x + 1) + 2$.

   a. $y = 1 + \frac{1}{x}$ b. $y = 8x^3$

   c. $y = \frac{1}{2x - 2}$ d. $y = \sqrt{6 - x}$

   e. $y = 3 - \sqrt{x}$ f. $y = (2x)^3$

5. Use a calculator to graph $y = x$, $y = x^2$, and $y = x^3$. Set a viewing window of $0 \le x \le 1$ and $0 \le y \le 1$. Write one or two sentences to compare the graphs.

6. Use a calculator to graph $y = x$ and $y = \sqrt{x}$. Set a viewing window of $0 \le x \le 1$ and $0 \le y \le 1$. Write one or two sentences to compare the graphs.

7. The graphs of several toolkit functions exhibit symmetry. The ability to identify symmetries will help you as you learn to graph other functions that do not belong to the toolkit.

    a. Notice when looking at the function $f(x) = x^2$, the expression for $f(-x)$ is identical to that for $f(x)$.

    $$f(-x) = (-x)^2 = x^2 = f(x)$$

    Since $f(-x) = f(x)$, for every point (x, y) on the graph of f there is also a point $(-x, y)$. The fact that these points have the same y-coordinate paired with opposite x-coordinates means that the graph of f is *symmetric about the y-axis*. A function that has this type of symmetry is called an *even function*. Which other toolkit functions are even functions?

    b. Now look at the toolkit function $m(x) = x^3$. Note that $m(-x)$ is not equal to $m(x)$. But notice that $m(-x)$ is equal to the opposite of $m(x)$.

    $$m(-x) = (-x)^3 = -x^3 = -m(x)$$

    Since $m(-x) = -m(x)$, for every point (x, y) that is on the graph of m, the point $(-x, -y)$ is also on the graph. The graph of m is *symmetric about the point* $(0, 0)$. A function with this type of symmetry is called an *odd function*. Which other toolkit functions are odd functions?

8. Why are the terms "even" and "odd" reasonable choices to describe functions whose graphs display the symmetries described in Exercise 7? (Hint: Think about power functions.)

9. Can the graph of a function be both even and odd? Explain.

10. Must a function be either even or odd? Are there toolkit functions that are neither? Do they exhibit other types of symmetry? Explain your answers and give examples to support your conclusions.

even function

odd function

11. Discuss the symmetry of the graph of $f(x) = 1 + \frac{1}{x-1}$ shown in Figure 13.

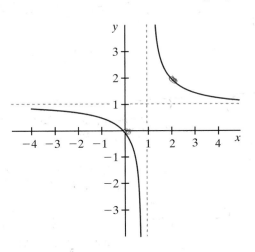

Figure 13   Graph of $f(x) = 1 + \frac{1}{x-1}$

12. Figure 14 shows the graph of a function $y = h(x)$ for $0 \le x \le 3$.

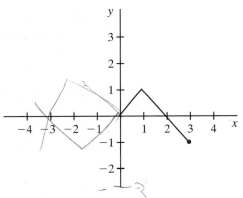

Figure 14   Graph of h for Exercise 12

  a. Sketch the graph of h from -3 to 3 if h is an even function.

  b. Sketch the graph of h from -3 to 3 if h is an odd function.

13. Use a graphing calculator to graph the functions $f(x) = x^2$ and $g(x) = 2^x$. How many times do these graphs intersect? Find the coordinates of all points of intersection.

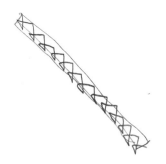

14. In defining the exponential function $f(x) = a \cdot b^x$, we stated that the base b must be greater than zero. What problems are associated with negative values of b? To help answer this question, consider $m(x) = (-2)^x$ and make a table of values for $x = -3, -2, -1, -0.5, 0, 0.5, 1, 1.5, 2$, and 3. (What happens if you try to sketch a graph of $y = m(x)$?)

15. Sketch a graph of each piecewise-defined function.

    a. $f(x) = \begin{cases} x & \text{if } x \geq -2 \\ |x| & \text{if } x < -2 \end{cases}$ **b.** $f(x) = \begin{cases} x^3 & \text{if } x < 0 \\ 2^x & \text{if } x \geq 0 \end{cases}$

    c. $f(x) = \begin{cases} x+2 & \text{if } x \leq 0 \\ 2 & \text{if } 0 < x < 4 \\ \sqrt{x} & \text{if } x \geq 4 \end{cases}$ **d.** $f(x) = \begin{cases} 1 & \text{if } |x| < 1 \\ x^2 & \text{if } x \geq 1 \end{cases}$

16. The symbol $\lfloor x \rfloor$ denotes the greatest integer that is less than or equal to x.

greatest integer function

For example, $\lfloor 11.02 \rfloor = 11, \lfloor 4.9 \rfloor = 4, \lfloor 2 \rfloor = 2, \lfloor -2.4 \rfloor = -3$. The function $f(x) = \lfloor x \rfloor$ is called the *greatest integer* or *floor function*. For any x-value between 0 and 1, $f(x) = 0$, so over this part of the domain f behaves like a constant function. Similarly, for any x-value between 1 and 2, $f(x) = 1$. Thus, the graph of f consists of a collection of horizontal line segments, or steps. The graph is shown in Figure 15. Note that it is discontinuous at every integer value of x. Identify the domain and the range of the function $f(x) = \lfloor x \rfloor$.

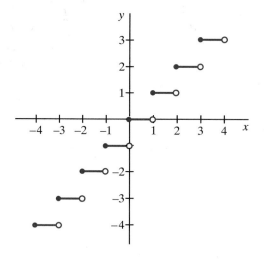

Figure 15   Graph of greatest integer function $f(x) = \lfloor x \rfloor$

17. Suppose $f(x)$ is a linear function, $f(0) = 4$, and $f(x) - f(x + 2) = -3$ for all values of x. Write an expression for f. (Hint: Information about the slope of the function is given by the expression $f(x) - f(x + 2) = -3$.)

SECTION 2.4   Finding the Domain of a Function

Each time you work with a function you may be interested in its domain. To see an example of what information the domain provides, refer back to the table of postage rates given in Figure 2 on page 59. Weights and postage rates are given in the columns of this table. Since the postage rate depends on the weight, the weight is the independent variable. We can think about a function that consists of the ordered pairs (*weight, postage rate*). The table shows postage weights from 0 to 16 ounces, so the domain consists of all real numbers from 0 to 16.

Throughout this book only real numbers are used for the domain and the range of functions. Whenever the domain for a function is not specified or implied by the context of the function used as a model, you should assume the domain is the largest set of real numbers that produce function values that are also real numbers. Restrictions on the domain are often necessary to avoid division by zero and even roots of negative numbers. When using technology, failure to recognize appropriate domain restrictions can result in a blank coordinate system or in error messages. More importantly, it can also lead to unreasonable conclusions about the phenomenon that the function is modeling.

EXAMPLE 1   Find the domains of $f(x) = \dfrac{3x - 1}{x^2 - 4}$ and $g(x) = \sqrt{6 - 7x}$.

Solution   The domain of f cannot include x-values that result in a zero in the denominator. Since $x^2 - 4$ is equal to 0 when $x = 2$ and when $x = -2$, the domain consists of all real numbers except 2 and -2.

To ensure real values in the range, the domain of g can include only x-values for which $6 - 7x$ is greater than or equal to 0. Solving the inequality $6 - 7x \geq 0$ gives $x \leq \dfrac{6}{7}$; therefore, the domain of g is $x \leq \dfrac{6}{7}$. ■

EXAMPLE 2   Find the domain of the function $y = \sqrt{x^2 - x - 12}$.

Solution   The domain of this function includes only those x-values for which $x^2 - x - 12$ is greater than or equal to zero. We will use the expression under the radical sign to define a new function: $f(x) = x^2 - x - 12$. Then to find the domain we need to find where $f(x)$ is greater than or equal to zero. Since $f(x) = x^2 - x - 12 = (x - 4)(x + 3)$, the factors of f are $(x - 4)$ and $(x + 3)$, and the zeros of the function are $x = 4$ and $x = -3$. The sign of $f(x)$ is determined by the signs of its factors; $f(x)$ is greater than zero when both factors are positive and also when both factors are negative. Figure 16 on the next page shows a useful way to display the signs of the factors on a number line.

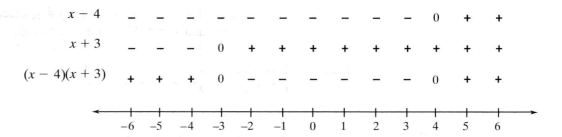

Figure 16 Signs of factors and product on a number line

We see that both factors are positive for *x*-values greater than 4 and both are negative for *x*-values less than −3. Therefore, the domain of $y = \sqrt{x^2 - x - 12}$ is $x \geq 4$ or $x \leq -3$. ∎

The technique just presented depends on being able to write the function as a product, on finding zeros, and on deciding whether or not the function changes sign at these *x*-values. For instance, the function $g(x) = (x - 5)^2$ has a zero at $x = 5$, but it does not change sign here. In fact, $g(x)$ is nonnegative for all values of *x*. Note that this technique requires that the inequality be written so that an expression is compared to zero, since zero is the unique number that separates positive and negative real numbers.

We can also approach the question in Example 2 by using a computer or graphing calculator to graph $y = x^2 - x - 12$ and then examine the graph to find *x*-values that produce nonnegative *y*-values. When using graphing technology, you need to choose the interval of *x*-values over which you want the function graphed. With some software and graphing calculators, the interval of *y*-values is automatically chosen to include all possible *y*-values that result when the function acts on the *x*-values. Other software and graphing calculators require that the user

viewing window

choose an interval of *y*-values. In either case, the result is a *viewing window* that shows the graph of a function in one particular area.

The graph of $y = x^2 - x - 12$ on the left in Figure 17 has a viewing window of $-10 \leq x \leq 10$ and $-15 \leq y \leq 40$. Keeping in mind that we cannot take the square root of a negative number, we are looking for the *x*-values associated with the portion of the graph that has *y*-values greater than or equal to zero.

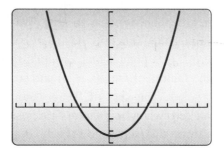 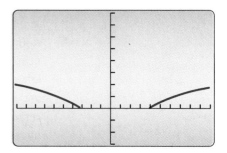

Figure 17 Graphs of $f(x) = x^2 - x - 12$ and $y = \sqrt{f(x)}$ with viewing windows of $-10 \leq x \leq -10$ and $-15 \leq y \leq 40$

Using the zero-finding option of your graphing calculator, you can find the zeros of f to be $x = -3$ and $x = 4$. Since we want to identify the x-values that are associated with the part of the graph on or above the x-axis, we choose x-values such that $x \le -3$ or $x \ge 4$. The graph on the right in Figure 17 shows $y = \sqrt{f(x)}$. Note that this graph does not exist for $-3 < x < 4$.

CLASS PRACTICE

1. Find the domain of each function.

   a. $f(x) = \sqrt{x - x^2}$

   b. $h(x) = \sqrt{x^2 - x}$

2. How are the domains of f and h related to each other?

3. Find the domain of $y = \dfrac{1}{\sin x}$.

EXAMPLE 3 Find the domain of $y = \sqrt{5x^3 - 5x^2 + 0.6}$.

Solution The domain of this function includes only those x-values for which $5x^3 - 5x^2 + 0.6$ is greater than or equal to 0. There are no good algebraic techniques for solving the inequality $5x^3 - 5x^2 + 0.6 \ge 0$, but you can define a new function $h(x) = 5x^3 - 5x^2 + 0.6$. Then use a calculator graph of h to find all x-values where $h(x) \ge 0$. Figure 18 shows the graph of h with a viewing window of $-10 \le x \le 10$ and $-10 \le y \le 10$. This viewing window shows the global behavior of the graph, but the details that we need to observe are obscured.

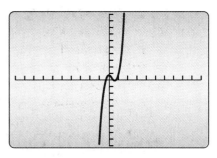

Figure 18 Graph of $h(x) = 5x^3 - 5x^2 + 0.6$ with a viewing window of $-10 \le x \le 10$ and $-10 \le y \le 10$

We need a smaller viewing window to observe the local behavior of the graph of h close to the origin. Figure 19 shows the graph of h with a viewing window of $-2 \leq x \leq 2$ and $-2 \leq y \leq 2$.

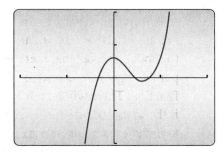

Figure 19 Graph of $h(x) = 5x^3 - 5x^2 + 0.6$ with a viewing
window of $-2 \leq x \leq 2$ and $-2 \leq y \leq 2$

We can now see that the graph of h crosses the x-axis once between $x = -1$ and $x = 0$. It also crosses the x-axis twice between $x = 0$ and $x = 1$. The values of these zeros are approximately -0.30, 0.48, and 0.82. So the values of x for which $h(x) \geq 0$ are approximately $-0.30 \leq x \leq 0.48$ and $x \geq 0.82$. Thus, the approximate domain of the original function is all x-values such that $-0.30 \leq x \leq 0.48$ or $x \geq 0.82$.

Figure 20 shows $y = \sqrt{h(x)}$ for $-2 \leq x \leq 2$ and $-2 \leq y \leq 2$. This graph confirms our statement about the domain of $y = \sqrt{5x^3 - 5x^2 + 0.6}$. The graph your calculator creates may show gaps between $y = \sqrt{h(x)}$ and the x-axis because the graph is nearly vertical there.

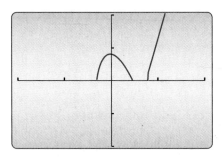

Figure 20 Graph of $y = \sqrt{h(x)}$ with a viewing
window of $-2 \leq x \leq 2$ and $-2 \leq y \leq 2$ ■

In the three previous examples, the domain of each function consists of intervals of real numbers. In most cases we expressed these intervals using familiar inequality notation. A more efficient way for communicating intervals of real numbers is to use *interval notation*. For example, the interval of real numbers indicated by $a \leq x \leq b$ can be written in interval notation as $[a, b]$. This is a *closed interval*. The interval of real numbers denoted by $a < x < b$ is written as (a, b). This is an *open interval*. The interval notation $[a, b)$ indicates a *half-open interval* and consists of the real numbers $a \leq x < b$. To denote all real numbers to the left or right of some number, we use the infinity symbol ∞ within parentheses. For example, the set of real numbers $x < a$ is indicated by $(-\infty, a)$; the set of real numbers $x > b$ is indicated by (b, ∞).

interval notation

closed interval

open interval

half-open interval

1. Use interval notation to represent the domain of each function in Examples 1, 2, and 3.

In the preceding examples, the concern with finding the domain of a function was not associated with any particular context. In the next section, when we use functions to model phenomena, there will be some restrictions on the domain that arise out of purely algebraic concerns. These are usually apparent from the equation of the function. There will be other restrictions that are implied by the situation. For example, if the independent variable represents a number of objects or people, you will not want to include fractions or negative numbers in the domain. In other situations the physical aspects of the problem impose additional limitations on the domain.

Exercise Set 2.4

1. Use algebra to find the domain of each function.

   a. $y = \sqrt{3 - x}$

   b. $y = \dfrac{1}{x - 3}$

   c. $y = \sqrt{(x - 1)(x + 2)x}$

   d. $y = \dfrac{1}{x^2 + 4}$

   e. $y = \dfrac{1}{x^2 - 5x + 6}$

   f. $y = \sqrt{x^2 - 9}$

   g. $y = \dfrac{1}{2^x - 2}$

   h. $y = \sqrt{\dfrac{x^2 - 4}{x + 7}}$

   i. $y = \sqrt{x^3 + 3x^2}$

2. Use algebra or graphical techniques to find the domain of each function.

  a. $w(t) = \dfrac{7}{t^3 - 3t - 8}$

  b. $r(x) = \sqrt{\dfrac{1}{x} + \dfrac{1}{x-1}}$

  c. $f(t) = \dfrac{1}{\sqrt{t^2 - 4}}$

  d. $h(r) = \sqrt{4r^4 - 4r^2 + 1}$

3. The two functions f and g defined below appear to be identical. However, they are identical functions only if they have the same domain and the same range. Compare the domains of f and g to help you decide if they are really identical.

$$f(x) = \dfrac{\sqrt{x+1}}{\sqrt{x-2}} \qquad g(x) = \sqrt{\dfrac{x+1}{x-2}}$$

SECTION 2.5 Functions as Mathematical Models Revisited

Now that you are familiar with the simple functions that make up the toolkit, you are ready to explore more complex and interesting functions. In this section you will analyze functions used to model real-world phenomena. Different attributes of the graphs of these functions (such as the basic shape of the graph, where the graph is increasing and where it is decreasing, the domain, the range, the zeros, and the turning points) provide information about the behavior of the phenomenon that the function models. Graphing calculators can plot many points quickly, which makes them efficient tools for graphing complex functions. However, the power of graphing calculators must be combined with your knowledge and common sense; if not, you may be misled by the information they provide.

EXAMPLE 1 Open Box

Suppose that you have a rectangular piece of cardboard that measures 18 inches by 24 inches. You can form an open box by cutting congruent squares from each of the corners of the cardboard and folding up the flaps. (See Figure 21.) What is the largest box you can make?

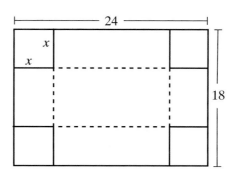

Figure 21 Cuts to form box

Solution We will assume that the largest box is the one with maximum volume. If you cut a square that measures x by x from each corner of the cardboard, then the dimensions of the open box will be x by $18 - 2x$ by $24 - 2x$. The volume of the box is given by the function $V(x) = x(18 - 2x)(24 - 2x)$.

The physical aspects of the problem impose limitations on the domain of V that are not imposed by the algebraic representation of the function. That is, without a context, the domain of V would be all real numbers. In this context, however, x must be a positive number that cannot exceed 9 inches. Thus, when you consider V as a model, the domain is all real numbers between 0 and 9, which is written $0 < x < 9$, or $(0, 9)$ in interval notation.

To get information about values of volume for squares of various sizes, we can generate a table of values of $V(x)$. Figure 22 shows that values of $V(x)$ increase for $0 \leq x \leq 3$. Somewhere between $x = 3$ and $x = 4$, the values of $V(x)$ reach a maximum and then begin to decrease.

Figure 22 Table of values for $V(x)$

x	$V(x)$	x	$V(x)$
0	0	5.0	560.0
0.5	195.5	5.5	500.5
1.0	352.0	6.0	432.0
1.5	472.5	6.5	357.5
2.0	560.0	7.0	280.0
2.5	617.5	7.5	202.5
3.0	648.0	8.0	128.0
3.5	654.5	8.5	59.5
4.0	640.0	9.0	0
4.5	607.5		

You can determine the location of the maximum volume more precisely by making another table on the interval between $x = 3.0$ and $x = 4.0$. Figure 23 shows that when you measure the square cutout to the nearest tenth of an inch, the maximum volume of 654.976 cubic inches occurs when x is 3.4 inches.

Figure 23 Refined table of values for $V(x)$

x	$V(x)$	x	$V(x)$
3.0	648.000	3.6	653.184
3.1	651.124	3.7	651.052
3.2	653.312	3.8	648.128
3.3	654.588	3.9	644.436
3.4	654.976	4.0	640.000
3.5	654.500		

Note that you could make another table to show even more values in the interval between 3.3 and 3.5. This would increase the precision of your answer but would not be useful unless you could actually measure and cut a square to the nearest hundredth of an inch.

You can also solve this problem graphically. Figure 24 shows a graph of V over the interval $[0, 9]$. The graph shows that for x-values between 0 and 9 the y-values increase, reach a maximum, and then decrease. The presence of this turning point indicates that there is a maximum box volume that we can create. The x-coordinate of this turning point tells us what size square to cut from the corners in order to create the box with the greatest possible volume, and the y-coordinate is the maximum volume.

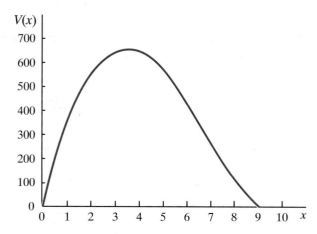

Figure 24   Graph of $V(x) = x(18 - 2x)(24 - 2x)$

According to one graphing calculator, the y-coordinate of the turning point is approximately 654.977, which occurs when $x \approx 3.394$. This information confirms what the table in Figure 23 indicated. The largest possible box has volume of approximately 655 cubic inches, and this volume is achieved when a square with side about 3.4 inches is cut from each corner.

EXAMPLE 2   Seismic Reflection

Geologists and other scientists are interested in the nature of rock formations beneath the surface of Earth. They can determine the depth and composition of layers under the surface without resorting to costly, time-consuming, and ecologically destructive drilling.

One procedure they use, called seismic reflection, is based on the fact that sound waves travel at different rates in different media. For example, sound travels at about 330 meters per second in air, 1450 meters per second in water, and 5000 meters per second in granite. In general, the more compact a medium, the faster waves will travel through it. Figure 25 on the next page shows a layer of earth of uniform thickness on top of a layer of sandstone. A seismologist will set up a vibration source at point A and a receiver at point B. Waves will radiate from A in all directions. Some waves will travel to B by traveling through the top layer of earth, along the interface between the two layers, and then back up to B. There is one path from A to B that requires the least amount of time; waves that follow this path will be the first to reach the receiver at B. We will try to find this path that takes the least time. Notice that the path of least time is not necessarily the same as the path of least distance. This is because the elapsed time depends on both the distance traveled and the speed.

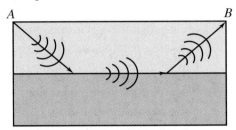

Figure 25   Earth and sandstone

Assume that waves travel through the upper layer of earth at 150 meters per second and through the sandstone layer along the interface at 1500 meters per second, that A and B are 1000 meters apart, and that the top layer of earth is 200 meters deep. We will also assume that the first wave to reach B travels through the upper layer, across the sandstone, and then back through the upper layer. (See Figure 26.) Under these assumptions, we can write a function to show the relationship between x (as shown in Figure 26) and the elapsed time. We will need to use the Pythagorean theorem and the fact that elapsed time is equal to distance divided by velocity.

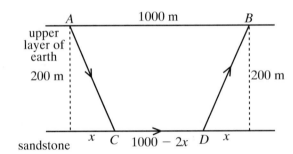

Figure 26   Path from A to B

The least-time path through the earth and along the interface consists of three sections. Two of these sections, $\overline{AC}$ and $\overline{BD}$, have length $\sqrt{40{,}000 + x^2}$. The elapsed time for these two sections in seconds is

$$2\,\frac{\sqrt{40{,}000 + x^2}}{150}.$$

The horizontal section of this path, $\overline{CD}$, has length $1000 - 2x$. The elapsed time for this section is

$$\frac{1000 - 2x}{1500}.$$

Thus, the total elapsed time $T(x)$ depends on x in the following way:

$$T(x) = 2\,\frac{\sqrt{40{,}000 + x^2}}{150} + \frac{1000 - 2x}{1500}.$$

This is quite a complicated function. Without a context, the algebraic representation of $T(x)$ allows all real numbers to be used in the domain of the function since no x-value would yield a negative value under the radical or a zero in the denominator of the function. However, in this example the distance from C to D is non-negative, so $1000 - 2x \geq$. This, in conjunction with the requirement that x is non-negative, gives a domain of $0 \leq x \leq 500$. Figure 27 shows a graph of T for $0 \leq x \leq 500$.

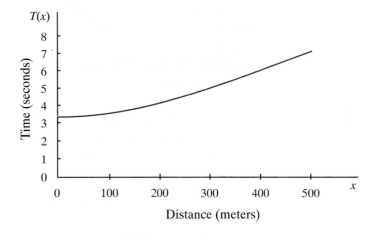

Figure 27   Graph of T for Example 2

This graph illustrates that shorter times are associated with small values of x, but it does not clearly show a turning point. Now that we have a global view of this function, we can get more information by zooming in on the portion of the graph representing $0 \leq x \leq 100$. You should adjust the viewing window to show $T(x)$-values between 3.25 and 3.5, which gives a more local view of the function and clearly shows the turning point. (See Figure 28.) The coordinates of this turning point are approximately (20.10, 3.32).

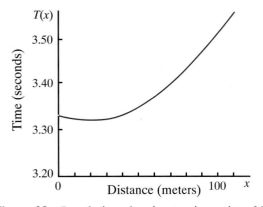

Figure 28   Local view showing turning point of T

Therefore, the wave requiring the least time reaches the interface between the two layers about 20 meters from a point directly below A. (See Figure 29.) It takes the waves about 3.3 seconds to travel the entire path from A to B.

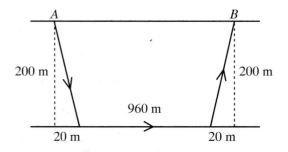

Figure 29   Least-time path ▪

EXAMPLE 3   Mortgage Payments

A real estate agent has a client who needs to borrow $65,000 to purchase a house. The client wants a 30-year variable rate mortgage, with a monthly mortgage payment of no more than $500. The equation

$$P(r) = \frac{Ar(1 + r)^n}{(1 + r)^n - 1}$$

expresses the monthly payment $P(r)$ in terms of A, the amount borrowed, r, the monthly interest rate expressed as a decimal (such as 0.01 for 1%), and n, the duration of the loan in months. What annual interest rates will result in an affordable monthly payment?

Solution   In this problem, $A = \$65,000$ and $n = 360$ months. We need to find the values of r for which P will be less than or equal to $500. That is, we need to solve the following inequality for r:

$$\frac{65,000r(1 + r)^{360}}{(1 + r)^{360} - 1} \le 500.$$

Technology can be of assistance in solving this inequality. First, define the function

$$P(r) = \frac{65,000r(1 + r)^{360}}{(1 + r)^{360} - 1}.$$

Then generate a list of values for the function. The table will show when function values fall below $500. What r-values should be included in the table? Mortgage interest rates are quoted as annual rates that are generally between 6% and 18%. Since r represents a monthly interest rate, divide the annual rates by 12. Therefore, the table should include r-values between 0.005 (which is $\frac{0.06}{12}$) and 0.015. (See Figure 30.)

Figure 30 Table of values for monthly mortgage payments

r	P	r	P
0.005	389.71	0.011	729.20
0.006	441.21	0.012	790.79
0.007	495.19	0.013	853.16
0.008	551.30	0.014	916.14
0.009	609.21	0.015	979.61
0.010	668.60		

The table shows that r-values of 0.007 and smaller result in values of $P(r)$ that are less than 500. A second table over a smaller interval shows that $P(r)$ is less than 500 when r is less than 0.00709. This means that the solution to the inequality

$$\frac{65,000r\,(1 + r)^{360}}{(1 + r)^{360} - 1} \leq 500$$

is $r < 0.00709$. A monthly rate of 0.00709 corresponds to an annual interest rate a little over 8.5%. Thus, the client will have a monthly payment below $500 whenever the annual interest rate on the variable mortgage is less than or equal to 8.5%.

You can also use technology to solve this problem graphically. Graph the function P and the constant function $f(r) = 500$ in the same viewing window. Solutions of the inequality occur wherever $P(r)$ is less than or equal to $f(r)$. Figure 31 shows P and f graphed on the same coordinate axes. The r-values for which $P(r) \leq f(r)$ are the solutions to the inequality

$$\frac{65,000r\,(1 + r)^{360}}{(1 + r)^{360} - 1} \leq 500.$$

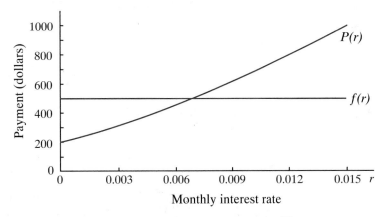

Figure 31 Graphs of $P(r) = \dfrac{65,000r(1 + r)^{360}}{(1 + r)^{360} - 1}$ and $f(r) = 500$

All r-values to the left of the point of intersection represent interest rates for which the monthly payment is less than $500. The graphs intersect at an r-value approximately equal to 0.00709. So, you can again conclude that an annual interest rate of 8.5% or lower will keep the monthly payment below $500. ■

EXAMPLE 4   The Bakery Problem

The Fresh Bakery currently sells 1000 huge chocolate chip cookies every week for $0.50 each. The bakery would like to increase its revenue from the cookie sales by increasing this price. A survey has convinced the manager that for every $0.01 increase, the bakery would sell 7 fewer cookies each week. At what price should the cookies be sold to maximize the revenue to the bakery?

Solution   If x represents the number of $0.01 increases, then $1000 - 7x$ represents the number of cookies sold. The price for each cookie is $0.50 + 0.01x$. Since revenue is the product of the number of cookies sold and the price per cookie, the function $R(x) = (1000 - 7x)(0.50 + 0.01x)$ can be used to generate revenue values. In this situation, x must be between 0 and $\frac{1000}{7}$ in order to have a positive value for revenue.

Since R is a quadratic function in factored form, we can find the maximum revenue without technology. The zeros of the function are $x = -50$ and $\frac{1000}{7}$. If we multiply the factors, the quadratic term is $-0.07x^2$, which indicates that the graph of R is a parabola opening downward. The parabola's x-intercepts are -50 and $\frac{1000}{7}$. A graph is provided in Figure 32.

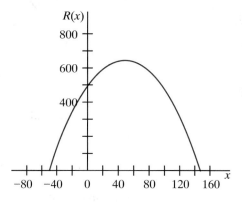

Figure 32   Graph of $R(x) = (1000 - 7x)(0.50 + 0.01x)$

We want to determine the number of $0.01 increases that maximizes the value of R. This number is the x-coordinate of the vertex of the parabola. The symmetry of the parabola implies that the x-coordinate of the vertex is midway between the x-intercepts. So, the x-coordinate of the vertex is $\frac{1}{2}(-50 + \frac{1000}{7}) \approx 46$. Recall that x represents the number of $0.01 increases, which implies that x must be an integer. Forty-six $0.01 price increases mean that each cookie will sell for $0.96 and the revenue will be $R(46) = \$650.88$. If the bakery were instead to sell each cookie for $0.97, then revenue, $R(47) = \$650.87$, would decrease. Since $R(46)$ produces the maximum revenue, the bakery should sell the cookies for $0.96 each. ∎

1. When a bottle rocket is shot into the sky, the height it reaches over time is modeled by the function $h(t) = -16t^2 + 117t + 5$, where t represents the number of seconds that have elapsed and $h(t)$ represents height in feet.

   a. Identify the domain and range of the function h. Your answers should reflect the fact that the function is used to model the bottle rocket's flight.

   b. According to the model, how high will the bottle rocket go?

   c. When will the bottle rocket be at least 175 feet in the air?

   d. Write a few sentences detailing how you determined your answer to Part c. Include any tables or graphs you may have used.

   e. According to the model, how long will the bottle rocket stay in the air?

2. The amount of power that can be generated by an undershot waterwheel in a stream depends on several factors, including the velocity of the waterwheel and the velocity of the stream. When the velocity of the stream is 4.1 feet per second, the power generated, P, is a function of the wheel's velocity, v:

$$P(v) = 1600v(4.1 - v)^2.$$

   a. Is a domain of all real numbers reasonable for the function P in the context of the problem? Explain your answer.

   b. Find the maximum power that can be generated by the waterwheel.

3. Imagine that a rectangular coordinate system is superimposed over the area where a submarine is traveling. In this system, a sonar buoy is located at the point with coordinates $(1, -1.5)$ and the submarine's path is along the graph of the function $y = x^2$. Write a function to express the distance between the submarine and the buoy in terms of the submarine's x-coordinate.

   a. Where must the submarine be located so that its distance from the sonar buoy is minimized?

   b. Where can the submarine be located so that its distance from the buoy is less than 4 units?

4. An open box is constructed by cutting congruent squares from the corners of a 30 inch by 20 inch piece of aluminum. What are the dimensions of the largest box that can be constructed?

5. The owners of a theme park know that an average of 50,000 people visit the park each day. They are presently charging $15.00 for an admission ticket. Each time in the past that they have raised the admission price, an average of 2500 fewer people have come to the park for each $1.00 increase in ticket price.

    a. Using x to represent the number of $1.00 price increases, write a function to express the relationship between x and the revenue from ticket sales.

    b. Identify the domain of this function. Your answer should reflect the fact that the function is used to model the revenue from ticket sales.

    c. What ticket price will maximize the revenue from ticket sales?

6. A museum hires a graphics company to produce a poster for its upcoming exhibit. The graphics company charges $1000 for production and design work and an additional $2.00 for each poster printed. The museum decides to sell the posters for $7.50 each.

    a. Write a function to represent the profit the museum realizes when it sells x posters.

    b. Identify the domain of this function. Your answer should reflect the fact that the function is used to model the profit from poster sales.

    c. How many posters should the museum sell to maximize profit?

7. Suppose a customer wants to borrow $10,000 and that the annual interest rate is 12%. This rate is equivalent to 1% per month. The monthly payment, P, is related to the duration of the loan in months, n, by the equation below.

    $$P(n) = \frac{10{,}000(0.01)(1.01)^n}{(1.01)^n - 1}$$

    a. Determine the monthly payment for loans of various duration: one month, one year, three years, and four years ($n = 1$, $n = 12$, $n = 36$, $n = 48$).

    b. Identify the domain and the range of the function P.

    c. The value $P = 100$ is not in the range of the function. Explain why this makes sense. What would happen to the outstanding balance on the loan if the monthly payment were $100?

8. Refer to the preceding exercise about loan repayment. Write a function to express the total amount the customer repays for the loan in terms of the duration of the loan. Use a computer or calculator to investigate how the total amount repaid varies as the duration of the loan increases from 1 year to 5 years. What implications does this have for borrowers and lenders?

9. Focusing a camera involves moving the lens so that it is an appropriate distance away from the film. When a camera is focused on a faraway object, the lens is at the infinity setting. The distance, f, that the lens must move from the infinity setting depends on two things: (1) the distance, x, from the object to the lens, and (2) the focal length, F, of the lens. (See Figure 33.) In particular,

$$f = \frac{F^2}{x - F}.$$

Since the focal length of any lens is a constant, f is a function of x. Use a calculator to investigate the relationship between f and x for a 50 mm lens ($F = 50$) between $x = 55$ mm and $x = 500$ mm. Compare the way f varies for close objects (small x-values) with the way it varies for distant objects (large x-values). Explain why close-up photographs are difficult to focus.

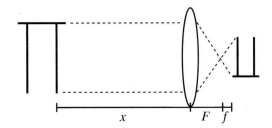

Figure 33   Camera diagram

10. The Roadrunner is standing at the edge of a cliff 800 meters above the ground. Wile E. Coyote is 42 meters above the ground on a ledge directly below the Roadrunner. Wile E. has purchased a spring catapult from the Acme Catapult Company. This catapult will propel Wile E. upward with an initial velocity of 117 meters per second. The equation that gives the height, h, of Wile E. above the ground t seconds after he releases the catapult is given by $h(t) = -4.9t^2 + 117t + 42$.

   a. Does Wile E. reach the Roadrunner? How high does he go?

   b. When is Wile E. at the same height as he was when he catapulted?

   c. When will Wile E. hit the ground?

   d. Wile E. goes back to the Acme Company to get a stronger catapult. Catapults come in integer strengths; each increase of one meter per second initial velocity costs $100. How much will Wile E. have to pay for a catapult strong enough to reach the Roadrunner?

   Source: Presentation by Wally Dodge at the National Council of Teachers of Mathematics Northeastern Regional meeting in Nashua, NH, November 1991.

11. A piece of wire 33 inches long may be cut to form a circle and/or a square. How would you cut the wire so that the resulting figure(s) would enclose the greatest area?

12. The Robinson and Barrett Oil Company wants to build a distribution center along an established, straight pipeline that runs near three cities as shown below. (See Figure 34.) To minimize the cost, the distribution center should be located so that the total length of pipe running from the distribution center to the three cities is as small as possible. Where should the distribution center be located?

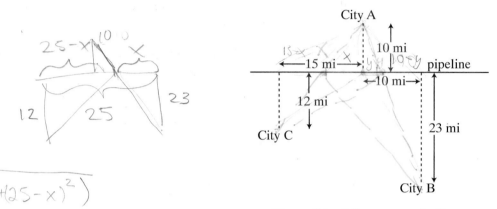

Figure 34   Oil company pipeline

$$\sqrt{\left(12^2 + (25 - x)^2\right)}$$

$$\sqrt{23^2 + x^2}$$

13. An underground telephone cable is to be laid between two boathouses on opposite banks of a straight river. One boathouse is 600 meters downstream from the other. The river is 200 meters wide. (See Figure 35.) If the cost of laying cable is $150 per meter underwater and $80 per meter on land, how should the cable be laid to minimize cost?

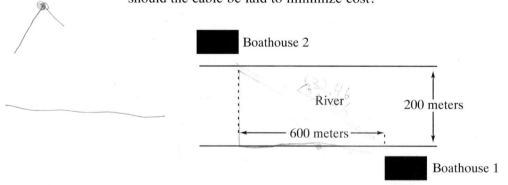

Figure 35   Underground telephone cable

14. In the Baja Desert Dune Buggy Race, participants travel in pairs consisting of a driver and a navigator. The race lasts several days and covers several hundred miles. There are periodic required check-in spots throughout the route. At one of the check-ins, a navigator is found to be dehydrated and needs to be transported to the nearest hospital. The check-in spot is located 25 miles from a point on the main road that is 70 miles from the hospital. The navigator will be driven by a medic to the main road to meet an ambulance; the ambulance will be on a road that runs directly to the hospital. The goal is to get the navigator to the hospital as quickly as possible.

a. If the dune buggy transporting the navigator travels 40 miles per hour and the ambulance travels 60 miles per hour, where along the road should the medic and the ambulance meet in order to minimize the time that the navigator must spend traveling?

b. If the situation requires that the medic get the navigator to the hospital in less than 1 hour and 45 minutes, where along the road could the dune buggy and the ambulance meet?

c. Look back over your work from Parts a and b. List at least three simplifying assumptions you made to answer these questions.

15. A delivery service handles only containers whose weight does not exceed 50 pounds. The minimum charge for a delivery is $5. A charge of 10 cents per pound is added for each full pound in excess of 5 pounds. Write an equation for the cost, c, as a function of weight, x. Identify the domain and the range of this function in the context of this problem.

16. A piece of fencing 120 feet long will be used to enclose three sides of a rectangular field. The fourth side of the field will be the wall of a barn. Let L be the length of a side of the field; let A be the area of the field.

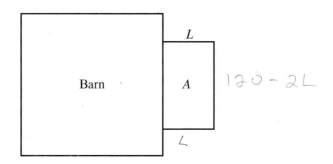

Figure 36 Fencing three sides of a field

a. Express A as a function of L.

b. What is the domain of A in the context of this problem?

c. Find the value of L that maximizes the area of the field.

17. Refer to Example 3 on page 84. At what annual interest rates will a 30-year loan for $80,000 require a monthly payment of less than $750?

18. The formula

$$d = 0.05s^2 + s$$

expresses an approximate relationship between the distance, d, required to stop a car and the speed, s, of the car (where d is measured in feet and s in miles per hour). At what speeds does it take more than 100 feet to stop?

19. Suppose you invest p dollars each month in an account that pays r percent interest per month, where r is expressed as a decimal. If you invest like this for n months, the balance $T(n)$ in the account is given by

$$T(n) = \frac{p(1 + r)((1 + r)^n - 1)}{r}.$$

A student invests $50 each month in an account that pays 6% annual interest. How long will it take until the balance in the account exceeds $1000?

SECTION 2.6 | Investigating a Conical Container

Water cups and popcorn and cotton candy containers are examples of handheld containers made in the shape of a cone. Usually the construction begins with a circular piece of material, and the cone is made by making a cut in the circle from the edge to the center along a radius and then overlapping the material. A small region of overlap results in a low, wide cone, while a larger overlap results in a tall, narrow cone. In this investigation you will determine how to construct the cone with largest possible volume from a circular material of given radius.

1. As an initial activity, cut a circle of radius 10 centimeters from a sheet of paper. Then cut a slit in the paper from the edge to the center along a radius as shown in Figure 37.

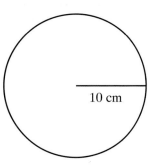

10 cm

Figure 37

Experiment with your circle by overlapping the sections along the slit until you have made the cone that appears to have maximum volume. Tape your cone so that it maintains its shape. (See Figure 38.) Now measure the cone and calculate the volume. Recall that the formula for the volume of a cone is $V = \frac{1}{3}\pi r^2 h$.

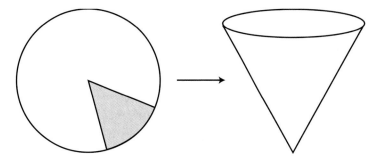

Figure 38

2. Compare the volume of your cone to those of other classmates. What dimensions produced the largest cone? Are you convinced that this is the cone with greatest possible volume? If so, how could we confirm this result analytically? If not, how could we determine the result analytically?

3. With a partner, find a function for the volume of the cone in terms of either its radius or its height. Use technology to find the dimensions of the cone with greatest volume. How close to the largest possible cone was the largest cone made by the class?

   Now begin with a circle having a different radius, and again make the cone with greatest possible volume. We could write a function for the cone's volume and use technology to find the dimensions needed for maximum volume. However, if we expected to do this over and over again, it might be more efficient to try to generalize and develop a rule for constructing a cone of maximum volume when given a circular piece of material. The question to investigate is the following: Given a circle of radius R, what are the radius and the height that will maximize the cone's volume? You will answer this question by gathering more data.

4. For at least six more circles of different radii, R, determine the cone's radius and/or height that produce the maximum volume of the cone. Then make a scatter plot of either the cone's radius or the cone's height that maximizes volume versus the radius of the circular material, R.

5. Determine a function to model this data set. This model will enable you to input the radius, $R,$ of the circular material with which you begin and output the radius or height that maximizes the volume of the cone. Test this model by using it to determine the dimensions of the largest possible cone that could be made from a 10 centimeter circle.

6. If the cone is constructed with maximum volume, how does the radius of the cone compare with the height of the cone? Is this shape consistent with the shape of an ice cream cone or conical water cup? If not, why do you think people choose different dimensions?

SECTION 2.7   Investigating Graphs of Transformations of Functions

Each group of functions below focuses on a particular type of function transformation. You should work in pairs to investigate and make generalizations about these transformations.

INVESTIGATION 1

Consider the following group of modified toolkit functions.

a. $g(x) = (x + 2)^2$ **b.** $g(x) = |x - 3|$ **c.** $g(x) = \sin(x + \frac{\pi}{2})$ **d.** $g(x) = \dfrac{1}{x - 8}$

Answer questions 1–3 for each function.

1. Use function notation to show how each function, g, is related to one of the toolkit functions, T. For example, if $g(x) = (x + 9)^2$ you would write: $g(x) = T(x + 9)$, where $T(x) = x^2$.

2. Choose an appropriate viewing window and use a graphing calculator to graph the function g and its related toolkit function, T, together. Sketch the graphs on your own paper, labeling carefully.

3. Carefully examine how the graph of g is different from its related toolkit function, T. Write specific notes about your observations.

4. Write a general statement explaining how the graph of $T(x \pm c)$, $c > 0$, differs from the graph of $T(x)$. ▨

INVESTIGATION 2

Consider the following group of modified toolkit functions.

e. $g(x) = x^3 + 1$ **f.** $g(x) = \sqrt{x} - 4$ **g.** $g(x) = \sin x + 10$ **h.** $g(x) = 2^x - 3$

5–7. Answer questions 1–3 from Investigation 1.

8. Write a general statement explaining how the graph of $T(x) \pm c$, $c > 0$, differs from the graph of $T(x)$. ▨

INVESTIGATION 3

Consider the following group of modified toolkit functions.

i. $g(x) = 4|x|$ **j.** $g(x) = 2x^3$ **k.** $g(x) = \frac{1}{2}\sqrt{x}$ **l.** $g(x) = 3\sin x$

9–11. Answer questions 1–3 from Investigation 1.

12. Write a general statement telling how the graph of $cT(x)$, $c > 0$, differs from the graph of $T(x)$. ▧

INVESTIGATION 4

Consider the following group of modified toolkit functions.

m. $g(x) = (3x)^2$ **n.** $g(x) = \frac{1}{4x}$ **o.** $g(x) = \sin(\frac{1}{2}x)$ **p.** $g(x) = 2^{3x}$

13–15. Answer questions 1–3 from Investigation 1.

16. Write a general statement explaining how the graph of $T(cx)$, $c > 0$, differs from the graph of $T(x)$. ▧

INVESTIGATION 5

Consider the following group of modified toolkit functions.

q. $g(x) = \sqrt{x - 5} + 7$ **r.** $g(x) = (x + 1)^2 - 3$

s. $g(x) = 3\sin(2x)$ **t.** $g(x) = 3|x - 4| + 2$

17–19. Answer questions 1–3 from Investigation 1. Notice that each function has several of the transformations investigated above. ▧

Basic Transformations of Functions

There are many functions that are not simple toolkit functions but nonetheless lend themselves to paper-and-pencil graphing. In many instances, sketching a graph of such a function by hand actually takes less time than using technology.

The examples in this section will provide you with information that will make sketching graphs faster and easier. The graphs you sketch by hand do not require pinpoint accuracy but rather should show characteristics such as intercepts, asymptotes, turning points, and general shape.

EXAMPLE 1   Sketch a graph of $g(x) = 2^x - 1$.

Solution   The equation of this function suggests that it is related to the toolkit function $T(x) = 2^x$. Using function notation, you can write $g(x) = T(x) - 1$. For each x-value, the y-value of g will be 1 less than the y-value of T. Thus, the graph of $g(x) = 2^x - 1$ can be obtained by shifting the toolkit graph $T(x) = 2^x$ down 1 unit. The result is the shifted exponential toolkit function with asymptote $y = -1$ shown in Figure 39.

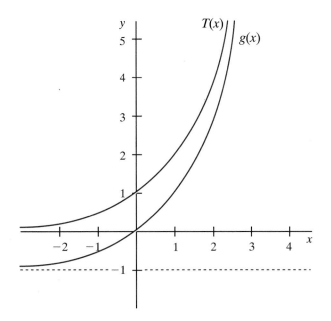

Figure 39   Graphs of $g(x) = 2^x - 1$ and $T(x) = 2^x$

EXAMPLE 2   Graph $g(x) = \sqrt{x - 3}$.

Solution   This function is a variation of the toolkit function $T(x) = \sqrt{x}$. We can write $g(x) = T(x - 3)$. Note that the function's argument, the value for which the function is being evaluated, is $x - 3$; x-values are decreased by 3 before the square root function acts on them to produce an output.

Therefore, to obtain a particular y-value on the graph of g, the x-value must be 3 more than it was to obtain the same y-value on the graph of T. For example, on the graph of $T(x) = \sqrt{x}$, the y-value is 2 when x is 4. To obtain a y-value of 2 from function g, the x-value has to be 7. Thus, the point $(4, 2)$ on $T(x) = \sqrt{x}$ is mapped to the point $(7, 2)$ on $g(x) = \sqrt{x - 3}$. This explains why the graph of $T(x) = \sqrt{x}$ must be shifted 3 units to the right to get the graph of $g(x) = \sqrt{x - 3}$ as shown in Figure 40.

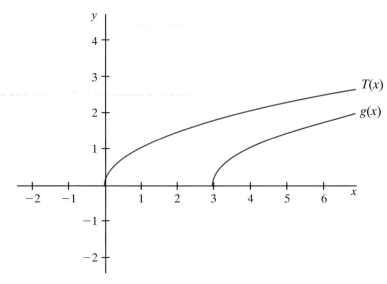

Figure 40   Graphs of $g(x) = \sqrt{x - 3}$ and $T(x) = \sqrt{x}$

There is an important difference between $y = f(x - k)$ and $y = f(x) - k$, where k is a constant. In the former, the argument of the function is $x - k$, which means that k is subtracted from x before the function acts on its input. In the latter, the argument of the function is x, and k is subtracted after the function acts on its input.

We will use the function $f(x) = x^3$ to review what was observed in the preceding examples. First, consider the function defined by $g(x) = (x - 1)^3 = f(x - 1)$. For any fixed y-value, the x-value will be one more than it was for $y = f(x)$, and the graph will be shifted one unit to the right. This is an example of a *horizontal translation*. Second, consider the function defined by $h(x) = x^3 + 1 = f(x) + 1$. For any fixed x-value, the y-value is one more than for $y = f(x)$, and the graph has been shifted one unit up. This is an example of a *vertical translation*.

horizontal translation

vertical translation

CLASS PRACTICE   1. Sketch a graph of $g(x) = |x| + 1$.

2. Sketch a graph of $g(x) = (x + 1)^2$.

EXAMPLE 3   Sketch a graph of $g(x) = 2\sin x$.

Solution   This function is related to the sine function in the toolkit. If we define $T(x) = \sin x$, then the function we want to graph can be written as $g(x) = 2T(x)$. For any fixed x-value, the y-coordinate on the graph of $g(x) = 2T(x)$ will be twice as large as the y-coordinate on the graph of $y = T(x)$. Thus, the points $(0, 0)$, $(\frac{\pi}{2}, 1)$, $(\pi, 0)$, and $(\frac{3\pi}{2}, -1)$ on the graph of $T(x) = \sin x$ are mapped to $(0, 0)$, $(\frac{\pi}{2}, 2)$, $(\pi, 0)$, and $(\frac{3\pi}{2}, -2)$ on the graph of $g(x) = 2\sin x$. These points and your familiarity with the graph of $T(x) = \sin x$ lead to the graph shown on the left in Figure 41. You should compare this graph to that of $T(x) = \sin x$, which is shown on the right in Figure 41. What relationships can you see between the domains and the ranges of the two functions?

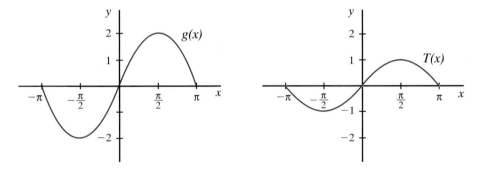

Figure 41   Graphs of $g(x) = 2\sin x$ and $T(x) = \sin x$

EXAMPLE 4   Sketch a graph of $h(x) = \sin(2x)$.

Solution   This is another function that is related to the sine function. Note that the argument of this function is $2x$; x-values are doubled before the function acts on them to produce an output. Therefore, to obtain a particular y-value on the graph of h, the x-value must be only half the x-value needed to obtain the same y-value on the graph of $T(x) = \sin x$. For example, on $T(x) = \sin x$, the y-value is 1 when x is $\frac{\pi}{2}$. To obtain a y-value of 1 from the function h, you must have $2x = \frac{\pi}{2}$ or $x = \frac{\pi}{4}$. Thus, the point $(\frac{\pi}{2}, 1)$ on $T(x) = \sin x$ is mapped to the point $(\frac{\pi}{4}, 1)$ on $h(x) = \sin(2x)$. The special values in Figure 42 on the next page show that the cycle of the sine curve over $0 \le x \le 2\pi$ is completed in an interval half as long, that is $0 \le x \le \pi$ for $h(x) = \sin(2x)$.

Figure 42 Table of values

y = sin x		y = sin (2x)		
x	y	x	2x	y
0	0	0	0	0
$\frac{\pi}{2}$	1	$\frac{\pi}{4}$	$\frac{\pi}{2}$	1
π	0	$\frac{\pi}{2}$	π	0
$\frac{3\pi}{2}$	−1	$\frac{3\pi}{4}$	$\frac{3\pi}{2}$	−1
2π	0	π	2π	0

If we continue this analysis, we get the graph of h shown on the left in Figure 43. Compare this graph to that of the toolkit function $T(x) = \sin x$, which is shown on the right in Figure 43. Notice that the period of $h(x) = \sin(2x)$ is π, which is half the period of $T(x) = \sin x$.

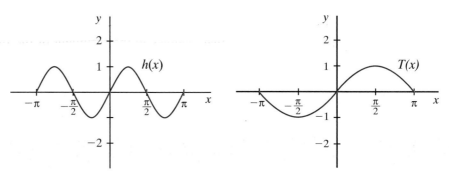

Figure 43 Graphs of $h(x) = \sin(2x)$ and $T(x) = \sin x$

vertical stretching

horizontal compression

The graphs of $g(x) = 2\sin x$ and $h(x) = \sin(2x)$ are both closely related to the graph of $T(x) = \sin x$, but the 2 in each equation has a different effect. The 2 in $g(x) = 2\sin x$ doubles y-values and causes a *vertical stretching* about the x-axis. The 2 in the argument of $h(x) = \sin(2x)$ halves x-values and causes a *horizontal compression* about the y-axis.

CLASS PRACTICE

1. Sketch a graph of $g(x) = \frac{1}{2}\sin x$. Describe the effect of "$\frac{1}{2}$" on the graph.

2. Sketch a graph of $g(x) = \sin\left(\frac{1}{2}x\right)$. Describe the effect of "$\frac{1}{2}$" on the graph.

You have now seen several different ways that constants and coefficients in the equation of a function can affect the graph. These effects are summarized in the table that follows.

Graphing Transformations of Functions, $c > 0$	
$g(x) = T(x) + c$	graph is translated c units upward
$g(x) = T(x) - c$	graph is translated c units downward
$g(x) = T(x + c)$	graph is translated c units to the left
$g(x) = T(x - c)$	graph is translated c units to the right
$g(x) = cT(x)$	graph is stretched or compressed vertically by a factor of c
$g(x) = T(cx)$	graph is stretched or compressed horizontally by a factor of c

Note that changes in the argument of the function alter the graph horizontally by influencing the domain, whereas changes outside the argument alter the graph vertically by influencing the range.

Exercise Set 2.8

1. Write a sentence to describe how the graph of each function is related to that of a toolkit function. Sketch a graph of each function. Label coordinates of two or three key points and any asymptotes to help the reader understand the transformation.

   a. $g(x) = |x| + 5$ b. $g(x) = x^3 - 2$ c. $g(x) = \frac{1}{x + 2}$

   d. $g(x) = 2^x + 3$ e. $g(x) = (x - 4)^3$ f. $g(x) = |x + 3| - 3$

   g. $g(x) = \sin(x + \pi)$ h. $g(x) = \sqrt{9x}$ i. $g(x) = \frac{1}{5}x^2$

   j. $g(x) = \frac{2}{x}$ k. $g(x) = 2^{\frac{x}{3}}$

2. How can the graph of the toolkit function $T(x) = \sqrt{x}$ be transformed to give the graphs of $g(x) = \sqrt{4x} = T(4x)$ and $h(x) = 2\sqrt{x} = 2T(x)$? Sketch both graphs and write a sentence or two describing how horizontal and vertical compressions and stretches are related for this particular function.

3. How can the graph of the toolkit function $T(x) = 2^x$ be transformed to give the graphs of $g(x) = 2^{(x+1)} = T(x + 1)$ and $h(x) = 2 \cdot 2^x = 2T(x)$? Sketch both graphs and write a sentence or two describing how shifts and stretches are related for this particular function.

4. Figure 44 shows the graph of a function g. The domain is $-3 \le x \le 4$ and the range $-1 \le y \le 2$. Make a graph of each transformation of g below and state the domain and range for each.

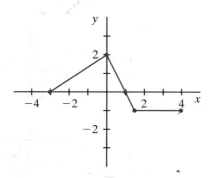

Figure 44   Graph of g for Exercise 4

a. $y = g(x) - 3$ **b.** $y = \frac{1}{3}g(x)$

c. $y = g(\frac{1}{3}x)$ **d.** $y = g(x + 3)$

5. Sketch a graph of each piecewise-defined function.

   a. $y = \begin{cases} (x - 1)^2 & \text{if } x \ge 0 \\ x^3 - 2 & \text{if } x < 0 \end{cases}$ **b.** $y = \begin{cases} \sqrt{\frac{1}{2}x} & \text{if } x > 0 \\ \sin(3x) & \text{if } x \le 0 \end{cases}$

   c. $y = \begin{cases} |x - 4| & \text{if } x \ge 2 \\ \sqrt{x + 7} & \text{if } -3 < x < 2 \\ \frac{3}{x} & \text{if } x \le -3 \end{cases}$

6. Use your knowledge of toolkit functions and your understanding of transformations to determine the domain and range of each function. You may want to look at a graph if you have difficulty.

   a. $f(x) = x^2 - 1$ **b.** $f(x) = \sin x - 2$

   c. $f(x) = \sqrt{x - 3}$ **d.** $f(x) = 2^x + 5$

   e. $f(x) = 0.01x^3$ **f.** $f(x) = |x + 3|$

   g. $f(x) = |x| + 3$ **h.** $f(x) = 3\sin x$

i. $f(x) = \dfrac{1}{x+4}$ **j.** $f(x) = 3 \cdot 2^x$

7. Each graph in Figures 45–48 is the transformation of a toolkit function. (In each graph, the scale on both axes is 1.) Write an equation for each graph.

a.

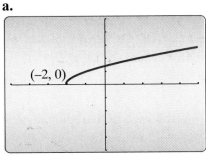

Figure 45

b.

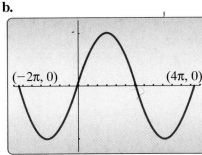

Figure 46

c.

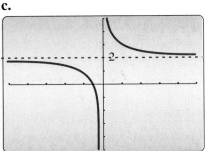

Figure 47

d.

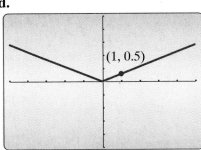

Figure 48

8. Investigate the effects on the graphs of the functions $g(x) = cT(x)$ and $h(x) = T(cx)$ for $c < 0$. Describe how negative c-values affect the graphs.

9. There are noticeable differences in the effects of c-values between 0 and 1 and c-values greater than 1 on the graphs of functions of the form $g(x) = cT(x)$. Describe these differences.

10. There are noticeable differences in the effects of c-values between 0 and 1 and c-values greater than 1 on the graphs of functions of the form $h(x) = T(cx)$. Describe these differences.

11. Write a sentence describing how the graph of each function is related to that of $h(x) = \lfloor x \rfloor$ (the greatest integer function). Sketch a graph of each function. Label axes carefully.

a. $y = \lfloor 4x \rfloor$ **b.** $y = \lfloor x \rfloor - 2$

SECTION 2.9   Combinations of Transformations

The generalizations about transformations of the form $g(x) = cT(x)$ and $g(x) = T(cx)$ listed in the summary table in the previous section required positive c-values. When c-values are negative, stretches and compressions occur in a slightly different way as the next two examples will illustrate.

EXAMPLE 1   Graph $f(x) = -\sqrt{x}$ and $g(x) = \sqrt{-x}$.

Solution   To graph each of these functions, we could either plot points or use technology. However, both functions are closely related to the toolkit function $T(x) = \sqrt{x}$ and are easily graphed by hand with the benefit of knowledge about transformations.

We will graph $f(x) = -\sqrt{x}$ first. For each fixed x-value, the y-value on the graph of f will be the opposite of the y-value on the graph of the toolkit function. For every point (x, y) on the graph of $T(x) = \sqrt{x}$, there is a point $(x, -y)$ on the graph of f. Thus, the graph of the toolkit function can be reflected about the x-axis to produce the graph of $f(x) = -\sqrt{x}$. The graph of f is shown on the left in Figure 49.

Now consider $g(x) = \sqrt{-x}$. Since the domain of the square root function is the nonnegative real numbers, x-values in the domain of g must satisfy the inequality $-x \geq 0$. This means that the domain is $x \leq 0$. Another way to look at this is to see that for each fixed y-value, the x-value on the graph of g will be the opposite of the x-value on the graph of the toolkit function. In either case, for every point (x, y) on the graph of $T(x) = \sqrt{x}$, there is a point $(-x, y)$ on the graph of g. This means that the graph of the toolkit function can be reflected about the y-axis to produce the graph of $g(x) = \sqrt{-x}$. The graph of g is shown on the right in Figure 49. You should verify the coordinates of a few points on the graphs of f and g to help you understand why their graphs look the way they do.

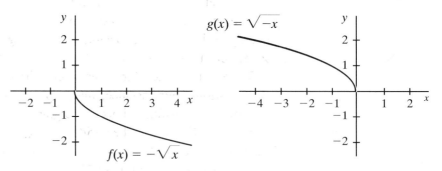

Figure 49   Graphs of $f(x) = -\sqrt{x}$ and $g(x) = \sqrt{-x}$

EXAMPLE 2   Sketch a graph of $h(x) = \sqrt{-2x}$.

Solution   The function h combines a horizontal compression with a reflection about the y-axis. Notice that the x-values in the domain of h must satisfy the inequality $-2x \geq 0$, so the domain is $x \leq 0$.

How is this graph of h related to that of the toolkit function $T(x) = \sqrt{x}$? Since the constant -2 is in the argument of h, you can best analyze the relationship between the graphs by thinking about what x-value is needed to produce a particular y-value. On the toolkit function, $y = 1$ when $x = 1$. On h, $y = 1$ when $-2x = 1$ or $x = -\frac{1}{2}$. Similarly, the point $(9, 3)$ on the toolkit function is mapped to the point $\left(-\frac{9}{2}, 3\right)$ on h. These observations suggest that the graph of h can be obtained by reflecting the toolkit graph about the y-axis and compressing it horizontally. Figure 50 shows the graphs of h and T together.

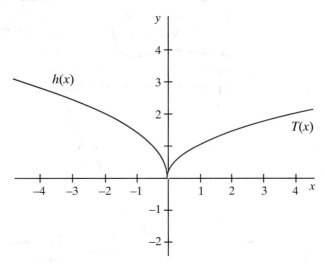

Figure 50   Graphs of $T(x) = \sqrt{x}$ and $h(x) = \sqrt{-2x}$

In general, a negative value for c in $g(x) = cT(x)$ or $g(x) = T(cx)$ has an effect similar to one of those observed in Example 1. The absolute value of c will determine the compression or stretch factor, and the negative sign will result in a reflection about either the x- or y-axis.

More complicated functions often exhibit a combination of the four types of transformations. For instance, the graph of $g(x) = \frac{1}{2}|x - 1| - 4$ will incorporate three different transformations of the graph of the toolkit function $T(x) = |x|$. Since

vertical compression

$g(x) = \frac{1}{2}|x - 1| - 4$ can be written as $g(x) = \frac{1}{2}T(x - 1) - 4$, the toolkit graph should undergo a *vertical compression*, be translated 1 unit to the right, and be translated 4 units down. (See Figure 51 on the next page.)

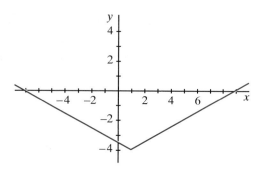

Figure 51 Graph of $g(x) = \frac{1}{2}|x - 1| - 4$

EXAMPLE 3 Refer to the function f whose graph is shown in Figure 52. Sketch the graphs of $g(x) = f(\frac{1}{2}x)$ and $h(x) = f(\frac{1}{2}x + 1)$.

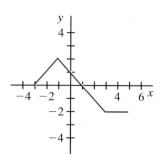

Figure 52 Graph of f for Example 3

What does the graph of

$g(x) = f\left(\frac{1}{2}x\right)$ *look like?*

horizontal stretch

Solution Notice that the domain of f is $-3 \le x \le 5$. The point $(-3, 0)$ on the graph indicates that $f(-3) = 0$; similarly $f(-1) = 2, f(1) = 0$, and so on.

What does the graph of $g(x) = f(\frac{1}{2}x)$ look like? You should be able to predict that the $\frac{1}{2}$ in the argument of the function causes a *horizontal stretch* by a factor of 2. Therefore, the domain of g is $-6 \le x \le 10$. Another way to reach this conclusion is to reason that the argument, a, of the function f must satisfy the inequality

$$-3 \le a \le 5.$$

Therefore, the argument $\frac{1}{2}x$ of the function $f(\frac{1}{2}x)$ must satisfy the inequality

$$-3 \le \frac{1}{2}x \le 5,$$

which is equivalent to

$$-6 \le x \le 10.$$

Points on the graph of g can be determined by choosing a particular y-value and then asking yourself what x-value will produce that y-value as output. For instance, on the graph of f the y-value is 2 when $x = -1$. On g the x-value must be twice as big to produce the same y-value, so the graph of g contains the point $(-2, 2)$. Similarly, the graph of g contains the points $(-6, 0)$, $(2, 0)$, $(6, -2)$, and $(10, -2)$. The complete graph of g is shown in Figure 53.

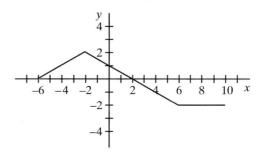

Figure 53   Graph of $g(x) = f(\frac{1}{2}x)$

Can you predict what the graph of $h(x) = f\left(\frac{1}{2}x + 1\right)$ *will look like?*

Can you predict what the graph of $h(x) = f(\frac{1}{2}x + 1)$ will look like? You may expect to translate the graph of $g(x) = f(\frac{1}{2}x)$ one unit to the left. But, after you find the domain and plot some points you will see the error in this prediction.

Since the argument of h is $\frac{1}{2}x + 1$, values in the domain of h must satisfy the inequality

$$-3 \le \tfrac{1}{2}x + 1 \le 5,$$

which is equivalent to

$$-4 \le \tfrac{1}{2}x \le 4$$

or

$$-8 \le x \le 8.$$

You can also determine what happens to particular points on the graph of f. For example, the graph of f contains the point $(-1, 2)$. What x-value will produce a y-value of 2 on $f(\frac{1}{2}x + 1)$? Since $f(-1) = 2$, the argument $\frac{1}{2}x + 1$ must be equal to -1. Solving for x gives $x = -4$, so the graph of h contains the point $(-4, 2)$. The complete graph of $h(x) = f(\frac{1}{2}x + 1)$ over the domain $-8 \le x \le 8$ is shown in Figure 54.

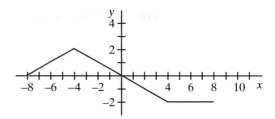

Figure 54   Graph of $h(x) = f(\frac{1}{2}x + 1)$

Comparing the graph of h in Figure 54 to that of g in Figure 53 indicates that the graph of $h(x)$ is obtained by shifting $g(x)$ to the left 2 units. Why is the shift 2 units instead of 1 unit? One way to answer this question is to reason that the horizontal stretch caused one horizontal unit on the graph of f to become 2 units on the graph of g.

You can help yourself remember this slightly surprising result in the future by writing the expression for h as $h(x) = f(\frac{1}{2}(x + 2))$. This form indicates that the graph of f has been stretched horizontally by a factor of 2 and then translated 2 units to the left. ■

Sometimes it is difficult to recognize the transformations that are involved in the graph of a function. The next example reviews an algebraic technique to help you recognize such transformations.

EXAMPLE 4   Sketch a graph of $y = 2x^2 + 12x + 19$.

completing the square

Solution   This equation contains an x^2-term, an x-term, and a constant term. It is not immediately obvious how the toolkit function $T(x) = x^2$ has been transformed. Fortunately, the equation can be rewritten using an algebraic technique called *completing the square*. We assume you learned this technique in a previous course. This example is included to refresh your memory and to illustrate the value of completing the square in graphing quadratic functions.

$$y = 2x^2 + 12x + 19$$

$$y = 2(x^2 + 6x) + 19$$

$$y = 2(x^2 + 6x + \text{?}) + 19$$

$$y = 2\left(x^2 + 6x + \left(\frac{6}{2}\right)^2\right) + 19 - 18$$

The "square was completed" by adding 9 to make a perfect square trinomial inside the parentheses. When we add 9 inside the parentheses, we are in effect adding 18 to the right side of the equation, so 18 is subtracted to maintain equality. Simplifying the final equation above gives

$$y = 2(x + 3)^2 + 1.$$

With the equation written in this form, our function can be expressed as $y = 2T(x + 3) + 1$, where $T(x) = x^2$. So the toolkit parabola has been stretched vertically by a factor of 2, shifted 3 units to the left, and shifted 1 unit up. The graph is shown in Figure 55.

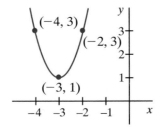

Figure 55   Graph of $y = 2x^2 + 12x + 19$ ■

Exercise Set 2.9

1. Write a sentence to describe how the graph of each function is related to that of a toolkit function. Sketch a graph of each function and label coordinates of two or three points.

   a. $f(x) = -2\sqrt{x}$

   b. $f(x) = x^2 - 2x - 24$

   c. $f(x) = 3(x + 1)^3 - 2$

   d. $f(x) = -\sin(-x)$

   e. $f(x) = 4x^2 + 28x + 53$

   f. $f(x) = \sqrt{5 - x}$

   g. $f(x) = -2\sin(x + \pi)$

   h. $f(x) = |3x - 6|$

   i. $f(x) = -\frac{1}{5}x^2$

   j. $f(x) = 4 - |x|$

   k. $f(x) = -x^2 + 7x - 12$

   l. $f(x) = -2^{x+3}$

   m. $f(x) = \frac{4}{x - 3}$

   n. $f(x) = \frac{1}{2x + 4}$

   o. $f(x) = -2\sqrt{x - 4} + 3$

   p. $f(x) = \frac{-2}{x + 1}$

   q. $f(x) = 2^{-x+1}$

   r. $f(x) = \sin(2x - \pi)$

   s. $f(x) = -\sqrt{2 - x}$

   t. $f(x) = -2|x - 4|$

2. The graph of a piecewise-defined function f is shown in Figure 56. For each transformation of f, sketch a graph and identify the domain and the range.

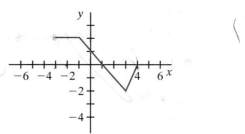

Figure 56   Graph of f for Exercise 2

   a. $g(x) = f(x + 3)$

   b. $g(x) = 2f(x) + 3$

   c. $g(x) = f(-2x)$

   d. $g(x) = -2f(x)$

   e. $g(x) = f(-x + 1)$

   f. $g(x) = \frac{1}{2}f(x) - 4$

   g. $g(x) = f(2x - 2)$

3.   Your study partner is graphing the functions below on his or her calculator with a viewing window of $-10 \le x \le 10$ and $-10 \le y \le 10$. When each of the functions is graphed, the calculator screen is blank. Explain to your study partner why the screen is blank, and then suggest a better viewing window for each.

a. $y = 2^x + 11$ b. $y = 14 + \frac{9}{x+12}$

c. $y = \sqrt{-x - 50}$

4.   Each graph in Figures 57–59 is the transformation of a toolkit function. Write an equation for each graph.

a.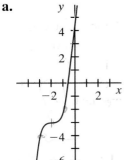

b.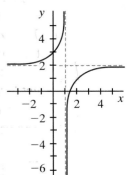

c.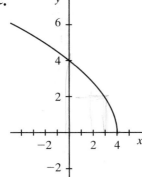

Figure 57   **Figure 58**   **Figure 59**

5.   For each of the situations in i–iv, make a scatter plot of the data and answer the following questions:

a.   What basic toolkit function does the scatter plot resemble? There may be information about the phenomenon that helps you determine what type of function to expect. For example, you might expect the amount of daylight each day of the year to be modeled by a *sinusoidal function*.

sinusoidal function

b.   How would the basic toolkit function need to be transformed to be more correctly placed and/or shaped? For example, would it be a better match if the toolkit function $T(x) = \sin x$ were moved to the left 2 units? stretched horizontally?

c.   Write a function that incorporates the changes you described in Part b and graph it with the data set.

i.   *Ball off the Roof* A ball is thrown upward from the top of an 80-foot building. Its height above ground at various times is given in Figure 60.

Figure 60   Height of thrown ball

Time (sec)	0	0.2	0.6	1.0	1.2	1.5	2.0	2.5	2.8	3.4	3.8	4.5
Height (ft)	80	92	110	130	134	142	144	140	132	112	90	44

ii. Stopping Distance The distance it takes to stop a car has two parts: thinking distance and braking distance, which when added together result in stopping distance. Figure 61 gives some approximate figures for thinking and braking distances. Three sets of ordered pairs can be compiled from this information; in each data set, speed is the independent variable. Investigate each relationship.

Figure 61   Stopping distances

Miles per hour	25	35	45	55	65
Thinking distance (ft)	27	38	49	60	71
Braking distance (ft)	34.7	68	112.5	168	234.7

iii. Tree Volume People who work in forestry need an easy way to estimate the volume of a tree so that they can determine the amount of timber available in a region of forest. Since volume cannot be directly measured until a tree is felled, foresters need to predict volume on the basis of other measurements. Figure 62 contains measurements of diameter and volume.

Figure 62   Diameter and volume of trees

Diameter (inches)	17	19	20	23	25	28	32	38	39	41
Volume (100s of board ft)	19	25	32	57	71	113	123	252	259	294

Source: Reprinted with permission from *Data Analysis*, © 1988 by the National Council of Teachers of Mathematics. All rights reserved.

iv. Focal Length The focal length setting on a camera is related to the length of a meterstick visible through the lens. As the camera "zooms in," a smaller and smaller portion of the meterstick is visible. Sample data are given in Figure 63.

Figure 63   Focal length and length of meterstick visible

Focal length	35	40	45	50	55	60	65	70
Length of meterstick visible	93	84	74	67.5	62.5	57	53	50

SECTION 2.10 Composition of Functions

Last week Lee and Paul went swimming in a quarry near their home. They took turns trying to dive to the bottom of the quarry but were never able to hold their breath long enough to make it all the way down. During the dives, Lee realized that it was much darker at his deepest point than at the surface. After the swim, he decided to search for a mathematical model that would relate the amount of underwater sunlight and the time he had been descending.

Lee did some research and found that the amount of sunlight reaching to a depth d decreases according to the function $f(d) = 2^{-d/15}$. In this function, d is the depth of the water in feet and $f(d)$-values represent the proportion of full sunlight at a particular depth. Lee estimated that he descended at a rate of 1.5 feet per second. Therefore, his depth in feet after t seconds is given by the function $g(t) = 1.5t$, where t is in seconds and $g(t)$-values are in feet. Both functions are illustrated below in terms of their input and output.

$$[\text{time}] \xrightarrow{g} [\text{depth}]$$

$$[\text{depth}] \xrightarrow{f} [\text{proportion of sunlight}]$$

Lee is interested in a model for the relationship between the proportion of sunlight and the time since beginning to descend. Since the proportion of sunlight is a function of depth and depth is a function of time, the amount of sunlight is also a function of time. The flow of values through g and then f is shown below.

$$[\text{time}] \xrightarrow{g} [\text{depth}] \xrightarrow{f} [\text{proportion of sunlight}]$$

To create the new function N that directly expresses the relationship between time and the proportion of sunlight, we will first restate the relationships between time, depth, and sunlight more symbolically. Since $g(t)$ can be used to determine the depth after a given time, t, $g(t)$ can be used as an input to the function f. Then $f(g(t))$ gives the proportion of sunlight available at the depth given by $g(t)$. This is illustrated by

$$N : t \xrightarrow{g} g(t) \xrightarrow{f} f(g(t)).$$

The function that relates time and sunlight directly is $N(t) = f(g(t))$. The symbol $f(g(t))$ is read "f of g of t." To find an algebraic expression for N, simply work from the inside, replacing function names with their algebraic expressions:

$$N(t) = f(g(t)) = f(1.5t) = 2^{(-1.5t/15)}.$$

composition

inner function

outer function

This function N is called the *composition* of the functions f and g. For any two functions f and g, the symbol $f \circ g$ is used to denote the composition of f and g and can be read "f composed with g." The notation $f \circ g(t)$ is equivalent to $f(g(t))$. In both notations, g is called the *inner function* and it acts on t, whereas f is called the *outer function* and it acts on the output of g, namely $g(t)$.

EXAMPLE 1 Let $f(x) = x^2$ and $g(x) = 2x + 1$. Find $f(g(2))$, $g(f(2))$, an expression for the composition $f \circ g$, and an expression for the composition $g \circ f$.

Solution

$$f(g(2)) = f(5) = 25$$
$$g(f(2)) = g(4) = 9$$
$$f \circ g(x) = f(g(x)) = f(2x + 1) = (2x + 1)^2 = 4x^2 + 4x + 1$$
$$g \circ f(x) = g(f(x)) = g(x^2) = 2(x^2) + 1 = 2x^2 + 1 \quad \blacksquare$$

Whenever two functions are composed, there are important relationships between their domains and ranges. In the composition $g \circ f$, values in the range of f provide possible domain values for g. Therefore, the range of f and the domain of g must have at least one value in common in order for $g \circ f$ to have meaning. In particular, a real number x is in the domain of the composition $g \circ f$ only if it satisfies two conditions. First, x must be in the domain of the inner function f. Second, x must produce an output $f(x)$ that is in the domain of the outer function g.

EXAMPLE 2 Let $f(x) = \frac{1}{x}$ and $g(x) = \frac{1}{x-2} + 1$. Find an expression for $g \circ f$ and identify the domain of the composition.

Solution It is useful to begin by examining the domain and range of f and g.

Domain of f: all real numbers except $x = 0$

Range of f: all real numbers except $f(x) = 0$

Domain of g: all real numbers except $x = 2$

Range of g: all real numbers except $g(x) = 1$

Since a number can be in the domain of $g \circ f$ only if it is in the domain of the inner function f, the domain of $g \circ f$ cannot include $x = 0$.

The first step in finding the new function $g \circ f$ is to find its algebraic expression:

$$g(f(x)) = g\left(\frac{1}{x}\right) = \frac{1}{\frac{1}{x} - 2} + 1.$$

Notice that if $x = \frac{1}{2}$, the denominator of the first term becomes zero. Therefore, the domain of this new function $g \circ f$ is all real numbers except $x = 0$ and $x = \frac{1}{2}$.

We can simplify the expression for $g \circ f$ as follows:

$$g(f(x)) = \frac{1}{\frac{1}{x} - 2} + 1 = \frac{1}{\frac{1 - 2x}{x}} + 1 = \frac{x}{1 - 2x} + \frac{1 - 2x}{1 - 2x} = \frac{1 - x}{1 - 2x}, \quad x \neq 0, x \neq \frac{1}{2}$$

Therefore,

$$g \circ f(x) = \frac{1 - x}{1 - 2x}, \qquad x \neq 0.$$

Notice that the final expression for $g \circ f$ implies that $x = \frac{1}{2}$ is not in the domain of $g \circ f$, but states directly that $x = 0$ is not in the domain. The fact that $x = 0$ is not in the domain of $g \circ f$ is evident in the unsimplified algebraic expression for $g \circ f$ but is no longer apparent when that expression is simplified. ∎

EXAMPLE 3   Let $f(x) = \sqrt{x}$ and $g(x) = x^2 - 9$. Find $f \circ g$, $g \circ f$, and the domain of each composition.

Solution   We begin by identifying the domain and range of f and g.

Domain of f: $x \geq 0$ Range of f: $f(x) \geq 0$

Domain of g: all real numbers   Range of g: $g(x) \geq -9$

Now find an expression for $f \circ g$.

$$f(g(x)) = f(x^2 - 9) = \sqrt{x^2 - 9}$$

To find the domain of $f \circ g$, first consider the domain of the inner function. Since the domain of g is all real numbers, at this point we have no restrictions. However, the expression for $f \circ g$ makes it clear that its domain must be restricted to values of x that satisfy $x^2 - 9 \geq 0$. One way to solve this inequality is to factor $x^2 - 9$, and then find values of x so that $(x + 3)(x - 3) \geq 0$. Recall that a product of two factors is positive when both factors are positive or when both factors are negative. Figure 64 shows the signs of the factors on a number line.

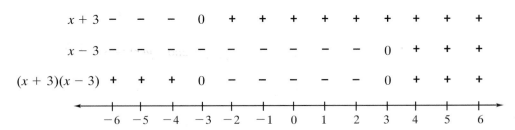

Figure 64   Signs of factors and product on a number line

Notice that $x^2 - 9 \geq 0$ when $x \leq -3$ or $x \geq 3$, so the domain of $f \circ g$ is $x \leq -3$ or $x \geq 3$.

Now consider $g \circ f$ and its domain. Combining the expressions for f and g, we see that

$$g(f(x)) = g(\sqrt{x}) = (\sqrt{x})^2 - 9.$$

The inner function, f, has a restricted domain of $x \geq 0$. Now, since $(\sqrt{x})^2 = x$ for nonnegative x-values, we can simplify the algebraic expression for $g \circ f$.

$$g \circ f(x) = x - 9, \quad x \geq 0 \quad \blacksquare$$

When you write an expression for a composition, be sure to indicate that the function has domain restrictions if they are not obvious in the final expression. In Example 3, it would be incorrect to write $g(f(x)) = x - 9$ without stating that x must be greater than or equal to zero, since the simplified expression implies a domain of all real numbers.

Exercise Set 2.10

1. Write and simplify an expression for $f \circ g$ and $g \circ f$. State the domain of each composition. Use exact values when possible.

   a. $f(x) = \frac{1}{x - 1}$ and $g(x) = \frac{1}{x + 1}$

   b. $f(x) = x^2 - 5$ and $g(x) = \sqrt{4x - 5}$

   c. $f(x) = \sqrt{x - 5}$ and $g(x) = \frac{1}{x^2}$

   d. $f(x) = x^2 + 1$ and $g(x) = x - 4$

   e. $f(x) = |x + 1|$ and $g(x) = \sqrt{x - 1}$

   f. $f(x) = \frac{1}{x - 2}$ and $g(x) = x + 3$

   g. $f(x) = \frac{x}{x - 1}$ and $g(x) = \sqrt{x + 2}$

   h. $f(x) = x^3 - 3x - 1$ and $g(x) = \frac{1}{x}$

   i. $f(x) = \sqrt{x}$ and $g(x) = \sin x$

2. Each of these functions is a composition of two simpler functions. Find f and g so that $h(x) = f(g(x))$. For example, if $h(x) = \frac{1}{x^3 - 1}$ we might choose $f(x) = \frac{1}{x - 1}$ and $g(x) = x^3$ or $f(x) = \frac{1}{x}$ and $g(x) = x^3 - 1$.

   a. $h(x) = \frac{1}{x - 11}$ b. $h(x) = \sqrt{x^2 - 5x - 14}$

   c. $h(x) = |\sin x|$ d. $h(x) = \sin|x|$

3. Suppose $g(x) = \frac{x + 3}{2}$. Evaluate $g(g(1))$ and $g(g(g(1)))$.

4. Consider the function f defined below. Evaluate
$$f(f(f(f(f(f(f(f(f(6))))))))).$$

$$f(x) = \begin{cases} f(x - 2) & \text{if } x > 5 \\ x^2 - 3 & \text{if } x = 5 \\ x^2 + 1 & \text{if } x < 5 \end{cases}$$

5. Consider the function f defined below.

$$f(x) = \begin{cases} x - 10 & \text{if } x > 100 \text{ and } x \text{ is an integer} \\ f(f(x + 11)) & \text{if } x \leq 100 \text{ and } x \text{ is an integer} \end{cases}$$

The domain of this function consists of the set of integers. You may be surprised to discover what the range is. Evaluate each of the following and then identify the range.

a. $f(102)$ b. $f(80)$

c. $f(97)$ d. $f(-10)$

e. $f(301)$

6. Suppose a group of students collected and analyzed data and found that the number of hours per week that a college student studies and the student's final grade point average (GPA) can be modeled by the equation

$$gpa(h) = \sqrt{0.5h + 0.9}.$$

In this equation h represents number of hours studied per week and can take on any value from 0 hours to 30 hours. Another group found that the relationship between students' final GPAs and their starting salaries after college can be modeled by

$$sal(g) = 6(2.7g + 3.8)^3 + 10{,}500.$$

In this equation g represents the GPA and can take on any value from 0 to 4.

a. In the context of the problem, explain the meaning of the composition of these two functions. Find an expression for the composition.

b. Find the domain of the composition.

c. Create and answer three questions about the situation modeled by the composition.

SECTION 2.11 Investigating Graphs of Compositions of Functions

Each part of this activity focuses on a particular type of function composition. You should work in pairs to investigate and make generalizations about these types of compositions.

INVESTIGATION 1   Consider the following group of function pairs.

    a. $f(x) = x^2 - 9$; $g(x) = |x^2 - 9|$

    b. $f(x) = 2 \sin x - 1$; $g(x) = |2 \sin x - 1|$

    c. $f(x) = 1 + \frac{1}{x}$; $g(x) = |1 + \frac{1}{x}|$

Answer questions 1–3 for each pair of functions.

    1. Use function notation to show how g is a composition involving function f. For example, if $g(x) = |2^x - 4|$ and $f(x) = 2^x - 4$, you would write $g(x) = |f(x)|$.

    2. Choose an appropriate viewing window and use a graphing calculator to graph the functions f and g together. Sketch the graphs and label key points carefully.

    3. Compare the graph of g to the graph of f. Note any changes in the domain. Write a few sentences about your observations.

    4. Write a general statement explaining how the graph of $g(x) = |f(x)|$ compares to the graph of f. Explain why this result is reasonable.

INVESTIGATION 2   Consider the following group of function pairs.

    d. $f(x) = x^2 - 9$; $g(x) = \sqrt{x^2 - 9}$

    e. $f(x) = |x - 1| - 3$; $g(x) = \sqrt{|x - 1|} - 3$

    f. $f(x) = 2^x$; $g(x) = \sqrt{2^x}$

    5–7. Answer questions 1–3 from Investigation 1.

8. Write a general statement explaining how the graph of $g(x) = \sqrt{f(x)}$ compares to the graph of f. ▨

INVESTIGATION 3   Consider the following group of function pairs.

g. $f(x) = x^2 - 9$; $g(x) = \dfrac{1}{x^2 - 9}$

h. $f(x) = \sin x$; $g(x) = \dfrac{1}{\sin x}$

i. $f(x) = |x - 1| - 3$; $g(x) = \dfrac{1}{|x - 1| - 3}$

9–11. Answer questions 1–3 from Investigation 1.

12. Write a general statement explaining how the graph of $g(x) = \dfrac{1}{f(x)}$ compares to the graph of f. ▨

INVESTIGATION 4   Consider the following group of function pairs.

j. $f(x) = (x - 3)^2 + 5$; $g(x) = (|x| - 3)^2 + 5$

k. $f(x) = \sin x - 2$; $g(x) = \sin|x| - 2$

l. $f(x) = 2^x$; $g(x) = 2^{|x|}$

13–15. Answer questions 1–3 from Investigation 1.

16.   Write a general statement explaining how the graph of $g(x) = f(|x|)$ compares to the graph of f. ▨

SECTION 2.12 Composition as a Graphing Tool

The concept of function composition is often useful when graphing complicated functions that can be decomposed into two simpler functions.

For instance, the function $h(x) = |x^2 - 9|$ can be expressed as $f \circ g$, where $f(x) = |x|$ and $g(x) = x^2 - 9$.

CLASS PRACTICE   Express each of these functions as a composition of two functions, where $f(x) = 5x^3 - 2$, $a(x) = |x|$, and $s(x) = \sqrt{x}$.

   1. $y = \sqrt{5x^3 - 2}$

   2. $y = 5|x|^3 - 2$

   3. $y = |5x^3 - 2|$

EXAMPLE 1   Sketch a graph of $h(x) = |3\sin x - 2|$ using the concept of function composition.

Solution   If we express h as $f \circ g$ with $f(x) = |x|$ and $g(x) = 3\sin x - 2$, then we will be able to take advantage of the fact that f is a simple toolkit function and g is a transformation of a toolkit function. The flow of the composition $f \circ g$ can be illustrated by

$$x \xrightarrow{\ g\ } 3\sin x - 2 \xrightarrow{\ f\ } |3\sin x - 2|.$$

Based on our knowledge of transformations, we know that the graph of g is a sine curve that is stretched vertically by a factor of 3 and shifted down 2 units. Its maximum y-value is 1, and its minimum is -5. The graph of g is shown on the left in Figure 65 on the next page.

To create the graph of $h(x) = |g(x)|$, consider what happens to each ordered pair $(a, g(a))$ on the graph of g when $g(a)$ is used as the input to the absolute value function. Since $|y| = y$ whenever $y \geq 0$, the point $(a, g(a))$ is the same as the point $(a, |g(a)|)$ whenever $g(a) \geq 0$. So, points on the graph of g that have nonnegative second coordinates will also be points on the graph of $|g|$. This means that the graph of $|g|$ is identical to the graph of g wherever the graph of g is on or above the x-axis. Now consider a point $(a, g(a))$ for which $g(a) < 0$. We know that $|y| = -y$ whenever $y < 0$, so the point $(a, -g(a))$ will be on the graph of $|g|$ whenever $g(a) < 0$. This means that any portion of the graph of g that lies below the x-axis will be reflected about the x-axis to produce the graph of $|g|$.

The complete graph of $h = f \circ g = |g|$ is shown on the right in Figure 65. How does the range of $h(x) = |3\sin x - 2|$ differ from the range of $g(x) = 3\sin x - 2$?

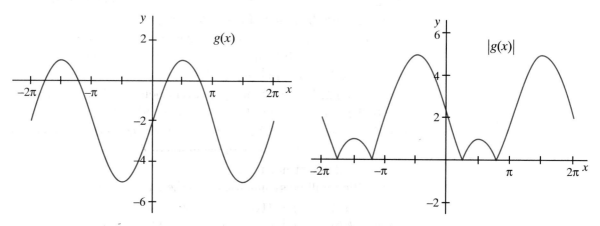

Figure 65 Graphs of $g(x) = 3 \sin x - 2$ and $h(x) = |3 \sin x - 2|$

The process of decomposing functions can produce many different answers. For example, if $h(x) = f(g(x)) = |3\sin x - 2|$, there are several pairs of functions, f and g, that can be composed to produce h:

$$f(x) = |3x - 2| \text{ and } g(x) = \sin x$$

or

$$f(x) = |x - 2| \text{ and } g(x) = 3\sin x$$

or

$$f(x) = |x| \text{ and } g(x) = 3 \sin x - 2.$$

If the purpose in decomposing h into $f \circ g$ is to simplify graphing h, then either f or g should be a simple toolkit function such as $y = |x|$, $y = \frac{1}{x}$, or $y = \sqrt{x}$.

EXAMPLE 2   Sketch a graph of $k(x) = (|x| - 3)^2 - 1$.

Solution   We can express k as $f \circ r$, where $f(x) = (x - 3)^2 - 1$ and $r(x) = |x|$. The graph of f is a transformation of the toolkit parabola, and r is the toolkit absolute value function. Notice that this composition differs from that in Example 1; here the absolute value function is the inner function in the composition. Writing this function as $k(x) = f(|x|)$ reminds us that the inner function is the absolute value function.

Consider the point $(a, k(a))$ on the graph of k. Since $|x| = x$ whenever $x \geq 0$, the point $(a, k(a)) = (a, f(|a|)) = (a, f(a))$ whenever $a \geq 0$. This means that the graph of k is identical to the graph of f whenever the first coordinate is nonnegative. So the graph of f and the graph of k are identical on the right side of the y-axis.

Now consider points on the graph of k for which first coordinates are negative. For clarity, let $a > 0$ be in the domain of f and consider a point whose x-coordinate is $-a$. Now $k(-a) = f(|-a|) = f(a)$, so the point $(-a, f(a))$ will be on the graph of the function k. In other words, for every point $(a, f(a))$ on the graph of f for which $a > 0$, the graph of k will have that point and also the point $(-a, f(a))$. This means that the graph of k will be symmetric about the y-axis. The complete graph of k is shown in Figure 66 along with the graph of f.

One effect of having absolute value as the inner function is that the resulting composition will be symmetric about the y-axis. This means that whenever you have a composition $k(x) = f(|x|)$, you can graph $y = f(x)$ for $x \geq 0$ and complete the graph of k by reflecting the right half about the y-axis.

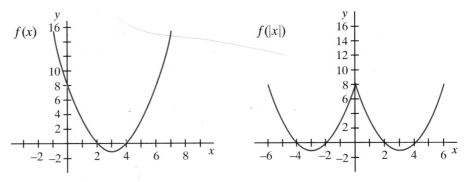

Figure 66 Graphs of $f(x) = (x - 3)^2 - 1$ and $k(x) = f(|x|)$

CLASS PRACTICE

1. Sketch a graph of $f(x) = |1 - x^3|$.

2. Sketch a graph of $g(x) = 1 - |x|^3$.

3. Write several sentences about how absolute value influences the graphs of
$$f(x) = |1 - x^3| \quad \text{and} \quad g(x) = 1 - |x|^3.$$

The functions we have graphed in the previous two examples have both involved composition with $y = |x|$. Notice that they have been graphed without technology and without plotting points. Instead, we analyzed the effect of taking the absolute value of an x-value or y-value and used this information as a graphing guide. The next two examples show how to use a similar approach, supplemented with some point plotting.

EXAMPLE 3   Sketch a graph of $r(x) = \sqrt{2x^2 - 12x + 10}$.

Solution   This function can be decomposed into $r(x) = g \circ f(x)$, where $g(x) = \sqrt{x}$ and $f(x) = 2x^2 - 12x + 10$. The graph of f is a transformation of $y = x^2$, but the transformations are not clear with f in its current form. We can complete the square to rewrite the function as $f(x) = 2(x - 3)^2 - 8$. The graph of f is a parabola that has been shifted 3 units to the right and 8 units down, giving a vertex at $(3, -8)$. We can also factor to solve the quadratic equation $0 = 2x^2 - 12x + 10$ and see that the zeros of f are $x = 1$ and $x = 5$. The graph of f is shown on the left in Figure 67.

The function r can be expressed as $r(x) = \sqrt{f(x)}$, and any point on the graph of r can be written as $(a, r(a)) = (a, \sqrt{f(a)})$. First, note that it is possible to take the square root of $f(a)$-values only when they are nonnegative. The graph of f in Figure 67 shows that $f(a)$ is greater than or equal to zero when $x \leq 1$ or when $x \geq 5$. Therefore, the domain of r is $x \leq 1$ or $x \geq 5$. Second, think about which points on the graph of f will also be on the graph of $r(x) = \sqrt{f(x)}$. The only numbers that are equal to their square roots are 0 and 1. So the points on f that have 0 or 1 as their y-coordinate are also on $r(x) = \sqrt{f(x)}$. In other words, if $f(a) = 0$ or $f(a) = 1$, then the point $(a, 0)$ or $(a, 1)$ is on both f and $r(x) = \sqrt{f(x)}$. Finally, consider the remaining x-values in the domain of r. Whenever the y-values on f are greater than 1, taking their square root decreases their size (since $\sqrt{a} < a$ for all $a > 1$). This means that the graph of $\sqrt{f}$ will lie below the graph of f for such x-values. Whenever the y-values on f are positive and less than 1, the y-values on $\sqrt{f}$ will lie above the y-values on f (because $\sqrt{a} > a$ when $0 < a < 1$). So the graph of $\sqrt{f}$ will lie above the graph of f for these x-values. The graph of r is shown on the right in Figure 67.

We can summarize the effect of composing the square root function with an inner function f as follows. The domain of the composition consists only of those numbers in the domain of f which produce nonnegative y-values. Points on f with y-values of 0 or 1 will be unchanged by the composition. Points on f with y-values between 0 and 1 will have larger y-values, and points with y-values greater than 1 will have smaller y-values.

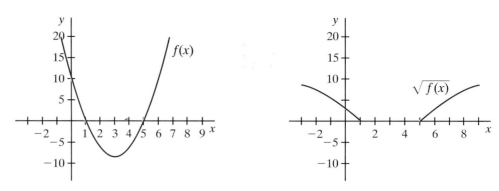

Figure 67   Graphs of $f(x) = 2x^2 - 12x + 10$ and $r(x) = \sqrt{f(x)}$ ■

EXAMPLE 4   Sketch a graph of $f(x) = \dfrac{1}{x^2 - 3x - 4}$.

Solution   This function can be decomposed as $f(x) = g \circ h$, where $g(x) = \frac{1}{x}$ and $h(x) = x^2 - 3x - 4$. The first step will be to sketch a graph of h. Factor h to find the zeros at $x = -1$ and $x = 4$. Using symmetry, we find the vertex of h is located at $x = \frac{3}{2}$ and has coordinates $(\frac{3}{2}, -\frac{25}{4})$. The graph of h is shown on the left in Figure 68.

Since f can be expressed as $f(x) = \frac{1}{h(x)}$, we can determine y-values on f by thinking about how taking the reciprocal affects the y-values on h. Taking the reciprocal of a number does not change its sign, so the y-values of f are positive wherever the y-values of h are positive, and the y-values of f are negative wherever the y-values of h are negative. We also know that we cannot take the reciprocal of zero, so the zeros of h, that is, $x = -1$ and $x = 4$, will not be in the domain of f. What happens to the graph of $f(x) = \frac{1}{x^2 - 3x - 4}$ as x-values get close to -1 and 4? As x-values get close to these numbers, y-values on h are getting very close to zero. The reciprocal of a number close to zero is a number of the same sign that is large in magnitude. So as the y-values on h get closer and closer to zero, the magnitude of their reciprocals, that is, of y-values on f, become larger and larger. As a result, the graph of $f(x) = \frac{1}{h(x)}$ has vertical asymptotes at $x = -1$ and $x = 4$.

What other observations can we make about the reciprocals of the y-values on h? For x-values where y-values of h are large in magnitude, y-values of f will be close to zero. Since the y-values of h increase without bound as x-values increase, the y-values on f will approach zero. Therefore, the x-axis is a horizontal asymptote for the graph of f. We also know that the reciprocal of 1 is 1 and the reciprocal of -1 is -1. So for all x-values where the y-value on h is equal to 1 or -1, f also has a y-value of 1 or -1. This means that the graphs of f and h will share points wherever the graph of h has a point with y-coordinate of ± 1. Finally, we can use the y-intercept on h to find the coordinates of another special point. Since the point $(0, -4)$ is on the graph of h, the point $(0, -\frac{1}{4})$ is on the graph of f. The graph of f is shown on the right in Figure 68.

What happens to the graph of $f(x) = \frac{1}{x^2 - 3x - 4}$ as x-values get close to −1 and 4?

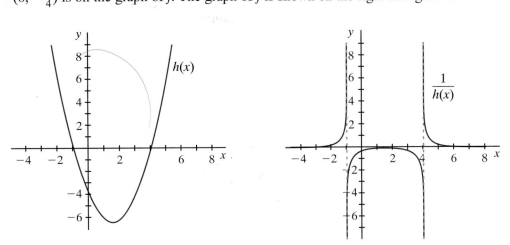

Figure 68   Graphs of $h(x) = x^2 - 3x - 4$ and $f(x) = \frac{1}{h(x)}$ ◼

EXAMPLE 5   Based on the graph of f in Figure 69, sketch a graph of $g(x) = \frac{1}{f(x)}$.

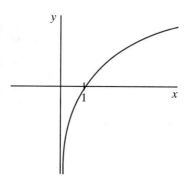

Figure 69   Graph of f for Example 5

Solution   This example differs from previous examples in that the equation for f has been left unspecified. The solution will rely solely on an analysis of tendencies in the graph of f.

Let's first define some notation that will help us describe the behavior of functions. The symbol " $\rightarrow$ " means "approaches" or "gets close to," and the symbol " ∞ " represents the concept of infinity. Therefore, " $x \rightarrow \infty$ " means " x-values increase without bound" and " $x \rightarrow -\infty$ " means " x-values decrease without bound." At times we want to indicate that x-values approach a specific value from a specific direction, so " $x \rightarrow 1^{+}$ " means that x approaches 1 by taking on values larger than 1 but getting closer and closer to the value of 1. For example, x might take on values like 1.1, 1.01, 1.001, etc. This notation is usually read " x approaches 1 from the right." Similarly, " $x \rightarrow 1^{-}$ " is read " x approaches 1 from the left."

This new notation will help describe the behavior of f and allow us to analyze the behavior of g. Recall the following observations from Example 4.

- The reciprocal of zero is undefined.

- The reciprocal of a number large in magnitude is close to zero.

- The reciprocal of a number close to zero is large in magnitude.

- The reciprocal of 1 is 1, and the reciprocal of -1 is -1.

- The sign of a number does not change when you take its reciprocal.

Applying these observations to the problem of graphing $g(x) = \frac{1}{f(x)}$ leads to the following conclusions.

Graphing the Reciprocal of a Function	
as $f(x) \to -\infty$	$\frac{1}{f(x)} \to 0^-$
as $f(x) \to 0^-$	$\frac{1}{f(x)} \to -\infty$
as $f(x) \to 0^+$	$\frac{1}{f(x)} \to \infty$
as $f(x) \to \infty$	$\frac{1}{f(x)} \to 0^+$

Information about the ordered pairs $(x, \frac{1}{f(x)})$ from above can be used to graph $g(x) = \frac{1}{f(x)}$. The graph of $g(x) = \frac{1}{f(x)}$ that incorporates this information is shown in Figure 70. Notice that there is little scale on either the x- or y-axis; the graph does not show specific points, but does show general tendencies.

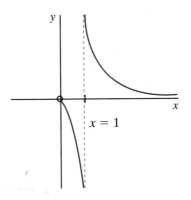

Figure 70   Graph of $g(x) = \frac{1}{f(x)}$ for Example 5   ▨

CLASS PRACTICE

1. Under what conditions is the domain of $g(x) = \dfrac{1}{ax^2 + bx + c}$ all real numbers? Sketch a graph to illustrate your answer.

2. Under what conditions is the domain of $h(x) = \sqrt{ax^2 + bx + c}$ all real numbers? Sketch a graph to illustrate your answer.

1. Identify the domain and the range of the functions f and g in Example 5.

2. Use the idea of decomposition to help you sketch a graph of each function without using technology. State f and g such that $y = f \circ g$. Note that you will use the concept of transformations of toolkit functions to graph some of the functions that were composed to produce the given function. Check your graphs using a graphing calculator.

    a. $y = \frac{1}{\sin x}$ **b.** $y = |(x - 2)^3|$ **c.** $y = (|x| - 2)^3$

    d. $y = \frac{1}{x^2 + 1}$ **e.** $y = \frac{1}{2^x - 1}$ **f.** $y = |2 + \frac{1}{x + 2}|$

    g. $y = \sqrt{|x - 4|}$ **h.** $y = \sqrt{|x - 4|} - 1$ **i.** $y = \sqrt{\sin x}$

    j. $y = -|\frac{1}{2}x^3 + 1|$ **k.** $y = \frac{1}{x^2 - 3x + 2}$ **l.** $y = ||x| - 4|$

    m. $y = (|x| - 2)^2 - 1$ **n.** $y = \frac{1}{|x| - 2} - 1$

3. Given the graph of $y = f(x)$, you can now sketch a graph of $y = |f(x)|$, $y = \sqrt{f(x)}$, $y = \frac{1}{f(x)}$, and $y = f(|x|)$. Follow the same reasoning as was presented for these four compositions to write instructions for sketching the graph of $y = (f(x))^2$ from the graph of $y = f(x)$.

4. The graph of g is shown in Figure 71. Sketch a graph of each of the following.

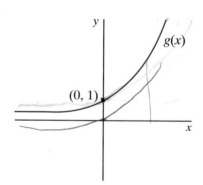

Figure 71   Graph of g for Exercise 4

    a. $y = \sqrt{g(x)}$ **b.** $y = |g(x)|$

    c. $y = \frac{1}{g(x)}$ **d.** $y = g(|x|)$

    e. $y = (g(x))^2$ **f.** $y = g(-x)$

5. You now have two powerful concepts to help you understand the graphs of complicated functions, namely transformations and compositions. In some cases, a complicated function may be rewritten strictly as a transformation or strictly as a composition. For each of the following, rewrite the given function first as a transformation involving a toolkit function and then as a composition. Lastly, sketch a graph. A discussion of Part a provides an example.

a. $f(x) = \frac{1}{x - 5}$

The function f is a transformation of the toolkit function $T(x) = \frac{1}{x}$, such that $f(x) = T(x - 5)$. Therefore, graph the toolkit reciprocal function and then move it 5 units to the right. As a composition, you could write $f(x) = \frac{1}{g(x)}$, where $g(x) = x - 5$. Graph $g(x) = x - 5$ and then follow the procedures for graphing the reciprocal of a known function. Notice that both methods produce the same graph, but the processes are very different. To make the processes visible, use a different set of axes for each method, and use different colors to indicate that the "first" graph you draw is used to produce the final graph.

b. $g(x) = |x| + 3$

c. $h(x) = |x + 3|$

d. $s(x) = \sqrt{x + 1}$

e. $p(x) = (x - 5)^2$

f. Now suppose you were asked to graph the function $w(x) = \sqrt{7 - \frac{3}{4}x}$. Would you rather consider this a transformation of the toolkit function $T(x) = \sqrt{x}$ or a composition of $f(x) = 7 - \frac{3}{4}x$ and $T(x) = \sqrt{x}$? Explain the reason for your choice and sketch a graph of w.

6. Graph $y = \sqrt{\frac{1}{4}x^2 - 2x + 4}$. Identify the domain. Write a sentence or two to explain why the graph looks the way it does.

In Exercises 7-10, recall that the symbol $\lfloor x \rfloor$ represents the greatest integer (or floor) function, which is the greatest integer less than or equal to x.

7. Suppose $p(x) = x^2 + 1$ and $q(x) = \lfloor x \rfloor$. Evaluate each of the following.

a. $p(q(3))$, $p(q(3.3))$, $p(q(3.5))$, and $p(q(3.9))$

b. $q(p(3))$, $q(p(3.3))$, $q(p(3.5))$, and $q(p(3.9))$

8. Sketch a graph of $y = \lfloor x^2 + 1 \rfloor$ over the domain $-2 \le x \le 2$.

9. Sketch a graph of $y = \lfloor x \rfloor^2 + 1$ over the domain $-2 \le x \le 2$.

10. Sketch a graph of $y = \lfloor \sin x \rfloor$.

SECTION 2.13 Inverses

A car salesperson advises potential customers that the monthly payment, p, on a new car depends on four factors: the price of the car, P, the size of the down payment, d, the monthly interest rate, r, and the number of months required to pay back the loan, n. These quantities are related by the formula

$$p = \frac{(P - d)r(1 + r)^n \cdot}{(1 + r)^n - 1}$$

In most instances the interest rate and the duration of the loan are determined by the seller, and the down payment is agreed upon by the buyer and seller. Since these three quantities do not vary, the monthly payment and the price of the car are the only true variables in the formula. The price of the car, P, determines the monthly payment, p, so price is usually thought of as the independent variable and the monthly payment is a function of the price.

Suppose a customer who has $2400 for a down payment wants to buy a $12,000 car. The annual interest rate is 9% (which is equivalent to 0.75% per month), and the term of the loan is 48 months. Using the formula given above, the salesperson determines that the monthly payment will be $238.90. Thus, the ordered pair (12000, 238.90) belongs to the function that contains ordered pairs of the form (*price of car, monthly payment*).

As often happens, this customer cannot afford a monthly payment of $238.90. The buyer's income and other expenses limit the monthly payment to $200. The buyer will have to shop around to find an affordable car. The price of a car the customer can afford with a monthly payment of $200 is found by solving the following equation for P.

$$200 = \frac{(P - 2400)(0.0075)(1.0075^{48})}{1.0075^{48} - 1}$$

The solution is $P = \$10,436.96$, so the customer can afford a car that costs about $10,437.

In this situation, the monthly payment is no longer the independent variable. Now the affordable price depends on the affordable monthly payment, and the roles of independent and dependent variables have been reversed. In effect, there is now a new function involved in this problem—it consists of ordered pairs of the form (*monthly payment, price of car*). This second function has a special relationship to the first function that consisted of ordered pairs (*price of car, monthly payment*).

inverses

These two functions are said to be *inverses* of each other; that is, one function "undoes" the other function. In the original function, the price of the car, P, is paired with the monthly payment, p, whereas in the inverse function, the monthly payment, p, is paired with the price of the car, P. The symbol f^{-1} is used to represent the inverse function of the function f.

The functions f and f^{-1} have interchanged the roles of independent and dependent variables and have reversed ordered pairs. So, if $f(a) = b$, then $f^{-1}(b) = a$. The domain of f is the same as the range of f^{-1}, and the range of f is the same as the domain of f^{-1}.

How is the graph of a function f related to the graph of f^{-1}?

How is the graph of a function f related to the graph of f^{-1}? For each point (a, b) on the graph of f, there is a point (b, a) on the graph of f^{-1}. The point (b, a) is the reflection about the line $y = x$ of the point (a, b). Thus, when f and f^{-1} are graphed on the same coordinate axes, the two graphs are symmetric about the line $y = x$. In fact, the graph of f can be reflected about $y = x$ to obtain the graph of f^{-1}.

EXAMPLE 1   Graph the inverse of $g(x) = x^3$.

Solution   A graph of g is shown in Figure 72. A graph of the inverse can be obtained by reflecting the graph of g about the line $y = x$. We have also graphed g^{-1} in Figure 72.

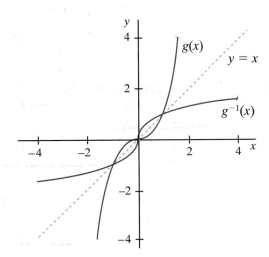

Figure 72   Graphs of g and g^{-1}

The function g contains ordered pairs of the form (x, x^3), so g^{-1} contains ordered pairs of the form (x^3, x) in which the first coordinate is the cube of the second coordinate. The form of the reversed ordered pairs can also be expressed as $(a, \sqrt[3]{a})$; this suggests that $g^{-1}(x) = \sqrt[3]{x}$. ■

EXAMPLE 2   Graph the inverse of $h(x) = \sqrt{x-1}$. Identify the domain and range of h and of h^{-1}.

Solution   The graph of h is a horizontal translation of the toolkit square root function. The graph of h^{-1} is obtained by reflecting the graph of h about the line $y = x$. The graphs of h and h^{-1} are shown in Figure 73.

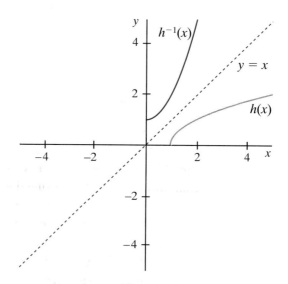

Figure 73   Graphs of h and h^{-1}

The domain and range of h^{-1} can be read from the graph. Notice that the domain of h is the same as the range of h^{-1} and vice versa.

Domain of h: $x \geq 1$ Range of h: $y \geq 0$

Domain of h^{-1}: $x \geq 0$ Range of h^{-1}: $y \geq 1$

The graph of h^{-1} in Figure 73 should suggest to you that $h^{-1}(x) = x^2 + 1, x \geq 0$. Think about why this is reasonable in light of the definition of h. Since h subtracts 1 from its input and then takes the square root, h^{-1} squares its input and then adds 1. ▨

CLASS PRACTICE   1.   Sketch graphs of f and f^{-1} for $f(x) = \frac{1}{3}x + 2$.

2.   Use the graph of f^{-1} to help you write an equation for f^{-1}.

In the preceding examples, we reflected the graph of a function about the line $y = x$ to obtain the graph of the function's inverse. In each example, the inverse was also a function. This will not always be true. For instance, the function $f(x) = x^2$ contains ordered pairs $(2, 4)$ and $(-2, 4)$. When the graph of f is reflected across $y = x$, the new graph contains ordered pairs $(4, 2)$ and $(4, -2)$. However, since 4 is paired with both 2 and -2, this relation is not a function. (Refer to Figure 74.)

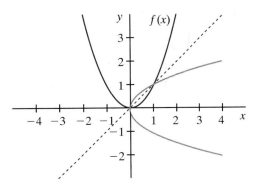

Figure 74   Graphs of $f(x) = x^2$ and its reflection across $y = x$

What condition must a function f satisfy for its inverse to be a function?

one-to-one function

What condition must a function f satisfy for its inverse to be a function? To avoid the problem encountered with the inverse of $f(x) = x^2$, two different x-values cannot be paired with the same y-value; in other words, each y-value of f must come from only one x-value. A function that has this property is called *one-to-one*. Every one-to-one function has an inverse that is a function. When we have a function like $f(x) = x^2$ that is not one-to-one, we frequently want to restrict the domain of f to make the resulting function a one-to-one function. When we choose domain restrictions, we do so in such a way that the range of the function is not changed. There are two convenient ways to restrict the domain of $f(x) = x^2$ to make it one-to-one. These are illustrated in Figure 75.

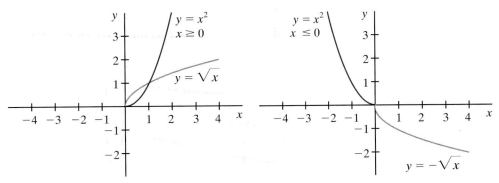

Figure 75   Graphs of $f(x) = x^2$, $x \geq 0$ and $f^{-1}(x) = \sqrt{x}$ (left)
and $f(x) = x^2$, $x \leq 0$ and $f^{-1}(x) = -\sqrt{x}$ (right)

If the domain of $f(x) = x^2$ is restricted to $x \geq 0$, then f is a one-to-one function and $f^{-1}(x) = \sqrt{x}$. If the domain of $f(x) = x^2$ is restricted to $x \leq 0$, then f is a one-to-one function and $f^{-1}(x) = -\sqrt{x}$.

A calculator is a useful tool for exploring the concept of inverse functions. Suppose you were doing some calculation and the number 37.958127 was on your calculator display. If you were to accidentally press the x^2 key, the new display would be 1440.819405. How could you retrieve your original display? Pushing the square root key will achieve this result, since taking the square root of a number "undoes" the effect of squaring. This sequence of events can be understood in terms of inverses. The x^2 key acts as a function that pairs the input 37.958127 with the output 1440.819405. The number 1440.819405 was then the input to a second function that was the inverse of the first; this function took the square root of its input, so the output 37.958127 was paired with the input 1440.819405. This "undoing" operation can be described mathematically with function composition. If functions f and g undo each other, then

$$g(f(x)) = x$$

for all x-values in the domain of f, and

$$f(g(x)) = x$$

inverse functions

for all x-values in the domain of g, and f and g are *inverse functions*.

Recall that the function $y = x$ is often called the identity function and consists of ordered pairs of the form (x, x). The composition of a function and its inverse is the identity function because $f : a \to b$ and $f^{-1} : b \to a$. Therefore, the composition $f^{-1} \circ f$ pairs a with a and $f \circ f^{-1}$ pairs b with b.

EXAMPLE 3   Graph $f(x) = x^2 - 3$ and its inverse. Put restrictions on the domain of f so that its inverse is a function. Identify the domain and range of f and f^{-1}. Confirm algebraically that $f\left(f^{-1}(x)\right) = x$ and $f^{-1}(f(x)) = x$.

Solution   Figure 76 shows the graphs of the function $f(x) = x^2 - 3$ and the reflection of f about $y = x$. The function f is not one-to-one and its reflection over $y = x$ is not a function. The task is to restrict the domain of f in such a way as to make f a one-to-one function. This restriction can be handled in two ways that retain as much of the graph of f as possible.

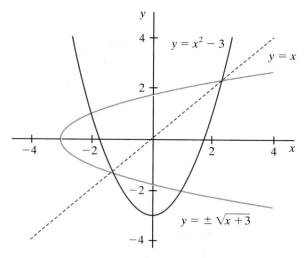

Figure 76   Graphs of $f(x) = x^2 - 3$ and its reflection across $y = x$

If you choose $x \geq 0$ as the domain of f, then the graph of the function will be the right half of a parabola and the graph of the inverse will be the upper half of a parabola. This choice is illustrated on the left in Figure 77. You should recognize from the graph that $f^{-1}(x) = \sqrt{x+3}$. The domain and range of f and f^{-1} follow.

$$\text{Domain of } f : x \geq 0 \qquad \text{Range of } f : y \geq -3$$

$$\text{Domain of } f^{-1} : x \geq -3 \qquad \text{Range of } f^{-1} : y \geq 0$$

On the other hand, you can also choose $x \leq 0$ as the domain of f. In this case, the graph of f will be the left half of a parabola, and the graph of f^{-1} will be the lower half of a parabola; see the graph on the right in Figure 77. With this domain restriction for f, the expression for the inverse is $f^{-1}(x) = -\sqrt{x+3}$. You should be able to identify the domain and range of f and f^{-1} based on this domain restriction.

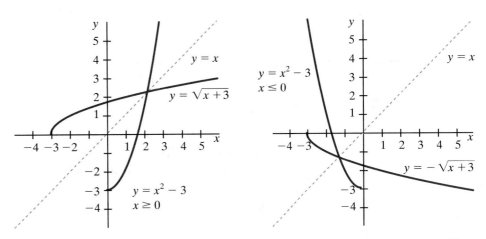

Figure 77   Two ways to restrict the domain to create a one-to-one function

The preceding discussion should make it clear why it is not correct to say that $f(x) = x^2 - 3$ and $g(x) = \sqrt{x+3}$ are inverse functions unless we specifically mention restrictions on the domain of f. It is correct to say that $f(x) = x^2 - 3$, $x \geq 0$, and $g(x) = \sqrt{x+3}$ are inverse functions.

Note that

$$f(g(x)) = f\left(\sqrt{x+3}\right) = \left(\sqrt{x+3}\right)^2 - 3 = x + 3 - 3 = x, \ x \geq -3$$

and

$$g(f(x)) = g\left(x^2 - 3\right) = \sqrt{\left(x^2 - 3\right) + 3} = \sqrt{x^2} = |x| = x, \ x \geq 0.$$

This confirms that with an appropriate restriction on the domain of f, $f(f^{-1}(x)) = x$ and $f^{-1}(f(x)) = x$.

EXAMPLE 4   If $f(x) = \frac{1}{2}x^3 - 4$, write an expression for f^{-1}.

Solution   The expression $f(x) = \frac{1}{2}x^3 - 4$ defines a cubic function that is one-to-one and therefore has an inverse that is a function. One way to analyze f is to show what

f does to each x-value to produce a y-value.

independent variable $\rightarrow$ cube $\rightarrow$ times $\frac{1}{2}$ $\rightarrow$ minus 4 = dependent variable

We know that f^{-1} must start with the dependent variable and end with the independent variable. Therefore, f^{-1} must proceed in the following manner.

dependent variable $\rightarrow$ add 4 $\rightarrow$ times 2 $\rightarrow$ cube root = independent variable

The preceding analysis suggests that $f^{-1}(x) = \sqrt[3]{2(x+4)}$. This procedure can be used to find the inverse of a function f only if f is one-to-one and the independent variable occurs exactly once in the expression for f.

The work to find an expression for f^{-1} can be organized as follows. Given the function $f(x) = \frac{1}{2}x^3 - 4$, we can write

$$y = \frac{1}{2}x^3 - 4.$$

Now solve for x:

$$y + 4 = \frac{1}{2}x^3$$

$$2(y + 4) = x^3$$

$$\sqrt[3]{2(y + 4)} = x.$$

The final equation above gives all the information needed to write the expression for f^{-1}. It shows that f^{-1} adds 4 to its input, multiplies by 2, and then takes the cube root. Since we usually name the input x, we write

$$f^{-1}(x) = \sqrt[3]{2(x + 4)}. \quad \blacksquare$$

EXAMPLE 5   If $g(x) = x^2 - 4x + 5$, write an expression for g^{-1}.

Solution   We begin by completing the square on g: $g(x) = (x - 2)^2 + 1$. So, the graph of g is a parabola with vertex at (2, 1). This graph is shown on the left in Figure 78. Since g is not one-to-one, we need to restrict the domain of g so that g^{-1} is a function. We can choose $x \geq 2$ as the domain to make g one-to-one and retain as much of its graph as possible.

Now to find the expression for g^{-1}, solve for the independent variable in the expression for g.

$$y = (x - 2)^2 + 1$$

$$y - 1 = (x - 2)^2$$

$$\pm\sqrt{y - 1} = (x - 2).$$

Either $(x - 2)$ is equal to $+\sqrt{y - 1}$, or it is equal to $-\sqrt{y - 1}$. Which should we choose? The domain of g provides information to help you make this decision. Since the domain of g is $x \geq 2$, it must be the case that $x - 2 \geq 0$. Since $x - 2$ is 0 or positive, it cannot be equal to $-\sqrt{y - 1}$ (since this is less than zero). This implies that $x - 2$ must be equal to the positive square root of $y - 1$.

$$\sqrt{y - 1} = x - 2$$
$$2 + \sqrt{y - 1} = x$$

Therefore, $g^{-1}(x) = 2 + \sqrt{x - 1}$. This is the inverse when the domain of g is restricted to $x \geq 2$. The graphs of g and g^{-1} with the domain of g restricted to $x \geq 2$ are shown on the right in Figure 78.

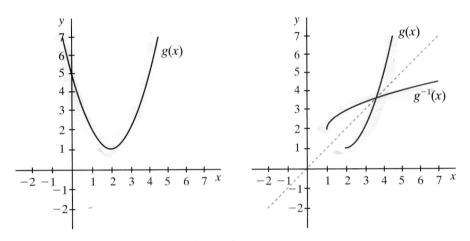

Figure 78 Graphs of $g(x) = x^2 - 4x + 5$ (left) and
$g(x) = x^2 - 4x + 5$, $x \geq 2$ (right) with g^{-1}

CLASS PRACTICE

1. In Example 5, what would be the expression for g^{-1} if the domain of g were restricted to $x \leq 2$?

2. Sketch the graphs of g and g^{-1} with this restricted domain.

EXAMPLE 6   If $g(x) = (x - 2)^2 + 1$, $x \geq 2$, we know that $g^{-1}(x) = 2 + \sqrt{x - 1}$. Write an expression for the compositions $g^{-1} \circ g$ and $g \circ g^{-1}$.

Solution $g^{-1}(g(x)) = g^{-1}((x - 2)^2 + 1)$

$$= 2 + \sqrt{(x - 2)^2 + 1 - 1}$$
$$= 2 + \sqrt{(x - 2)^2}$$
$$= 2 + |x - 2|$$

Since the domain of g is $x \geq 2$, $|x - 2| = x - 2$. Therefore,

$$g^{-1}(g(x)) = 2 + |x - 2| = 2 + (x - 2)$$

$$= x, \quad x \geq 2.$$

$$g(g^{-1}(x)) = g(2 + \sqrt{x - 1})$$

$$= (2 + \sqrt{x - 1} - 2)^2 + 1$$

$$= (\sqrt{x - 1})^2 + 1.$$

Since the domain of g^{-1} is $x \geq 1$, $(\sqrt{x - 1})^2 = x - 1$. Therefore,

$$g(g^{-1}(x)) = (\sqrt{x - 1})^2 + 1 = x - 1 + 1 = x, \quad x \geq 1.$$

This shows that $g^{-1}(g(x)) = x$ and $g(g^{-1}(x)) = x$, which confirms an important property of inverse functions. However, notice that the domain of the first composition is $x \geq 2$, but the domain of the second composition is $x \geq 1$. ∎

Exercise Set 2.13

1. Determine an equation for f^{-1}.

   a. $f(x) = (x - 1)^3 + 2$

   b. $f(x) = \dfrac{1}{x + 1}$

   c. $f(x) = \sqrt{5 - x}$

   d. $f(x) = 5 - 2x$

   e. $f(x) = \sqrt{9 - x^2}, \quad x \geq 0$

   f. $f(x) = \dfrac{x}{x - 1}$

   g. $f(x) = 6x - x^2, \quad x \leq 3$

   h. $f(x) = -\sqrt{x + 1}$

2. For each function in Exercise 1, graph f and f^{-1} on the same coordinate axes.

3. For each function in Exercise 1, identify the domain of f and of f^{-1}.

4. Let $p(x) = -\sqrt{x + 3}$.

   a. Find an equation for p^{-1}.

   b. Graph p and p^{-1} on the same coordinate axes.

   c. Identify the domain of p and of p^{-1}.

   d. For what values of x is it true that $p(p^{-1}(x)) = x$?

   e. For what values of x is it true that $p^{-1}(p(x)) = x$?

f. Graph $p \circ p^{-1}$. Be sure to use the appropriate domain.

g. Graph $p^{-1} \circ p$. Be sure to use the appropriate domain.

5. Find an equation for h^{-1}. Based on your answer, what symmetry must be present in the graph of h?

   a. $h(x) = \sqrt[3]{1 - x^3}$

   b. $h(x) = \sqrt{1 - x^2}, \ x \geq 0$

   c. $h(x) = \dfrac{3x - 1}{2x - 3}$

6. What happens when you try to find the inverse of the function $f(x) = \dfrac{x^2}{x - 3}$? What does this tell you about the graph of f?

7. Find the inverse of each function. When necessary, give a restriction on the domain of f so that f^{-1} will be a function. A sketch of f and f^{-1} will be helpful.

   a. $f(x) = x(x - 2)$

   b. $f(x) = |x - 1| + 1$

   c. $f(x) = -\sqrt{4 - x^2}$

8. Let $f(x) = \sin x$.

   a. Sketch a graph of the inverse of f.

   b. Suggest a domain restriction on f so that the inverse of f will be a function.

9. Write an equation for the inverse of g. A sketch of the graph will be helpful.

   a. $g(x) = \begin{cases} x - 2 & x \leq 2 \\ 2x - 4 & x > 2 \end{cases}$

   b. $g(x) = \begin{cases} -(x - 1)^2 + 3 & \text{if } x \leq 1 \\ \dfrac{1}{2}x + \dfrac{5}{2} & \text{if } 1 < x < 7 \\ \sqrt{x - 7} + 6 & \text{if } x \geq 7 \end{cases}$

SECTION 2.14 Using Inverses to Straighten Curves

When you try to fit a curve to a set of data, your goal is to find a function whose graph fits the data well. The function you choose may be indicated by theory or by past experience. Often you must choose a function based on the scatter plot of the data. If the scatter plot indicates that a straight line will not provide a satisfactory fit, you must decide just how "curved" the graph of the data is. In Exercise Set 2.9 toolkit functions and transformations were used to find models for nonlinear data sets. You probably realized that it is often difficult to determine what function the curvature indicates just by looking at the scatter plot. For example, you probably cannot easily distinguish the difference in the curvature of certain sections of quadratic, cubic, and exponential graphs. In contrast, to see whether or not a graph is linear requires only a check to decide whether or not the points line up.

re-express the data

A commonly used technique for fitting curves to data is to first transform or *re-express the data* to make them linear. That is, rather than working directly with the given data, we generate a new data set by performing a mathematical operation on one or both coordinates. Then we fit a line to the new re-expressed data set. If the fit is satisfactory, then later we can "undo" the transformation and change the variables back to their original state.

How does one go about re-expressing the data to straighten a scatter plot?

How does one go about re-expressing the data to straighten a scatter plot? Several techniques are available, and one of the most common is based on the ideas of inverse functions and compositions. If we form a conjecture about the kind of function that models the data in a scatter plot, then we can use the inverse of that function to "undo" the curvature. For example, if there is a quadratic relationship between two variables, there is curvature in the graph because the y-values result from squaring the x-values. Taking the square root of the y-values will have the effect of composing a square root function with a quadratic function. This composition results in a linear relationship between the original x-values and the transformed y-values. The process of using an inverse function to *linearize* data will be illustrated in the following example.

linearize

EXAMPLE 1 Free Fall

Students in a physics class are studying free-fall to determine the relationship between the distance an object has fallen and the length of time since it was released. They collect the data in Figure 79 and create a scatter plot with amount of time since release, t, graphed on the horizontal axis and distance fallen by the object, d, graphed on the vertical axis.

Figure 79   Free-fall data

Time (sec)	Distance (cm)	Time (sec)	Distance (cm)
0.16	12.1	0.57	150.2
0.24	29.8	0.61	182.2
0.25	32.7	0.61	189.4
0.30	42.8	0.68	220.4
0.30	44.2	0.72	254.0
0.32	55.8	0.72	261.0
0.36	63.5	0.83	334.6
0.36	65.1	0.88	375.5
0.50	124.6	0.89	399.1
0.50	129.7		

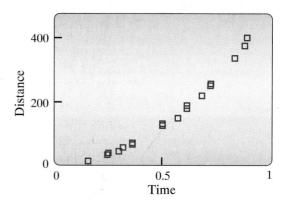

Figure 80   Scatter plot of free-fall data

The scatter plot in Figure 80 indicates that distance depends on time in a non-linear way. The points show upward curvature so we will need a model that is concave up. This information could indicate a quadratic relationship between the variables. If the underlying relationship can be modeled by a quadratic function of the form $d = a \cdot t^2$, then

$$\sqrt{d} = \sqrt{a \cdot t^2} = \sqrt{a} \cdot t, \quad t \ge 0.$$

If $\sqrt{d} = \sqrt{a} \cdot t$, the ordered pairs $\left(t, \sqrt{d}\right)$ can be modeled by a linear function. It is useful to think of taking the square root as "undoing" the squaring that was done by a quadratic function.

The scatter plot in Figure 82 shows time on the horizontal axis and the square root of the distance measurements on the vertical axis. The square root of the distance values are rounded to two decimal places in the table in Figure 81, but more decimal places are retained for the calculations that follow.

Figure 81   Re-expressed free-fall data

Time (sec)	$\sqrt{Distance}$ (cm)	Time (sec)	$\sqrt{Distance}$ (cm)
0.16	3.48	0.57	12.26
0.24	5.46	0.61	13.50
0.25	5.72	0.61	13.76
0.30	6.54	0.68	14.85
0.30	6.65	0.72	15.94
0.32	7.47	0.72	16.16
0.36	7.97	0.83	18.29
0.36	8.07	0.88	19.38
0.50	11.16	0.89	19.98
0.50	11.39		

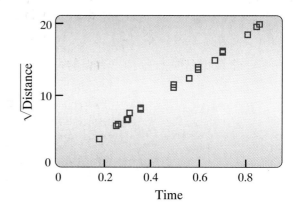

Figure 82   Scatter plot of re-expressed free-fall data

These points lie along a line, so we can proceed to fit a median-median or least squares line to the transformed data. Fitting a median-median line through points of the form $(t, \sqrt{d})$ yields the following equation with constants rounded to four decimal places:

$$\sqrt{d} = 22.1870t + 0.1176.$$

This equation relates the square root of the distance that an object falls to the time since it was released. The linear fit and corresponding residual plot are shown in Figure 83.

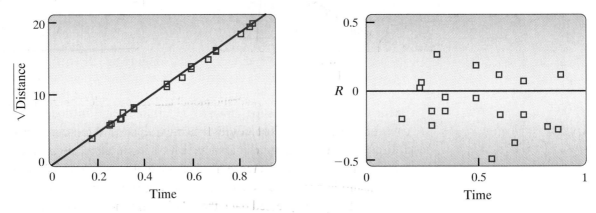

Figure 83   Linear fit and residual plot for re-expressed free-fall data

Since the residuals are scattered randomly about the horizontal axis, we have additional evidence that the square root function has successfully straightened the original data. This information helps confirm our conjecture that the relationship between these variables is quadratic. We can now proceed to transform the variables back to the original ones. To show how the actual distance fallen, not the square root of the distance, is related to time, square both sides of the equation above. Doing so yields the equation

$$d = (22.1870t + 0.1176)^2 = 492.26t^2 + 5.22t + 0.01.$$

Note that rounding constants to two decimal places in the final expression enables you to predict distance values to the nearest tenth of a centimeter, which corresponds to the accuracy of the distance measurements in the original data.

How well does the graph of this equation fit the original data? To answer this question graph the parabola on the original scatter plot and then examine a residual plot. These plots are provided in Figure 84.

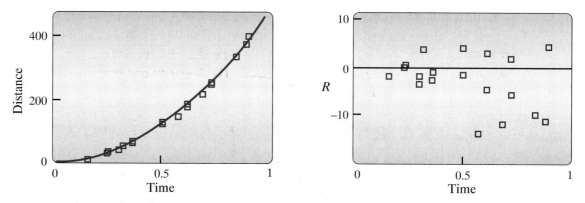

Figure 84 Quadratic model and residual plot for free-fall data

The residuals indicate a satisfactory fit. They are randomly scattered with some positive and some negative values appearing throughout the plot. Though the size of the residuals seems to increase as t-values increase, this is often the case when the relationship is quadratic. Since this trend was not observed in the residual plot from the linear fit to re-expressed data, the relative size of residuals is consistent throughout. You should be aware of this trend in using the model to extrapolate, however. ▨

As you attempt to fit models to data, you use the shape of a scatter plot, your knowledge of mathematical functions, and your understanding of the phenomenon to help decide how to re-express the data. Re-expressions may involve powers, roots, reciprocals, combinations of these, or other functions yet to be studied. Educated trial and error is often required in searching for the mathematical relationship between the variables. As you experiment with different re-expressions, you will find that information from the residual plot is very helpful as you evaluate the success of your re-expression.

The residual plots in Figure 83 and Figure 84 provide different, but important, information. In Figure 83 the residuals are associated with the square root of the d-values. This residual plot helps you decide whether your choice of re-expression is a good one. The residuals are randomly distributed and indicate that taking the square root of d-values is a good choice. If the re-expression technique does not succeed in making the data linear, there will usually be a noticeable pattern in this residual plot. Once you are satisfied with the re-expression, examine the model superimposed on the original data as we have done in Figure 84. Notice that the pattern of the residuals is the same in both residual plots. However, the size of the residuals in Figure 84 provides additional information. Since these residuals are associated with the actual d-values, they indicate the magnitude of the error to be expected when we use the model to make predictions.

Example 1 shows an important technique in data analysis. Do you understand why changing from ordered pairs (t, d) to ordered pairs $(t, \sqrt{d})$ makes the data linear? The upward curvature in the original scatter plot indicated that the d-values needed to be lowered. There are many ways to re-express the d-values to pull the points down—square roots, cube roots, or fourth roots all have this effect. Since the relationship between these variables is quadratic, taking the square root lowers the points just enough so that they will lie along a line.

Exercise Set 2.14

1. Suppose an object slides without friction down an inclined plane in a laboratory. Sparks are generated on paper tape every tenth of a second to mark the position of the object. The distance between sparks can be measured to generate a data set that relates position of the object with time since release. A data set is provided in Figure 85.

Figure 85   Inclined plane data

Time (sec)	Position (cm)	Time (sec)	Position (cm)
0	0.0	1.0	12.5
0.1	0.2	1.1	15.0
0.2	0.6	1.2	17.7
0.3	1.3	1.3	20.6
0.4	2.2	1.4	23.8
0.5	3.3	1.5	27.2
0.6	4.7	1.6	30.8
0.7	6.3	1.7	34.8
0.8	8.1	1.8	38.9
0.9	10.2	1.9	43.1

a. Plot time on the horizontal axis and position on the vertical axis. What function might be a good model for this relationship?

b. Based on your answer to Part a, what re-expression will linearize the data? Perform this re-expression and examine a graph. Are you satisfied with the re-expression?

c. Find a model that describes the relationship between position and time.

d. Is this a good model? Write several sentences to support your answer.

2. Suppose you thought that the scatter plot of the free-fall data in Figure 80 on page 139 indicated a cubic relationship. If you had performed a cube root re-expression of these data, what would you expect to have observed in the new scatter plot? Explain your answer. Verify your response by doing the re-expression.

3. The scatter plot in Figure 86 is concave down. Suggest ways to re-express the data set to linearize them.

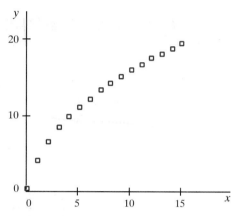

Figure 86 Scatter plot for Exercise 3

4. a. Make a scatter plot of the data set in Figure 87.

Figure 87   Data set for Exercise 4

x	2	6	10	14	20
y	40	1080	5000	13,720	40,000
x	24	40	50	55	
y	69,120	320,000	625,000	831,875	

b. Predict what will happen if you fit a line to the data set in Figure 87. Check your predictions by fitting a least squares line to the data set and examining the residuals.

c. If the data set is quadratic, doing the re-expression $(x, \sqrt{y})$ will linearize the data. Do this re-expression and fit a least squares line to the re-expressed data set. Examine the residuals. Does this appear to be a good model? Why or why not?

d. The ordered pairs in Figure 87 were generated by a third-degree power function. Knowing this, what re-expression technique will linearize the data? Re-express the data and fit a least squares line to these new re-expressed data. Solve your equation for y and graph the final model with the original data. Does this appear to be a good model? Why?

5. In 1969, John S. Rinehart published the article "Waterfall-Generated Earth Vibrations" in the June issue of *Science* magazine. In this article, he stated that a waterfall typically produces a vibration in which one frequency is predominant.

   a. From the data given in Figure 88, find a model for the relationship between the predominant frequency and the height of the indicated waterfall.

   Figure 88   Waterfall heights and frequencies

Waterfall	Height (m)	Frequency (Hz)
Lower Yellowstone	100	5
Yosemite	77	3
Canadian Niagara	50	6
American Niagara	50	8
Upper Yellowstone	37	9
Lower Gullfoss	24	6
Firehole	13	19
Godafoss	11	21
Upper Gullfoss	8	40
Fort Greely	5	40

   Source: Reprinted with permission from *Data Analysis*, copyright 1988 by the National Council of Teachers of Mathematics. All rights reserved.

   b. What should be the predominant frequency of a waterfall that is 42 meters high? How confident are you in this prediction?

   c. Does your model appear to have less error for tall waterfalls or for short waterfalls? Explain.

6. Reconsider the data provided in Figure 62 (page 111) concerning the diameter and volume of a group of trees. Use re-expression to find a model for the relationship between diameter and volume.

7. Large bags of dry dog food provide daily feeding instructions based on the weight of the dog. The data set in Figure 89 was developed from a chart provided on one dog food bag. Use these data to determine the mathematical model for the relationship between the weight of a dog and the daily amount of food recommended by the manufacturer. You may need to experiment with several re-expressions until you find one that seems to straighten the data. Document your attempts by sketching the scatter plot or residual plot associated with each of the re-expressions you try.

Figure 89   Dog food data

Weight of dog (lb)	8	19	36	57	79	102	130	159	190
Amount of food (cups)	1	2	3	4	5	6	7	8	9

8. We linearized the free-fall data provided in Figure 79 on page 139 by taking the square root of the d-values. Given that the functional relationship between distance fallen and time is quadratic, consider other ways that you could re-express the data to make the data set linear.

    a. Square the values of time and examine a scatter plot of the ordered pairs (t^2, d). Find the equation of the median-median line through these points. Examine the residual plot to determine whether this re-expression has successfully linearized the data.

    b. Divide the distance values by corresponding time values to create ordered pairs $\left(t, \frac{d}{t}\right)$. Find the equation of the median-median line through these points. Examine the residual plot to determine whether this re-expression has successfully linearized the data.

    c. Based on the re-expressions in Parts a and b, find models for the ordered pairs (t, d). Graph these with the original data and examine residual plots for each model.

    d. Compare the three different models you found for this data set. Which model do you feel is best? Why?

9. Calculators and computers with curve-fitting software often have a quadratic regression (or quadratic least squares) option.

    a. Use this feature of your calculator or computer software to find a model for the inclined plane data provided in Figure 85 on page 142.

    b. Graph this model with the data and examine a residual plot.

    c. Compare this model to the other model you found for the inclined plane data. (Quadratic regression programs produce the quadratic function that minimizes the sum of the squared residuals. They do not re-express the data but instead use formulas to directly find the coefficients a, b, and c in the quadratic model $y = ax^2 + bx + c$.)

Investigations Using Data Collection

The following investigations produce nonlinear data sets that can be linearized using re-expression techniques discussed in the previous section. You should work in groups of three or four while doing these investigations. Since it is not practical for each of you to perform all the investigations, it is suggested that you share the results of your investigations with the class. Those of you taking measurements should agree on accuracy and measuring techniques. Measurement error involved with the data collection should be explained whenever possible.

For your assigned investigations below, you should make a scatter plot of the data and complete the following items.

a. What basic toolkit function does the scatter plot resemble? (There may be information about the phenomenon that helps you know what type of function to expect.)

b. Use re-expression to find a model for the relationship between the variables.

c. Is this a good model? Write several sentences to support your answer.

INVESTIGATION 1   **Water Flow**

Question   What is the relationship between the level of water in a cylindrical container and the time since the water began draining?

Equipment   A 2- or 3-liter soda bottle, masking tape, device for puncturing the bottle, ruler, stopwatch

Data Collection   Fill the soda bottle to the top of the cylindrical part. Place a strip of masking tape vertically on the bottle from the bottom to the top of the cylindrical portion. Mark the tape at uniform intervals. Puncture the bottle below the cylindrical part, and allow the water to begin to drain at the instant you start a stopwatch. When the water level reaches one of your predetermined marks, record the time. Continue this process until the water level falls to the base of the cylindrical part of the bottle. Eight to ten ordered pairs (*time, height of water*) should be sufficient.   ■

INVESTIGATION 2   **Pendulum**

This investigation is also described in Section 1.4.

Question   What is the relationship between the period of a pendulum and its length?

Equipment   Long string (two or more meters), small nut or several washers, stopwatch, tape measure or meterstick

Data Collection Construct a pendulum by tying a small nut or several washers onto a string at least two meters long. Vary the length of the string by about 20 centimeters from one trial to the next and measure the period of the pendulum (time to complete one swing across and back). To measure the period, hang the string so that one end is stable, pull the weight slightly (about 20°) to one side, let the weight make 10 complete swings, record the time, and divide the time by 10. Be sure to include some long lengths as well as some short lengths. Ten ordered pairs (*length of string, period*) should be sufficient.

INVESTIGATION 3 Spring

Question What is the relationship between the period of oscillation and the mass of an oscillating object?

Equipment Meterstick, hook (for suspending spring from meterstick), masking tape, spring, hook (to support masses), various masses, stopwatch or a data collection device with a motion sensor

Data Collection Tape a hook to the end of a meterstick. Suspend the spring with the attached mass from the hook. Record the total mass you are using; be sure to include the mass of the hook apparatus. Carefully push the mass straight up and then release it to allow it to oscillate. Find the period of oscillation and record the period. Vary the mass and repeat the steps above. One method for collecting these data would be to use a stopwatch to time 10 oscillations and divide by 10 to get the period of oscillation. A second method would be to use a motion sensor with a computer or graphing calculator to determine the period. Eight to ten ordered pairs (*mass, period*) should be sufficient.

INVESTIGATION 4 Pressure

Question What is the relationship between the pressure and the volume of a gas held at constant temperature?

Equipment A data collection device with a pressure sensor

Data Collection The data can be collected using a pressure sensor and a computer or graphing calculator. Vary the volume and record the pressure at each volume. Six to eight ordered pairs (*volume, pressure*) should be sufficient.

INVESTIGATION 5 Ball Toss

Question What is the relationship between the height of a ball tossed in the air and the elapsed time?

Equipment A ball, a data collection device with a motion sensor or a stopwatch with tape measures or metersticks

Data Collection Toss a ball into the air and use one of the following methods to record the height of the ball at various times. The best data result from using a motion sensor with a computer or graphing calculator to gather time and height data. If this equipment is not available, have one student toss a ball near a wall on which you have already marked several heights. Students with stopwatches should then note the times as the ball passes the marked heights. If this experiment is conducted without a data collection device, six to eight ordered pairs (*time, height*) should be sufficient. ▨

SECTION 2.16 Introduction to Parametric Equations

During the winter of 1993 a blizzard struck the East Coast. A man who lived in the Washington, D.C. area went outside during the storm to dump some trash. Just as he stepped out the door, a gust of wind blew his cap from his head. The cap flew over a fence 4 feet high that was 25 feet away. We want to determine how far away the cap landed and for how long the cap traveled before hitting the ground. It will be helpful to know that the man is 6 feet 3 inches tall.

Before proceeding, we must make some assumptions about the forces acting on the cap. Two obvious forces were at work: gravity and the wind. The height of a freely falling object is given by $h = \frac{1}{2}gt^2 + v_0t + h_0$, where $g = -32$ ft/sec^2, v_0 is the initial velocity, h_0 is the initial height, t is time in seconds, and h is height in feet. Assume that $v_0 = 0$ and, since the man is 6 feet 3 inches tall, $h_0 = 6.25$ ft. Also assume that the wind was blowing horizontally across the yard. The weather person reported gusts up to 40 miles per hour that day, so assume that this was the speed of the wind that carried the cap away. Using this fact, the horizontal position of the cap can be determined by the familiar $d = r \cdot t$ formula, where $r = 40$ miles per hour, t is in hours, and d is in miles. Since these units are not practical for a cap blowing across a yard, we will convert 40 miles per hour to feet per second as follows:

$$\frac{40 \text{ miles}}{\text{hour}} \cdot \frac{1 \text{ hour}}{3600 \text{ sec}} \cdot \frac{5280 \text{ feet}}{1 \text{ mile}} = 58.\overline{6} \text{ feet/second.}$$

We now have two functions to model the path of the cap. The cap moves in the horizontal direction according to $x(t) = 58.\overline{6}t$ and in the vertical direction according to $y(t) = -16t^2 + 6.25$. In both functions t is in seconds and distance traveled is in feet.

Using these functions we can substitute values for time and determine the location of the cap. For example, when $t = 0$, $x = 0$ and $y = 6.25$. So the initial position is 0 feet away from the man in the horizontal direction and 6.25 feet above ground level; that is, at $t = 0$ the cap is on the man's head. (See Figure 90.)

6.25

Figure 90   Man with cap

After 1 second, $x = 58.\overline{6}$ ft and $y = -9.75$ ft. The negative y-value indicates that the cap has already landed, so it is necessary to calculate distances for some intermediate time values to determine the x-coordinate when $y = 0$. The following table in Figure 91 shows the position of the cap every tenth of a second.

Figure 91   Cap positions

Time (sec)	x (ft)	y (ft)
0	0	6.25
0.1	5.87	6.09
0.2	11.73	5.61
0.3	17.60	4.81
0.4	23.47	3.69
0.5	29.33	2.25
0.6	35.20	0.49
0.7	41.07	-1.59
0.8	46.93	-3.99
0.9	52.80	-6.71
1.0	58.67	-9.75

Source: Presentation by Alan Bellman at Teaching Contemporary Mathematics Conference, Durham, NC, February 1994.

You can use the information in Figure 91 to analyze the path of the cap according to the functions x and y. When $t = 0.4$ second, the cap has traveled approximately 23.47 feet in the horizontal direction and when $t = 0.5$ second, $x = 29.33$ feet. Therefore, at some time between $t = 0.4$ and $t = 0.5$, the cap reached the fence, which was 25 feet away. Since the corresponding y-values are less than 4 (the height of the fence), the model indicates that the cap will hit the fence before it hits the ground.

CLASS PRACTICE

1. Use the information in Figure 91 to create ordered pairs (x, y) and plot these ordered pairs to get a graphical representation of the path of the cap. Include the fence in your graph by placing a vertical line segment 4 units high on the horizontal axis where $x = 25$.

parametric equations

parameter

The equations $x(t) = 58.\overline{6}t$ and $y(t) = -16t^2 + 6.25$ are examples of *parametric equations*. Parametric equations are very useful whenever the vertical and horizontal components of position can be described as functions of time, since they specify the position of an object in the coordinate plane at any instant of time. The variable t is said to be a *parameter*. You might think of t as the variable that determines the values of x and y. It is often the case, as in this example, that t represents time, while $x(t)$ and $y(t)$ represent the position of an object in the xy-plane at time t. Previously you found a direct relationship between x and y, usually in the form $y = f(x)$, to describe similar scenarios. A function, $y = f(x)$, allows you to easily determine y-coordinates for given x-coordinates, but $y = f(x)$ cannot relate position in the xy-plane directly to time.

Graphing calculators typically have a parametric mode that allows us to create tables of values and graphs based on pairs of parametric equations. Use the parametric mode on your calculator to repeat the work you did to make the table of values and graph representing locations of the cap. You will need to enter equations for both x and y and specify the t-values for the table calculations. Let the minimum t-value equal 0, the maximum t-value equal 1, and the increment or step size equal 0.1. You will also need to specify the x- and y-dimensions of a viewing window for your graph. To get more detailed information about the path of the cap, change the increment or step size for t to 0.05 or 0.01.

According to the data in Figure 91, the man's cap hit the fence and landed in his yard. But the man reported that the cap cleared the fence. Where did we go wrong?

Several possibilities exist. There may have been a mistake somewhere in the calculations. Or there may have been a mistake in the assumptions leading to the model. The equation for x assumed that the cap was being moved horizontally by a 40 mph wind. We have no way to verify this assumption, but it seems unlikely that the wind would have been stronger than what was reported (and a stronger wind would be needed for the cap to clear the fence). The equation for y was based on a formula for the position of a freely falling body. But this formula applies only to bodies falling in a vacuum where there is no air resistance. The cap was not falling in a vacuum and by the nature of its shape, falls somewhat like a parachute. Thus, the equation $y(t) = -16t^2 + 6.25$ is probably not correct. In this situation, acceleration is actually the result of two components: gravity and air resistance; so the coefficient of t^2 should be smaller in magnitude than -16. In Exercise 1 at the end of this section, you will modify the function y and reconsider the path of the cap.

Consider the parametric equations $x(t) = 2t + 1$ and $y(t) = 4t$. If we make a table of x- and y-values for both positive and negative values of t, we have the values shown in Figure 92.

Figure 92 Table of values for $x(t) = 2t + 1$ and $y(t) = 4t$

t	-3	-2	-1	0	1	2	3
x	-5	-3	-1	1	3	5	7
y	-12	-8	-4	0	4	8	12

What do you think the graph of the ordered pairs (x, y) will look like? When t changes by a constant, both the x-value and the y-value also change by a constant. The y-values all change by 4, and the x-values all change by 2, so the ratio of the change in y to the change in x is constant, $\frac{4}{2} = 2$. Thus, the graph of the ordered pairs (x, y) will be a line with slope 2. What is the equation of this line in terms of x and y?

To write the equation in terms of x and y, eliminate the parameter t from the system $x = 2t + 1$ and $y = 4t$. If

$$x = 2t + 1,$$

then

$$t = \frac{x - 1}{2},$$

and

$$y = 4t = 4\left(\frac{x - 1}{2}\right) = 2x - 2.$$

The parametric equations $x = 2t + 1$ and $y = 4t$ describe the same graph as $y = 2x - 2$.

EXAMPLE 1 Make a prediction about the shape of the graph of the parametrically defined curve, $x(t) = 2t$ and $y(t) = t^2$. Use your calculator to confirm or modify your guess and make a sketch of the graph.

Solution A table of x- and y-values for both positive and negative values of t is shown in Figure 93.

Figure 93 Table of values for $x(t) = 2t$ and $y(t) = t^2$

t	−3	−2	−1	0	1	2	3
x	−6	−4	−2	0	2	4	6
y	9	4	1	0	1	4	9

We should recognize the y-values in the table as some of the y-values of the quadratic toolkit function. Plotting the points from the x- and y-coordinates in Figure 93 along with the corresponding points from the quadratic toolkit function shows the graph is definitely a transformation of the toolkit function $f(x) = x^2$. (See Figure 94.) The graph of $f(x) = x^2$ is shown along with the points from the parametrically defined curve (represented by dots).

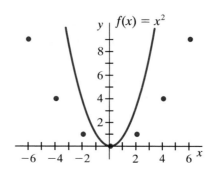

Figure 94 Graph of $f(x) = x^2$ and points on the parametric curve $x(t) = 2t$ and $y(t) = t^2$

Since the y-values in the table are paired with x-values twice as large as those they are paired with in the quadratic toolkit function, we conclude that the graph of $x(t) = 2t$, $y(t) = t^2$ looks like $f(x) = x^2$ stretched horizontally by a factor of 2. Therefore, $y = f(\frac{1}{2}x) = (\frac{1}{2}x)^2 = \frac{1}{4}x^2$ should be an equivalent representation of the parametric equations. We can confirm the equivalence of these functions by eliminating the parameter t from the equations $x = 2t$ and $y = t^2$ as follows:

$$x = 2t$$

$$t = \frac{x}{2}$$

$$y = t^2 = \left(\frac{x}{2}\right)^2 = \frac{x^2}{4}$$

Exercise Set 2.16

1. A motion sensor can be used with a calculator or computer laboratory system to determine a better model for the height of the cap described at the beginning of this section. If you have access to this equipment, drop a baseball cap over the motion sensor and fit a quadratic curve to the data representing the height of the falling cap. The coefficient of the quadratic term from this equation can be used to replace the -16 in $y(t)$ to obtain an equation for vertical displacement that more accurately represents the combined effects of gravity and air resistance. If you do not have access to equipment for gathering your own data, assume that the new coefficient is approximately -9. Now repeat the analysis described on pages 149–151. You will see that with most modifications to $y(t)$, the new model is consistent with the cap's clearing the fence. Now investigate the original questions: How far and for how long did the cap move in the horizontal direction before hitting the ground?

2. Are you convinced that your new model for the cap's position is correct? Write a paragraph expressing how confident you are in the final model and in the predictions you made in Exercise 1.

3. Consider the parametric equations $x(t) = 2t$ and $y(t) = 3t$. Make a table of x- and y-values for various values of t. Use your table to predict the shape of the graph associated with these equations; then use your calculator to confirm or modify your guess. Write a few sentences to describe the graph and to explain why it looks as it does.

4. In the following exercises, make a prediction about the shape of the graph of the parametrically defined curves. Use your calculator to confirm or modify your prediction; then make a sketch of the graph. Be sure to label your axes and make notes about where some *t*-values are located on your graph. When possible, describe why the graph looks as it does.

    a. $x(t) = 2t + 1$ $y(t) = 3t - 2$

    b. $x(t) = t$ $y(t) = 2t^2$

    c. $x(t) = t^2$ $y(t) = t$

    d. $x(t) = t^2$ $y(t) = t^2$

    e. $x(t) = t^2$ $y(t) = \sin t$

    f. $x(t) = \sin(t - \frac{\pi}{2})$ $y(t) = \sin t$

    g. $x(t) = t - \sin t$ $y(t) = 1 - \sin(t + \frac{\pi}{2})$, $t \in [0, 30]$, $\triangle t = 0.3$

5. In Parts a–d of Exercise 4, eliminate the parameter *t* and express *y* in terms of *x*. In some cases your result may be a relation that is not a function.

6. An air traffic controller is monitoring airplanes on a radar screen. The radar sweeps out a circle every second, showing the location of planes flying in the vicinity. The screen is a 50 cm by 50 cm square. The controller notices that at a certain instant, one plane, *A*, is located along the left edge of the screen 40 cm from the bottom left corner. Another plane, *B*, is located 45 cm to the right and 6 cm above the lower left corner. On the next rotation (1 second later), the controller observes that *A* is located at position 0.65 cm to the right and 0.30 cm below its previous position and *B* is 0.55 cm to the left and 0.45 cm above its previous position. Assume that the planes continue to fly in the same direction at a constant speed while they are visible on the screen.

    a. Write parametric equations to describe the horizontal and vertical positions from the bottom left corner for each plane as time passes.

    b. Assuming that the two planes are flying at the same altitude, will they collide? Explain.

    c. If the planes do not collide, how close do they get to each other in terms of centimeters on the radar screen?

    d. If one centimeter on the screen corresponds to 800 feet in the air, what is the minimum distance between the planes?

7. Coco, the wonder cat, is watching a spider on a wall that is 18 feet wide and 12 feet tall. The spider moves for 10 seconds along a path described by the parametric equations

$$x(t) = 0.3(t - 2.5)^2 + 2 \quad \text{and} \quad y(t) = 1.5 \sin t + 6,$$

if the lower left corner of the wall is taken as the origin. Coco can reach 3 feet up and can move freely from side to side.

a. What is the closest that Coco can come to catching the spider? When does this happen? Give answers to the nearest tenth. Document your work and clearly explain your reasoning.

b. How close does the spider get to the origin?

c. Watching the path of the spider, we see it changes its vertical direction several times. When does this happen? What aspect of the equations indicate that this change in direction will occur?

8. At 6:00 A.M. the Princess of the Sea reports that she is located 230 miles due north of Hamilton, Bermuda, traveling south at 15 mph. At the same time, the Atlanic Sovereign is 25 miles due west of Hamilton, traveling west at 20 mph.

a. Assuming the ships continue in the same directions at the same speeds, write parametric equations that model the paths of the two ships as they would appear on radar located at Hamilton.

b. Let d be the distance between the two ships at time t. Graph distance versus time for the next 12 hours of travel.

Chapter 2 Review Exercises

1. The local theater is putting on a play and is offering an incentive for student groups to attend. Each of the first 10 tickets costs $10. If more than 10 students but less than or equal to 20 students attend, each student pays $9 per ticket. If more than 20 students attend, then each ticket costs $7.50. Sketch a graph of the total cost T of purchasing s tickets.

2. Find the domain of the following functions.

   a. $f(x) = \dfrac{x}{x^2 - 5}$ **b.** $f(x) = \sqrt{x^2 + 5x + 6}$ **c.** $f(x) = \sqrt[3]{\dfrac{x+1}{x-2}}$

3. Suggest an appropriate domain restriction so that the inverse of $f(x) = 2x^2 - 4x - 10$ will be a function. Find f^{-1}.

4. Each graph in Figures 95–98 is the transformation or composition involving a toolkit function. Write an appropriate function for each graph.

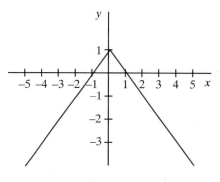

Figure 95

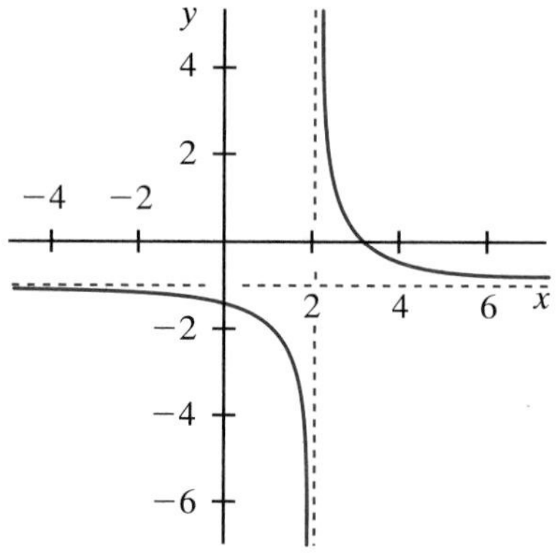

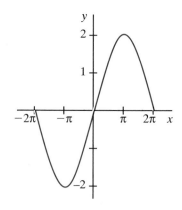

Figure 96

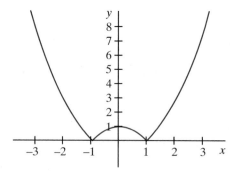

Figure 97

Figure 98

5. The driver of a delivery truck earns $10 per hour. The operating cost is $0.5 + \frac{x}{40}$ dollars per mile when the truck travels at an average speed of x miles per hour. What is the optimal speed for the truck as it travels 240 miles around town?

6. The sketch of $y = f(x)$ is shown in Figure 99.

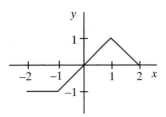

Figure 99   Graph of f for Exercise 6

a. State the domain and range of f.

b. Sketch graphs of the following.

   i. $g(x) = |f(x)| + 2$ **ii.** $g(x) = 2f(\frac{1}{2}x)$ **iii.** $g(x) = f(-x)$

   iv. $g(x) = f(|x|)$ **v.** $g(x) = f(2x - 3)$

7. After breakfast, a cup of coffee is left on the kitchen counter. A graph of the temperature of the coffee (in degrees Celsius) over time (in minutes) is shown. Use correct mathematical language to describe the graph.

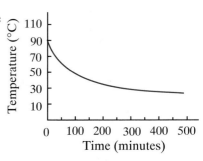

Figure 100   The cooling of hot coffee

8. a. Describe the graph defined by the parametric equations:

$$x(t) = 5\sin t + 1 \qquad y(t) = 5\sin\left(t + \tfrac{\pi}{2}\right) - 2$$

b. Why is the graph reasonable in light of the definitions of x and y?

c. Modify the equations so that the graph is moved to the left 2 and up 1 from its current location.

9. Sketch an even function that is increasing on the interval $-3 \le x \le 0$ and concave down on the interval $-1 \le x \le 3$.

10. Let $f(x) = \frac{1}{x^2 - 4}$ and $g(x) = \sqrt{x + 2}$. Find $f \circ g$ and give the domain.

11. Explain how to use composition to obtain a graph of $f(x) = \frac{1}{x^2 - 9}$.

12. Explain how you can use the concept of function inverses to straighten data that appear nonlinear.

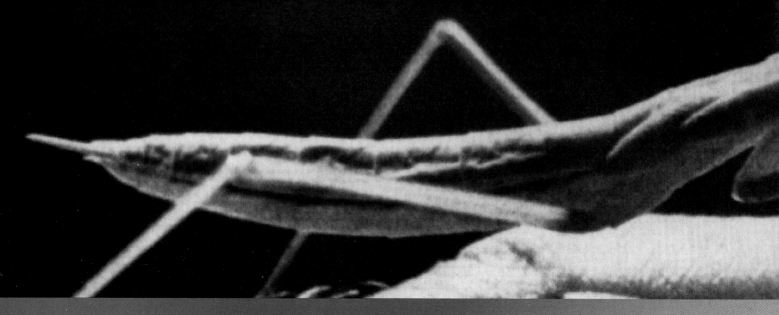

3 Exponential and Logarithmic Functions

3.1	Functions Defined Recursively	160
3.2	Loans and the Binary Search Process	167
3.3	Geometric Growth Models	173
3.4	Investigating the Mantid Problem	180
3.5	Geometric Series: Summing Geometric Growth	182
3.6	Investigating Garbage Disposal	188
3.7	Compound Interest	190
3.8	Graphing Exponential Functions	197
3.9	Introduction to Logarithms	207
3.10	Graphing Logarithmic Functions	214
3.11	Solving Exponential and Logarithmic Equations	220
3.12	Logarithmic Scales	229
3.13	Data Analysis with Exponential and Power Functions	235
3.14	Error Bounds in Re-expressed Data	251
3.15	Investigating More Data Collection	255
3.16	Investigation: Assessing Your Model	262
Chapter 3 Review Exercises		264

The Mantid Problem

A mantid is an insect that we commonly refer to as a praying mantis. Mantids are often used in biological studies because they are the insect version of a sloth. They rarely move, so it is very easy to keep track of them. Researchers have been studying the relationship between the distance a mantid will move to seek food and the amount of food already in the mantid's stomach. The distance is measured in millimeters, and the amount of food is measured in centigrams. In the research, food was placed progressively nearer to a mantid. The distance at which the mantid began to move toward the food was labeled the maximal distance of reaction (R). The amount of food in the mantid's stomach was measured. This amount was called the degree of satiation (S). Measurements for 15 mantids are given in the table below.

S (cg)	11	18	23	31	35	40	46	53	59	66	70	72	75	86	90
R (mm)	65	52	44	42	34	23	23	8	4	0	0	0	0	0	0

Source: Wildlife, Field-Test Version, COMAP, Inc., Lexington, MA, 1992.

Can you find an appropriate function to model the relationship between these variables?

SECTION 3.1 | Functions Defined Recursively

closed form
or explicit expressions

recursive functions

In the previous chapter, we learned that functions are special sets of ordered pairs. In most of the examples in the previous chapter, functions were described by an algebraic expression that can be evaluated for a particular input value resulting in a single output value. Such algebraic expressions are called *closed form* or *explicit expressions*. For these functions, the equation $y = f(x)$ was used to relate a y-value to a given x-value. In this section, we will investigate functions that are defined *recursively*. The domain values for *recursive functions* are whole numbers, and the range values are defined in terms of preceding range values.

EXAMPLE 1

Joan has a headache and decides to take a 200-mg ibuprofen tablet to relieve it. The ibuprofen is absorbed into her system and stays in her system until it is metabolized and filtered out by the liver and kidneys. Every four hours, Joan's body removes 67% of the ibuprofen that was in her body at the beginning of that four-hour time period. How much of the ibuprofen will remain in her system 24 hours after taking the 200-mg tablet?

iterative process

Solution We can generate values for the amount of ibuprofen in Joan's system by using an *iterative process*. In any iterative process, the current value of a variable is used to determine the next value. Therefore, each new amount of ibuprofen can be determined by decreasing the current amount by the amount of ibuprofen filtered out of Joan's system during each four-hour period. Since Joan begins with 200 mg of ibuprofen, we write

$$D_0 = 200,$$

where D_0 is used to represent the initial value.

We will use D_1 to represent the amount of ibuprofen left after four hours. The subscripts are used to count the steps in the iterative process. In this example, the subscripts also represent the number of four-hour periods since taking the tablet. In four hours, her kidneys have filtered 67% of the drug from her bloodstream, so

$$D_1 = D_0 - 0.67D_0 = 66.$$

After a second four-hour time period, the amount of drug in her body, represented by D_2, is given by

$$D_2 = D_1 - 0.67D_1 = 21.78.$$

Similarly, successive quantities of ibuprofen in her body can be generated as follows:

$$D_3 = D_2 - 0.67D_2 = 7.187,$$

$$D_4 = D_3 - 0.67D_3 = 2.372,$$

and, in general,

$$D_n = D_{n-1} - 0.67D_{n-1}, \quad n = 1, 2, 3,\dots.$$

A spreadsheet or calculator can generate successive values of D_n as shown in Figure 1. Note that values in the table have been rounded to three decimal places but that more decimal places were used in all computations. The amount of ibuprofen in Joan's body drops to less than 1 mg during the fifth time interval. If she takes one 200-mg dose, Joan will have only about 0.258 mg remaining in her body 24 hours later.

n	0	1	2	3	4	5	6
D_n	200	66	21.78	7.187	2.372	0.783	0.258

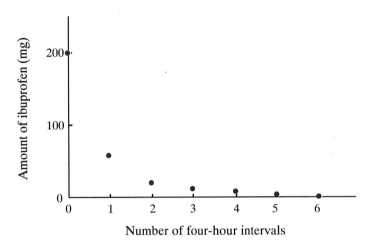

Figure 1   Amount of ibuprofen in Joan's body (one 200-mg dose) over time

The graph in Figure 1 shows ordered pairs (n, D_n) generated by the recursive system

$$D_0 = 200$$

$$D_n = D_{n-1} - 0.67D_{n-1}, \quad n = 1, 2, 3, \ldots.$$

Each point on the graph represents the amount of ibuprofen in Joan's body at the end of a particular four-hour time period. Notice that there is obvious curvature in this graph. The amount of ibuprofen in Joan's body does not decrease by the same number of milligrams during each time period. ▨

CLASS PRACTICE

1. Modify the recursive system in Example 1 to answer the following questions.

    a. Suppose the tablet Joan took was 250 mg. How much ibuprofen would be in her body after 4, 8, 12, 16, 20, and 24 hours?

    b. Suppose Joan's kidneys filter only 50% of the ibuprofen in four hours. If she takes a 200-mg tablet, how many milligrams would Joan have in her body after 4, 8, 12, 16, 20, and 24 hours?

2. Write one or two sentences to interpret the following recursive systems in the context of Example 1 about filtering ibuprofen. Find how much ibuprofen remains after 24 hours. In each exercise, the domain of D_n is $n = 1, 2, 3,\dots$.

a. $D_0 = 300,$ $\quad D_n = D_{n-1} - 0.8D_{n-1}$

b. $D_0 = 150,$ $\quad D_n = D_{n-1} - 0.2D_{n-1}$

c. $D_0 = 500,$ $\quad D_n = 0.2D_{n-1}$

d. $D_0 = 500,$ $\quad D_n = 0.5D_{n-1}$

EXAMPLE 2   Joan strained her knee playing tennis and her doctor has prescribed ibuprofen to reduce the inflammation and control pain. Joan is instructed to take two 200-mg ibuprofen tablets every four hours for three days. Joan doesn't like taking medicine, so she decides to take only one tablet every four hours for six days. After the six days, Joan's knee has not responded to the medication. Naturally, she knew that her knee would take longer to respond to the reduced treatment, but she did not expect no response at all. What could have happened?

Solution   This situation is similar to Example 1 except that Joan now repeats the dosage of 200 mg every four hours. At the end of n four-hour periods, Joan's body has filtered 67% of the ibuprofen that was in her body after $n - 1$ four-hour periods. In addition, 200 mg from a new tablet have been added. The recursive system representing the amount of ibuprofen in Joan's body if she takes 200 mg every four hours is

$$D_0 = 200$$
$$D_n = D_{n-1} - 0.67D_{n-1} + 200, \quad n = 1, 2, 3,\dots \quad \textbf{(1)}$$

The subscript n represents the number of four-hour time periods that have elapsed since Joan took the first dose. The results of iterating the recursive system (1), rounded to two decimal places, are shown together with a graph in Figure 2.

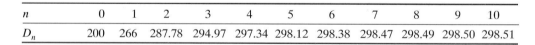

n	0	1	2	3	4	5	6	7	8	9	10
D_n	200	266	287.78	294.97	297.34	298.12	298.38	298.47	298.49	298.50	298.51

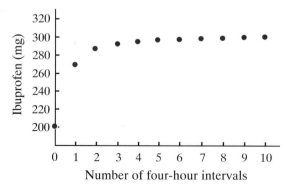

Figure 2   Amount of ibuprofen in Joan's body (one tablet every four hours)

The points shown in Figure 2 represent the amount of ibuprofen in Joan's body just after she takes a 200-mg tablet. Between each two points the level of the ibuprofen declines. We assume that the level "jumps" at the moment she takes another tablet, and the recursive system enables us to compute these levels. If we record the level only after she takes a tablet, then these values reach an *equilibrium* of approximately 298.51 mg. To understand why equilibrium has been reached at 298.51 mg, consider how much of the ibuprofen will be filtered out in four hours. Joan's kidneys will filter out 67% of the 298.51 mg in her body, or approximately 200 mg, which will be replaced when she takes the next tablet. That is, equilibrium occurs when the amount of ibuprofen taken into the body is the same as the amount filtered prior to taking the next tablet.

equilibrium

Suppose the ibuprofen has a therapeutic level of 450 mg. This means that there must be at least 450 mg in Joan's body in order for her to receive the benefits of the medicine. No wonder she thought the medicine was not working. It wasn't! ▨

Exercise Set 3.1

1. Modify the recursive system in equation (1) to represent the amount of ibuprofen in Joan's body if she takes two 200-mg tablets every four hours. Generate values of D_n to verify that this dosage achieves the therapeutic level (450 mg).

   a. How long does it take the ibuprofen to reach the therapeutic level?

   b. What is the equilibrium level of the ibuprofen in Joan's body if she takes two 200-mg tablets every four hours over a long period of time?

2. If Joan takes her medication every four hours, determine the amount of ibuprofen in her body after two days (12 iterations) and the equilibrium level resulting from each of the following recursive systems. Plot the ordered pairs you generate on a graph. In some cases the initial dosage differs from subsequent doses. In each exercise, the domain of D_n is $n = 1, 2, 3, \ldots$.

   a. $D_0 = 200,$ $\quad D_n = D_{n-1} - 0.4D_{n-1} + 200$

   b. $D_0 = 800,$ $\quad D_n = 0.6D_{n-1} + 200$

   c. $D_0 = 600,$ $\quad D_n = 0.4D_{n-1} + 200$

   d. $D_0 = 600,$ $\quad D_n = D_{n-1} - 0.4D_{n-1} + 200$

   e. $D_0 = 600,$ $\quad D_n = 0.6D_{n-1} + 300$

   f. $D_0 = 600,$ $\quad D_n = 0.4D_{n-1} + 300$

3. Each of the recursive systems in Exercise 2 can be written in the form

$$D_0 = a, \quad D_n = (1 - r)D_{n-1} + b, \quad n = 1, 2, 3,\ldots.$$

a. What does r represent? Why is r between 0 and 1?

b. By looking back at the results of Exercise 2 and by trying other variations, determine the effect of changing a, r, and b on the amount of ibuprofen in Joan's body after five days.

c. Determine the equilibrium level in terms of a, r, and b. Recall that $D_n = D_{n-1}$ when equilibrium is reached.

d. Use the result of Part c to determine the equilibrium level of any medicine if you take 200 mg every four hours and your kidneys filter out 50% of the medicine in your body every four hours.

4. A company has $10,000 worth of equipment and, for tax purposes, the company wants to figure the depreciation of the equipment over a 10-year time period. One method is to reduce the value by the same amount each year. A second method is to decrease the value of the equipment by the same percent of the current value each year.

a. Using the first method, generate a table and graph for the value of the equipment if it is decreased each year by $1000.

b. Using the second method, generate a table and graph for the value of the equipment if it is decreased by 20% each year.

5. One of the primary responsibilities of the manager of a swimming pool is to maintain the proper concentration of chlorine in the pool. The concentration should be between 1 and 2 parts per million (ppm). If the concentration gets as high as 3 ppm, swimmers experience burning eyes. If the concentration drops below 1 ppm, the water will become cloudy, which is unappealing. If it drops below 0.5 ppm, algae begin to grow. During a period of one day, 15% of the chlorine present in the pool dissipates (mainly because of the sun).

a. If the chlorine content starts at 2.5 ppm and no additional chlorine is added, how long will it be before the water becomes cloudy?

b. If the chlorine content starts at 2.5 ppm and 0.5 ppm of chlorine is added daily, what will happen to the level of chlorine in the long run?

c. If the chlorine content starts at 2.5 ppm and 0.1 ppm of chlorine is added daily, what will happen to the level of chlorine in the long run?

d. How much chlorine must be added daily for the chlorine level to stabilize at 1.8 ppm?

6. The Fish and Wildlife Division monitors the trout population in a stream that is under its jurisdiction. Its research indicates that natural predators, together with pollution and fishing, are causing the trout population to decrease at a rate of 20% per month. The Division proposes to introduce additional trout each month to replenish the stream. Assume the current population is 300. Use tables and graphs to investigate the following questions.

   a. What will happen to the trout population over the next 10 months with no replenishment program?

   b. What is the long-term result of introducing 100 trout into the stream each month?

   c. Investigate the result of changing the number of trout introduced each month. What is the impact of the number of trout added each month?

   d. Investigate the impact of changing the initial population on the long-term behavior of the population. What is the effect of the initial population?

   e. What is the impact of the rate of decrease in the population during the replenishment program?

   f. There are three parameters in this problem: the initial number of trout, the rate of decrease, and the number of trout added each month. Which parameter is most important? Why?

7. Medicines generally have a therapeutic range rather than a single therapeutic level. In other words, a medicine is effective if its level in the body is between two values. At concentrations below this range, too little is present to have a measurable effect, and concentrations above this range may be toxic. The level in the body peaks just after the medicine is taken, and the level is at a minimum just before a dose is taken. Suppose Joan takes anti-inflammatory medicine at the prescribed dosage of 440 mg every 12 hours and her kidneys filter 60% of the medicine from her body every 12 hours. Use tables and graphs to complete the following tasks.

   a. Generate a sequence of values for the level of medicine in Joan's body just before each dose.

   b. In the long run, the level of medicine in Joan's body will range between what two values?

   c. Suppose the therapeutic range of the anti-inflammatory medicine is between 300 mg and 800 mg. What adjustment, if any, needs to be made in Joan's dosage to stay within this range in the long run?

8. Suppose Bill wants to buy a compact disc player that costs $249. He has a part-time job, and he is able to save $20.00 each week. Suppose he accumulates the money at home.

   a. Write a recursive system that can be used to determine how much Bill has saved over time.

   b. Use the recursive system to generate values for the amount of money Bill has saved in 13 weeks.

   c. Make a graph to show the amount Bill has saved versus the number of weeks he has been saving.

   d. The points graphed in Part c should appear linear. Explain why these ordered pairs are linear.

   e. Write an explicit function $A = f(t)$ that can be used to generate the same ordered pairs you graphed in Part c.

   f. State the domain and the range of the explicit function within the context of this problem. Compare them to the domain and range of the recursively defined function.

Now suppose that Bill deposits his savings in a bank that will pay 0.1% interest each week.

   g. Write a recursive system that can be used to determine how much Bill has saved over time.

   h. Use the recursive system to generate values for the amount of money Bill has saved in 13 weeks.

   i. Make a graph to show the amount Bill has saved versus the number of weeks he has been saving.

   j. Are the ordered pairs graphed in Part i linear? Explain why or why not.

   k. State the domain and range of the recursively defined function.

SECTION 3.2 Loans and the Binary Search Process

Loans

Iterative procedures are useful not only to generate successive values of a function, but also to find solutions to various types of problems. Suppose you are interested in purchasing a car and need a $5000 loan. The lending agency is going to charge you interest each month and you are going to make a payment each month. You plan to pay $125 each month until the loan is paid off. Suppose the interest rate is 0.75% per month. How long will it take you to repay the loan? What is the total amount you will have to repay?

When repaying a loan, interest is charged on all the money owed at the end of each month. In this example, you will owe $(0.0075)(\$5000) = \$37.50$ in interest on the loan at the end of the first month. After making the first payment, you will owe $\$5000 + \$37.50 - \$125 = \$4912.50$ on the loan. The amount you still owe on the loan at the end of a month is the old amount plus interest, minus the amount of your payment. The process of paying off a loan is known as *amortizing* the loan, or *loan amortization*. We can write a recursive system for the amount that you owe on the loan, which is

loan amortization

$$L_0 = 5000, \quad L_n = L_{n-1} + (0.0075)L_{n-1} - 125, \quad n = 1, 2, 3,...,$$

principal

where L_n is the amount you owe on the loan after n months, also known as the *principal*. If we iterate this system, the values we generate represent the principal at the end of each month.

$$L_1 = 5000 + (0.0075)5000 - 125 = 4912.50,$$

$$L_2 = 4912.50 + (0.0075)4912.50 - 125 = 4824.34,$$

$$L_3 = 4824.34 + (0.0075)4824.34 - 125 = 4735.53.$$

Note that values have been rounded to the nearest cent. All the decimal places on the calculator were retained in the computations.

After three months, you will owe $4735.53 on the loan. Notice that you have paid $375 (three payments of $125), but only $\$5000 - \$4735.53 = \$264.47$ was applied toward the principal of the loan. The remaining $110.53 was interest on the loan.

We can continue generating values of L using a more compact form of the iterative equation,

$$L_n = (1.0075)L_{n-1} - 125,$$

which yields

$$L_4 = (1.0075)4735.53 - 125 = 4646.04,$$

$$L_5 = (1.0075)4646.04 - 125 = 4555.89,$$

$$L_6 = (1.0075)4555.89 - 125 = 4465.06.$$

After six months, you owe $4465.06 from the original $5000 principal. You have paid $750, and these payments have reduced the debt by $534.94. More than $215 was interest on the loan.

We are interested in the length of time it will take to repay this loan and the total amount you will have to repay. If you continue payments of $125, the final payment will probably exceed the outstanding balance. While it is possible to compute loan payments so that all the payments are equal (or within a few cents), it is not unusual for the lender to make all the payments, except for the last one, a whole dollar amount, or even round these payments to the nearest five- or ten-dollar amounts. Doing this will almost certainly make the final payment different from the rest. This final payment is known as the *balloon payment*.

balloon payment

The graph and partial table in Figure 3 show that it will take 47 payments to get the balance down to $91.25. After one additional month, the balance will be $91.25(1.0075) = $91.93. Thus, the balloon payment will be $91.93.

n	0	1	2	3	4	...	44	45	46	47
Balance ($)	5000	4912.50	4824.34	4735.53	4646.04	...	458.68	337.12	214.64	91.25

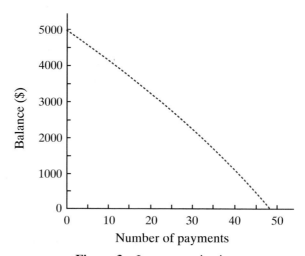

Figure 3 Loan amortization

The total amount repaid to the lender is (47 · $125) + $91.93 = $5966.93. We see that it costs $966.93 to borrow the $5000 for 48 months.

We can generalize the iterative system used to determine the total amount repaid as follows. If we borrow an amount A and let I represent the interest rate and P the amount of the payment during each time period, we can describe the amount owed on a loan after n time periods with the recursive system

$$L_0 = A, \quad L_n = L_{n-1} + I \cdot L_{n-1} - P, \quad n = 1, 2, 3,...$$

or

$$L_0 = A, \quad L_n = (1 + I)L_{n-1} - P, \quad n = 1, 2, 3,....$$

The Binary Search Process

Typically with a loan we know the amount we want to borrow, the interest rate we will have to pay, and the length of time we have to pay off the loan. We want to determine the payment that will allow us to pay off the loan in the required time.

EXAMPLE 1   Suppose you buy a car and take out a loan of $10,000 at 8% annual interest to be paid back over four years. What is the yearly payment you must make to pay off the loan in four equal payments?

Solution   If there were no interest, you would have to pay $2500 each year to repay the $10,000. Since you must pay interest on the loan, $2500 per year is obviously not enough. The interest for the first year is $800 (8% of $10,000), so you might guess the payment will be $2500 + $800 = $3300. If you repay $3300 the first year, the outstanding balance will be less than $10,000 and interest for the second year will be less than $800; so a payment of $3300 will result in an overpayment for the last three years. Since $2500 is too low and $3300 is too high, your next guess might be the average of these two payments, or $2900. It is not obvious what the balance will be after four payments, so we need to do the recursive calculations.

The outstanding principal is modeled by the recursive system

$$L_0 = 10,000, \quad L_n = L_{n-1} + 0.08 L_{n-1} - 2900, \quad n = 1, 2, 3, 4.$$

The successive values are

$$L_0 = 10,000,$$
$$L_1 = 10,000 + 0.08(10,000) - 2900 = 7900,$$
$$L_2 = 7900 + 0.08(7900) - 2900 = 5632,$$
$$L_3 = 5632 + 0.08(5632) - 2900 = 3182.56,$$
$$L_4 = 3182.56 + 0.08(3182.56) - 2900 = 537.16.$$

An annual payment of $2900 is not sufficient to pay off the loan in four years, as a balance of $537.16 remains after the four payments. If we again select the average of two payments, one that is too large and one that is too small, we will have arrived at a better guess. We know $3300 is too large and $2900 is too small, so the next guess will be $\frac{\$2900 + \$3300}{2} = \$3100$. The successive balances are now

$$L_0 = 10,000,$$
$$L_1 = 10,000 + 0.08(10,000) - 3100 = 7700,$$
$$L_2 = 7700 + 0.08(7700) - 3100 = 5216,$$
$$L_3 = 5216 + 0.08(5216) - 3100 = 2533.28,$$
$$L_4 = 2533.28 + 0.08(2533.28) - 3100 = -364.06.$$

We see that a payment of $3100 is too large, since $L_4 = -364.06$ implies that you pay $364.06 more than necessary.

After initially considering payments of $2500 and $3300, which were too small and too large, respectively, we made two additional calculations for the loan payment. The first estimated payment, $2900, was too small, and the second, $3100, was too large. The correct loan payment is some value between $2900 and $3100, so we use the midpoint, which is $3000. This results in a positive balance of $86.55 after four years, so this payment is too small. Our next estimate will be midway between $3000 (with a positive balance) and $3100 (with a negative balance). This midpoint is $3050, and it leaves a balance of $-$138.75$ after four years. Now we know that the correct payment is between $3000 and $3050. We can continue this process of guessing until we get a final balance of $0 or as close to $0 as possible with rounding the payment to the nearest cent. The result of these calculations reveals that you need to pay $3019.21 per year to pay off the loan in four years. The computations below verify that this yearly payment is correct.

$$L_0 = 10,000,$$
$$L_1 = 10,000 + 0.08(10,000) - 3019.21 = 7780.79,$$
$$L_2 = 7780.79 + 0.08(7780.79) - 3019.21 = 5384.04,$$
$$L_3 = 5384.04 + 0.08(5384.04) - 3019.21 = 2795.56,$$
$$L_4 = 2795.56 + 0.08(2795.56) - 3019.21 = -0.009.$$ ▪

binary search

The process we used to find the loan payment is called a *binary search*. The binary search is an iterative process in that it uses the same procedure again and again, but with a new search area, or feasible interval, at each step.

Summary of the Binary Search Procedure Applied to Loan Payment

To determine a feasible interval, use a payment that is too low as the lower bound of the interval and one that is too high as the upper bound. Find the midpoint of this interval and determine the final balance for this payment. If there is a negative balance after the required number of payments, use the midpoint as the new upper bound of the interval. Otherwise, use the midpoint as the lower bound of the interval. Repeat the process using the new smaller interval and stop as soon as the balance is zero or as close to zero as required.

The first eight steps of the binary search applied to the loan payment problem are summarized in Figure 4.

Figure 4   Binary search applied to loan payment

Low guess	High guess	Midpoint	Principal after 4 yrs.
$2500	$3300	$2900	$537.16
$2900	$3300	$3100	$-$364.06
$2900	$3100	$3000	$86.55
$3000	$3100	$3050	$-$138.75
$3000	$3050	$3025	$-$26.10
$3000	$3025	$3012.50	$30.23
$3012.50	$3025	$3018.75	$2.06
$3018.75	$3025	$3021.88	$-$12.04

1. How long would it take to amortize a loan and how much will the loan cost under each of the following conditions? In each case, L_0, I, and P represent the initial amount borrowed, the monthly interest rate, and the monthly payment, respectively.

   a. $L_0 = \$5000$, $I = 1\%$, $P = \$200$

   b. $L_0 = \$5000$, $I = 1.5\%$, $P = \$200$

   c. $L_0 = \$5000$, $I = 0.5\%$, $P = \$200$

   d. $L_0 = \$5000$, $I = 1\%$, $P = \$250$

2. Maisha opens a retirement account on her 35th birthday with a deposit of $2400. Each year on her birthday, she plans to deposit an additional $2400. The account earns interest at a rate of 10% annually.

   a. How much will Maisha have saved by the time she retires at age 65?

   b. Suppose Maisha wants to have $1 million in the account by age 65. To the nearest hundred dollars, how much should she deposit each year? Assume the initial deposit remains $2400.

   c. Suppose Maisha starts saving 10 years earlier, at age 25. To the nearest hundred dollars, how much should she deposit each year to have $1 million at retirement? Assume the initial deposit remains $2400.

3. Compute and compare the annual payments on a five-year loan at 6% annual interest for the following cars:

   a. Mazda Miata priced at $21,000

   b. Chevrolet Camaro priced at $26,000

   c. Jeep Cherokee priced at $28,000

4. Isaac wants to buy a car and is shopping for a four-year loan.

   a. If he needs to borrow $15,000 and the loan is at 8% annual interest, what is his annual payment?

   b. If he can afford to pay no more than $3600 a year on a loan, how much can he borrow at 8% annual interest?

   c. If he wants to borrow $14,000 and he can afford to pay no more than $4000 a year on a loan, what is the highest interest rate he can accept?

5. You and your parents need to borrow $50,000 to pay for your college tuition. The Village Bank offers a 15-year loan at 8% annual interest. The Hometown Bank offers a 20-year loan at 7% annual interest. Which loan is better? Explain what you mean by "better."

6. You are going to use a binary search procedure to find a word that your friend has chosen from the dictionary.

   a. Outline the steps in your process.

   b. If the dictionary contains 234,931 words, how many iterations of the procedure are needed to find the word?

SECTION 3.3 Geometric Growth Models

geometric growth

Some simple, recursively defined functions have important applications, one of which is representing *geometric growth*. The first example from Section 3.1, the amount of ibuprofen in Joan's system at time n if she takes only one 200-mg tablet, is an example of a geometric growth model. In that case, we used the recursive system

$$D_0 = 200, \quad D_n = D_{n-1} - 0.67D_{n-1},$$

which can be written as

$$D_0 = 200, \quad D_n = 0.33D_{n-1}, \quad n = 1, 2, 3,\dots.$$

The amount of ibuprofen remaining after n time intervals is always 33% of the amount remaining after $n - 1$ time intervals. In a geometric growth model, the value of the function at time n is directly proportional to the value at time $n - 1$. While this suggests a recursive definition for geometric growth, we can also find an explicit or closed form representation.

EXAMPLE 1 Blue jeans fade when they are washed. Suppose a pair of jeans loses 2% of its color with each washing. How much of the original color is left after 50 washes?

Solution The recursive system is

$$C_0 = 1, \quad C_n = 0.98C_{n-1}, \quad n = 1, 2, 3,\dots,$$

where $C_0 = 1$ is the initial color at 100% and C_n is the amount of color remaining in the jeans after n washings.

Since losing 2% is the same as retaining 98%, the coefficient in the iterative model is 0.98. With 50 iterations of the equation for C_n, the result is that the jeans have about 36% of the original color left after 50 washings. ▨

EXAMPLE 2 The consumption of electrical power in a community increases by 5% per year. This year, the community used 500 thousand kilowatt-hours of electrical power. How long will it take before the annual electrical power consumption doubles?

Solution This situation can be represented by the recursive system

$$P_0 = 500, \quad P_n = 1.05P_{n-1}, \quad n = 1, 2, 3,\dots.$$

Since the amount of power used increases by 5% each year, next year the community will use 105% of what it used this year. Iterating the equation $P_n = 1.05P_{n-1}$ shows that after 14 years annual consumption will be about 990 thousand kWh and after 15 years it will be about 1,039 thousand kWh. The amount of power required by the community will double in approximately 15 years, if the need for electrical power continues to increase at 5% per year. ▨

EXAMPLE 3   Potassium-42 is a radioactive element that is often used in biological experiments as a tracer element. Potassium-42, like all radioactive elements, decays into a non-radioactive form at a rate proportional to the amount present. Potassium-42 loses 5.545% of its mass every hour. If 1 milligram is initially present in an animal, when will only 0.1 milligram be present?

Solution   Using the recursive system

$$P_0 = 1, \quad P_n = 0.94455P_{n-1}, \quad n = 1, 2, 3,\ldots,$$

we need to determine the values of n for which P_n is less than or equal to 0.1. Iterating the equation for P_n shows that $P_{40} = 0.1021$ mg and $P_{41} = 0.0964$ mg. Sometime between 40 and 41 hours, we expect to have only 0.1 mg remaining.

In this case, the amount of radioactive material does not remain constant for an hour and then jump to a new value. Rather, the amount changes at every instant. During each hour, 5.545% of the mass decays from radioactive to nonradioactive material. We know that there is some "smooth" curve that connects the discrete points generated by the recursive model and that the amount of potassium-42 changes between the 40th and 41st hours. ▪

Explicit Functions for Geometric Growth

Each of the previous examples uses a recursive model that can be written as

$$Y_0 = a, \quad Y_n = (1 + k)Y_{n-1}, \quad n = 1, 2, 3,\ldots,$$

growth rate

decay rate

How can geometric growth be described by an explicit function?

where k is the *growth rate* if $k > 0$ or k is the *decay rate* if $k < 0$. The values of Y_n increase if $k > 0$, and the values of Y_n decrease if $k < 0$. In each case, the value of Y_n depends entirely on the value of k and the value of Y_{n-1}. As usual, n represents the number of iterations.

We know that geometric growth can be described by a recursive system. How can geometric growth be described by an explicit function? To find an explicit function, we begin by iterating the equation $Y_n = (1 + k)Y_{n-1}$. In Example 2 concerning electrical power consumption, we find that

$$P_0 = 500,$$
$$P_1 = (1 + 0.05)P_0,$$
$$P_2 = (1 + 0.05)P_1,$$
$$P_3 = (1 + 0.05)P_2,$$
$$P_4 = (1 + 0.05)P_3.$$

We can rewrite each of these equations in terms of P_0, which yields

$$P_1 = (1 + 0.05)P_0,$$
$$P_2 = (1 + 0.05)P_1 = (1 + 0.05)^2 P_0,$$
$$P_3 = (1 + 0.05)P_2 = (1 + 0.05)^3 P_0,$$
$$P_4 = (1 + 0.05)P_3 = (1 + 0.05)^4 P_0.$$

The nth term is given by

$$P_n = (1 + 0.05)^n P_0.$$

In general, we can convert recursive equations for geometric growth to explicit functions in terms of Y_0, k, and n, as follows:

$$Y_1 = (1 + k)Y_0,$$

$$Y_2 = (1 + k)Y_1 = (1 + k)^2 Y_0,$$

$$Y_3 = (1 + k)Y_2 = (1 + k)^3 Y_0,$$

$$Y_4 = (1 + k)Y_3 = (1 + k)^4 Y_0.$$

The nth term is given by

$$Y_n = (1 + k)^n Y_0.$$

Since n can be thought of as the independent variable and Y the dependent variable, this equation can be rewritten in the more traditional functional form,

$$Y(n) = (1 + k)^n Y_0. \tag{2}$$

In equation (2) the independent variable n is the exponent, so this function is an exponential function. The exponential function $Y(n) = (1 + k)^n Y_0$ is the explicit form of the recursive growth model $Y_0 = a$, $Y_n = (1 + k)Y_{n-1}$. In the recursive representation, n is a discrete variable with positive integer values representing the number of time intervals. According to equation (2), it appears that n could be a continuous variable representing the time elapsed since we started measuring the growth. In fact, if we plot the discrete values generated by the recursive system and graph the continuous function $Y(t) = (1 + k)^t Y_0$, the graphs will coincide for nonnegative integer values of t.

EXAMPLE 4 Suppose you plan to make a one-time deposit into a bank account that will earn 0.75% monthly interest. How large must this deposit be so that you will have a college fund of $25,000 available after 18 years, which is 216 months?

Solution We can solve this problem using the recursive system $S_0 = a$, $S_n = (1.0075)S_{n-1}$. To do so, we have to guess and check to find the appropriate value of S_0 that gives $S_{216} = 25,000$. Using the explicit function $S(n) = (1.0075)^n S_0$, we need to find the value of S_0 such that $25,000 = (1.0075)^{216} S_0$. Solving for S_0 gives the equation

$$S_0 = \frac{25,000}{(1.0075)^{216}},$$

or $S_0 = \$4977.46$. We see that the explicit function is useful when we do not need all the intermediate values that the recursive system generates. ∎

CLASS PRACTICE

1. Find the explicit function for Example 1 on page 173.

2. Find the explicit function for Example 2 on page 173.

3. Find the explicit function for Example 3 on page 174.

Note that the explicit form can be used to generate values for nonnegative integer values of n, and these values are the same as values generated using recursive equations. However, the explicit form can also be used to generate values for noninteger values of n. Sometimes these values are meaningful; in other contexts noninteger values do not have meaning.

Discrete and Continuous Exponential Growth

The first three examples in this section illustrate both the differences and similarities that exist in discrete and continuous functions. In Example 1, jeans fade only when they are washed. In the time between washings, we assume that there is no fading. In Example 2, the amount of power used does change between one year and the next. However, the net effect of many smaller changes throughout the year was a 5% gain at the end of the year. In Example 3, the amount of radioactive material at the end of each hour was found using a recursive equation. In this situation, the decay happens at every instant. In all three examples, we can find both a recursive model and an explicit model.

Recursive models are, by their very nature, discrete. Explicitly defined functions are often associated with continuous phenomena. By restricting the domain of a function, we can use an explicit function to generate discrete values. Mathematicians and scientists have long used continuous functions to model phenomena that are really discrete, and likewise have used discrete forms to model continuous phenomena. When an explicit function is used for a discrete-valued function, it is wise to include a restriction on the domain. If the domain restriction is omitted, it may leave the mistaken impression that the function is continuous.

We want to be able to use both the recursive and explicit forms to generate values of a function. Furthermore, we want to be able to decide which is more appropriate for a given situation.

EXAMPLE 5   Population Doubling Time

Suppose the population of a certain country is known to grow geometrically and to double in 24 years. Find a model for the population at any given time. Use your model to estimate the population after 30 years.

Solution   Since the problem involves a quantity that doubles every 24 years, a natural choice for the base of the exponential function is 2. The function $y = 2^x$

doubles whenever the value of x increases by 1 unit. The explicit function has the form $f(t) = a \cdot 2^{kt}$. Since $f(0) = a$, we know that a is the initial population. We need to determine the value of k so that the population at time $t = 24$ is $2a$. We need to solve the equation $2a = a \cdot 2^{24k}$ for k.

$$2a = a \cdot 2^{24k}$$

$$2 = 2^{24k}$$

$$2^1 = 2^{24k}$$

$$1 = 24k$$

$$k = \frac{1}{24}$$

Thus, the function that models the population is $f(t) = a \cdot 2^{\frac{1}{24}t}$, where t is any real number greater than or equal to zero. Since $f(30) = a \cdot 2^{\frac{30}{24}} \approx 2.4a$, the population will have increased from its original value by a factor of approximately 2.4 after 30 years.

In this situation, the recursive system $A_0 = a$, $A_n = 2A_{n-1}$, $n = 1, 2, 3, \ldots$, where n represents one 24-year time period, would provide the same population values at the end of each 24-year period. The recursive model indicates very little about the intervening years. To determine the population after 30 years, the explicit form is more useful, since we can evaluate $f(t) = a \cdot 2^{kt}$ at $t = 30$ to estimate the population after 30 years. ▩

EXAMPLE 6   Rebecca starts working for a company at a salary of $30,000 per year. Based on the company's history, she can expect raises of 5% each year on the anniversary of her employment. When will she first earn at least $40,000 per year?

Solution   We can use the recursive system $A_0 = 30,000$, $A_n = 1.05A_{n-1}$ to determine when Rebecca's salary will equal or exceed $40,000. Her first pay raise will result in a salary of $31,500. Her fifth raise will bring her salary up to $38,288, and her sixth raise will put her pay over $40,000 at $40,203. We can also use an explicit function to find the time at which her salary will reach $40,000. This function is $S(t) = 30,000(1.05)^t$. By evaluating this function, we find that $S(5) = 38,288$ and $S(6) = 40,203$. Both the recursive model and the explicit model show that Rebecca will make at least $40,000 with her sixth raise. ▩

In many cases, either the recursive system or the explicit function can be used to arrive at the same answer. If we wanted to know how much Rebecca might earn annually if she stayed with this company for 30 years, it would be easier to use the explicit function and substitute $t = 30$. In such cases, it is more efficient to use the explicit function when we need to predict far into the future. In cases in which the intermediate values are important, such as the balance due on a loan after each payment, a recursive system is preferable. Even if we use the explicit function in Example 6, the context of the problem implies that the domain is limited to integer values.

In Exercises 1 through 4, state the recursive system and the explicit function, when possible. Use the most appropriate form to answer each question. Discuss whether the growth (or decay) is discrete or continuous.

1. Research City is growing by 14% each year. If the population of the city is approximately one million people and the rate of growth continues at 14% annually, write a function to model the population over the next 10 years.

2. The population of Coastal City grows by 3% each year as a result of only births and deaths among current residents. The population is currently one million. Each year 15,000 more people move into the city than move out. Write a function to model the population over the next 10 years.

3. Each year the population of rabbits in a meadow decreases by 5%. If the year begins with 230 rabbits and the population continues to decrease by 5% each year, describe what will happen to the population over the next ten years.

4. Each year the population of rabbits in a meadow decreases by 5%. Farmer Dan decides to augment the rabbit population by releasing five new rabbits into the meadow each year. If the year begins with 230 rabbits, describe what will happen to the population over the next 10 years.

5. According to the North Carolina HIV/STD Control Branch, a total of 7422 cases of AIDS had been diagnosed in North Carolina for all years up through the end of 1996. This compares to 6712 cases by the end of 1995. If the number of cases continues to grow at the same rate, how many cases will be diagnosed this year? How many cases in 2005?

   Source: HIV/STD Control Section, Epidemiology & Special Studies Branch, First Quarter 1997 HIV/AIDS Surveillance Report, NC Dept. of Environment, Health, & Natural Resources, Division of Epidemiology.

6. An annual inflation rate of $k\%$ means that items will cost $k\%$ more next year than they did this year. Based on a yearly inflation rate of 3%, use the current cost given below to estimate the cost of the following items in 10, 20, 30, and 40 years.

Jeans	$35.00
Hamburger	$1.90
Car	$19,000.00
Home	$120,000.00
Textbook	$45.00
Movie Ticket	$7.00

7. How much money would you need to invest now in an account that earns 0.5% monthly interest so that in 20 years you would have $50,000?

8. The population of a certain organism triples every hour. Write a function that models this growth. By what factor does the population grow in one-half hour?

half-life

9. The *half-life* of a radioactive material is the period of time required for it to decay to one-half of its original mass. The half-life of thorium-234 is only 25 days. Write a model for the amount of thorium-234 left after t days. What percent of an original amount is left after 300 days?

10. The half-life of thorium-232 is 14 billion years. What percent of an initial quantity will be left after 2000 years?

11. The population of the People's Republic of China in 1990 was approximately 1.2 billion and was growing at a rate of 1.3% annually.

   a. Find an explicit exponential function to model the population.

   b. To the nearest year, how long will it take the population to double?

   c. Using the information in Part b, write an alternate function for the population.

   d. Write a recursive system to model the population.

   e. In 1997, Great Britain returned Hong Kong to the People's Republic of China. This resulted in a one-time increase of 6 million people. How does this affect the year in which the population doubles the 1990 figure?

12. The cane toad, a native of South America, was introduced to Australia in 1935 as a way to control the destructive sugar cane beetle. It has no natural predators in Australia (cane toads secrete a poison when squeezed or bitten), and it reproduces rapidly. There were 101 cane toads brought to Australia in June 1935. By March 1937, the population of cane toads had grown to 64,500. Write a short newspaper article that explains to people in Australia how the cane toad population is growing. Include factual information that they will understand and find significant. Graphs and tables are helpful but should not be the only information given. Also include a fact sheet of your work so the editor can verify the accuracy of your story. This fact sheet will not be included in the article.

   Source: Smithsonian, October 1990, pp. 138–144.

13. The recursive system $P_0 = a$, $P_n = (1 + k)P_{n-1}$, $n = 1, 2, 3, \ldots$ can be written as an explicit exponential function $P(n) = a(1 + k)^n$. What recursive system can be written as an explicit linear function of the form $y(n) = a + kn$?

SECTION 3.4 | Investigating The Mantid Problem

A mantid is an insect that we commonly refer to as a praying mantis. Mantids are often used in biological studies because they are the insect version of a sloth. They rarely move, so it is very easy to keep track of them. Researchers have been studying the relationship between the distance a mantid will move to seek food and the amount of food already in the mantid's stomach. The distance is measured in millimeters, and the amount of food is measured in centigrams. In the research, food was placed progressively nearer to a mantid. The distance at which the mantid began to move toward the food was labeled the maximal distance of reaction (R). The amount of food in the mantid's stomach was measured. This amount was called the degree of satiation (S). Measurements for 15 mantids are given in Figure 5.

Figure 5 Satiation and reaction distance

S (cg)	11	18	23	31	35	40	46	53	59	66	70	72	75	86	90
R (mm)	65	52	44	42	34	23	23	8	4	0	0	0	0	0	0

Source: Wildlife, Field-Test Version, COMAP, Inc., Lexington, MA, 1992.

1. Determine a piecewise–defined function to model the relationship between the satiation of a mantid and the distance it will move for food.

2. According to your model, what is the greatest distance that a mantid will move for food?

hunger threshold

3. Biologists call the level of satiation at which an animal will not seek food the *hunger threshold.* What is the hunger threshold for the mantid?

Researchers have also studied the digestion of the mantid. They have gathered data on the degree of satiation and the time that had passed since the mantid had last eaten its fill. By combining the measurements of a number of mantids, biologists have a fairly accurate picture of how quickly the digestive system of the mantid works. The data below compare the length of time that a mantid is deprived of food in hours (T) and the amount of food in its stomach at that time (S). For example, one hour after a mantid has eaten its fill, its stomach contains 90 cg of food.

Figure 6 Time and satiation

Time T (hr)	0	1	2	3	4	5	6	8	10
Satiation S (cg)	94	90	85	82	88	83	70	66	68
Time T (hr)	12	16	19	20	24	28	36	48	72
Satiation S (cg)	50	46	51	41	32	29	14	17	8

4. Explain why it makes sense that S should decrease as T increases.

5. The biologists assume that the mantid will digest a fixed percentage of the food in its stomach each hour. Since the amount of food decreases by a fixed percentage each hour, the recursive equation $S_n = p \cdot S_{n-1}$ should fit reasonably well for some value of p. Use the binary search process to determine the value of p that best fits the data. In the recursive model, what does n represent?

6. Find an explicit function $S = f(T)$ that gives the same values as the recursive model from Problem 5.

7. Suppose a mantid has been without food for 40 hours. How far do you estimate it will travel seeking food? How did you determine your answer?

8. Suppose a mantid is willing to travel 47 mm for food. Approximately how long has it gone without eating? How did you determine your answer?

9. Sketch a graph relating the time since eating to the distance a mantid will travel seeking food. Be sure to include the times when the mantid is past the hunger threshold.

10. Write a paragraph to describe, using appropriate mathematical terminology, what you have learned about the eating habits of mantids.

Geometric Series: Summing Geometric Growth

The worldwide consumption of oil in 1995 was about 25 billion barrels, and consumption was increasing at a rate of 2.2% per year. If this rate of increase continues, how much oil will be needed to meet the demands of society in 2025? The explicit function, $A(n) = 25(1.022)^n$, where n represents the number of years since 1995, can be used to show that the amount of oil needed in 2025 is $A(30) \approx 48$ billion barrels. That is certainly a large amount, but knowing the amount in any one year does not really tell the whole story. A more important question is the following: What is the total amount of oil that will be used between 1995 and 2025? This sum is the quantity of oil that is being depleted from world oil reserves over the next 31 years.

To determine the total amount of oil used from 1995 to 2025, we want to find the sum

$$T = A(0) + A(1) + A(2) + \cdots + A(30).$$

We can rewrite the equation for T as

$$T = A(0) + (1.022)A(0) + (1.022)^2 A(0) + (1.022)^3 A(0) + \cdots + (1.022)^{30} A(0).$$

geometric series

Notice that each term in the sum is 1.022 times the previous term. This sum is an example of a *geometric series* in which the successive addends are found by multiplying the previous term by some fixed value.

The general form of a geometric series is

$$S = a + ar + ar^2 + ar^3 + \cdots + ar^n. \tag{3}$$

common ratio

Is there an efficient way to calculate the sum of a geometric series?

The ratio of two successive terms in the geometric series in equation (3) is the constant r, which is known as the *common ratio*. Is there an efficient way to calculate the sum of a geometric series? A formula for the sum of a geometric series can be found by using an algebraic trick. First, multiply both sides of equation (3) by r, which yields

$$Sr = ar + ar^2 + ar^3 + ar^4 + \cdots + ar^{n+1}. \tag{4}$$

Now subtract equation (4) from equation (3). That is, subtract the left side of (4) from the left side of (3) and the same with the right sides. Note that most of the terms on the right sides of equations (3) and (4) are the same. The resulting equation is

$$S - Sr = a - ar^{n+1}.$$

Solving for S yields

$$S = \frac{a - ar^{n+1}}{1 - r}. \tag{5}$$

Equation (5) represents the sum of the first $n + 1$ terms, a through ar^n, of a geometric series with common ratio r, assuming $r \neq 1$. If $r = 1$, the series

$$S = a + ar + ar^2 + ar^3 + \cdots + ar^n$$

is equivalent to

$$S = a + a + a + a + \cdots + a$$

and the sum of this series is simply $S = (n + 1)a$.

Returning to the question regarding world oil consumption from 1995 through 2025, we need to find the sum of terms 0 through 30 of a geometric series with initial term 25 and common ratio 1.022. This sum is given by

$$T = \frac{25 - 25(1.022)^{31}}{1 - 1.022} \approx 1095 \text{ billion barrels of oil.}$$

Estimates of proven (as of 1995) oil reserves recoverable with 1995 technology are in the range of 1000 to 1100 billion barrels worldwide. If oil consumption continues to increase at the 1995 rate, oil reserves will be exhausted by 2025.

Source: Energy Information Administration, U.S. Dept. of Energy, International Energy Database, *International Energy Outlook 1996.*

Summing geometric series happens frequently enough in mathematics that either the formula for the sum of a geometric series, or the method used here to arrive at that sum, should be a part of the mathematical tools you have at your disposal. Below are two additional examples from the financial world where this concept is used.

EXAMPLE 1 **Balance of an Annuity**

annuity

Suppose your parents want to set aside money for college tuition for your younger sister. They begin saving when she is 12 by opening an account with an initial deposit of $100. For six years thereafter, they deposit an additional $100 at the beginning of each month. An account into which regular payments are made (or from which regular withdrawals are made) is called an *annuity*. The account into which your parents place the money earns 0.5% monthly interest, which is added to the account at the end of each month. How much money will be in the account at the end of six years?

Solution The initial deposit earns interest for 72 months, which means that the initial deposit grows geometrically to $100(1.005)^{72}$. The money deposited at the beginning of the second month earns interest for 71 months and grows to $100(1.005)^{71}$. Each successive deposit earns interest for one month less than the previous deposit. For simplicity, assume that the parents close the account on the day that they make the final deposit of $100 so that the final deposit earns no interest. The time line in Figure 7 shows each deposit together with its value when the account is closed.

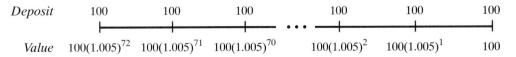

Figure 7 Time line for interest and deposit in an annuity

The balance of the annuity after six years is the sum of the values of all 73 deposits, which is

$$B = 100 + 100(1.005) + 100(1.005)^2 + \cdots + 100(1.005)^{71} + 100(1.005)^{72}.$$

This is a geometric series with initial term 100 and common ratio 1.005, so we can use equation (5) to write the sum of this series as

$$B = \frac{100 - 100(1.005)^{73}}{1 - 1.005},$$

which equals \$8784.09. This is the balance of the annuity after six years. Note that only \$7300 was deposited, but compounded interest brought the balance up to \$8784.09. ■

EXAMPLE 2 **Loan Payment**

Suppose you borrow \$20,000 to buy a car, and you have to pay back the loan over 48 months at 1% interest per month. What payment P is required to pay off the loan in 48 equal monthly payments?

Solution   This problem can be modeled with the recursive system

$$L_0 = 20{,}000, \quad L_n = 1.01L_{n-1} - P, \quad n = 1, 2, 3, \ldots, 48,$$

where L_n is the amount still owed on the loan after n months. This problem is similar to those solved in Section 3.2 using a binary search. Now we will demonstrate a more direct algebraic solution using geometric series.

Using the recursive system, we know that after one month,

$$L_1 = 1.01(20{,}000) - P.$$

After two months, the amount still owed is given by

$$L_2 = 1.01L_1 - P,$$

which by substitution is equivalent to

$$L_2 = 1.01^2(20{,}000) - 1.01P - P.$$

Similarly, since $L_3 = 1.01L_2 - P$, we have

$$L_3 = 1.01^3(20{,}000) - 1.01^2P - 1.01P - P.$$

We can continue to iterate for 48 months, which yields

$$L_{48} = 1.01^{48}(20{,}000) - 1.01^{47}P - 1.01^{46}P - \cdots - 1.01P - P.$$

In general, for any value of n,

$$L_n = 1.01^n(20{,}000) - P(1.01^{n-1} + 1.01^{n-2} + \cdots + 1.01^2 + 1.01 + 1). \quad \textbf{(6)}$$

Embedded in the right side of equation (6) is the geometric series

$$1 + 1.01 + 1.01^2 + \cdots + 1.01^{n-2} + 1.01^{n-1},$$

which has first term 1 and common ratio 1.01. Using equation (5) to simplify this sum results in an explicit function:

$$L(n) = 1.01^n(20,000) - P\left(\frac{1 - 1.01^n}{1 - 1.01}\right).$$

The explicit form can be used to find the payment, P, required to pay off the loan in 48 months. We want $L(48) = 0$. This means that

$$0 = 1.01^{48}(20,000) - P\left(\frac{1 - 1.01^{48}}{1 - 1.01}\right),$$

which can be rewritten as

$$P\left(\frac{1 - 1.01^{48}}{-0.01}\right) = 1.01^{48}(20,000).$$

We can then solve for P,

$$P = \frac{1.01^{48}(20,000)(-0.01)}{1 - 1.01^{48}},$$

or, equivalently,

$$P = \frac{1.01^{48}(20,000)(0.01)}{1.01^{48} - 1},$$

which means that $P = 526.68$. The loan will be paid off in 48 months with a monthly payment of \$526.68. You can verify using the recursive system $L_0 = 20,000$ and $L_n = L_{n-1}(1.01) - 526.68$ that the balance after 48 payments, L_{48}, is zero. (Due to rounding, your answer may not be exactly zero.) The final balloon payment may be a few cents more or less than \$526.68. ■

Exercise Set 3.5

1. Determine the sum of the first 10, 100, and 1000 terms of the following geometric series with initial term Y_0 and common ratio r.

   a. $Y_0 = 100$, $r = 1.05$ b. $Y_0 = 100$, $r = 0.95$

   c. $Y_0 = 1$, $r = \frac{1}{2}$ d. $Y_0 = 1$, $r = -\frac{1}{2}$

   e. $Y_0 = \frac{3}{10}$, $r = \frac{1}{10}$ f. $Y_0 = 1$, $r = -\frac{3}{2}$

   g. How does the value of r seem to affect the long-term behavior of the sums in these problems?

2. In the oil consumption problem at the beginning of this section, the consumption of oil was increasing at a rate of 2.2% per year. By how much would the 31-year total consumption of oil be reduced if the annual rate of increase were reduced to 1.1%?

3. You have an annuity with an initial balance of $1000 and you deposit $150 each month. This annuity pays 10% annual interest, compounded monthly.

   a. Write a recursive model for the situation. What does the model describe?

   b. How much money will you have after 3 years?

   c. Write an explicit equation in terms of A_0 to calculate the amount in the account after 4 months if the initial deposit is A_0.

   d. Write an explicit equation in terms of A_0 and n to calculate the amount in the account after n months if the initial deposit is A_0.

4. The Forestry Department states that the number of bass in a lake is growing at a rate of 1.5% a year. At the present time, there are 200 bass in the lake. If this rate of growth continues, how many bass will there be in 15 years?

5. The number of new cars sold at a car dealer each year grows at a rate of 1.5% a year. At the present time, the dealer is selling 200 cars each year. If this rate of growth continues, how many cars will be sold in 15 years?

6. Explain the difference between Exercises 4 and 5.

7. Tim and Tom are twins. They both went to work at the age of 20 in identical jobs. On their birthday of each year, they received identical bonuses of $2000. However, there were some differences. For example, Tim was conservative early in life, and each year he invested his $2000 bonus in a savings program earning 9% interest annually. At age 30, Tim decided to have some fun in life and he began spending his $2000 bonuses on vacations in Hawaii. The money he had already saved stayed in the savings plan and earned interest. This continued until he was 65 years old. Tom, on the other hand, believed in his youth that life was too short to be concerned about saving for the future. For nine years, he spent his $2000 bonuses on vacations in Hawaii. At age 30, he realized that someday he might not be able to work and then he would need savings to support himself after retiring. He began investing his $2000 bonuses in a savings account earning 9% interest annually. This continued until he was 65. Compare the amounts of money that Tim and Tom have in the bank on their 65th birthday.

   Source: Smith, Keith, *"Tim and Tom's Financial Adventure" Pull-Out,* **Consortium**, Number 38, Fall 1991, COMAP, Inc., Lexington, MA.

8. Use the method for summing a geometric series to arrive at a generalized explicit formula for an annuity with an initial deposit of D and regular deposits of D for n time periods, with an interest rate i for each time period. The recursive system is

$$D_0 = D, \quad D_n = D_{n-1}(1 + i) + D.$$

Note that the balance D_n represents a total of $n + 1$ deposits of D and n interest calculations.

9. Use the method for summing a geometric series to arrive at a general explicit formula for the balance of a loan with an initial balance B, equal periodic payments of P, n payments, with an interest rate i each time period. The recursive system is

$$B_0 = B, \quad B_n = B_{n-1}(1 + i) - P.$$

10. Use the method for summing a geometric series to arrive at a general explicit formula for the recursive system

$$Y_0 = A, \quad Y_n = aY_{n-1} + b.$$

11. The lucky winner of a grand sweepstakes will be given the choice of receiving a single lump sum payment the year she wins the sweepstakes or receiving install-ment payments for a period of 10 years. Suppose you are the lucky winner, and you are offered installment payments of $50,000 per year for 10 years.

   a. If you assume an annual interest rate of 5% per year, how large would the lump sum award need to be for you to accept the one-time award? Ignore any taxes associated with the winnings. Document how you arrive at the payment you choose.

   b. How large would the annual interest rate need to be for a lump sum award of $350,000 to be comparable to the $50,000 yearly installment payments over 10 years?

SECTION 3.6 Investigating Garbage Disposal

Garbage Barge Returns in Search of a Dump

After more than eight weeks of wandering in the Caribbean and up and down the East Coast in a futile search for a final resting place, Long Island's outcast garbage was anchored off Brooklyn yesterday, awaiting the outcome of another round of legal and political wrangles.

Rejected by six states and three countries, the trash and garbage—3100 tons of baled refuse piled atop the barge *Mobro* and towed by the tug *Break of Dawn*—had an invitation to return to Islip, Long Island, where its 60-day 6000-mile odyssey began.

Source: Robert D. McFadden, *New York Times*, May 18, 1987, page A1. Copyright© 1987 by The New York Times. Reprinted by permission.

Most of the refuse discarded by Americans winds up in landfill sites. According to the U. S. Environmental Protection Agency, each American discards an average of 4.1 pounds of garbage a day. One cubic foot of compacted garbage weighs approximately 70 pounds. Some densely populated states already have a shortage of landfill sites and ship garbage out of state to less densely populated areas. For example, California uses landfill sites in Arizona, while New York uses sites in Vermont and Texas.

1. How much garbage, both in weight and volume, was discarded in the United States in 1995? (According to the *U.S. Bureau of the Census*, the U.S. population in 1995 was 263.8 million.)

2. It has been reported that the weight of garbage discarded per person each year has remained relatively stable over the last 50 years. If the amount of garbage discarded per person remains the same for the next 50 years, how many cubic feet of garbage will be discarded in 2045? (The population of the United States is currently growing at about 1% per year.)

3. Of course, the problem isn't just the amount discarded in a particular year. Once the garbage is discarded, it doesn't go away! Determine the amount of accumulated garbage from 1995 to 2045.

4. The EPA has recommended that Americans reduce their per capita output of garbage by 10%. According to the agency's estimates, the breakdown of the common constituents of municipal solid waste is the following:

Disposable diapers	2%
Large appliances	2%
Food and yard waste	7%
Plastic bottles	1%
All paper	40%
Construction debris	12%

Where can the largest reductions be made? What would be the effect of reducing the per capita production of garbage by 10% on the accumulated waste from 1995 to 2045? By how much must the amount of paper waste be reduced to produce a 10% reduction in total garbage over this time period?

5. The landfill and recycling center at Sturbridge, Massachusetts, is a model for the country. Paper, tires, glass, plastics, aluminum cans, and used oil all get recycled. The result is that only 40 cubic yards of compacted garbage are deposited in the landfill site each week. The population of Sturbridge in 1995 was 8563 and is expected to rise to 9360 by the year 2000.

a. Assume that the residents of Sturbridge were at one time typical Americans who each discarded 4.1 pounds of garbage per day. By what percentage have they reduced their production of garbage through recycling?

b. Sturbridge's present landfill has 3 acres of space with a depth of 50 feet available. If the growth rate of the population remains constant, how long will the landfill last?

6. Students in Precalculus are doing research on the county landfill that opened in 1996. According to the county manager, the landfill has a capacity of 600 million tons and an expected life span of 100 years. The population of the county in 1990 was 110,000. If the residents generate garbage at the rate of 4.1 pounds per day, and the growth rate remains constant for the next 100 years, what is the growth rate for the population of the country?

Source: Growth, Field Test Version COMAP, Inc., Lexington, MA, 1992.

SECTION 3.7 Compound Interest

In previous sections of this chapter, we encountered situations in which money was deposited into an interest-bearing account, and the interest was added to the account at regular intervals. For example, suppose Maria deposits $1000 in an account that earns interest at a rate of 8% annually. After one year, the balance of the account will increase to $1080. After two years, the balance will be $1166.40, and so on, according to the recursive equation

$$A_n = A_{n-1} + 0.08A_{n-1},$$

future value

compound interest

where A_n is the account balance after n years. A_n is known as the *future value* of the initial deposit of $1000 after n years. The interest in Maria's account is *compounded annually*, which means that the interest is credited to the account at the end of each year. Instead of adding all of the annual interest at the end of each year, financial institutions often use quarterly, monthly, or continuous compounding. For example, a bank that compounds quarterly would add one-fourth of the annual interest to an account four times a year, or every three months.

What effect does the frequency of compounding have on future value?

What effect does the frequency of compounding have on future value? Suppose Maria deposits $1000 in an account that pays annual interest of 8% with quarterly compounding. Using the formula $L = pIt$, where L is the amount of interest, p the principal, I the interest rate, and t the time, the amount of interest after three months is $L = 1000(0.08)\frac{1}{4}$. The account balance after one quarter is

$$\$1000 + \$1000(0.08)\frac{1}{4} = \$1000(1.02) = \$1020.$$

Earning 8% interest compounded quarterly is equivalent to multiplying the previous balance by 1.02 each quarter. At the end of the second quarter, the account has a balance of

$$\$1020(1.02) = \$1000(1.02)^2 = \$1040.40.$$

At the end of four compounding periods, the balance will be

$$\$1040.40(1.02)(1.02) = \$1000(1.02)^4 = \$1082.43.$$

Notice that this amount is larger than the corresponding balance of $1080.00 at the end of one year using annual compounding. Over time, quarterly compounding can have a significant influence on the balance compared to annual compounding.

CLASS PRACTICE

1. **a.** Write a recursive system and an explicit function for the future value, F, of an initial deposit, A_0, after N years with quarterly compounding. Let the annual interest rate be represented by I.

   b. Write a recursive system and an explicit function as in Part a, but use monthly compounding.

c. Write a recursive system and an explicit function that can be used to find the future value, F, when interest is compounded k times each year. Use A_0 as the initial deposit, N as the number of years, and I as the annual interest rate.

2. Study Figure 8 to compare the balance in Maria's account for various compounding frequencies. Use an annual interest rate of 8% and an initial deposit of $1000 to verify several entries in the table. (All table entries are rounded to the nearest dollar.) Write a paragraph describing how the future value changes as the frequency of compounding increases.

Figure 8 **Future value of $1000 for various compounding frequencies with 8% annual interest rate (values rounded to the nearest dollar)**

Years(N)	Annually $k = 1$	Quarterly $k = 4$	Monthly $k = 12$	Weekly $k = 52$	Daily $k = 365$	Hourly $k = 8760$	By the minute $k = 525{,}600$
5	$1469	$1486	$1490	$1491	$1492	$1492	$1492
20	$4661	$4875	$4927	$4947	$4952	$4953	$4953
30	$10,063	$10,765	$10,936	$11,003	$11,020	$11,023	$11,023
50	$46,902	$52,485	$53,878	$54,431	$54,574	$54,597	$54,598
75	$321,205	$380,235	$395,475	$401,573	$403,164	$403,418	$403,429

By reading across any row of Figure 8, you can observe that the balance seems to level off as the frequency of compounding increases. For each length of deposit, there appears to be an upper limit to the future value of the account balance. Each entry in Figure 8 can be calculated with the recursive formula

$$A_n = A_{n-1}\left(1 + \frac{I}{k}\right),$$

where A_n is the future value after the interest is compounded n times, or with the explicit formula

$$A(N) = A_0\left(1 + \frac{I}{k}\right)^{kN},$$

where $A(N)$ is the future value after N years. Note that in this situation, N must be a positive integer. In each case, A_0 represents the original deposit, I is the annual interest rate, and k is the number of compounding periods in a year. In Figure 8 above, $A_0 = \$1000$ and $r = 0.08$, so the values are based on the formula

$$\text{Future value} = \$1000\left(1 + \frac{0.08}{k}\right)^{kN} = \$1000\left(\left(1 + \frac{0.08}{k}\right)^k\right)^N.$$

The value of N is constant in each row of Figure 8, so increases in the future values in a particular row result solely from the quantity $\left(1 + \frac{0.08}{k}\right)^k$. Since the balances associated with each value of N appear to have a limit, we suspect that the quantity $\left(1 + \frac{0.08}{k}\right)^k$ has some limiting value as k becomes very large. Figure 9 on the next page provides values of this quantity for k-values associated with increasing frequency of compounding. Correct to six decimal places, the limiting value appears to be 1.083287.

Figure 9 Values of $\left(1 + \frac{0.08}{k}\right)^k$ rounded to six decimal places

| Annually | Quarterly | Monthly | Weekly | Daily | Hourly | By the minute |
$k = 1$	$k = 4$	$k = 12$	$k = 52$	$k = 365$	$k = 8760$	$k = 525{,}600$
1.080000	1.082432	1.083000	1.083220	1.083278	1.083287	1.083287

Using a calculator to compute the value of $e^{0.08}$, we find that the value of $\left(1 + \frac{0.08}{k}\right)^k$ approaches the value of $e^{0.08}$ for large values of k. Does this relationship hold for other values of I, the annual interest rate? Look at the case where $I = 1$. We use a calculator to evaluate $\left(1 + \frac{1}{k}\right)^k$ for increasing values of k, and the results are displayed in Figure 10. The quantity $\left(1 + \frac{1}{k}\right)^k$ seems to have a limiting value of about 2.718280 as k becomes very large. Using a calculator, we find that the value of e^1 equals 2.71828 rounded to five decimal places.

Figure 10 Values of $\left(1 + \frac{1}{k}\right)^k$ rounded to six decimal places

$k = 10$	$k = 100$	$k = 1{,}000$	$k = 10{,}000$	$k = 100{,}000$	$k = 1{,}000{,}000$
2.593742	2.704814	2.716924	2.718146	2.718268	2.718280

Mathematicians began exploring the behavior of $\left(1 + \frac{1}{k}\right)^k$ early in the eighteenth century. In essence, they found that the value of $\left(1 + \frac{1}{k}\right)^k$ approaches a limiting value as k gets larger and larger. Mathematicians define this limiting value, which is 2.71828..., as the number e in honor of the Swiss mathematician Leonhard Euler. In mathematical terms, we can write

$$\lim_{k \to \infty}\left(1 + \frac{1}{k}\right)^k = e,$$

which is read, "The limit of $\left(1 + \frac{1}{k}\right)^k$ as k increases without bound is the number e."

Now consider the value of $\lim_{k \to \infty}\left(1 + \frac{I}{k}\right)^k$ for any value of I. We have already found that $\lim_{k \to \infty}\left(1 + \frac{0.08}{k}\right)^k = e^{0.08}$. We can also use a calculator to compare the values of $e^{0.06}$ and $\left(1 + \frac{0.06}{k}\right)^k$ for large values of k, as well as several other values of I. What we find supports the result

$$\lim_{k \to \infty}\left(1 + \frac{I}{k}\right)^k = e^I.$$

Thus, the exponential function with base e can be used to approximate future values when interest is compounded frequently. For example, if 8% annual interest is compounded frequently, the future value in N years of an initial deposit of $1000 is equal to $\$1000\left(1 + \frac{0.08}{k}\right)^{kN}$, which can be approximated by $\$1000e^{0.08N}$. The value of the expression $\left(1 + \frac{0.08}{k}\right)^k$ approaches $e^{0.08}$ as the frequency of compounding increases. When this limiting value is used, the future value computations represent the result

continuous compounding   of *continuous compounding,* or compounding at every instant. The equation

$$\text{Future value} = A_0(e^I)^N = A_0 e^{IN}$$

gives the balance after N years in an account with continuous compounding, initial deposit A_0, and annual interest rate I. Since the compounding is continuous, N can be any positive real number, and is not restricted to positive integers. This equation gives a reasonable approximation of future value for quarterly compounding and is even more accurate for monthly and daily compounding.

The function $F(N) = A_0 e^{IN}$ describing the future value of an account with continuous compounding is an example of an exponential function. All exponential functions can be written as transformations $y = e^x$. This important function can be used to describe many real-world situations, from the cooling of coffee sitting on a table to the increase of carbon dioxide in the atmosphere. In the next several sections, you will become more acquainted with this important function and the situations that it can be used to model.

Exercise Set 3.7

1. Verify that the entries in Figure 8 can be obtained by raising the corresponding entries in Figure 9 to the Nth power (where N represents the number of years) and multiplying by $1000.

2. Compute the balance that results when $2000 is deposited for one year in an account paying 8% annual interest compounded quarterly. How does this balance compare to an approximation using continuous compounding?

3. If Jack invests $250 at an annual interest rate of 7.5%, what is the future value after two years if the interest is compounded quarterly? monthly? weekly? daily? continuously?

4. **a.** Which has the greater future value after 5 years, $1000 invested at 8% with annual compounding or $1000 invested at 7.75% with quarterly compounding? Use graphs to determine if the number of years affects which deposit has a greater future value.

    b. Which has the greater future value after 5 years, $1000 invested at 8% compounded annually or $800 invested at 9% compounded annually? Does the number of years affect which deposit has the greater future value?

5. It has been said that the island of Manhattan was purchased for $24 in 1626. Suppose the $24 had been invested in 1626 at 6% annual interest compounded quarterly. What would it be worth today?

6. Banks usually offer a variety of investment accounts from which their customers can choose. Interest rates and frequency of compounding vary from bank to bank, and even within banks, for different types of accounts. The following questions compare alternative investments.

   a. If you deposit $100 in an account that pays 8% interest compounded quarterly, what is the balance in the account at the end of one year?

   b. If a competing bank compounds only once each year, what would its interest rate need to be to yield the same balance on your $100 deposit as in Part a?

effective annual yield

   c. The answer to Part b is called the *effective annual yield*, or *effective annual interest rate*. It is the interest rate that if compounded only once would yield the same balance after one year as a one-year account with more frequent compounding. The effective rate provides a way to compare different interest rates that have different compounding frequencies. Suppose another bank offers 7.75% interest compounded monthly. What is the effective annual yield?

7. If we denote the effective annual rate by R, then R satisfies the equation

$$A(1 + R) = A\left(1 + \frac{I}{k}\right)^k,$$

where A is the initial deposit, k is the number of compounding periods per year, and I is the annual interest rate.

   a. Verify algebraically that $R = \left(1 + \frac{I}{k}\right)^k - 1$.

   b. Use effective annual yield to compare a 7.25% certificate with quarterly compounding to a 7% certificate with monthly compounding. Which option will provide a better return on your investment?

8. Suppose that when you were born your parents estimated they would need $50,000 for college expenses. The best interest rate they could find was offered on a certificate of deposit paying 6% annual interest compounded monthly.

   a. What is the effective annual interest rate for this account?

   b. How much money should your parents have invested to have a balance of $50,000 on your eighteenth birthday?

present value

9. The amount of money that would have to be invested today to yield some specified amount in the future is called the *present value* of that future amount. (In Exercise 8, you were asked to calculate the present value of $50,000 to be paid in eighteen years.) Suppose a professional athlete signs a one-year contract for $2,000,000 and agrees to be paid over a period of five years. At the beginning of each of the next five years, he or she will be paid $400,000. What is the present value of this contract assuming a 9% interest rate compounded annually? Another way to ask this question is, how much money should the team management deposit in an account earning 9% annual interest when the contract is signed to guarantee that they can pay this five-year deal? Since the athlete is paid at the beginning of each year, assume that the present value of the first payment of $400,000 is $400,000.

10. Suppose the interest rate is 7% in the athletic contract discussed in Exercise 9. What is the present value of this contract? Does this make sense compared to the answer to Exercise 9? What would you expect to be true for an interest rate of 11%?

11. Describe precisely how present value can be used to compare two contracts similar to those described in Exercise 9, in which one contract is paid immediately and the other is paid at the beginning of each year over several years.

12. a. Julie estimates that she will need $5000 in four years to buy her first car. She has found that she can purchase a certificate of deposit that will pay 6.5% compounded annually. How large should the certificate be to enable her to purchase the car?

    b. Julie finds that another bank offers a 6.25% certificate with quarterly compounding. How large does this certificate need to be? Which certificate is the better investment?

    c. Describe precisely how the concept of present value can be used to compare investment options like those Julie faces. In such a comparison, does the higher or lower present value represent a better option?

13. A court settlement requires Ms. Jones to pay Mr. Murphy $4000 within six months. Ms. Jones plans to wait until the end of the six months to pay Mr. Murphy, but Mr. Murphy is anxious to settle the case and will accept less money if Ms. Jones will pay immediately. What is the least amount of money Mr. Murphy should accept for an immediate settlement? Assume an interest rate of 5% compounded monthly.

14. On the same day, Chris deposits $500 in account A that pays 8% interest compounded annually and $1000 in account B that earns 6% interest compounded quarterly.

    a. Use a computer or calculator to draw graphs to illustrate the balances in the two accounts over time.

    b. After how many years will the two accounts have the same balance?

15. Verify the following statement: The number e is the balance that results when $1 is invested for one year at an interest rate of 100% compounded continuously.

16. Imagine that you have been selected as the winner of the Wonderful Mathematics Student Award and now you have to choose between two prize options. The first option is to receive a $50,000 certificate of deposit that pays 8% interest compounded continuously. The second option is to place one cent in a fund that is guaranteed to double every six months.

    a. Which option would you choose if you actually will receive the award five years from now?

    b. Which option would you choose if you will receive the award in 20 years?

    c. Find the time at which the two options are equivalent.

SECTION 3.8   Graphing Exponential Functions

base

In Chapter 2 we graphed the toolkit function $f(x) = 2^x$ and in Section 3.7 we introduced the exponential function $f(x) = e^x$. Both of these functions are examples of exponential functions of the form $f(x) = b^x$, where $b > 0$ and $b \neq 1$. The positive real number b is called the *base* of the exponential function. The graph of $f(x) = 2^x$ passes through the points $(0, 1)$ and $(1, 2)$, and the graph of $f(x) = e^x$ passes through the points $(0, 1)$ and $(1, e)$. The graphs of all exponential functions $f(x) = b^x$ will pass through the point $(0, 1)$ since $b^0 = 1$ for all $b > 0$. Likewise, since $b^1 = b$ and $b^{-1} = \frac{1}{b}$, these graphs will also contain the points $(1, b)$ and $(-1, \frac{1}{b})$. In addition, all the graphs of $f(x) = b^x$ will have the x-axis as an asymptote. Figure 11 shows a graph of $f(x) = e^x$ along with the toolkit function $g(x) = 2^x$.

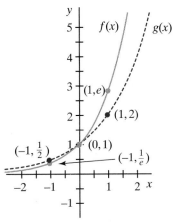

Figure 11   Graphs of the functions $f(x) = e^x$ and $g(x) = 2^x$

You should recall the laws of exponents $b^{m+n} = b^m \cdot b^n$ and $(b^m)^n = b^{mn}$. These laws can be used to show why exponential functions whose equations appear quite different may, in fact, be the same function and therefore have the same graph. The following examples show two such cases.

EXAMPLE 1

Graph the function $f(x) = 3^{-x}$ and the function $g(x) = \left(\frac{1}{3}\right)^x$. Explain why the graphs are identical.

Solution

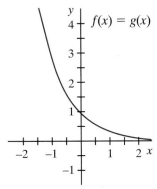

Figure 12   Graph of $f(x) = 3^{-x}$ and $g(x) = \left(\frac{1}{3}\right)^x$

The functions shown in Figure 12 have graphs that coincide. To understand why $f(x) = g(x)$ for all x, recall that $3^{-1} = \frac{1}{3}$, so we can write $(3^{-1})^x = \left(\frac{1}{3}\right)^x$. Because $(3^{-1})^x = 3^{-1 \cdot x}$, we have $3^{-x} = \left(\frac{1}{3}\right)^x$ and so $f(x) = g(x)$. Recall from Chapter 2 that the graph of $f(-x)$ is the reflection about the y-axis of the graph of the function $f(x)$. Thus, the graph of the function $f(x) = 3^{-x}$ is the reflection of the graph of $h(x) = 3^x$ about the y-axis. The graph of $f(x)$ and $g(x)$ is decreasing and has the positive x-axis as a horizontal asymptote. Exponential functions with a base between 0 and 1, or those with a base greater than 1 but with a negative coefficient in the exponent, are used to model *exponential decay*. ■

exponential decay

EXAMPLE 2 Use a graphing calculator to graph $y = 4 \cdot 2^x$ and $y = 2^{x+2}$. Use your knowledge of transformations to explain why these graphs coincide?

Solution The graph of $y = 4 \cdot 2^x$ is a vertical stretch of the toolkit function $f(x) = 2^x$. The point $(0, 1)$ moves to $(0, 4)$ and the point $(1, 2)$ moves to $(1, 8)$. The graph of $y = 2^{x+2}$ is a horizontal shift 2 units to the left of $f(x) = 2^x$. The point $(0, 1)$ moves to $(-2, 1)$, and the point $(1, 2)$ moves to $(-1, 2)$. If we use the fact that $b^{m+n} = b^m b^n$, we see that $2^{x+2} = 2^x \cdot 2^2 = 4 \cdot 2^x$. Therefore, for this exponential function, the vertical stretch and the horizontal shift produce the same graph. Although the expressions look different, they are actually two different ways of writing the same function. ■

Other transformations of exponential functions, such as horizontal compressions and vertical shifts, result in curves that are important in modeling various types of growth and decay.

EXAMPLE 3 In a laboratory experiment, the growth of a tumor in a mouse was monitored over time. Using data analysis techniques, the function $s = 0.32e^{0.11d}$ was found to be a good model for the growth. In this equation, d represents the number of days since monitoring began, and s represents the size of the tumor in cubic centimeters. Graph the function over an appropriate domain. What is the initial size of the tumor? Determine the size after 10 days and after 20 days.

Solution The graph of $s = e^d$ has been compressed vertically by a factor of 0.32 and stretched horizontally by a factor of $\frac{1}{0.11} \approx 9$ to obtain the graph shown in Figure 13. The vertical compression moves the point $(0, 1)$ to $(0, 0.32)$ and moves the point $(1, e)$ to $(1, 0.32e)$. The horizontal stretch moves $(1, 0.32e)$ to $(9, 0.32e)$ or approximately $(9, 0.87)$. Based on the context of this example, the appropriate domain is nonnegative values of d. Although we expect there to be physical limitations on the size of the tumor, it is not known how large the tumor can grow, so we have not attempted to show this in the graph.

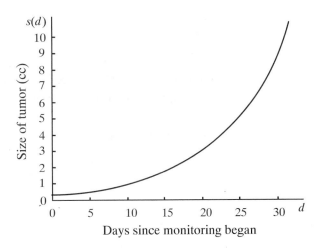

s(d)

Size of tumor (cc)

Days since monitoring began

Figure 13 Graph of $s = 0.32e^{0.11d}$

The initial size of the tumor is 0.32 cc since this is the s-value associated with $d = 0$. Substituting $d = 10$ and $d = 20$ into the given equation reveals that the size of the tumor is about 0.96 cc after 10 days and about 2.89 cc after 20 days. ◾

EXAMPLE 4 When a cup of hot coffee sits undisturbed, the coffee begins to cool and will continue to do so until it reaches room temperature. If the initial temperature of the coffee is 85°C and the room temperature is 24°C, find an exponential function that can be used to predict the temperature at any time.

Solution The cooling process can be modeled reasonably well modeled using exponential decay. We can use the function $y = e^{-x}$ to model exponential decay, but we must transform the curve so that it has a horizontal asymptote at 24, not 0, and a y-intercept of 85, not 1. Adding 24 to e^{-x} shifts the entire graph up 24. We must also stretch the graph vertically so that it passes through the point $(0, 85)$. This can be done by multiplying e^{-x} by 61. The graph of the function $y = 61e^{-x} + 24$ is shown in Figure 14.

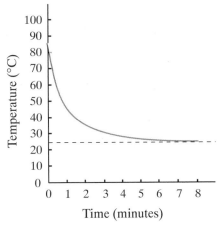

Temperature (°C)

Time (minutes)

Figure 14 Graph of a cooling curve

Objects cool at different rates, depending on their composition and shape. In graphing the function in Figure 14, we did not take into account how quickly the coffee cooled to room temperature. We can see from the graph that it seemed to cool very quickly, nearly reaching room temperature in less than 4 minutes, which is unrealistically fast. In real life, coffee takes much longer to cool. To model this slower cooling process, we need to apply a horizontal stretch to the function $y = 61e^{-x} + 24$. This will result in a model of the form $y = 61e^{-kx} + 24$, where $0 < k < 1$. Finding the particular value of k in a given situation requires that we know more than just the initial and room temperatures. ■

CLASS PRACTICE

1. Use your calculator to graph three possible cooling curves: $f(x) = 61e^{-0.5x} + 24$, $g(x) = 61e^{-0.2x} + 24$, and $h(x) = 61e^{-0.1x} + 24$.

2. Compare the temperatures of f, g, and h at 4 minutes.

3. Compare the times required for f, g, and h to reach a temperature of 30°C.

EXAMPLE 5

learning curves

When you start to learn a new skill, such as typing, your proficiency begins at a rather low level and grows toward some maximum level. At first, your proficiency grows quickly, but as you near the maximum level, your skill grows more slowly. Exponential curves that model this behavior are called *learning curves*. Such functions can be modeled with transformed exponential functions. Consider the function $f(x) = 45 - 35e^{-0.3x}$, where x represents the number of weeks of practice and $f(x)$ represents typing speed in words per minute. Sketch a graph of this function and use the graph to describe the specific characteristics of this learning curve.

Solution   The domain of this function, based on the situation described above, is $x \geq 0$. To sketch this function, begin with the graph of $g(x) = 35e^{-0.3x}$, which is a decreasing exponential function with a y-intercept of 35. (See Figure 15.)

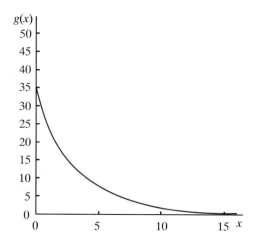

Figure 15   Graph of $g(x) = 35e^{-0.3x}$

Our next step is to flip the graph of g about the x-axis to get the graph of $-g(x) = -35e^{-0.3x}$, as shown in Figure 16.

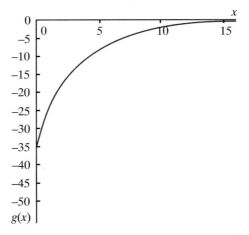

Figure 16 Graph of $-g(x) = -35e^{-0.3x}$

Finally, there is a vertical translation of $y = -g(x)$ up 45 units. The resulting graph is that of the desired function $f(x) = 45 - 35e^{-0.3x}$, shown in Figure 17.

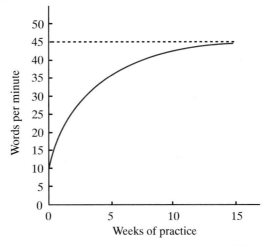

Figure 17 Graph of $f(x) = 45 - 35e^{-0.3x}$

When $x = 0$, the value of the function is 10, so you start typing at 10 words per minute. In the first five weeks of practice, you improve to about 37 words per minute, but in the next five weeks you improve only 6 more words per minute. At this point, you type a little over 43 words per minute but gain very little with additional practice. The horizontal asymptote at 45 indicates that there is an upper limit of 45 words per minute for your typing proficiency. ■

EXAMPLE 6   Use the ideas of composition to sketch the graph of $f(x) = e^{\frac{1}{x-1}}$.

Solution   The function f is a composition of $g(x) = \frac{1}{x-1}$ and $h(x) = e^x$. The graphs of $g(x) = \frac{1}{x-1}$ and $h(x) = e^x$ are shown in Figure 18.

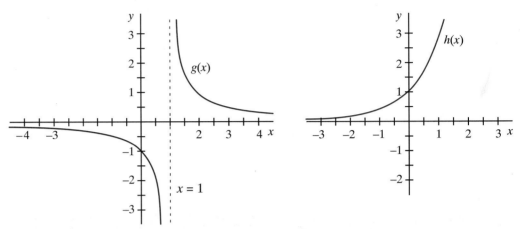

Figure 18   Graphs of $g(x) = \frac{1}{x-1}$ and $h(x) = e^x$

Since $f(x) = h(g(x))$ and h has no domain restrictions, the domain of the composite function f is the same as the domain of g. Therefore, x cannot equal 1. Also, since $h(x) > 0$ for all values of x, we know that $h(g(x)) > 0$ for all x in the domain of g.

Since values of $g(x)$ never equal zero, the input to h is never zero, so the y-value of $h(g(x))$ is never 1. It is also important to note that the graph of g has $y = 0$ as a horizontal asymptote. As x-values increase or decrease without bound, $g(x) = \frac{1}{x-1}$ approaches zero, so $f(x) = e^{\frac{1}{x-1}}$ approaches e^0, or 1. This produces a horizontal asymptote for the graph of f at $y = 1$.

Now consider the behavior of f close to $x = 1$. As x approaches 1 from the left, $\frac{1}{x-1}$ decreases without bound, and the value of $f(x) = e^{\frac{1}{x-1}}$ approaches 0 from above. The open circle on the graph of the composition signifies that $y \to 0^+$ as $x \to 1^-$. As x approaches 1 from the right, $\frac{1}{x-1}$ increases without bound, and the value of $e^{\frac{1}{x-1}}$ also increases without bound. The graph of the composition $f(x) = e^{\frac{1}{x-1}}$ is shown in Figure 19.

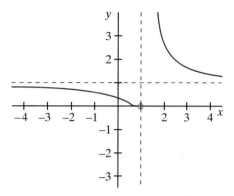

Figure 19 Graph of $f(x) = e^{\frac{1}{x-1}}$ ▨

Exercise Set 3.8

1. As illustrated in Examples 1 and 2, the laws of exponents can often be applied to simplify graphing exponential functions. Describe two different ways to graph the following equations as transformations of an exponential function. Which way seems easier to you? Rewrite each equation to make your choice of transformation obvious, and sketch a graph. Identify the horizontal asymptote and important points.

    a. $y = e^{2x}$ **b.** $y = 4 \cdot 2^{x-1}$

    c. $y = (-4)2^x$ **d.** $y = 4^{-x-2}$

2. **a.** Explain why these expressions are all equal:

$$3^{3x-2} = 3^{3(x-\frac{2}{3})} = 27^{x-\frac{2}{3}} = \tfrac{1}{9}(27)^x.$$

    b. Graph $y = 3^{3x-2}$ using one of the expressions for 3^{3x-2}. Label the coordinates of the points that correspond to $(0, 1)$ and $(1, b)$.

    c. Which form did you choose to graph? Why?

3. Graph the following functions. Label asymptote(s) and important points.

    a. $y = 5^{x-7}$ **b.** $y = e^{1.61x}$

    c. $y = 5 + 3^{2x}$ **d.** $y = 1 + 2e^{-x}$

    e. $y = -8 + 0.5^x$ **f.** $y = 3 - 6^x$

    g. $y = 8^{2x-2}$ **h.** $y = 7(3^x)$

    i. $y = -3(2^{2x}) + 1$ **j.** $y = |-5 + 3^{x+2}|$

4. A typical worker at a supermarket bakery can decorate $f(t)$ cakes per hour after t days on the job, where

$$f(t) = 10\left(1 - e^{-\frac{1}{4}t}\right).$$

a. Sketch a graph of f. Restrict the domain to meaningful values of t.

b. How many cakes can a newly employed worker decorate in an hour?

c. After eight days, how many cakes can a worker decorate in an hour?

d. Based on this graph, after a worker has decorated cakes for a very long time, how many cakes can he or she decorate in an hour?

5. Rheumatoid arthritis patients are treated with large doses of aspirin. Research has shown that the concentration of aspirin in the bloodstream increases for a short period of time after the drug is administered and then decreases exponentially. For a typical patient, this relationship is given by $a = 14.91e^{-0.18t}$, where t represents the number of hours since peak concentration and a represents the concentration of aspirin measured in milligrams per cubic centimeter of blood.

a. Graph the function over an appropriate domain. Label the coordinates of the points that correspond to $t = 0$ and $t = 1$.

b. Determine the peak concentration of aspirin.

c. Determine the amount of aspirin remaining four hours after peak concentration.

d. Use graphing technology to determine the time at which the concentration of aspirin is 5 mg per cc of blood.

6. In a classroom experiment, students made fudge the old-fashioned way. They cooked sugar, milk, and chocolate until it reached 234° Fahrenheit. Then they removed the mixture from the heat, added butter and vanilla, and let the mixture cool to 110°. As they monitored the temperature, they kept records and used data analysis to determine a mathematical model for the temperature of the fudge as a function of time. Their model is $F = 154e^{-0.00063t} + 77$, where F is the temperature of the fudge in degrees Fahrenheit and t represents the number of seconds elapsed since stirring in the butter.

a. Graph this function over an appropriate domain. Interpret the y-intercept and the horizontal asymptote.

b. Determine the temperature of the mixture after 30 minutes.

c. Use graphing technology to determine how long it takes for the temperature to reach 110°.

7. McKenzie invests $1000 in a local bank at an annual interest rate of 5.14% compounded quarterly.

   a. Use a graph to determine when her investment will be worth $1500.

   b. Suppose the 5.14% is compounded continuously. When will she have $1500?

8. Use the ideas of composition to sketch a graph of each function without resorting to graphing technology.

   a. $f(x) = \dfrac{1}{e^x}$

   b. $f(x) = 3^{|x|}$

   c. $f(x) = 2^{\frac{1}{x}}$

   d. $f(x) = 2^{(x^2)}$

   e. $f(x) = e^{-x^2 - 2x}$

   f. $f(x) = 3^{\sin x}$

   g. $f(x) = e^{-x^2}$

   h. $f(x) = \dfrac{1}{1 + 2e^{-x}}$

9. Newton's Law of Cooling states that the change in temperature of an object as it cools is proportional to the difference between the current temperature and the ambient temperature. Suppose the initial temperature of a cup of hot coffee was 92°C and the ambient temperature was 3°C. The temperature of this coffee can be modeled with the following recursive system:

$$T_0 = 92, \quad T_n = T_{n-1} - k(T_{n-1} - 3).$$

In this situation, n represents the number of minutes that have elapsed, and k is a constant.

   a. Given the initial temperature of 92°C and a value of $k = 0.02$, use your calculator to generate values of the temperature over time and to produce a graph of these values. Explain, in terms of the recursive system, why the graph is shaped the way it is.

   b. There are several steps involved in constructing an explicit function to model this situation. Start with the decreasing exponential function $f(x) = e^{-x}$. Sketch a graph of this function. Identify its y-intercept and its horizontal asymptote.

   c. Transform f to produce a function g whose graph has a horizontal asymptote of $y = 3$ and a y-intercept of 92.

   and goes through (20,50)

   d. Now transform g so that it decreases less quickly. This transformed function should produce a graph similar to the one you generated in Part a.

10. Suppose that because of limitations in food and living space, a population of field mice experiences constrained growth. This means that the population cannot increase beyond a certain maximum number, M. The change in population is directly proportional to the product of the current population and the difference between the maximum population, M, and the current population.

This can be expressed by the following recursive system:

$$P_0 = P, \quad P_n = P_{n-1} + kP_{n-1}(M - P_{n-1}),$$

where P represents the initial population, n the number of months, and k the constant of proportionality. Populations that grow in this manner are said to exhibit *logistic growth*.

logistic growth

a. Assume that the initial population is 100, the maximum population is 1000, and $k = 0.0006$. Use your calculator to generate values of the population over time and to produce a graph of these values. Explain, in terms of the recursive system, why the graph is shaped the way it is.

b. The graph from Part a can be used to find an explicit function to model logistic growth. Begin with the function $f(x) = 1 + e^{-x}$. Sketch a graph of this function. What is the range of this function? Identify the y-intercept and the horizontal asymptote.

c. Now graph $g(x) = \frac{1}{1 + e^{-x}}$. What is the range of this function? Identify the y-intercept and the horizontal asymptote.

d. Transform g so that it has a y-intercept of 100 and grows less quickly. The horizontal asymptote should be $y = 1000$. This transformed function should produce a graph similar to the one you generated in Part a.

SECTION 3.9 Introduction to Logarithms

In Exercise 5 on page 204, you were given the function $a = 14.91e^{-0.18t}$ and asked to use a graph to determine the time at which the aspirin concentration is 5 mg per cc of blood. What happens when you attempt to solve the equation $5 = 14.91e^{-0.18t}$ algebraically? Dividing both sides by 14.91 gives approximately $0.335 = e^{-0.18t}$. How do we isolate the variable t? We can say that t is the number that when multiplied by -0.18 and used as the exponent on e produces 0.335. Our goal is to write this statement using mathematical symbols.

logarithms

To rewrite $0.335 = e^{-0.18t}$ so that t is isolated, we need to define *logarithms*. Logarithms allow us to rewrite an exponential equation so that the exponent is isolated. Specifically, if $a = b^c$, then c is the logarithm of a with base b, which is written as $c = \log_b a$. Thus, $\log_b a$ is by definition the number c so that $b^c = a$. Using logarithms we can rewrite $0.335 = e^{-0.18t}$ as

$$\log_e 0.335 = -0.18t.$$

natural logarithms

Logarithms with base e are referred to as *natural logarithms* and written with the special notation ln (read "el en"). We can use a calculator to find the value of $\log_e(0.335) = \ln(0.335) \approx -1.09$ and solve the resulting equation

$$-1.09 = -0.18t$$

to show that the aspirin level is 5 mg per cc of blood after approximately 6.1 hours.

In this exercise, we were given an output value of an exponential function and asked to find the corresponding input value. This process is the same as the one we use to find inverse functions. To find the inverse of an exponential function $y = b^x$, we must solve for x. According to the definition of logarithm, we can rewrite $y = b^x$ as $x = \log_b y$ for $b > 0$, $b \neq 1$. So, given $f(x) = b^x$, the inverse function is $f^{-1}(x) = \log_b x$.

Figure 20 shows the graphs of the exponential function $f(x) = e^x$ and its inverse, $g(x) = \log_e x = \ln x$, which can be obtained by reflecting the graph of $y = e^x$ about the line $y = x$.

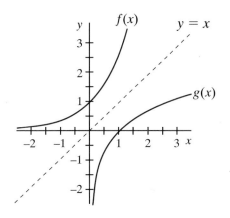

Figure 20 Exponential and logarithmic functions as inverses

The domain of the logarithmic function is the same as the range of the exponential function, and the range of the logarithmic function is the domain of the exponential function. Thus, the logarithmic function has the positive real numbers as its domain and all real numbers as its range.

Most calculators and computers have two keys or functions for logarithms: one marked "log," which represents $\log_{10} x$, and a second marked "ln," which represents $\log_e x$ or $\ln x$. The logarithm with base 10 is called the *common logarithm* and often is written with no base. While any positive base (not equal to one) could be used for a logarithmic function, most of our work with logarithmic functions will be with either base 10 or base e.

common logarithm

EXAMPLE 1   Convert each equation to its equivalent logarithmic or exponential form.

    a. $10^{-3} = 0.001$ **b.** $x^y = z$

    c. $\log_2 \frac{1}{16} = -4$ **d.** $\log_a b = c$

Solution   We use the definition of logarithm to rewrite each of these equations.

    a. $10^{-3} = 0.001$ is equivalent to $\log_{10}(0.001) = -3$.

    b. $x^y = z$ is equivalent to $\log_x z = y$.

    c. $\log_2 \frac{1}{16} = -4$ is equivalent to $2^{-4} = \frac{1}{16}$.

    d. $\log_a b = c$ is equivalent to $a^c = b$.

Notice that the value of the logarithm is the exponent when the equation is expressed in exponential form. ▩

EXAMPLE 2   Find the inverse of $f(x) = 10^x + 3$. Specify the domain and range of $f(x)$ and $f^{-1}(x)$.

Solution   The function $f(x)$ is one-to-one with domain all real numbers and range $y > 3$. To find $f^{-1}(x)$, we first solve for x in the equation $y = 10^x + 3$ as follows:

$$y = 10^x + 3$$
$$y - 3 = 10^x$$
$$\log_{10}(y - 3) = x.$$

Note that we had to subtract 3 from both sides of the equation before using the definition of logarithm to isolate x. Now we can write

$$f^{-1}(x) = \log_{10}(x - 3) = \log(x - 3).$$

The domain of $f^{-1}(x)$ is the range of $f(x)$, or $x > 3$; the range of $f^{-1}(x)$ is the domain of $f(x)$, or all real numbers. Both functions have been graphed in Figure 21. The graph of $y = \log(x - 3)$ is the reflection of the graph of $y = 10^x + 3$ across the line $y = x$.

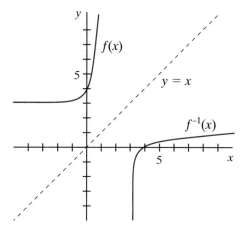

Figure 21   Graphs of $f(x) = 10^x + 3$ and its inverse, $f^{-1}(x) = \log_{10}(x - 3)$ ■

Compositions of Exponential and Logarithmic Functions

The composition of an exponential and a logarithmic function can be simplified since the functions are inverses of one another and both are one-to-one functions. If $f(x) = 10^x$, then $f^{-1}(x) = \log x$, and we expect that

$$f(f^{-1}(x)) = x \text{ and } f^{-1}(f(x)) = x.$$

We know, however, that some restrictions on x might be needed. For what x-values are these equations true? First, consider $f(f^{-1}(x)) = 10^{\log x}$. This is a composition with domain restrictions on the inner function $f^{-1}(x) = \log x$. For $f^{-1}(x) = \log x$ to be defined, it must be true that $x > 0$. There are no domain restrictions on the outer function, so

$$f(f^{-1}(x)) = 10^{\log x} = x, \text{ if } x > 0.$$

According to this statement, $10^{\log 100} = 100$, but $10^{\log(-2)} \neq -2$. In the first equation $\log 100 = 2$, so $10^{\log 100} = 10^2 = 100$ as expected. In the second equation, to evaluate $\log(-2)$ we would need to find the value a for which $10^a = -2$. However, no power of 10 is negative. Therefore, $\log(-2)$ does not exist and consequently $10^{\log(-2)} \neq -2$.

CLASS PRACTICE

1. Use your calculator to verify that $e^{\ln 4} = 4$.

2. Use your calculator to graph the functions $y = 10^{\log(x)}$ and $y = e^{\ln(x)}$. What are the domain and range of each function?

Now consider

$$f^{-1}(f(x)) = \log(10^x).$$

There are no domain restrictions on the inner function $f(x) = 10^x$ of this composition because $f(x) = 10^x$ is defined for all real numbers. Furthermore, since values of 10^x are always positive, we can always take the common logarithm of any number of the form 10^x. So,

$$f^{-1}(f(x)) = \log(10^x) = x, \quad \text{if } x \text{ is any real number.}$$

According to this statement, $\log(10^3) = 3$ and $\log(10^{-2}) = -2$. We can verify these easily since $\log(10^3) = \log(1000) = 3$ and $\log(10^{-2}) = \log\frac{1}{100} = -2$.

CLASS PRACTICE

1. Use your calculator to verify that $\ln(e^5) = 5$.

2. Use your calculator to graph the functions $y = \log(10^x)$ and $y = \ln(e^x)$. What are the domain and range of each function?

Now use your calculator to evaluate $\log(e^3)$. Are you surprised that the result is not 3? The bases of the logarithmic function and the exponential function must be the same for their composition to be the identity function. In general,

$$b^{\log_b x} = x, \quad \text{if } x > 0$$

and

$$\log_b b^x = x, \quad \text{if } x \text{ is any real number.}$$

Since the values of logarithms are themselves exponents, it is not surprising there are laws of logarithms that are closely related to the laws of exponents. The laws of exponents and the corresponding laws of logarithms are stated here and discussed below.

Summary of the Laws of Exponents and the Laws of Logarithms		
Product $b^r \cdot b^s = b^{r+s}, \quad b > 0$	$\log_b(rs) = \log_b r + \log_b s,$	$r, s, b > 0, b \neq 1$
Quotient $\frac{b^r}{b^s} = b^{r-s}, \quad b > 0$	$\log_b\left(\frac{r}{s}\right) = \log_b r - \log_b s,$	$r, s, b > 0, b \neq 1$
Power $(b^r)^s = b^{r \cdot s}, \quad b > 0$	$\log_b(r^s) = s \cdot \log_b r,$	$r, b > 0, b \neq 1$

To verify the first law of logarithms, which states that $\log_b(rs) = \log_b r + \log_b s$, first define a new constant k such that $k = \log_b r + \log_b s$. Since the expressions k and $\log_b r + \log_b s$ are defined to be equal, we can use them as exponents of b and the resulting expressions will still be equal. So,

$$\begin{aligned} b^k &= b^{\log_b r + \log_b s} \\ &= b^{\log_b r} \cdot b^{\log_b s} \\ &= rs. \end{aligned}$$

Now using $b^k = rs$ and the definition of logarithms, we know that $k = \log_b(rs)$. Since k was defined to be $\log_b r + \log_b s$, we have $\log_b(rs) = \log_b r + \log_b s$.

Another law of logarithms that will be used extensively is $\log_b(r^s) = s \cdot \log_b r$. The verification of this property again involves a law of exponents and composition. Define $q = \log_b r$. The exponential statement equivalent to this is $b^q = r$. Now if we raise both sides of this equation to the s power, we have $(b^q)^s = r^s$. We know that $(b^q)^s = b^{qs}$, so $b^{qs} = r^s$. Using the definition of logarithm, $b^{qs} = r^s$ is equivalent to $\log_b(r^s) = qs$. Since $q = \log_b r$, $sq = s \cdot \log_b r$. Therefore, $\log_b(r^s) = s\log_b r$, since both sides are equal to sq.

We have proven the laws of logarithms that involve products and powers. You will be asked in the exercises to prove the law involving quotients and the additional property, $\log_b r = \log_b s$ if and only if $r = s$. It is worth noting here that these properties of logarithms make difficult products, quotients, and powers much easier to compute. Products are turned into sums, quotients into differences, and powers into products. Scientists, starting with seventeenth-century astronomers, used logarithms to do difficult and lengthy calculations needed for many of the important discoveries they made. Imagine the time it would take to divide a number like 8497.231 by 0.00097388 without a calculator. With tables of logarithms, this division problem can be turned into a subtraction problem. Some of the most widely published books for nearly three centuries were tables of logarithms, and one of the first applications of the modern computer was to generate more accurate and precise tables of logarithms. The slide rule, which was based on logarithms, was the equivalent of a calculator for generations of students. Of course, with the spread of computers and calculators, very few people still use logarithms for computational purposes. Nonetheless, they remain important to mathematicians and scientists for other reasons. You will learn about some of these reasons later in this chapter.

Exercise Set 3.9

1. Convert each equation to exponential form.

    a. $\log_4 64 = 3$

    b. $\log_5 1 = 0$

    c. $\log 10{,}000 = 4$

    d. $\log_{1/2} 8 = -3$

2. Convert each equation to logarithmic form.

    a. $e^1 = e$

    b. $99^0 = 1$

    c. $\sqrt{16} = 4$

    d. $\sqrt[3]{64} = 4$

3. Evaluate the following without a calculator.

    a. $\log_3 81$

    b. $\ln \frac{1}{e^3}$

    c. $\log_2 \sqrt[5]{2^3}$

    d. $\log(0.01)$

4. By thinking about various powers of the base, determine between which two integers the value of the logarithm falls. For example, $\log_2 5$ is between 2 and 3 since $\log_2 4 = 2$ and $\log_2 8 = 3$ and $4 < 5 < 8$.

   a. $\log 12$ **b.** $\log 1.2$

   c. $\log_4 5$ **d.** $\log_7 21$

   e. $\log_{0.5}\left(\frac{1}{7}\right)$ **f.** $\log_5 1000$

   g. $\ln 27$ **h.** $\ln\left(\frac{1}{3}\right)$

5. Find the inverse of the following functions and specify the domain of each.

   a. $f(x) = e^{-5x}$ **b.** $f(x) = 3^{x-1}$

   c. $f(x) = \ln(2x)$ **d.** $f(x) = \log_2(x - 3)$

6. Prove the following laws of logarithms.

   a. $\log_b\left(\frac{r}{s}\right) = \log_b r - \log_b s, \qquad r, s, b > 0, b \neq 1$

   b. $\log_b\left(\frac{1}{a}\right) = -\log_b a, \qquad a, b > 0, b \neq 1$

7. **a.** What attribute of the logarithmic function allows us to say that if $\ln r = \ln s$, then $r = s$ for $r, s > 0$?

   b. What attribute of the logarithmic function allows us to say that if $r = s$, then $\ln r = \ln s$ for $r, s > 0$?

   c. For other toolkit functions, is it true that if $f(r) = f(s)$, then $r = s$? Explain your answer.

8. Computer scientists compare the speeds of some algorithms by finding the time required to complete a task as a function of the number of input items. Suppose there are three algorithms for solving one type of problem. The amount of time each of the algorithms A, B, and C takes is a function of the number of input items n.

   $$T_A(n) = 10{,}000 + 2\ln(n)$$

   $$T_B(n) = 100 + 3n^2$$

   $$T_C(n) = (0.001)e^n$$

   a. Rank the speed of the algorithms for $n = 10$ and $n = 300$.

   b. When these algorithms are used in industry, they often have thousands of input items. Which algorithm would be best for industrial use? Write a few sentences to justify your answer.

9. Solve the following equations for x. Give exact values whenever possible.

   a. $\log_x 27 = -3$ **b.** $\log_{64} x = \frac{2}{3}$

c. $x = 5^{2\log_5 6}$

d. $\log_6(x^2) = -2$

e. $\log_2(x^2 + 5x + 10) = 4$

f. $\log_{1/x} 9 = 2$

10. Evaluate each of the following. State any restrictions on values of the variables.

a. $\log_2 2^5$

b. $7^{\log_7(a+3)}$

c. $\log_{64} 8$

d. $(-3)^{\log_{-3} 5}$

11. Use a computer or calculator to graph $y = \log x$ and $y = \ln x$ on the same axes.

a. Write a paragraph in which you compare these graphs. What special features do they share? In what ways are they different?

b. Use graphs as needed to answer the following questions.

  i. For what values of x are values of $\log x$ negative? zero? positive?

  ii. For what values of x are values of $\ln x$ negative? zero? positive?

  iii. For what values of x are values of $\log x$ between zero and one? equal to one? greater than one?

  iv. For what values of x are values of $\ln x$ between zero and one? equal to one? greater than one?

12. Use the entries in Figure 22 to compare the rate of growth of $f(x) = \sqrt{x}$ and $g(x) = \ln x$.

Figure 22   Comparison of $f(x)$ and $g(x)$

x	$f(x) = \sqrt{x}$	$g(x) = \ln x$
0.1	0.316	−2.303
0.7	0.837	−0.357
1.0	1.000	0
1.5	1.225	0.405
5	2.236	1.609
10	3.162	2.303
25	5.000	3.219
100	10.000	4.605
1000	31.623	6.908
1,000,000	1000.000	13.816

The functions $f(x) = \sqrt{x}$ and $g(x) = \ln x$ are the inverses of the functions $h(x) = x^2$ and $k(x) = e^x$, respectively. Graph both $h(x)$ and $k(x)$ on the same axes and compare the rate at which they are increasing when $x > 0$. Use these graphs to support the conclusions you made when comparing the rate of growth of $f(x) = \sqrt{x}$ and $g(x) = \ln x$.

SECTION 3.10 Graphing Logarithmic Functions

We have already seen that the graph of $y = \log_b x$ can be obtained by first graphing its inverse $y = b^x$ ($b > 0$, $b \neq 1$) and then reflecting this graph about the line $y = x$.

When we consider exponential functions with bases greater than 1, we know that exponential functions with larger bases grow more rapidly than those with smaller bases. Since the graph of a logarithmic function is the reflection of an exponential graph about the line $y = x$, we expect logarithmic functions with larger bases to grow more slowly than those with smaller bases. Figure 23 shows graphs of $y = 3^x$ and $y = \log_3 x$ on the left, and graphs of $y = 10^x$ and $y = \log_{10} x$ on the right. For the logarithmic functions you can see that the larger base produces a curve that increases more slowly.

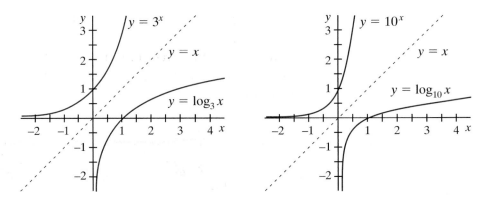

Figure 23   Exponential and logarithmic functions with base 3 and base 10

Notice that the graph of $y = \log_3 x$ appears to be a vertical stretch of the graph of $y = \log_{10} x$ and the graph of $y = 3^x$ appears to be a horizontal stretch of $y = 10^x$. You will see more about this relationship in the exercises at the end of this section.

We know that the base for a logarithm must be positive, but so far we have looked only at graphs of logarithmic functions with bases greater than 1. Now we will examine the graph of $y = \log_{1/3} x$. The function $g(x) = \left(\frac{1}{3}\right)^x$ is a decreasing exponential function. The graph of this function contains the points $(0, 1)$ and $(1, \frac{1}{3})$, and the positive x-axis is a horizontal asymptote for the graph. The inverse of this function is $g^{-1}(x) = \log_{1/3} x$, which therefore contains the points $(1, 0)$ and $(\frac{1}{3}, 1)$ and has the positive y-axis as a vertical asymptote. The graphs of these functions are shown in Figure 24.

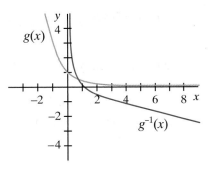

Figure 24 Graphs of $g(x) = \left(\frac{1}{3}\right)^x$ and $g^{-1}(x) = \log_{1/3} x$

The logarithmic function $f(x) = \log_b x$, $b > 0$, $b \neq 1$ should now be added to your function toolkit.

To graph a function that is a transformation of $y = \log_b x$, we will apply the same techniques used most recently for exponential functions. The most useful characteristics for graphing transformation $y = \log_b x$ are the points $(1, 0)$ and $(b, 1)$ and the vertical asymptote $x = 0$. The special characteristics of logarithmic graphs are listed below.

Characteristics of Logarithmic Functions $y = \log_b x$, $\quad b > 0, b \neq 1$
1. The point $(1, 0)$ is on every graph.
2. The point $(b, 1)$ is on every graph.
3. The y-axis $(x = 0)$ is an asymptote for each graph.
4. The domain of each function is the set of real numbers greater than zero.
5. The range of each function is the set of real numbers.
6. Each function is one-to-one.

EXAMPLE 1   Graph $f(x) = \ln x + 1$.

Solution   The graph is provided in Figure 25 on the next page. It has exactly the same shape as the graph of $y = \ln x$, but the graph has been shifted up one unit. The vertical asymptote remains the same. However, the special point $(1, 0)$ moves to $(1, 1)$, and the point $(e, 1)$ is shifted up to $(e, 2)$.

The expression $\ln x + 1$ can be rewritten as $\ln x + \ln e$ and this can be further simplified to $\ln(ex)$ by using the laws of logarithms. The graph of $f(x) = \ln x + 1$ is the same as the graph of $f(x) = \ln(ex)$. More generally, $\ln x + k$ is equivalent to $\ln x + \ln(e^k)$, which is equal to $\ln(e^k x)$. So a vertical shift of k units produces exactly the same graph as a horizontal compression (or stretch) using e^k.

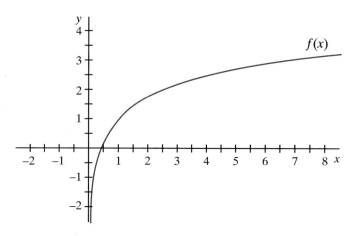

Figure 25   Graph of $f(x) = \ln(x) + 1 = \ln(ex)$ ▨

EXAMPLE 2   Graph $f(x) = \log_2(x^2)$. Note that this function can be rewritten as $g(x) = 2\log_2 x$ using the laws of logarithms. Are the functions f and g truly identical?

Solution   Notice first that x^2 is greater than zero for all $x \neq 0$. This means that the domain of $f(x) = \log_2(x^2)$ will be all real numbers except zero. Notice also that $f(-x) = f(x)$, so the graph of this function is symmetric about the y-axis. We might be tempted to simplify $\log_2(x^2)$ and rewrite it as $2\log_2 x$. However, when we do this, we change the domain and now can use only positive values of x. For positive values of x, the graphs of $f(x) = \log_2(x^2)$ and $g(x) = 2\log_2 x$ are identical. Both graphs contain the points $(1, 0)$ and $(2, 2)$. The graph of $f(x) = \log_2(x^2)$ also contains the points $(-1, 0)$ and $(-2, 2)$. The functions f and g are not identical because they have different domains. (See Figure 26.)

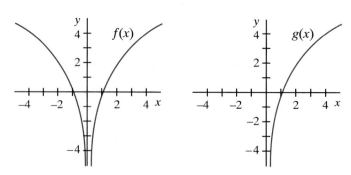

Figure 26   Graphs of $f(x) = \log_2(x^2)$ and $g(x) = 2\log_2 x$ ▨

EXAMPLE 3   Sketch the graph of $f(x) = \ln(x^2 - 3x - 4)$.

Solution   We can approach this graph as the composition $f(x) = g(h(x))$ where $h(x) = x^2 - 3x - 4$ and $g(x) = \ln x$. Domain plays an important role in this composition. The graph of $h(x)$ is shown in Figure 27 and can be used to analyze the domain of $f(x)$ and the flow of the composition. Since $\ln x$ is defined only for $x > 0$, the domain of the composition is restricted to those values of x for which $h(x) = x^2 - 3x - 4$ is positive. That is, the domain of f is restricted to those values of x for which the graph of $h(x) = x^2 - 3x - 4$ is above the x-axis. This happens when $x > 4$ or $x < -1$. For $g(x) = \ln x$, as x gets close to zero, the y-values tend toward negative infinity, making the y-axis a vertical asymptote. The values of the inner function $h(x)$ approach zero as x gets close to 4 and to -1. Therefore, the graph of f will have vertical asymptotes at $x = 4$ and at $x = -1$.

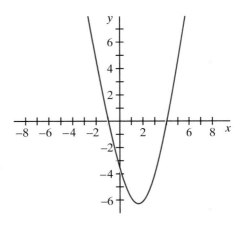

Figure 27   Graph of $h(x) = x^2 - 3x - 4$

Since $\ln 1 = 0$, the values of x for which $x^2 - 3x - 4 = 1$ are zeros of f; these are $x = \frac{3 \pm \sqrt{29}}{2}$, or $x \approx -1.2$ and $x \approx 4.2$. Notice that the y-values of f are negative for those x-values where $0 < x^2 - 3x - 4 < 1$. Since the natural logarithm of any number a is less than a itself, the graph of $f(x) = \ln(x^2 - 3x - 4)$ is below the graph of $h(x) = x^2 - 3x - 4$. The graph of $h(x) = x^2 - 3x - 4$ is symmetric about $x = \frac{3}{2}$. This symmetry is also present in the graph of f. The graphs of f and h are shown in Figure 28 on the next page.

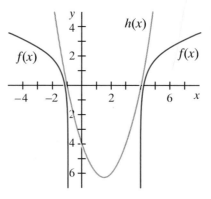

Figure 28 Graphs of $f(x) = \ln(x^2 - 3x - 4)$ and $h(x) = x^2 - 3x - 4$ ▨

Exercise Set 3.10

1. For what values of b is $f(x) = \log_b x$ an increasing function? a decreasing function?

2. Graph each function below. Label the points that correspond to $(1, 0)$ and $(b, 1)$ and the asymptote that corresponds to $x = 0$. Identify the domain and range of each.

a. $y = 3 + \log x$ **b.** $y = \log_{1/2} x$

c. $y = 4\log_3 x$ **d.** $y = \ln(x - 5)$

e. $y = \ln(5x)$ **f.** $y = \log(-x)$

g. $y = |\log_4 x|$ **h.** $y = \ln(4x - 3)$

i. $y = -3\log x + 5$ **j.** $y = -(3\log x + 5)$

3. Consider the function $f(x) = \log_2 \sqrt{\frac{x}{8}}$.

a. Graph f by rewriting it as $f(x) = \frac{1}{2}\log_2\left(\frac{x}{8}\right)$.

b. We can use another law of logarithms to write this function in the form

$$f(x) = \frac{1}{2}(\log_2 x - \log_2 8) = \frac{1}{2}\log_2(x) - \frac{1}{2}(3) = \frac{1}{2}\log_2(x) - \frac{3}{2}.$$

Use this form of the equation to graph the function.

c. Which form seems easier for graphing by hand?

4. Use the laws of logarithms or composition of functions to sketch a graph of each of the following. Identify the property you use or the two functions that compose to form f.

a. $f(x) = \log\left(\frac{1}{x}\right)$

b. $f(x) = \ln \sqrt[3]{x}$

c. $f(x) = \frac{1}{\ln x}$

d. $f(x) = \log|x|$

e. $f(x) = \ln(x^2 - 5)$

f. $f(x) = \ln(\sin x)$

g. $f(x) = \log(10^{-x^2+2})$

h. $f(x) = 10^{\log(-x^2+2)}$

5. Graph $f(x) = \ln x$ and $g(x) = \log_{10} x$ on the same x- and y-axes. Use transformations to explain how the graphs are related to each other.

SECTION 3.11 Solving Exponential and Logarithmic Equations

Solving equations that involve exponents or logarithms often requires techniques that involve composition as well as laws of exponents and logarithms.

EXAMPLE 1 The population of the world, as of August 1, 1996, was estimated by the U.S. Census Bureau to be about 5.779 billion and was growing at an annual rate of about 1.38%. If this rate were to remain constant, how long would it take for the population of the world to reach 10 billion?

Solution If we assume the rate of growth is a constant percent (of the current population), then the function that models this growth is $P(t) = 5.779(1.0138)^t$, where t represents the number of years since August 1, 1996. To determine the value of t for which $P(t) = 10$, we need to solve the equation $10 = 5.779(1.0138)^t$. Taking the common logarithm of both sides of this equation yields the following:

$$\log 10 = \log(5.779(1.0138)^t)$$

$$\log 10 = \log(5.779) + t\log(1.0138).$$

Solving for t,

$$t = \frac{\log 10 - \log(5.779)}{\log(1.0138)} \approx 40.009.$$

Therefore, the population of the world will reach 10 billion in just a little more than 40 years, sometime in August, 2036. It would also have been possible to first divide both sides of the equation $10 = 5.779(1.0138)^t$ by 5.779 and then take the logarithm of both sides of the resulting equation. The same answer can be obtained by taking the natural logarithm of both sides as follows:

$$10 = 5.779(1.0138)^t$$

$$\frac{10}{5.779} = (1.0138)^t$$

$$\ln\left(\frac{10}{5.779}\right) = \ln(1.0138^t)$$

$$\ln\left(\frac{10}{5.779}\right) = t\ln(1.0138).$$

Solving for t,

$$t = \frac{\ln\left(\dfrac{10}{5.779}\right)}{\ln(1.0138)} \approx 40.009. \quad \blacksquare$$

EXAMPLE 2   If a group of people is asked to memorize a list of nonsense words, then the percent of those who remember all of the words after t hours can be approximated by the logarithmic function $f(t) = 1 - k \ln(t + 1)$, where k is a constant that depends on the length of the list of words and other factors. Determine the value of k based on the observation that after 5 hours only 20% of the members of a group can remember all of the words. Using this value of k, determine what percent remember all the words after 1 hour. At what time will no one in the group be able to remember all the words?

Solution   Initially, 100% of the people know all the words, so $f(0) = 1.00$. Since 20% of the people can remember all the words after 5 hours, $f(5) = 0.20$. To determine the value of k, we need to solve the equation $0.20 = 1 - k \ln(5 + 1)$ for k.

$$0.20 = 1 - k \ln(5 + 1)$$

$$-0.80 = -k \ln 6$$

$$k = \frac{0.80}{\ln 6} \approx 0.446.$$

Using this value of k, we now need to determine the value of $f(1)$.

$$f(1) = 1 - 0.446 \ln(1 + 1)$$

$$= 1 - 0.446 \ln(2)$$

$$\approx 0.691.$$

Since $f(1) \approx 0.691$, after one hour about 69% of the group still remember all the words. Finally, to determine how long it will take until no one in the group can remember all the words, we need to find t such that $f(t) = 0$. We need to solve the following equation for t:

$$0 = 1 - 0.446 \ln(t + 1)$$

$$-1 = -0.446 \ln(t + 1)$$

$$\frac{1}{0.446} = \ln(t + 1).$$

We can now use both sides of the preceding equation as exponents for base e, so that

$$e^{(1/0.446)} = e^{\ln(t+1)}$$

$$e^{(1/0.446)} = t + 1$$

$$t = e^{(1/0.446)} - 1$$

$$t \approx 8.4$$

Thus, after about 8.4 hours, no one in the group can remember all the words. For values of t larger than 8.4, the function f would produce negative values for the percent, which would be meaningless in this context. ▨

Change of Base

Mara was writing a report on the life of Marie Curie. She found that in 1921, President Warren G. Harding presented Marie Curie with 1 gram of radium on behalf of the women of the United States. Although not essential to her report, two of Mara's references gave a function to describe the decay of radium. One text gave the function describing the amount of radium left after t years to be $R_1(t) = 2^{-0.0006039t}$ and the other gave $R_2(t) = e^{-0.0004186t}$. Mara was curious about the two different functions given in her references. After thinking about the two functions for awhile, Mara realized that they were the same function expressed with two different bases.

How do we change from one exponential base to another?

We know that $3^4 = 9^2$ and that $2^{-2} = \left(\frac{1}{16}\right)^{\frac{1}{2}}$. Using a calculator, we can verify that $5^2 \approx e^{3.22}$ and $2^{-0.0006039} \approx e^{-0.0004186}$. How do we change from one exponential base to another? When is it useful to do so? These questions are very important when we write exponential functions as models of real-world phenomena. Equivalent functions may look different because they are written with different bases.

How can we express $f(x) = 2^x$ as an exponential function with base e? We need to express 2 as a power of e, say, e^k. So we must solve the equation $2 = e^k$ to find k. Both sides of this equation are positive, so we can use the technique of taking logarithms of both sides of the equation as follows:

$$2 = e^k$$

$$\ln 2 = \ln(e^k)$$

$$\ln 2 = k \ln e$$

$$\ln 2 = k$$

$$0.693 \approx k.$$

So,

$$2 = e^k = e^{\ln 2} \approx e^{0.693},$$

and we can write

$$f(x) = 2^x \approx (e^{0.693})^x \approx e^{0.693x}.$$

This helps to explain the difference in the two functions Mara found, since

$$R_1(t) = 2^{-0.0006039t} = (e^{\ln 2})^{(-0.0006039)t} = e^{(-0.0006039)(\ln 2)t} = e^{-0.0004186t} = R_2(t).$$

Since exponential and logarithmic functions are inverses, you might suspect that we can also change the base of logarithmic functions. Let's try to rewrite $g(x) = \log_2(x)$ with a natural logarithm. Let

$$g(x) = y = \log_2(x)$$

and then rewrite this equation as

$$2^y = x.$$

We now take the natural logarithm of both sides (which are positive),

$$\ln(2^y) = \ln x$$

$$y \ln 2 = \ln x$$

$$y = \frac{\ln x}{\ln 2} \approx 1.44 \ln x.$$

Therefore,

$$g(x) = \log_2 x \approx 1.44 \ln x.$$

The fact that $g(x) = \log_2 x \approx 1.44 \ln x$ means that the graph of $y = \ln x$ can be stretched vertically by a factor of about 1.44 to produce the graph of $g(x) = \log_2 x$.

EXAMPLE 3   Evaluate $\log_3 7$.

Solution   Neither the calculator nor the computer can directly evaluate $\log_3 7$, but we can use the techniques above to change the base and then evaluate using a known base. Let $y = \log_3 7$, and rewrite this equation in exponential form, $3^y = 7$.

Since we now have an exponential equation, we can take either the common logarithm or the natural logarithm of both sides of the equation as follows.

$$3^y = 7 \qquad\qquad\qquad 3^y = 7$$

$$\log(3^y) = \log 7 \qquad\qquad \ln(3^y) = \ln 7$$

$$y \log 3 = \log 7 \qquad\qquad y \ln 3 = \ln 7$$

$$y = \frac{\log 7}{\log 3} \qquad\qquad y = \frac{\ln 7}{\ln 3}$$

$$y \approx 1.77 \qquad\qquad\qquad y \approx 1.77$$

The technique illustrated in the preceding example can be used to change the base of any logarithm or to evaluate any logarithm. To find $\log_a b$, let

$$y = \log_a b.$$

Then,

$$a^y = b$$

$$\log_c a^y = \log_c b$$

$$y \log_c a = \log_c b$$

$$y = \frac{\log_c b}{\log_c a}.$$

So,

$$\log_a b = \frac{\log_c b}{\log_c a}.$$

EXAMPLE 4   In recent years, the population of Japan has been growing at an annual rate of about 0.3%. If this rate remains constant, how long will it take for the population to double? How long will it take for the population to quadruple?

Solution   When we solved a problem similar to this one in Section 3.3, we did not have an analytical way to determine the doubling time. Logarithms provide such a way. Since the population is assumed to grow at a constant annual growth rate, each year the population is 1.003 times the previous year's population. If the current population of Japan is some number N, when the population doubles it will be $2N$. We need to solve the equation $2N = N(1.003)^t$, where t is the number of years necessary for doubling. This yields

$$2N = N(1.003)^t$$

$$2 = (1.003)^t$$

$$\ln 2 = \ln(1.003)^t$$

$$\ln 2 = t\ln(1.003)$$

$$t = \frac{\ln 2}{\ln(1.003)}$$

$$t \approx 231.4.$$

The population of Japan will double in approximately 231.4 years, assuming a continuation of the current annual growth rate of 0.3%. If this growth rate does not change, the population should quadruple in two doubling times, or approximately 462.8 years. We can verify this result algebraically by solving the equation $4N = N(1.003)^t$ for t.   ■

CLASS PRACTICE

1. Rework Example 4 by first writing a function for the population of Japan in the form $P(t) = ae^{kt}$ and then solving for the doubling time.

2. Verify from your work in the preceding exercise that $\frac{\ln 2}{k}$ is a general expression for the doubling time when the population function is written in the form $P(t) = ae^{kt}$.

3. Let g represent the annual growth rate of a population expressed in decimal form. Explain why the expression $\frac{\ln 2}{g}$ provides a good approximation of doubling time whenever g is small (less than 20%).

4. Write a function for the population of Japan using base 2. Explain why this base is useful for solving doubling-time problems.

EXAMPLE 5

Suppose an archaeologist finds a piece of human bone fragment that contains 43% of the amount of radioactive carbon-14 normally found in the bone of a living person. She knows that the half-life of carbon-14 is 5730 years. How old is the bone fragment?

Solution   Since less than half of the carbon-14 remains, we know that the bone fragment is more than 5730 years old. To find a more precise answer, we should first determine the general equation $P(t) = P_0 e^{kt}$ and then determine the value of k. Since the half-life of carbon-14 is 5730 years, the initial value of P_0 will have decayed to $0.5P_0$ when $t = 5730$. Therefore, we can write the equation $0.5P_0 = P_0 e^{5730k}$, and solve for k as follows:

$$0.5P_0 = P_0 e^{5730k}$$

$$0.5 = e^{5730k}$$

$$\ln(0.5) = 5730k$$

$$k = \frac{\ln(0.5)}{5730}$$

$$k \approx -0.000121.$$

We can now write the equation $P(t) = P_0 e^{-0.000121t}$. We want to determine the value of t when $P(t) = 0.43P_0$, so we need to solve

$$0.43P_0 = P_0 e^{-0.000121t}$$

$$0.43 = e^{-0.000121t}$$

$$\ln(0.43) = -0.000121t$$

$$t = \frac{\ln(0.43)}{-0.000121}$$

$$t \approx 6975.$$

We conclude, therefore, that the bone is about 6975 years old. ▪

CLASS PRACTICE

1. Solve the carbon-14 dating problem using base $\frac{1}{2}$. Explain why this is a simpler base for half-life problems.

Carbon-14 is useful in the dating of relatively old artifacts and fossils because its half-life is relatively long. However, if we wanted to measure the age of objects that were millions of years old, the carbon-14 dating process would be useless, because virtually all of the carbon-14 present in the organism at death would have decayed. Scientists use isotopes of uranium and thorium to date very old objects. Similarly, carbon-14 dating is not useful in dating material from the last few decades, since the abundance of carbon-14 in the material is virtually indistinguishable from the amount of carbon-14 in living matter. Tritium, with a half-life of 12.1 years, is often used for dating more recent matter such as wines.

1.   Rewrite $g(x) = \left(\frac{1}{2}\right)^x$ as an exponential function with base 10.

2.   Dot and Dan plan to invest \$5000 in an account that pays 9% annual interest compounded monthly. Write three distinct, explicit expressions for the future value of the account after N years. Use $\left(1 + \frac{0.09}{12}\right)$, 2, and e as bases.

3.   A house purchased four years ago for \$40,000 was later sold for \$60,000. If the value of the house continues to increase exponentially at the same rate, how much will it be worth next year?

4.   Find the continuously compounded interest rate that is equivalent to 10% compounded quarterly.

5.   **a.** Show that $\frac{\ln(0.5)}{k}$ is a general expression for half-life when the decay function is written as $P(t) = P_0 e^{kt}$.

     b. Find an expression for half-life if the decay function is written as $P(t) = P_0 2^{kt}$.

6.   When a certain medicine enters the bloodstream, it is absorbed gradually by body tissue, and its concentration decreases exponentially with a half-life of 3 days. If the initial concentration of the medicine in the bloodstream is A, what will the concentration be 30 days later? Assume that no additional doses are taken.

7.   Given a population of 10 million and an annual growth rate of 3%, how long will it take this population to double? What is the size of the population in triple the doubling time?

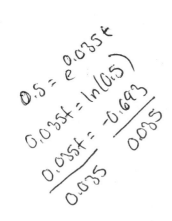

8.   A population of bacteria is growing exponentially. A researcher determines the population of the bacteria to be 2000 at 2:00 and to be 4000 at 2:10. What is the doubling time of the population? What is the estimated population at 2:05?

9.   Carbon-11 decays into boron at the rate of roughly 3.5% per minute. What is the half-life of carbon-11?

10. What exponential function of the form $y = ae^{bx}$ goes through the following points?

     a. $(0, 3)$ and $(5, 80)$

     b. $(3, 10)$ and $(6, 50)$

11. A local bank advertises that if you deposit at least \$5000 in an investment account, your money will double in 8 years. Assuming that interest is compounded monthly, find the interest rate on the account.

Chernobyl reactor

12. Cesium-137, one of the dangerous radionucleides produced in the fallout of the Chernobyl disaster, has a half-life of 30.3 years. How much time must pass until the radiation emitted by the radioactive cesium at Chernobyl is reduced to 10% of the initial radiation?

13. When money is deposited in a savings account, the balance increases exponentially. The time required for the balance to double is independent of the balance.

   a. To determine a general formula for doubling time, solve the equation $2A = A\left(1 + \frac{I}{k}\right)^{kN}$ for N in terms of I and k.

   b. Use this formula to verify several doubling times in Figure 29.

Figure 29   Doubling time in years

Annual interest rate, I	Annual compounding	Quarterly compounding	Daily compounding	Continuous compounding
0.05	14.21	13.95	13.86	13.86
0.08	9.01	8.75	8.67	8.66
0.10	7.27	7.02	6.93	6.93
0.12	6.12	5.86	5.78	5.78
0.15	4.96	4.71	4.62	4.62

   c. Describe how doubling time changes as interest rate and frequency of compounding change.

14. Bankers often approximate doubling time with the Rules of 69, 70, and 72. In the previous exercise, you should have found that exact doubling time is given by

$$N = \frac{\ln 2}{\ln\left(1 + \frac{I}{k}\right)^{k}},$$

where I is the annual percentage rate and k the number of compounding periods per year. In the case of continuous compounding, where $k \to \infty$ and $\left(1 + \frac{I}{k}\right)^{k} \to e^{I}$,

$$N = \frac{\ln 2}{\ln(e^{I})} = \frac{\ln 2}{I} \approx \frac{0.6931472}{I} \approx \frac{69}{100I}.$$

Thus, bankers can approximate doubling time with $N \approx \frac{69}{100I}$, which is commonly referred to as the Rule of 69.

   a. Use the Rule of 69 to estimate the doubling time for money invested at a rate of 8% compounded continuously.

   b. Verify several estimates for doubling time in Figure 30 on the next page.

Figure 30   Doubling time approximations in years

Annual interest rate, I	$\dfrac{69}{100I}$	$\dfrac{70}{100I}$	$\dfrac{72}{100I}$
0.05	13.80	14.00	14.40
0.08	8.63	8.75	9.00
0.10	6.90	7.00	7.20
0.12	5.75	5.83	6.00
0.15	4.60	4.67	4.80

    c. Compare the entries in Figure 29 and Figure 30 to determine which approximation is more appropriate for annual compounding. Which is more appropriate for quarterly compounding? for daily compounding?

    d. Sometimes the 69 in the numerator is replaced with 70 or 72. Explain when replacing 69 with 70 or 72 improves the accuracy of the approximation for the doubling time.

15. Change $\log_b a$ to an equivalent expression involving logarithms with base c.

16. Rewrite $\log x$ using the natural logarithm.

17. Show that $(\log_a b)(\log_b a) = 1$, for all $a > 0, b > 0, a \neq 1, b \neq 1$.

18. Solve the following equations for x. Give exact values whenever possible.

    a. $5^x = 3$ **b.** $5^{x-1} = e^x$

    c. $3^{x+1} = 9^x \cdot 27^{x+1}$ **d.** $\log_x 12 = 3$

    e. $\ln(\ln(x + 6)) = 0$ **f.** $10^{2x+1} = 4^{x-1}$

    g. $e^{x^2} = 10$ **h.** $5^{3x-2} = 3^{x+1}$

    i. $\log_2(x + 4) + \log_2(x + 2) = \log_2 3$ (Be careful with domain!)

    j. $\log(x^4) = (\log x)^3$ **k.** $5^{x^2-x} = 7$

    l. $5^{2x} - 7(5^x) + 12 = 0$

19. Solve for the indicated variable.

    a. t: $e^{rt} = 2$ **b.** k: $T = Ae^{-Bk} + C$

    c. t: $S = A\ln(Bt + C)$ **d.** c: $y = \dfrac{A}{1 + Be^{-Acx}}$

    e. y: $\ln(y - 10) = \ln(a) + bx$ **f.** y: $\ln y = \ln a + b\ln x$

20. Find the inverse of each of the following functions and identify the domain of each.

    a. $f(x) = 2^{x+1} - 3$ **b.** $f(x) = 3\ln(2x)$

    c. $f(x) = e^{2x} + 1$ **d.** $f(x) = \dfrac{\ln(x + 5)}{2}$

SECTION 3.12 Logarithmic Scales

In 1972, about 2.8 million ninth graders were enrolled in a math class. By the time these students reached their first year of college, only 250,000 of them were enrolled in a math class. In 1980, this group of students earned their bachelor's degrees, and only 10,000 of them were enrolled in a math class. In 1982, while earning master's degrees, 3000 students were enrolled in a math class. In 1986, while working on doctoral degrees, only 500 of them were enrolled in a math class. The data are given below in Figure 31.

Figure 31 Enrollment in mathematics classes over time for a single age group

Year	Math class enrollment
1972	2,800,000
1976	250,000
1980	10,000
1982	3000
1986	500

A scatter plot of the data is shown in Figure 32. Notice that the vertical scale required to show all the data makes it difficult to distinguish the smaller *y*-values from each other.

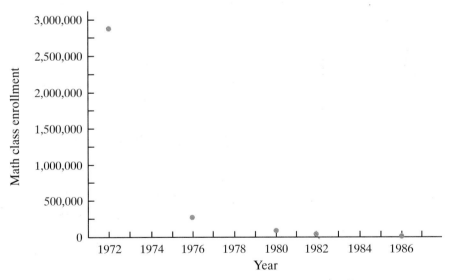

Figure 32 Scatter plot of mathematics class enrollment

The mathematics class enrollment for this group of students appears to be decreasing exponentially. To better understand the data, it is useful to see what happens when we take the common logarithm of the *y*-values and plot them against the year. These values are given in Figure 33 on the next page.

Figure 33   Enrollment and log(enrollment)

Year	Enrollment	Log(enrollment)
1972	2,800,000	6.4472
1976	250,000	5.3979
1980	10,000	4.0000
1982	3000	3.4771
1986	500	2.6990

The scatter plots in Figure 34 show two ways to represent the data. The graph on the left changes the vertical scale of the coordinate system using powers of 10 and plots the actual enrollments on this altered coordinate system. The graph on the right re-expresses the data by taking the logarithm of the enrollments and plots the re-expressed data on a standard vertical scale.

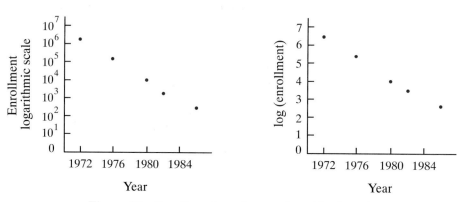

Figure 34   Enrollment graphs on a logarithmic scale

logarithmic scale

By altering the vertical scale on the graph at the left so that it has equal spacing between 10^1, 10^2, 10^3, and so on, we have produced a graph that appears linear. This new spacing creates what is called a *logarithmic scale*, in which equal differences on the x-axis correspond to equal ratios on the y-axis. The result is a linear relationship between the variables graphed on this logarithmic scale. This was a standard technique for analyzing exponential functions before the age of calculator and computer technology. Special graph paper that had the vertical scale rewritten as powers of 10 was used, and the data were plotted on this graph paper. Modern technology allows us to re-express the data instead of the coordinate system, as shown on the graph at the right.

Scientists commonly use logarithms to create a simple scale for data, when some values are many times larger than other values in the data set. For example, the pH of a solution is determined by the concentration of hydrogen ions in the solution, measured in moles per liter. Solutions with a high concentration of hydrogen ions are acidic, while solutions with a low concentration are basic. The concentration of hydrogen ions in household ammonia is approximately 0.0000000000005 moles/liter, while the concentration of hydrogen ions in lemon juice is ten billion times larger at approximately 0.005 moles/liter. To describe these vastly different values on the same scale, logarithms are used. The pH of a solution is defined to be pH $= -\log(H^+)$, where H^+ is the concentration of hydrogen ions. So we say that household ammonia has a pH of

$-\log(0.0000000000005) = 12.3$, while lemon juice has a pH of $-\log(0.005) = 2.3$. The pH of pure water is 7. Solutions with a pH less than 7 are acidic, while solutions with a pH greater than 7 are basic. By using logarithms, chemists take values that vary by huge amounts and place them on a simple scale from 0 to 14, which is much easier to understand.

Other logarithmic scales involve similarly disparate numerical values being collapsed onto a smaller, common scale. Astronomers use magnitudes to rate the apparent brightness of stars, physicists use decibels (dB) to measure the loudness of sound, and geologists use the Richter scale to rate the energy produced by an earthquake. In all of these settings, logarithmic scales make it convenient to investigate data with numerical values that otherwise would be very difficult to compare.

The Decibel Scale for Sound

The intensity of a sound is dependent on the actual energy carried in the wave and is measured in units of power per unit area, or watts/meter^2. The greater the intensity, the louder the perceived sound. The intensities of some common sounds are given in Figure 35. Physicists define a loudness level (measured in decibels, dB) that is related to the actual intensity, I, of the sound by the equation

$$\text{Loudness level (dB)} = 10 \log\left[\frac{I(\text{W/m}^2)}{10^{-12}(\text{W/m}^2)}\right].$$

The threshold of human hearing is a sound with intensity 10^{-12} W/m^2. It is defined to be of loudness level 0 dB and serves as a baseline by which other sounds are measured. You can use the equation above to verify that the loudness is 0 dB when the intensity is 10^{-12} W/m^2.

Figure 35   Intensity and loudness levels of some sounds

Sound	Intensity (W/m^2)	Loudness level (dB)
Threshold of hearing	10^{-12}	0
Normal breathing	10^{-11}	10
Rustling leaves	10^{-10}	20
Whisper	10^{-9}	30
Quiet library	10^{-8}	40
Quiet radio	10^{-7}	50
Ordinary conversation	10^{-6}	60
Busy street traffic	10^{-5}	70
Factory	10^{-4}	80
Niagara Falls	10^{-3}	90
Siren (at 30 m)	10^{-2}	100
Loud thunder	10^{-1}	110
Rock concert (at 2 m)	1	120
Jet plane takeoff (at 30 m)	10	130
Rupture of eardrum	10^4	160

A sound of 10 decibels has ten times the intensity of a 0 dB sound; a sound of 20 decibels has one hundred times the intensity of a 0 dB sound, or ten times the intensity of a 10 dB sound. Similarly, a sound of 30 dB has one thousand times the intensity of a 0 dB sound, or ten times the intensity of a 20 dB sound. When the loudness level increases by an increment of 10 decibels, the actual sound intensity is multiplied by a factor of ten. Therefore, the intensity of a 60 dB sound (ordinary conversation) is not double the intensity of a 30 dB sound (a whisper). Rather, the sound intensity of conversation is three factors of ten, or 1000, times the intensity of a whisper, and the intensity of a whisper is 1000 times that of the least audible sound. Equal differences in loudness level are equivalent to equal ratios of sound intensity.

EXAMPLE 1   Compare the decibel level of two sirens at a distance of 30 meters to the decibel level of a single siren at the same distance.

Solution

$$\text{Loudness level (2 sirens)} = 10 \log \frac{2 \cdot 10^{-2}}{10^{-12}}$$

$$= 10 \log(2 \cdot 10^{10})$$

$$= 10(\log 2 + 10) \approx 103 \text{ dB}$$

The total intensity of two simultaneous sounds is the sum of the individual intensities of the two sounds, but the individual decibel levels do not add to produce the total decibel level. When the intensity level doubles, the decibel level increases by $10 \log 2 \approx 3$ dB.

Exercise Set 3.12

1. At a party with 25 people, everyone is talking at once. What is the sound level in decibels?

2. The sound intensity level in large cities has been increasing by about one decibel annually. To what percent increase in intensity does one decibel correspond? If this annual increase continues, in how many years will the sound intensity double?

3. The Richter scale measures the magnitude of an earthquake in terms of the total energy released by the earthquake. One form of Richter's equation is

    $$M = \frac{2}{3} \log E - 2.9,$$

    where M is the magnitude and E is the energy in joules of the earthquake.

    a. If an earthquake releases 10^{13} joules of energy, what is its magnitude on the Richter scale?

b. A very powerful earthquake occurred in Colombia on January 31, 1906 and measured 8.6 on the Richter scale. Approximately how many joules of energy were released?

4. According to seismologists, an earthquake that registers 2 on the Richter scale is hardly perceptible, while an earthquake that measures 5 on this scale is capable of shattering windows and dishes and is generally classified as "minor." The San Francisco earthquake in 1989 caused great damage and registered 7.1 on the Richter scale.

San Francisco earthquake, 1989

    a. Compare the energy release of the San Francisco earthquake to a level 5 minor earthquake; i.e. calculate the ratio of their released energies.

    b. If an earthquake releases ten times as much energy as the San Francisco earthquake, what would it measure on the Richter scale?

    c. If an earthquake releases twice as much energy as the San Francisco earthquake, what would it measure on the Richter scale?

5. In the first edition of *Contemporary Precalculus Through Applications*, the authors erroneously stated that "each number on the Richter scale represents an earthquake 10 times as strong as one of the next lower number." Megan Bisk, a student at Wachusett Regional High School in Holden, Massachusetts, corrected us and proved that a 0.67 increase on the Richter scale represents an earthquake 10 times as strong.

    a. Prove that Megan was correct.

    b. Show that the correct statement is "Each number on the Richter scale represents an earthquake approximately 32 times as strong as one of the next lower number."

6. An empty auditorium has a sound level of 40 dB (due to heating, air-conditioning, and outside noise). On Saturday, 100 students are taking a college entrance exam. While they are working on the test, the only sounds are labored breathing and pencils rapidly moving across paper. The noise level then rises to 60 dB (not counting the groans). If each student contributes equally to the total noise, what would be the noise level if only 25 students were taking the test?

7. Because of dissolved carbon dioxide in the atmosphere, the pH of rain and snow is lower than that of pure water. The pH of pure water is 7, while the pH of rain and snow is about 5.6. Determine the concentration of hydrogen ions in pure water and in rain and snow. How many times greater is one than the other?

8. The basic astronomical unit of brightness is magnitude. The perceived, or apparent, magnitudes of two stars (m_1 and m_2) are related to their actual intensities (I_1 and I_2) by the equation

$$m_1 - m_2 = 2.5 \log\left(\frac{I_2}{I_1}\right).$$

If a bright star has intensity I_1 and a dim star has intensity I_2, then $I_2 < I_1$ and $0 < \frac{I_2}{I_1} < 1$. Since the logarithm of a number between 0 and 1 is negative, the quantity ($m_1 - m_2$) will be negative. Thus, $m_1 < m_2$, and this implies that the magnitude of the first, brighter star is smaller. We see, then, that small magnitudes are associated with brighter, more intense light sources. In 1856, Norman Pogson defined the magnitude scale such that the first magnitude, $m_1 = 1$, corresponds to the brightest-appearing stars and the sixth magnitude, $m_2 = 6$, corresponds to the faintest stars visible to the naked eye.

So the intensity of a star of magnitude 1 is 100 times the intensity of a star with magnitude 6. The magnitude system has been extended to negative magnitudes for bright objects like the sun and the moon and to positive magnitudes beyond 6 for stars visible only with a telescope. Figure 36 illustrates the magnitude scale.

Figure 36 Levels of brightness

Object	Apparent magnitude
Sun	−26
Full moon	−13
Venus at brightest	−4.6
Jupiter at brightest	−2.9
Mars at brightest	−2.6
Sirius, the brightest star	−1.5
Polaris, the North Star	2
Faintest star visible with binoculars	8
Faintest star visible with 8-inch telescope	14
Stars barely visible with largest telescopes	28

a. Using this table, determine how much brighter a full moon is than Venus.

b. To observe stars beyond the sixth magnitude, a telescope is required. Telescopes, however, also have limitations. The *limiting magnitude* of a telescope is the magnitude of the faintest star that can be seen with the telescope. A telescope with lens diameter D meters has a limiting magnitude, L, given by the formula

$$L = 17.1 + 5.1(\log D).$$

Find the lens diameter of a telescope with limiting magnitude of 11.1.

SECTION 3.13 Data Analysis with Exponential and Power Functions

In Chapter 2 we discussed straightening data that could be modeled by a quadratic function by using the inverse function, or the square root function. In this section we will extend the idea of re-expression to linearize other data sets. You will see that a data set that is best modeled by an exponential function can be straightened by using logarithms. This procedure is also based on the concept of inverse functions. We will also use logarithms to straighten data that are best modeled by a power function. This technique is not based on inverse functions, but rather on properties of logarithms.

Fitting Exponential Models to Data

EXAMPLE 1 The data in Figure 37 show how the population per square mile in the United States has changed over a period of years. On the scatter plot, the horizontal axis denotes the number of years since 1790. Find a function to model the population density over time.

Figure 37 Data and scatter plot of U.S. population per square mile

Year	Years since 1790	Population/mi^2	Year	Years since 1790	Population/mi^2
1790	0	4.5	1900	110	21.5
1800	10	6.1	1910	120	26.0
1810	20	4.3	1920	130	29.9
1820	30	5.5	1930	140	34.7
1830	40	7.4	1940	150	37.2
1840	50	9.8	1950	160	42.6
1850	60	7.9	1960	170	50.6
1860	70	10.6	1970	180	57.5
1870	80	10.9	1980	190	64.0
1880	90	14.2	1990	200	70.3
1890	100	17.8			

Source: Reprinted with permission from *The World Almanac and Book of Facts* Copyright© PRIMEDIA Reference Inc. All rights reserved.

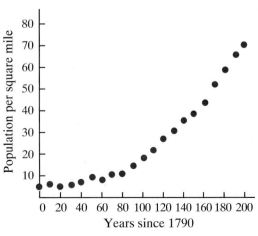

What nonlinear function fits the data?

Solution   It is clear that the scatter plot is not linear, but the curvature is difficult to judge. Just how curved is the graph? What nonlinear function fits the data? From our earlier work in this chapter, we know that populations often grow exponentially. If the data can be modeled by an exponential function of the form $y = ae^{bx}$, then

$$\ln y = \ln(ae^{bx}).$$

We can rewrite this equation using the laws of logarithms so that

$$\ln y = \ln a + \ln(e^{bx})$$

$$\ln y = \ln a + bx.$$

Since $\ln y = \ln a + bx$, the ordered pairs $(x, \ln y)$ will lie on a straight line. Similarly, it is also true—if the ordered pairs $(x, \ln y)$ are linear, then the ordered pairs (x, y) are exponential. Figure 38 shows the least squares line and the residuals for the re-expressed data points $(x, \ln y)$.

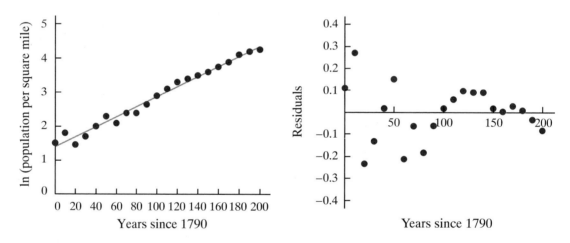

Figure 38   Least squares line and residuals for re-expressed population data

The re-expressed data look more linear than the original data. We can see this both in the plot of the re-expressed data with the linear model and in the residual plot. The residuals do not appear to have any pattern.

Fitting a least squares line to the re-expressed data yields

$$\ln y = 0.0147x + 1.3924.$$

Note that the constants have been rounded to four decimal places. To find a model for the original data (x, y), we express the equation $\ln y = 0.0147x + 1.3924$ in exponential form as follows:

$$y = e^{0.0147x + 1.3924}$$

$$y = e^{0.0147x} e^{1.3924}$$

$$y = 4.02e^{0.0147x}.$$

This exponential function models the relationship between the number of years since 1790 and the population per square mile. We have re-expressed the years to begin at zero in the scatter plot and will use these values in the data analysis that follows. In an exponential function, the independent variable, in this case, the number of years, is a factor in the exponent. Using large numbers, like 1790, as the exponent will cause unexpected errors unless we keep many more decimal places for our constants. If our model is in the form $P(t) = P_0 e^{kt}$ and k is rounded to too few decimal places, multiplying k by numbers close to 2000 would result in a very bad fit of the model to the data. In this example, we have kept more decimal places in our actual calculations than we show in the equations. Rounding, especially in the exponent, gives a fit with much larger residuals. Figure 39 shows the model graphed with the original data points.

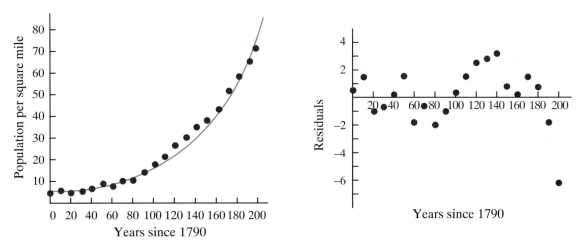

Figure 39   Exponential model and residuals for population data

Study the residual plots in Figures 38 and 39. The random pattern of the residuals from the least-squares line in Figure 38 indicates that taking the logarithm of the y-values successfully straightened out the curvature in the original data. The size of these residuals is deceptive, however. They represent errors in the logarithms of y-values rather than errors in actual y-values. The same random pattern of residuals can be seen in Figure 39. Positive residuals from the linear fit remain positive and negative residuals remain negative.   ▨

semi-log re-expression
semi-log plot

The technique of linearizing data by taking the logarithms of the y-values is frequently used. This method of re-expressing data is called *semi-log re-expression*, and a graph of ordered pairs $(x, \ln y)$ is called a *semi-log plot*. Semi-log re-expression will linearize any data set that can be modeled by an exponential function of the form

$$y = ae^{bx}.$$

1. Calculators and computers with curve fitting software often have an exponential regression (or exponential least squares) option.

    a. Use this feature of your calculator or computer software to find a model for the population data provided in Example 1 on page 235.

    b. Graph this model with the data and examine the residual plot.

    c. Compare this model to the one found in Example 1. Though the models may at first appear different, they are algebraically equivalent. Exponential regression programs use semi-log re-expression and linear least squares to produce the exponential model.

Fitting Power Functions to Data

A problem with using inverses to re-express data is that we often do not know what function relates the variables, so guessing its inverse is time consuming and difficult. While we might get lucky by using "educated guessing," in the long run this is an inefficient approach to finding a way to linearize ordered pairs of data. In the following example, we will discuss the use of logarithmic functions to straighten data that can be modeled by a power function of the form $y = ax^b$. This process takes much of the guessing out of the re-expression process.

EXAMPLE 2   The data in Figure 40 represent the age in years and volume in hundreds of board feet of several hardwood trees of the same species. Determine a model that fits this data set. Use your model to predict the volume of a 150-year-old tree.

Age (years)	Volume (100s of board feet)
20	1
40	6
80	33
100	56
120	88
160	182
200	320

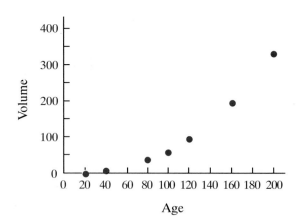

Figure 40   Tree volume data and scatter plot

Solution The scatter plot of the data shown in Figure 40 has the age of the tree plotted on the horizontal axis and the volume of wood on the vertical axis. (One board foot is 1 foot square by 1 inch thick.) Inspection of the scatter plot indicates that the relationship between these variables is not linear, but it is difficult to judge what kind of curvature is displayed. Since there is no underlying theory that would lead us to expect a particular relationship between age and volume, we have no choice but to guess. We have seen previously that taking the square roots of the y-values can straighten a curve if the functional relationship is quadratic, so we can try re-expressing the data and looking at a scatter plot of ordered pairs $(x, \sqrt{y})$. Figure 41 shows a least squares line fit to these re-expressed ordered pairs, as well as the corresponding residuals.

The scatter plot looks reasonably linear, but there is a clear pattern in the residual plot. This pattern of positive residuals at each end and negative residuals in the middle indicates that the ordered pairs $(x, \sqrt{y})$ are not linear, but actually curve upward. Since the re-expressed ordered pairs are not linear, the original ordered pairs (x, y) do not fit a simple quadratic function of the form $y = a(x - h)^2$. Our attempt to straighten the original scatter plot by taking the square roots of the y-values has not brought the y-values down far enough. There is either a vertical shift that we have not accounted for, or the relationship is not quadratic. Since the re-expressed data are still concave up, we might suspect that the original pairs (x, y) should be modeled by a function that grows more quickly than a parabola.

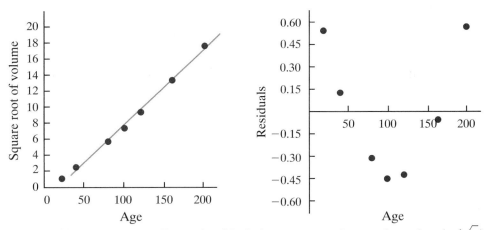

Figure 41 Least squares line and residuals for re-expressed tree volume data $(x, \sqrt{y})$

Since the graph of the original ordered pairs increases more quickly than the graph of a parabola, we might guess that the data are exponential. We perform a semi-log re-expression by taking the natural logarithm of the volume and plotting those values versus age, as shown in Figure 42 on the next page. This scatter plot is not linear, so the data are not exponential of the form $y = ae^{bx}$. In fact, the concavity in Figure 42 is the opposite of the concavity for the scatterplot of the original data. The logarithm pulls the y-values too far down, indicating that this re-expression is too strong. A possible model for the original data is somewhere between a quadratic and an exponential function.

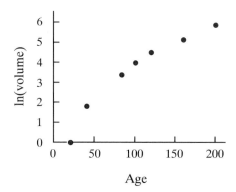

Figure 42   Semi-log re-expression of tree volume data

Perhaps the relationship can be modeled by a power function with an equation of the form $y = ax^p$. If p is greater than 2, the function increases more quickly than a quadratic function. In addition, this power function contains the point $(0, 0)$. This is consistent with the tree data since a tree initially has zero age and zero volume. Guessing the power of the function and then using the inverse to straighten the data would be time-consuming and tedious. A more efficient method uses logarithms.

If we take the natural logarithm of both sides of the equation of the power function

$$y = ax^p,$$

we have

$$\ln y = \ln(ax^p).$$

We can rewrite this equation using the laws of logarithms so that

$$\ln y = \ln a + \ln(x^p),$$

which simplifies to

$$\ln y = \ln a + p \ln x. \tag{7}$$

Equation (7) is a linear equation in $\ln x$ and $\ln y$. If we make a graph with $\ln x$ on the horizontal axis and $\ln y$ on the vertical axis, then the ordered pairs $(\ln x, \ln y)$ that satisfy equation (7) will lie on a line. The slope of this line is p, and the intercept on the vertical axis is $\ln a$.

We will try this technique on the tree volume data. We re-express the original ordered pairs (x, y) as the ordered pairs $(\ln x, \ln y)$. Figure 43 shows a least squares line fit to the re-expressed data and a scatter plot of the residuals. The residuals indicate that the linear fit for $(\ln x, \ln y)$ is good, so we conclude that the original ordered pairs (x, y) can be modeled by a power function.

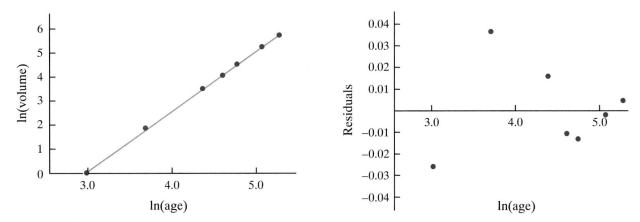

Figure 43   Least squares line and residuals for re-expressed tree volume data $(\ln x, \ln y)$

Fitting a least squares line to the re-expressed data $(\ln x, \ln y)$ yields the equation

$$\ln y = 2.4926 \ln(x) - 7.4415.$$

We can find the equation of the power function as follows:

$$\ln y = 2.4926 \ln x - 7.4415$$
$$y = e^{(2.4926 \ln x - 7.4415)}$$
$$y = e^{2.4926 \ln x} \, e^{-7.4415}$$
$$y = (e^{-7.4415}) e^{\ln(x^{2.4926})}$$
$$y = (0.0005864) x^{2.4926}. \qquad \textbf{(8)}$$

Figure 44 shows a graph of this power function superimposed on the original ordered pairs. The residuals indicate that this function does a good job of modeling the relationship between tree age and volume.

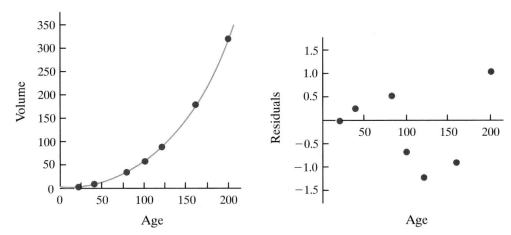

Figure 44   Power function model and residuals for tree volume data

To predict the volume of a 150-year-old tree, substitute $x = 150$ into equation (8). We get about 156, or approximately 15,600 board feet. ■

log-log re-expression

log-log plot

The technique of linearizing data by taking the logarithms of both the x-values and y-values is called *log-log re-expression*, and a graph of ordered pairs $(\ln x, \ln y)$ is called a *log-log plot*. Log-log re-expression will linearize any data set that can be modeled by a power function of the form $y = ax^p$.

An important observation regarding semi-log re-expression and log-log re-expression should be emphasized here. The linear equation obtained for the least squares line will involve re-expressed variables. Thus, the residuals from a semi-log plot or a log-log plot indicate only how well the re-expression has linearized the data. For example, the residuals on the right in Figure 43 are very small. However, they do not accurately represent the errors you may encounter when you use the model in equation (8) to make predictions about the original data. To determine the size of the errors to expect when using equation (8), you need to graph this function with the original ordered pairs. Then analyze the residuals associated with the original ordered pairs as shown in Figure 44.

CLASS PRACTICE

1. Calculators and computers with curve fitting software often have a power regression (or power least squares) option.

   a. Use this feature of your calculator or computer software to find a model for the tree volume data provided in Example 2 on page 238.

   b. Graph this model with the data and examine the residual plot.

   c. Compare this model to the one found in Example 2. Power regression programs use log-log re-expression and linear least squares to produce the power function model.

The preceding examples illustrate two techniques for straightening curved data: semi-log re-expression and log-log re-expression. These techniques have limitations. Re-expression to produce linearity is often an experimental process involving trial and error. Your knowledge of functions, inverses, special properties of functions, and some knowledge of the phenomena being modeled should guide you as you try different re-expressions and examine scatter plots.

EXAMPLE 3 A thermometer is placed in a cup of hot water, and the cup is placed in the refrigerator. The thermometer is checked periodically for the temperature of the water. The temperature of the refrigerator is set at 47.5°F. The data and a scatter plot are shown in Figure 45, with time measured in minutes since the thermometer was placed in the refrigerator and temperature measured in degrees Fahrenheit. Find a function that models the relationship between time and temperature and use it to estimate the temperature of the water at the beginning of the experiment.

Time (minutes)	Temperature (°F)
10	114
16	109
20	106
33	97
50	89
65	82
85	74
128	64
144	62
178	58
208	54
244	52
299	50
331	49

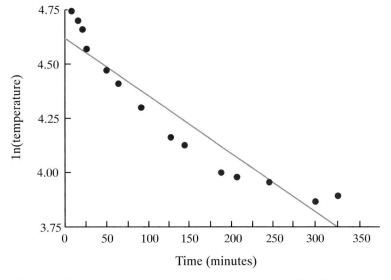

Figure 45   Data and scatter plot for the temperature of hot water over time

Solution   Since the refrigerator is set at 47.5°F, we expect the temperature of the hot water to level off at 47.5°F. Thus, the model for the (*time, temperature*) data should have a horizontal asymptote at 47.5. The functions that we know that have horizontal asymptotes are either a reciprocal function or a decaying exponential function. However, a reciprocal function also has a vertical asymptote, which is not consistent with the cooling phenomenon. Therefore, an exponential function has the characteristics we want for our model. To fit an exponential function to the data, we can try to linearize the data by using semi-log re-expression. Figure 46 contains a scatter plot of the ordered pairs (*time*, ln(*temperature*)) with a least squares line.

Figure 46   Scatter plot of semi-log re-expression of cooling data

The re-expressed data in Figure 46 have not been straightened by the semi-log re-expression. Recall that a semi-log re-expression will be successful only if the data can be modeled by an equation of the form $T = ae^{bt}$, where t is time and T is temperature. Functions of the form $T = ae^{bt}$ all have a horizontal asymptote at $T = 0$. However, the temperatures that we are modeling in this example will not approach zero but will level off at $T = 47.5$. If the cooling phenomenon can be modeled by an exponential function, its equation will need to account for a vertical shift of 47.5 units. The form of this equation is $T = ae^{kt} + 47.5$.

Using our knowledge of functions and transformations, we know that the ordered pairs $(t, T - 47.5)$ are shifted down 47.5 units and therefore will level off at $T = 0$. Thus, if the relationship between time and temperature can be modeled by a function of the form $T(t) = ae^{kt} + 47.5$, then a semi-log re-expression performed on the ordered pairs $(t, T - 47.5)$ will linearize the cooling data. Figure 47 shows the least squares line through the re-expressed data $(t, \ln(T - 47.5))$ and the corresponding residual plot. The least squares line appears to fit the re-expressed data well, and the residuals are random in sign.

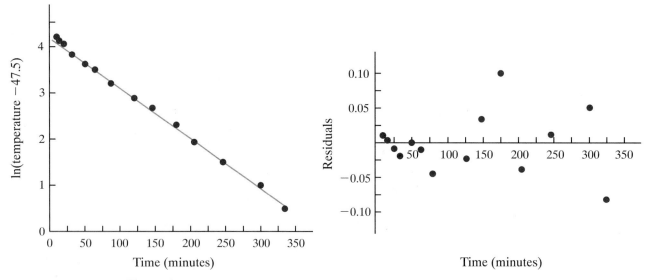

Figure 47 Least squares line and residuals for re-expressed cooling data

The equation of the least squares line through the ordered pairs $(t, \ln(T - 47.5))$ is

$$\ln(T - 47.5) = -0.01151t + 4.3006.$$

Solving for T gives

$$\ln(T - 47.5) = -0.01151t + 4.3006$$

$$T - 47.5 = e^{-0.01151t + 4.3006}$$

$$T - 47.5 = e^{4.3006}e^{-0.01151t}$$

$$T = 73.75e^{-0.01151t} + 47.5.$$

Figure 48 shows a graph of this exponential function superimposed on the original scatter plot. A residual plot is also provided. The residuals in Figure 48 are very small relative to the observed temperatures. As expected, there is no noticeable pattern in the residual plot, so we conclude that the model is a good one.

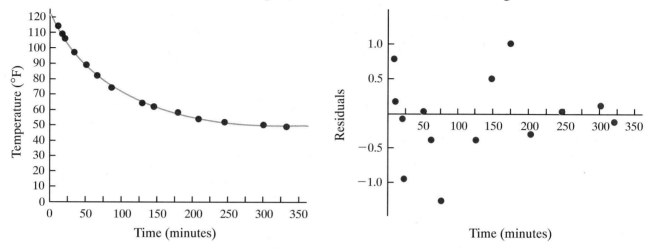

Figure 48 Exponential model and residuals for cooling data

The temperature was first recorded 10 minutes after the water started to cool. Had we observed the temperature as soon as we put the hot water into the cup, our model $T(t) = 73.75e^{-0.01151t} + 47.5$ suggests that the initial temperature would have been $T(0) = 73.75e^{-0.01151(0)} + 47.5 \approx 121$. Of course, we wouldn't expect the temperature to be exactly 121°F, but we would expect it to be somewhere in the neighborhood of this temperature. How close we would expect the value to be is discussed in the next section.

In our discussion of mathematical models, we have explained that a good model is one that shares the essential features of the phenomenon it describes. We expect there to be some deviation between the model and the observed data points, and our decision concerning the adequacy of a model is a subjective one. In this example, the model $T(t) = 73.75e^{-0.01151t} + 47.5$ seems to fit the general trend of the data, and we conclude that the decreasing exponential function is a satisfactory model for the cooling phenomenon.

Semi-log and log-log re-expression are tools for linearizing data sets that can be modeled by an exponential function or a power function. Each technique is based upon an assumption that the data are in "standard" position. For exponential functions, this means that the horizontal asymptote is the x-axis. For power functions, this means the point (0, 0) lies on the curve or the curve is asymptotic to the axes. Before performing a semi-log re-expression on exponential data, you may need to shift the data points horizontally so that the x-axis is the asymptote. Data to be modeled by a power function may need to be shifted both horizontally and vertically before performing a log-log re-expression. When performing a semi-log re-expression to find an exponential model, remove any point(s) with y-coordinate equal to zero. When performing a log-log re-expression to find a power function model, remove any point(s) in which either the x-coordinate or the y-coordinate is zero.

1. Fit a quadratic model to the population per square mile data in Example 1 on page 235 by performing an appropriate re-expression. How does the fit of the quadratic model compare to the fit of the exponential model in the solution for Example 1?

2. Study the residual plot for the cooling model in Figure 48. Notice that the residuals associated with later observations are smaller in magnitude than the residuals associated with earlier observations. How would you account for this?

3. In Example 1 we modeled U.S. population data by performing a semi-log re-expression. Suppose, instead of the natural logarithm, we had used the common logarithm. How would the new model for the data compare to the model using natural logarithms?

4. In a laboratory experiment, a tumor was induced in a plant, and the growth of the tumor was recorded over time. The data set is given in Figure 49. Use re-expression and least squares analysis to find a model for the data. At what rate is the tumor growing?

Figure 49   Growth of tumor

Number of days after inducing	14	19	23	26	28
Size of tumor (cc)	1.85	4.25	7.5	11	14.5
Number of days after inducing	30	33	35	37	41
Size of tumor (cc)	18.95	28.65	38	49.75	84.5

5. The following data set was collected by a precalculus class using a Calculator Based Laboratory™ (CBL™) and calculator. A probe that measures light intensity was aimed at a light source in a dark room. The light intensity was measured at specific distances from the probe. The distance is measured in centimeters, and the light intensity is measured in milliwatts per square centimeter (mW/cm^2). Find a function that models the data in Figure 50.

Figure 50   Light intensity data

Distance	120	130	140	150	160	170	180	190	200	210
Intensity	0.7085	0.6102	0.5415	0.4640	0.4032	0.3609	0.3216	0.2889	0.2570	0.2370

6. A group of students collected data relating the length of the side of an equilateral triangle with the area of the triangle. The data set is collected by drawing triangles on graph paper. The students counted the squares on the graph paper to estimate the area. Find a function that models the data in Figure 51.

Figure 51   Areas of equilateral triangles

Side	2	3	4	5	6	7	8	9	10	11	12	13	14	15
Area	1.75	4.0	7.0	11	15.5	20.5	27	36	43.5	52	62	74	86	97

Source: Adapted with permission from *Data Analysis*, ©1988 by the National Council of Teachers of Mathematics. All rights reserved.

7. The Highway Department uses a formula to determine how fast a car in an accident was traveling based on the length of the skid. The following data set was collected under controlled conditions (surface and weather). Find a function that models the data in Figure 52.

Figure 52   Skid length and speed

Speed (mph)	20	25	30	35	40	45	50	55	60	65	70	80
Skid length (ft)	15	23	33	45	59	75	93	112	113	156	181	240

8. Researchers in anthropology are interested in the way urban growth affects lifestyles and stress levels. They measured the average walking speed of people living in various cities and collected the data set shown in Figure 53.

Figure 53   Walking speed and population

Population (thousands)	341.9	5.5	0.4	78.2	867.0	14.0
Walking speed (ft/sec)	4.8	3.3	2.8	3.9	5.2	3.7
Population (thousands)	23.7	70.7	304.5	138.0	2602.0	
Walking speed (ft/sec)	3.3	4.3	4.4	4.4	5.1	

Source: *UMAP Module 551*, COMAP, Inc., Lexington, MA, 1983

a. Find an equation to model the relationship between walking speed and population. (First, re-express the data and find the equation of the least squares line.)

b. According to your model, in what size city will the walking speed be zero?

c. What does your model imply about walking speed as population increases? Be specific. Is there an upper bound to walking speed?

d. What does your model imply about the relationship between urban growth and stress?

9. The biology student who found a model for predicting temperature by listening to crickets stirred much interest among her friends. One friend decided to repeat the experiment, but he gathered his data in a different way. He used a stopwatch to determine the time required to hear 50 chirps. Figure 54 shows the number of seconds required to count 50 cricket chirps at various temperatures.

Figure 54   Cricket chirps and temperature

Number of seconds to count 50 chirps	94	59	42	36	32	32
Temperature (°F)	45	52	55	58	60	62
Number of seconds to count 50 chirps	26	24	21	20	17	16
Temperature (°F)	65	68	70	75	82	83

a. Make a scatter plot of the data and re-express the data to produce linearity.

b. Fit a least squares line to the linearized data. Write an equation that will allow you to predict the temperature based on a count of cricket chirps.

c. Suppose you hear a cricket chirping, and you need 25 seconds to count 50 chirps. Use your model to predict the temperature.

10. Rheumatoid arthritis patients are treated with large quantities of aspirin. The concentration of aspirin in the bloodstream increases for a period of time after the drug is administered and then decreases in such a way that the amount of aspirin remaining is a function of the amount of time that has elapsed since peak concentration. Figure 55 gives data for a particular arthritis patient after taking a large dose of aspirin. Find a model for these data and use it to predict how much aspirin remains in the patient's bloodstream ten hours after the 15-mg reading.

Figure 55   **Drug concentration data**

Hours elapsed since peak concentration	0	1	2	3	4
Aspirin (mg) per 100 cc of blood	15	12.5	10.5	8.7	7.3
Hours elapsed since peak concentration	5	6	7	8	9
Aspirin (mg) per 100 cc of blood	6.1	5.1	4.3	3.6	3.0

sidereal year

Kepler's third law

11. Figure 56 provides data for a planet's distance from the sun and the time it takes to complete an orbit. A planet's mean distance from the sun is measured in millions of miles. The number of years required for a planet to complete its revolution around the sun is called a *sidereal year*. Determine a model that fits this data set. (The relationship between distance from the sun and sidereal year is called *Kepler's third law*. It was discovered empirically by Johann Kepler in 1618. He experimented, without the aid of data analysis or computer technology, until he found a mathematical statement to describe this relationship.)

Figure 56   **Planet data**

Planet	Distance	Years
Mercury	36.0	0.241
Venus	67.0	0.615
Earth	93.0	1.000
Mars	141.5	1.880
Jupiter	483.0	11.900
Saturn	886.0	29.500
Uranus	1782.0	84.000
Neptune	2793.0	165.000
Pluto	3670.0	248.000

12. You can use the graph of the function $y = \sqrt{x}$ provided in Figure 57 to estimate the area of the region that is under the graph, above the x-axis, and between the vertical lines $x = 0$ and $x = 2$. This area appears to be a little less than two square units, since it contains one complete block and two partial blocks.

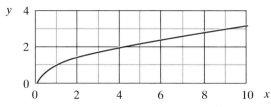

Figure 57 Graph of $y = \sqrt{x}$

a. Repeat this process for the regions between $x = 0$ and $x = t$ for integer values of t from 1 to 10. The data set you create should consist of ordered pairs of the form (t, A), where A represents the area under the graph between $x = 0$ and $x = t$ and above the x-axis.

b. Make a scatter plot of your ordered pairs. Use the scatter plot to make a guess about the relationship between t and A. Check your guess by re-expressing the data to make them linear. If you are not satisfied with your choice of re-expression, experiment with others until your data appear linear.

c. Find an equation of the least squares line through the linearized data. Solve your equation for the dependent variable A, and examine the model superimposed on the original data.

d. Techniques from calculus allow us to determine that the area under the curve $y = \sqrt{x}$ between $x = 0$ and $x = t$ is given exactly by the equation $A = \left(\frac{2}{3}\right)t^{\frac{3}{2}}$. How close to this equation is your empirical model?

e. How well does the theoretical model stated in Part d fit the data you gathered?

SECTION 3.14 Error Bounds in Re-expressed Data

In the previous section, we fit exponential and power models to nonlinear data by using semi-log and log-log re-expressions. We can use these models to make predictions for values of the dependent variable when given a value of the independent variable. We fit the exponential model $P(t) = 4.02e^{0.0147t}$ to the data representing the population per square mile and the year since 1790. If asked to make a prediction about the population per square mile for a given year, we would evaluate the function to produce our estimate. How accurate do you think our prediction would be? Do you think a predicted value for 1987 or 1995 would be off by more or less than a prediction for 1823?

Example 3 in the previous section considered the exponential decrease in temperature of hot water cooling in a refrigerator. The model we found for the relationship between time and temperature was $T(t) = 73.75e^{-0.01151t} + 47.5$. We used this model to estimate that the temperature at $t = 0$ was approximately 121°. Naturally, we do not expect that this was the exact value. We would not be surprised if the temperature was 119° or 123°, but we would be surprised to find that the initial temperature was 100° or 150°. We can extend our work with error bounds for linear data to create error bounds for nonlinear data. For the linearized data, we will use error bounds that are two standard deviations of the residuals either side of the linear model $y = mx + b$.

The linear model for the re-expressed cooling data is $y = -0.01151x + 4.3006$. The standard deviation of the residuals about this line is 0.04472. Thus, typical values of $\ln(T - 47.5)$ will fall between $-0.01151t + 4.2112$ and $-0.01151t + 4.3900$. These error bounds allow us to estimate error in nonlinear models in a way that takes into account the increased precision for exponential models as they approach their asymptote. The linear equations we have at this point are

$$\ln(T - 47.5) = -0.01151t + 4.3006 \pm 0.0894.$$

Re-expressing these equations in exponential form gives us

$$T - 47.5 = e^{-0.01151t + 4.3006 \pm 0.0894}.$$

Solving for T and simplifying, we find the two error bounds for temperature are

$$T_{\text{lower}}(t) = \left(e^{4.3006 - 0.0894}\right)e^{-0.01151t} + 47.5 = 67.44e^{-0.01151t} + 47.5$$

and

$$T_{\text{upper}}(t) = \left(e^{4.3006 + 0.0894}\right)e^{-0.01151t} + 47.5 = 80.65e^{-0.01151t} + 47.5.$$

Notice that these two error bounds both have the horizontal asymptote $T = 47.5$. They also have the same coefficient in the exponent. The functions differ in their leading coefficient. They are graphed in Figure 58 with the model

$$T(t) = 73.75e^{-0.01151t} + 47.5.$$

We still estimate that the initial temperature is $T(0) \approx 121$, but we can use the error bounds to quantify the uncertainty in this estimate. Based on these error bounds, we expect that the initial temperature was somewhere between $T_{\text{lower}}(0) \approx 115°$ and $T_{\text{upper}}(0) \approx 128°$. If we use our model to predict the temperature at 100 minutes, our guess would be $T(100) = 71°$. As before, we do not expect that this is exactly the temperature we would have obtained if we had measured the temperature at 100 minutes. We can, however, expect to find the temperature somewhere between $T_{\text{lower}}(100) \approx 69$ degrees and $T_{\text{upper}}(100) \approx 73°$. Notice that the error bounds suggest that we are more certain of the temperature as the time increases. After about five hours, both upper and lower error bounds give the same temperature rounded to the nearest degree.

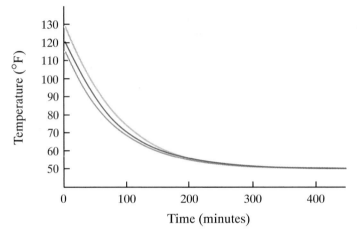

Figure 58   Exponential model and error bounds for cooling data

Error bounds can also be created for log-log re-expressions. If the re-expression $X = \ln x$ and $Y = \ln y$ linearizes the data, a least squares line of the form $Y = mX + b$ can be computed. Then we can determine the standard deviation of the residuals. If this value is s, then we would consider Y-values between $Y = mX + b - 2s$ and $Y = mX + b + 2s$ to be typical values for Y. Since $X = \ln x$ and $Y = \ln y$, this means that we would consider y-values between $y = e^{b-2s} x^m$ and $y = e^{b+2s} x^m$ as typical values for the dependent variable y. The derivation for the lower bound is shown below.

$$\ln y = m \ln(x) + (b - 2s)$$
$$e^{\ln y} = e^{m \ln(x) + (b - 2s)}$$
$$y = e^{m \ln(x)} \cdot e^{b-2s}$$
$$y = e^{\ln(x^m)} \cdot e^{b-2s}$$
$$y = e^{b-2s} \cdot x^m$$

In Example 2 on page 238, we found a model $V(A) = 0.0005864A^{2.4926}$ relating the age, A, in years, of hardwood trees and the volume, V, in hundreds of board feet, of wood contained in the tree. The data are repeated below.

Figure 59   **Data for tree age and volume**

Age (in years)	20	40	80	100	120	160	200
Volume (hundreds of board feet)	1	6	33	56	88	182	320

A log-log re-expression results in the linear model $Y = 2.4926X - 7.4415$, where $X = \ln A$ and $Y = \ln V$. The standard deviation of the residuals for this linear model is $s = 0.0214$. The re-expressed linear model has error bounds that are $Y = 2.4926X - 7.4843$ and $Y = 2.4926X - 7.3987$. The model, then, has error bounds of $V_{lower}(A) = 0.0005618A^{2.4926}$ and $V_{upper}(A) = 0.0006120A^{2.4926}$. The power model and error bounds are shown in Figure 60.

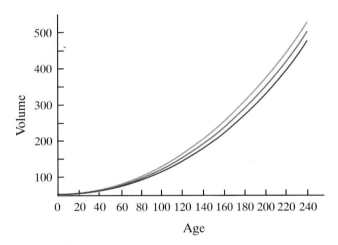

Figure 60   Power model and error bounds for tree growth

If we wanted to predict the volume of wood in a tree that is 150 years old, our prediction would be $V(150) \approx 156$, or 15,600 board feet, though we would expect the value would be somewhere between $V_{lower}(150) \approx 149$ and $V_{upper}(150) \approx 163$.

Exercise Set 3.14

1. Find error bounds for the population growth model in Example 1 in Section 3.13. Use the error bounds to give upper and lower bounds for the population in 1995 and 1823. Is the present population per square mile between the error bounds?

2. Saint Patrick's Cathedral in Dublin, Ireland, has 11 bells that were all hung before 1897. A plaque in the cathedral gives the diameter, in inches, and the weight, in pounds, of each of the 11 bells.

Figure 61 **Diameters and weights of cathedral bells**

Diameter (in)	29.5	31.5	34.0	35.0	36.5	38.0	42.0	47.0	49.5	55.0	62.0
Weight (lb)	801	925	1050	1116	1109	1253	1638	2122	2467	3339	5089

a. If a twelfth bell, 70 inches in diameter, were constructed similar to the others, what do you think would be the likely weight of that bell?

b. If a twelfth bell, 25 inches in diameter, were constructed similar to the others, what do you think would be the likely weight of that bell?

c. Which of the two "potential twelfth bells" in Parts a and b has the smaller range of possible weights? Explain your answer.

SECTION 3.15 Investigating More Data Collection

In Chapter 1 you collected some of your own sets of data that you analyzed with linear data analysis tools. Now that you are equipped with some nonlinear data analysis tools, such as semi-log and log-log re-expressions, you are ready to collect and thoroughly analyze more of your own data. A list of investigations, as well as a list of research topics, follows. Most of the investigations can be completed in a short amount of time with minimal equipment, but some will require a data-gathering device such as a CBL™. Ideally, you should work together in groups of two or three while doing the investigations. Since it is not practical for each student in a given class to perform all of the investigations in this section, it is suggested that you share the results of your investigations with the class. Investigating research topics for data should be done individually or in pairs.

As before, students taking measurements should agree on accuracy and measuring techniques. Measurement error involved with the data collection should be explained whenever possible. When you are asked to find a model, make use of all data analysis tools studied thus far and keep a detailed record of your work. Include an explanation for choosing a particular model and discuss the accuracy of that model, using the appropriate data analysis tools. You should also interpret the meaning of that model and different mathematical components of the model in the context of the problem.

INVESTIGATION 1   Rebound Height of a Bouncing Ball

As a child, you no doubt played with rubber balls. You most likely noticed that when you drop a rubber ball, it keeps bouncing, but the height to which the ball rebounds decreases with each bounce.

Question   What is the relationship between a ball's rebound height and the number of times the ball bounces?

Equipment   Data collection device with a motion detector; ring stand and utility clamp; rubber ball; hard, level surface (If a data collection device is not available, it is possible to get height estimates using a tape measure.)

Data Collection   Set up your data collection device and use the ring stand to hold the motion detector so that the sensor faces the hard, level surface. Begin running the appropriate program for your data collection device and hold the ball under the motion detector but not too close. When you are ready to start collecting data, signal the program to do so and then drop the ball. If the ball bounces outside the detector's range, check to be sure the hard surface being used is level and try again.

Analysis   Make a scatter plot of (*number of bounces, rebound height*). Describe the relationship. Is this the relationship you expected? Find a model for the data. Use the model to interpolate and/or extrapolate.

Other Questions to Consider Do all balls bounce the same way? You can try this investigation with different types of balls and make comparisons. ▨

INVESTIGATION 2 Dice Rolling

If you take 50 dice and roll them on the floor, how many would you expect to show a 2? If you remove all the 2s and roll the remaining dice, how many 2s do you expect to see on the second roll? What happens in the long run?

Question What is the relationship between the number of times you have rolled the dice and the number of dice remaining after removing the 2s?

Equipment Fifty dice, jar

Data Collection Take a jar containing 50 dice, shake the jar to mix the dice, and empty it onto the floor. Remove the dice that are showing 2s and record the number of dice remaining. Return the remaining dice to the jar, shake it well, empty the jar again, remove the dice that are showing 2s, and record the number of dice remaining. Continue this process until no dice remain.

Analysis Make a scatter plot of (*number of rolls, number of remaining dice*). Describe the relationship. Is this the relationship you expected? Find a model for the data. Use the model to interpolate and/or extrapolate. Compare your results to those obtained from Investigation 3: Pennies, in Chapter 1 on page 21. ▨

INVESTIGATION 3 Cooling

As soon as a cup of hot chocolate is poured, it begins to cool. The temperature drops over a period of time until it reaches room temperature. What is the nature of this cooling process?

Question What is the relationship between the temperature of a warm object and the amount of time the warm object has been cooling?

Equipment Data collection device with a temperature probe, thermometer, water, cup, microwave

Data Collection Set up your data collection device with the temperature probe. Heat water in the cup in the microwave to boiling. Use the thermometer to record the temperature of the water. Insert the temperature probe into the water for several seconds. Make sure your data collection device is on and ready to record data. Take the temperature probe out of the water and start the necessary program to begin collecting data. The temperature probe should remain exposed to the air while collecting data. You should collect about one data point per second for about one and a half minutes.

Analysis   Make a scatter plot of (*time, temperature*). Describe the relationship. Is this the relationship you expected? Find a model for the data. Use the model to interpolate and/or extrapolate.

Other Questions to Consider   What happens to a warm object that is put in ice? Try this investigation again by placing the warmed temperature probe in a can of ice water and compare the results.   ▨

INVESTIGATION 4   Warming

Everyone knows that you have to eat an ice cream cone quickly or it will melt all over you, but what is the nature of this warming process?

Question   What is the relationship between the temperature of a cool object and the amount of time the object has been warming?

Equipment   Data collection device with a temperature probe, thermometer, water, ice, can

Data Collection   Set up your data collection device with the temperature probe. Fill approximately two-thirds of the can with ice and then put in just enough cold water so that the ice can be stirred. Use the thermometer to record the temperature of the water. Insert the temperature probe into the water for several seconds. Make sure your data collection device is on and ready to record data. Take the temperature probe out of the water and start the necessary program to begin collecting data. The temperature probe should remain exposed to the air while collecting data. You should collect about one data point per second for about one and a half minutes.

Analysis   Make a scatter plot of (*time, temperature*). Describe the relationship. Is this the relationship you expected? Find a model for the data. Use the model to interpolate and/or extrapolate. Compare this model to the model found in Investigation 3: Cooling.

Other Questions to Consider   What happens to a cool object that is heated? Try this investigation again by heating the cooled temperature probe with a hair dryer and compare the results.   ▨

INVESTIGATION 5   Pressure

Pressure is the force exerted by an opposing body and is expressed in units of force per unit area. If you have ever played around with a bicycle pump, you've experienced the effects of pressure. You know that if you close the valve at the end that attaches to a tire and try to pump, it is very difficult, if not impossible, to push the pump all the way down.

Question What is the relationship between the pressure and volume of a gas held at a constant temperature?

Equipment Data collection device with a pressure sensor, plastic syringe

Data Collection Set up the data collection device with the pressure sensor and plastic syringe. Set the syringe to the desired setting and close the release valve. Run the appropriate program for the data collection device and collect 8 to 10 data points by decreasing the volume on the syringe by 2 or 3 cubic centimeters at a time and allowing the data collection device to record the corresponding pressure each time.

Analysis Make a scatter plot of (*volume, pressure*). Describe the relationship. Is this the relationship you expected? Find a model for the data. Use the model to interpolate and/or extrapolate. What is the trend as you decrease the volume on the syringe? ▨

INVESTIGATION 6 Light Intensity

Light intensity is the amount of force or energy of light per unit area. We know this as brightness and understand that the closer we are to a light source, the brighter it will be around us.

Question What is the relationship between the intensity of light and the distance from the source?

Equipment Data collection device with a light intensity sensor, light source, wooden block, tape, tape measures or metersticks

Data Collection Position the light source and sensor so that they are at the same level. Darken the room as much as possible. Set up the data collection device with the sensor and run the appropriate program. Collect 15 data points by moving the sensor farther away from the light source for each reading. Be sure that the shortest distance (in cm) between the sensor and light source exceeds the wattage of the bulb. For example, if using a 40-watt bulb, start with the sensor 50 cm from the light source.

Analysis Make a scatter plot of (*distance, intensity*). Describe the relationship. Is this the relationship you expected? Find a model for the data. Use the model to interpolate and/or extrapolate.

Other Questions to Consider What is the effect of using a different light source? Repeat this investigation using a different type of light source and compare the results. List several physical characteristics of a light source that you think might have an effect on the intensity of that light source at a fixed distance from the source. ▨

INVESTIGATION 7   Water Jug

The first person to get water out of a full water jug is able to fill his or her cup much faster than the last person to get water out of the jug. This makes sense because there is more pressure exerted on the water near the spigot when the jug is full.

Question   What is the relationship between the number of the cup being filled and the amount of time it takes to fill that cup?

Equipment   Water jug or similar liquid container with spigot near the bottom, cup, stopwatch

Data Collection   Designate one person to open and close the spigot and one person to run the stopwatch. Measure the time needed to fill the cup with water from the jug. Data should be recorded until no water comes out. Do not tilt the jug to get more water out.

Analysis   Make a scatter plot of (*cup number, time to fill that cup*). Describe the relationship. Is the relationship what you expected? Find a model for the data. Use the model to interpolate and/or extrapolate. What influence does gravity have on your results?

Other Questions to Consider   Does the shape of the container affect the investigation? Try using different shaped containers, repeat the investigation, and compare the results.

Research Topics

You will need access to a library or the Internet to pursue the following research topics. All the data that can be collected will have year values as the independent variable. Recall that it is a good idea to let the independent variable represent the number of years since the first year. For example, if the first few year values in a particular data set are 1915, 1917, and 1920, then you should let your independent variable represent the number of years after 1915. The variable values in this case would be 0, 2, and 5.

RESEARCH TOPIC 1   Populations

Example 1 on page 235 shows that population growth can often be modeled with an exponential function. In this example, we found an exponential model for the U.S. population per square mile since 1790.

Question   Are the populations of any other countries or continents growing exponentially?

Research   Find as much recent population data as you can. Start by looking for countries with large populations, such as Japan.

Analysis   Make a scatter plot of (*number of years after first year in data set, population*). Carefully examine each data set and describe the relationships. Find appropriate models for each set of data and explain your choices. Answer the question above. ▇

RESEARCH TOPIC 2   Medical Costs

According to the Consumer Price Index (CPI), it cost $2.34 in June 1997 to purchase the same type of medical care that cost $1.00 to purchase in 1982. The overall CPI for June 1997 is 160, which means that it cost $1.60 to purchase an item that cost only $1 in 1982. This means that medical costs have risen more than twice as much as the average commodity has risen.

Source: U.S. Department of Labor. Table 1. Consumer Price Index for All Urban Consumers (CPI-U): U.S. City Average, by expenditure category and commodity and service group. http://stats.bls.gov/news. release/cpi.t01.htm

Question   How are medical care costs growing, in general?

Research   Find as much recent data on medical care cost as you can.

Analysis   Make a scatter plot of (*number of years after first year in data set, medical care cost*). Carefully examine each data set and describe the relationships. Find appropriate models for each set of data and explain your choices. Answer the question above. ▇

RESEARCH TOPIC 3   College Costs

Tuition and fees for attending colleges across the United States increased approximately 5% from 1995–96 to 1996–97.

Source: College Board Online. http://www.collegeboard.org/press/html/1tables.html#anchor5243995

Question   How has the cost of attending college been changing across the country?

Research   Find as much recent college cost data as you can. Start by looking for data on four-year institutions.

Analysis   Make a scatter plot of (*number of years after first year in data set, cost of particular year*). Carefully examine each data set and describe the relationships. Find appropriate models for each set of data and explain your choices. Answer the question above. ▇

RESEARCH TOPIC 4   AIDS Cases

"By 15 December 1995, 1,291,810 cumulative cases of AIDS had been reported to the World Health Organization (WHO)—an increase of more than 25% over the 1,025,073 cases reported by the end of 1994."

Source: HIV/AIDS: Figures and Trends, Mid-1996 Estimates, http://.hiv.unaids.org/unaids/press/figures.html

Question   How has the number of AIDS cases in the United States been growing since 1980?

Research   Try to find estimates for the number of AIDS cases in the United States for every year beginning with 1980.

Analysis   Make a scatter plot of (*number of years after first year in data set, number of AIDS cases in the United States*). Carefully examine the data set and describe the relationship. Find an appropriate model for the set of data and explain your choice. Answer the question above.

Investigation: Assessing Your Model

Suppose you are given the data in Figure 62, which relates the focal length setting on a camera to the length (in centimeters) of a meterstick visible through the lens. As the camera "zooms in," a smaller and smaller portion of the meterstick is visible. If you do not know a theoretical relationship between these variables, you must rely on residual plots to help you find a good model for these data.

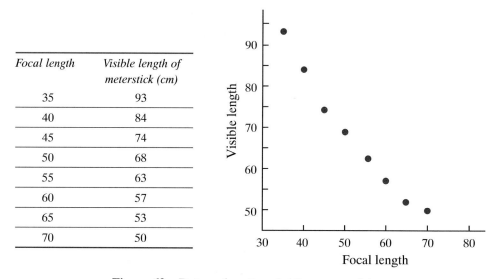

Focal length	Visible length of meterstick (cm)
35	93
40	84
45	74
50	68
55	63
60	57
65	53
70	50

Figure 62   Data and scatter plot for camera data

In this investigation, you will fit four different models to this data and use the residual plots to assess the quality of the fit.

1.   Fit each of the four models described to the data.

   a. Quadratic Model: Use your calculator to perform a quadratic fit.

   b. Reciprocal Model 1: Re-express the focal length as $\frac{1}{x}$ and fit a linear model to the re-expressed data $\left(\frac{1}{x}, y\right)$. Rewrite this model as a reciprocal function of the form $R_x(x) = \frac{A}{x} + B$.

   c. Reciprocal Model 2: Re-express the length of the meterstick visible as $\frac{1}{y}$ and fit a linear model to the re-expressed data $\left(x, \frac{1}{y}\right)$. Rewrite this model as a reciprocal function of the form

$$R_y(x) = \frac{1}{Ax + B} = \frac{\frac{1}{A}}{x + \frac{B}{A}}.$$

d. Power Model: Perform a log-log re-expression and find a power model for the relationship.

Which model is the best? In some situations, you will have information from the context of the problem that suggests a particular model. For example, population growth is generally modeled with exponential functions, and objects falling under the force of gravity are modeled with quadratic models. If you know something about the phenomenon, there is no need to create competing models. You should fit the model that you know describes the phenomenon. If you don't understand the phenomenon and are trying to get some information about what it might be, you need to assess the different models under consideration.

2. The first step in determining whether a model is reasonable is to look at the residual plots and the graphs plotted against the data. Compare the four residual plots for the four different models. Is one significantly better than another?

Data analysis is never a perfect process. All of the residual plots should be strikingly similar. While you can never truly prove one model is correct, you can gather evidence to support a particular model over other models. One way to do this is to create the model using only part of the data. Then use the model created with this portion of the data to predict the remaining values in the data set. In this example, you will use the first five data values to predict the last three data values.

Fit a model to the first five data values, and graph the resulting function against the whole data set to see how well the model captures the flow of the data. (Since you are not using all of the data, none of the fits will be as good as those seen earlier.)

Figure 63 Partial data for focal length

Focal length	35	40	45	50	55
Visible length of meterstick (cm)	93	84	74	68	63

3. Fit each of the four models in Problem 1 using only the first five data points. Once you have found each model, superimpose it on the entire data set and look at the residuals for the entire data set. Based on this analysis, what kind of function do you think best describes the relationship between focal length and the length of the meterstick visible?

Chapter 3 Review Exercises

1. Suppose you plan to save $10,000 to buy a car. Each month you will make $150 deposits into an account that pays 6% annual interest compounded monthly.

    a. Write a recursive system that describes this scenario.

    b. How long will it take to save $10,000?

    c. Write an explicit expression for this scenario.

2. What is the effective annual yield of an account that earns 6% annual interest compounded monthly? compounded continuously?

3. At what annual interest rate compounded daily would you have to invest your money if you wanted to double the initial deposit in nine years? (You make no additional deposits.)

4. Sketch graphs of the following functions. Label important features.

    a. $f(x) = e^{x-3}$

    b. $f(x) = 3^{2x} + 1$

    c. $y = \ln(x^2 - 4)$

    d. $y = \log_3(x + 2)$

    e. $y = \log(2x)$

    f. $y = \dfrac{1}{e^x - 1}$

5. Evaluate:

    a. $\log_{1/2} 8$

    b. $\ln e^{17}$

    c. $10^{\log x^2}$

6. Solve the following equations:

    a. $4^x = 12$

    b. $\log x + \log(x - 3) = 1$

    c. $6000 = 4000 e^{0.06t}$

    d. $\ln(\ln x) = 2$

7. If $\log_b x = 2$, what is $\log_b \frac{1}{x}$?

8. Find the inverse of the following functions. State the domain of each inverse function.

    a. $f(x) = e^{2x+1} - 2$

    b. $f(x) = 2\log(x - 4)$

9. The price of concert tickets was $22.50 last year, and the price is $25 this year. If the growth continues at the same exponential rate in the future, how long will it take for the price of the tickets to double? Give the answer to the nearest tenth of a year.

10. Suppose you record the temperature and corresponding time of a cup of tea that has been left outside on a summer day. You collect the data in Figure 64.

Figure 64

Time (minutes)	0	3	6	9	12	15	18	21	24	27	30	33
Temperature (°C)	74.9	69.2	65.1	60.8	57.4	54.8	51.4	49.2	46.3	45.0	43.3	41.6

    a. Find a model for the relationship between the elapsed time and the temperature of the tea. Assume that the outside temperature was 30°C.

    b. How long does it take for the temperature of the tea to fall to 35°C?

11. A person's typing speed is modeled by the function $W = 80(1 - e^{-0.08t})$, where W is the number of words per minute this person can type after t weeks of practice.

    a. Approximately how many weeks did it take to increase this person's typing speed to 60 words per minute?

    b. Is there an upper limit to this person's typing speed? If so, what is it?

4 Modeling

4.1	The Tape Erasure Problem	268
4.2	Radioactive Chains	273
4.3	Free Throw Percentages	275
4.4	Choosing the Best Product	279
4.5	The Tape Counter Problem	288
4.6	Developing a Mathematical Model	295
4.7	Some Problems to Model	297

Linear Irrigation

Suppose your uncle is designing an irrigation system for his farm in eastern North Carolina. His land is very flat, and the fields, which are rectangular in shape, are roughly 2200 feet long and 1000 feet wide. Your uncle is considering a linear irrigation system, which is essentially one long pipe to water each field. A linear irrigation system is generally set on wheels that keep it above the level of the plants. Nozzles are placed periodically along the pipe, and each nozzle sprays water in a circular region. The entire system moves slowly down the field, watering the plants beneath it as it moves.

There are 20 sprinkler nozzles available to your uncle. He has enough pumps to maintain water pressure so that each nozzle delivers a uniform spray to a circular region 50 feet in radius with a flow rate of 10 gallons per minute. How far apart should the nozzles be placed to produce the most uniform distribution of water on a field 1000 feet wide?

Sources: Adapted with permission from *Mathematical Modeling in the Secondary School Curriculum: A Resource Guide of Classroom Exercises* by the National Council of Teachers of Mathematics. All rights reserved.
"The Irrigation Problem," *Everybody's Problems,* **Consortium**, Number 63, Fall, 1997, COMAP, Inc., Lexington, MA.

The Tape Erasure Problem

There are many problems in the world around us that are solved with concepts and skills acquired through mathematical training. The solutions to some problems are direct and exact, but more often we are required to think more deeply and develop mathematical models in our search for solutions. Mathematical modeling is a process that requires creative thinking, experimentation, setting priorities, and making choices. We present several examples to illustrate methods for developing mathematical models to solve a variety of problems.

A tape recording is made of a meeting between two managers. Their conversation starts at the twenty-first minute on the tape, and it lasts for 8 minutes. The tape records for 60 minutes. While playing back the tape, one of the managers accidentally erases 15 consecutive minutes of the tape, but does not know which 15 minutes were erased.

Questions for the Tape Erasure Problem

1. What is the probability that the manager erased the entire conversation?

2. What is the probability that all or some part of the conversation was erased?

3. Suppose the exact position of the conversation on the tape is not known, except that it began sometime after the twenty-first minute. What is the probability that the entire conversation was erased?

You may have solved probability problems in previous mathematics courses by computing the ratio of the number of successful outcomes to the number of all possible outcomes. This approach is limited to those situations in which there are a finite number of outcomes. The questions above require a different approach since there are infinitely many locations for the erasure and, therefore, an infinite number of possible outcomes. We will use geometric models to find these probabilities.

To answer Question 1 above, we need to analyze the starting time of the 15-minute erasure. We can get acquainted with the problem by looking at a few specific starting times. Suppose the erasure begins 10 minutes from the beginning of the tape. Then the portion of the tape between the tenth and twenty-fifth minutes is erased; this section includes a portion of the conversation but not the entire conversation. What if the erasure begins at the fifteenth minute or the twenty-fifth minute? If the erasure begins at the fifteenth minute, it will end at the thirtieth minute and thus erase the entire conversation. If the erasure begins at the twenty-fifth minute, it will erase only a portion of the conversation.

sample space

If we let x be the number of minutes from the beginning of the tape to the start of the erasure, then x can be any real number from 0 to 45. Since x represents every possible way that the tape erasure can occur, we can analyze this problem by looking at x-values. Using a portion of a number line to represent these values, we can see that the set of all possible outcomes is an interval 45 units in length. This interval is called the *sample space*. An interval on the number line is a simple geometric model of this problem; every possible starting time for the erasure is represented by a point in this interval.

event space

The conversation lasts from the twenty-first minute to the twenty-ninth minute, so the erasure can start as early as the fourteenth minute and as late as the twenty-first minute and erase the entire conversation. Thus, the value of x must be between 14 and 21 to be in the region we call the *event space*. The event space is the part of the sample space that consists of all successful outcomes for the problem. An illustration of both the sample space and the event space is shown in Figure 1. In this case, a successful outcome is the erasure of the entire conversation. The *probability of a success* is the ratio of the length of the event space to the length of the sample space. The event space is 7 units long, and the sample space is 45 units long, so the probability that the entire conversation was erased is $\frac{7}{45}$, or about 0.16.

probability of a success

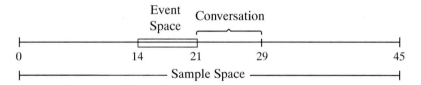

Figure 1   Model for Question 1 of the tape erasure problem

To answer Question 2, again let x represent the number of minutes from the beginning of the tape to the start of the erasure. The sample space is the same as it was in Question 1, but the event space is larger. The event space now includes values of x that result in any portion of the conversation being erased. You should find that a portion of the conversation will be erased if x is between 6 minutes and 29 minutes. So the event space is 23 units long, and the probability that at least some part of the conversation was erased is $\frac{23}{45}$, or about 0.51.

For Question 3, we need to consider both the starting time of the erasure and the starting time of the conversation. Let x represent the number of minutes until the start of the erasure, and let y be the number of minutes until the start of the conversation. Since the erasure is 15 minutes long, x can vary from 0 to 45. Since the conversation starts after the twenty-first minute and is 8 minutes long, y can vary from 21 to 52. The sample space consists of all ordered pairs (x, y) in the rectangle shown on the next page in Figure 2. Our sample space is now two-dimensional with an area of 1395 square units.

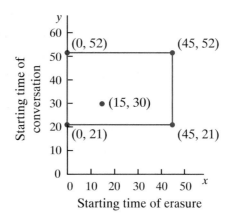

Figure 2 Sample space for Question 3 of the tape erasure problem

Which points in the sample space are in the event space?

Which points in the sample space are in the event space? One way to investigate the event space is by testing individual points. For example, the point (15, 30) corresponds to the erasure starting at the fifteenth minute and the conversation starting at the thirtieth minute. The erasure will be over before the conversation begins, so this point is not in the event space. By similar reasoning, you should be able to decide that (17, 23) and (20, 27) are in the event space, whereas (30, 25) is not. In general, the beginning time of the conversation must be later than the beginning time of the erasure. This is written mathematically as $y \geq x$. Also, the conversation must start when less than 7 minutes of the erasure have gone by, so there will still be 8 minutes of erasure time left to erase the 8-minute conversation. Therefore, we have $y \leq x + 7$. This means that a point is in the event space if its coordinates satisfy the inequalities $x \leq y \leq x + 7$, which is the region shaded in Figure 3. Its area is 192.5 square units. The probability that a point in the sample space is also in the event space is the ratio of the area of the event space to the area of the sample space, or

$$\frac{192.5}{1395} \approx 0.14.$$

This means that the probability that the entire conversation was erased is about 0.14 when we know only that the conversation started after the twenty-first minute.

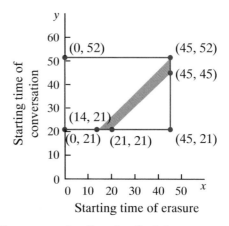

Figure 3 Event space for Question 3 of the tape erasure problem

A significant aspect of the tape erasure problem is the geometric model used to represent it. This modeling process can be applied to many probability problems where counting is not possible. Once the model is created and the sample and event spaces are identified, probabilities are calculated as the ratio of two geometric lengths or areas.

geometric probability

The following exercise set consists of more problems that can be solved using techniques of *geometric probability*. The geometric aspects of some problems are more obvious than others. The two keys to each problem are the identification of which events are occurring randomly and the establishment of a model that represents these events by points.

Exercise Set 4.1

1. The Inner Circle Sandwich Shop has a square dartboard measuring 18 inches by 18 inches that hangs on a wall. Customers can win free sandwiches by throwing darts at the board. The dartboard contains three concentric circles with their centers at the center of the square. The radii of the circles are 2 inches, 4 inches, and 6 inches, respectively. The prize for throwing a dart into the inner circle (the bull's eye) is a free large sandwich. Customers win a medium sandwich if their dart lands in the circle of radius 4 inches (excluding the bull's eye). The prize for throwing a dart inside the circle of radius 6 inches (excluding the two inner circles) is a free small sandwich. Customers win nothing if the dart lands outside the largest circle.

    a. If you want to find the probability of winning a sandwich, you need to make two assumptions. What are those assumptions?

    b. Find the probability that a customer will win a medium sandwich.

    c. The management has decided that too many people are winning sandwiches, so they want to reduce the probability of winning. What would the radius of the largest of the three circles have to be so that the probability of *not* winning a sandwich would be 0.95?

2. A surveillance company has detection devices that illuminate a region shaped like a quarter-circle with a radius of 100 yards. These devices will be distributed evenly along a border pass that is 1000 yards wide. (See Figure 4.)

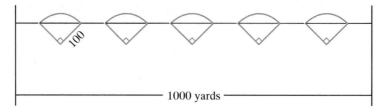

Figure 4   Surveillance problem

a.   Suppose five of the devices are used to survey the border. If an intruder enters the pass at a random point, what is the probability of detection?

b.   How many devices are needed so that the probability of detecting an intruder is greater than 0.9?

3.   Max and Yvonne want to meet at the ice cream shop to enjoy their favorite after-dinner treat. Both agree to arrive sometime between 8:00 and 8:30. They also agree that the first person to arrive will buy two cones and then wait for the other person. If the second person has not arrived within 12 minutes, the first person will start to eat that person's cone and will then continue to wait for the friend. If they arrive at the same time, each will buy his or her own cone. Find the probability that each person eats only one ice-cream cone.

SECTION 4.2 | Radioactive Chains

For some radioactive materials, one radioactive material decays into a second radioactive material, which then decays into a stable form. This is known as a radioactive chain with a length of two, since there are two radioactive materials in the chain. There are naturally occurring radioactive chains of lengths 13, 17, and 19. Suppose that each atom of radioactive material A decays into an atom of radioactive material B, and B then decays into a stable material. In a given time period, 15% of A and 50% of B decay. If you begin with 1000 atoms of material A, what is the relationship between the amounts of material A and material B present in the long run?

This problem can be modeled with iterative equations. Since material A decays at a rate of 15% per time period, the amount of A present after n time periods can be represented by the equations

$$A_0 = 1000, \quad A_n = 0.85A_{n-1}.$$

Representing the amount of material B is a little more complicated. Material B decays at a rate of 50% per time period. However, each atom of A becomes an atom of B, so the 15% decrease in A becomes an increase for B. Putting these two ideas together generates the iterative equations

$$B_0 = 0, \quad B_n = 0.5B_{n-1} + 0.15A_{n-1}.$$

We can use a graphing calculator to generate and compare the values of A_n and B_n over time. (See Figure 5.)

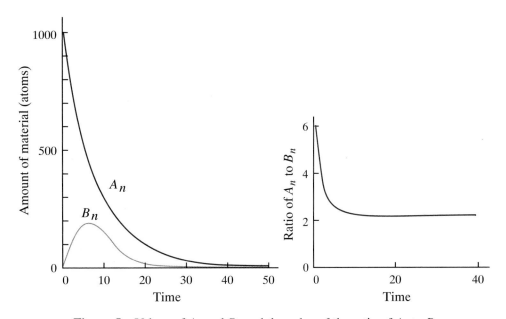

Figure 5 Values of A_n and B_n and the value of the ratio of A_n to B_n

By iterating the equations and comparing the values of A_n to B_n, we can see that after only a few time intervals, the ratio of $\frac{A_n}{B_n}$ stabilizes to approximately 2.33. In the long run, there will be a little over twice as much of material A as there will be of material B.

Exercise Set 4.2

1. Suppose that each year 5% of material A decays into material B, and 25% of B decays into a stable material. What is the long-run ratio of A to B? Does it depend on the initial amounts of A and B?

2. Suppose you have three radioactive materials X, Y, and Z. Material X decays into material Y, and material Y decays into material Z. Each hour, 2% of X decays into Y, 10% of Y decays into Z, and 5% of Z decays into lead.

    a. Determine the long-run ratios of Z to X and Z to Y if initially there are 100 grams of X, no Y, and no Z.

    b. Determine the long-run ratios of Z to X and Z to Y if initially there are 5, 25, and 10 grams of X, Y, and Z, respectively.

3. The life of the American bison can be broken down into three stages: calves (0–1 year old), yearlings (1–2 years old), and adults. The survival rate for each age group is the following: calves, 0.6; yearlings, 0.75; and adults, 0.95. Only adults reproduce, with a reproduction rate of 0.42. Only females are counted in these rates. If a herd is started with 200 adults (100 females and 100 males), what will be the long-term distribution among the three age groups?

    Source: Sandefur, James T. 1990. *Discrete Dynamical Systems: Theory and Applications,* page 282, by permission of Oxford University Press.

SECTION 4.3 Free Throw Percentages

Imagine that you are watching a televised game between the Chicago Bulls and the Boston Celtics. The Bulls' star player drives to the basket and is fouled. As he stands at the free throw line, the announcer states that he is hitting 78% of his free throws this year. He misses the first shot but makes the second. Later in the game, the player is fouled for the second time. As he moves to the free throw line, the announcer states that he has made 76% of his free throws so far this year. Can you determine how many free throws this player has attempted and how many he has made this year?

This appears to be a straightforward question. If we let s represent the number of shots successfully made and a the number of attempts, then we can represent the problem analytically using the system of equations

$$\frac{s}{a} = \frac{78}{100}$$

and

$$\frac{s+1}{a+2} = \frac{76}{100}.$$

Now we can solve for s and a. Since $100s = 78a$ and $100s + 100 = 76a + 152$, $s = 20.28$ and $a = 26$.

Since s and a represent numbers of shots, only integer values are meaningful. Do we round 20.28 down to 20 shots out of 26 attempts? Hitting 20 of 26 shots gives the player a percentage of 77, not the reported 78. Hitting 21 out of 26 gives a percentage of 81. Did the announcer make a mistake? If not, what is going on?

The key is the expression 78%. When writing the system of equations, we made the assumption that the percentages reported were exact. If the announcer followed general rounding practice, the value 78% can actually represent any percentage that is at least 77.5% and less than 78.5%. Similarly, a reported percentage of 76% means the actual percentage is at least 75.5% and less than 76.5%. With the assumption that the announcer reported rounded percentages, we really have a system of inequalities to solve, rather than a system of equations. Our analytic representation of the problem should be

$$\frac{775}{1000} \le \frac{s}{a} < \frac{785}{1000} \tag{1}$$

and

$$\frac{755}{1000} \le \frac{s+1}{a+2} < \frac{765}{1000}. \tag{2}$$

We are searching for a solution (or solutions) where s and a are both positive integers. We can restate the compound inequalities as follows:

$$\frac{775}{1000} \le \frac{s}{a} \quad \text{and} \quad \frac{s}{a} < \frac{785}{1000}$$

and

$$\frac{755}{1000} \le \frac{s+1}{a+2} \quad \text{and} \quad \frac{s+1}{a+2} < \frac{765}{1000}.$$

Since both s and a are positive, we can multiply across these inequalities and isolate s in each inequality to obtain

$$s \geq 0.775a \text{ and } s < 0.785a$$

and

$$s \geq 0.755(a + 2) - 1 \text{ and } s < 0.765(a + 2) - 1,$$

which simplifies to

$$s \geq 0.755a + 0.51 \text{ and } s < 0.765a + 0.53. \tag{3}$$

lattice points

We wish to find the *lattice points* (points whose coordinates are integers) in the region bounded by the four lines described by the inequalities above.

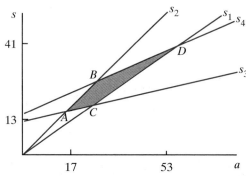

Figure 6 Region containing the solution

The graph in Figure 6 shows the general properties of the region we are interested in but is not a precise drawing. The intersection points of the lines are A: (17, 13.345), B: (26.5, 20.8025), C: (25.5, 19.7625), and D: (53, 41.075).

CLASS PRACTICE   **1.**   Verify the coordinates of points A, B, C, and D.

How can you find the lattice points in a region bounded by a quadrilateral?

How can you find the lattice points in a region bounded by a quadrilateral? To find the lattice points, you might examine a calculator-generated graph of the region bounded by the four lines. If you try this approach, you will see why it is not used. Identifying integer coordinates in the region is not possible from a calculator-generated graph because the slopes of the lines are almost identical and there is very little difference in the intercepts.

One efficient way to search for lattice points analytically is to examine the graph in Figure 6. Between points A and D, integer a-values are in the interval $17 \leq a \leq 53$. We are looking for integer s-values that satisfy both inequality (1) and inequality (2). We can accomplish this by using the equations

$$s_1 = 0.775a, \tag{4}$$

$$s_2 = 0.785a, \tag{5}$$

$$s_3 = 0.755a + 0.51, \tag{6}$$

and

$$s_4 = 0.765a + 0.53 \tag{7}$$

to determine s-values for a given integer a-value. A lattice point exists if there is an integer between the s-values produced by equations (4) and (5) and there is the same integer value between the s-values produced by equations (6) and (7). For example, substituting $a = 18$ into equations (4) and (5) yields values of $s_1 = 13.95$ and $s_2 = 14.13$. There is an integer value between 13.95 and 14.13, namely 14. Is the ordered pair (18, 14) also in the region bounded by inequality (3)? Substituting $a = 18$ into equations (6) and (7) yields values of $s_3 = 14.1$ and $s_4 = 14.3$. There is no integer between 14.1 and 14.3; therefore the ordered pair (18, 14) is not a lattice point in the region. We can generate a table of values for the four equations and search for integer values that satisfy both sets of related inequalities. A section of the table is shown in Figure 7.

Figure 7   Table of values showing one lattice point

a	s_1	s_2	s_3	s_4
20	15.5	15.7	15.61	15.83
21	16.275	16.485	16.365	16.595
22	17.05	17.27	17.12	17.36
23	17.825	18.055	17.875	18.125
24	18.6	18.84	18.63	18.89
25	19.375	19.625	19.385	19.655
26	20.15	20.41	20.14	20.42

Figure 7 shows that (23, 18) is a lattice point that satisfies both inequality (1) and inequality (2). We continue this process with appropriate a-values and boundary lines to identify the remaining lattice points. There are six lattice points in the region defined by our inequalities:

$$(23, 18), (27, 21), (32, 25), (36, 28), (40, 31), (49, 38).$$

Therefore, there is not just one answer to our original question, "How many times has the basketball player been to the free throw line?"

This example illustrates the important role assumptions play in the mathematical modeling process. We began this problem by assuming that the percentages given were exact. When this proved to be an incorrect assumption we had to ask ourselves questions and reinterpret the given information. We then modified the model to reflect this new assumption, which meant rewriting a system of equations as a system of inequalities. We also had to modify our strategy for finding the lattice points.

Initially, we thought we could look at a graph, but the slopes and intercepts of our boundary lines varied only slightly so that approach was not helpful. We considered tables of values instead. The abilities to persevere and to consider alternative strategies are important skills for the mathematical modeler.

Exercise Set 4.3

1. A basketball player is making 50% of her shots before a game, and after making 8 of 14 shots she is now hitting 55% of her shots. What is the fewest number of shots she could have taken so far this season?

2. Suppose that a baseball player is batting 0.299. The next time he is at bat, he gets a hit. Is it possible for his batting average now to be 0.306? Explain.

SECTION 4.4　Choosing the Best Product

People are often faced with the decision of which product to buy. Naturally, they like to choose the product that comes closest to meeting their needs. If one product satisfies all their requirements, then there is little to think about. However, it is often the case that no single product satisfies all their requirements, so they are faced with the difficulty of determining the product that comes closest to doing so. The following problem illustrates this situation.

A school board has decided that every mathematics classroom will have a computer for demonstrations. The principal has asked your class to help determine which type of computer to buy. The class investigates the different computers and finds a consumer magazine that rates the different computers from which each teacher can pick. The magazine rates the computers from 0 to 10 on performance and affordability. A computer with a score of $(0, 0)$ performs very poorly and is unaffordable, while one with a score of $(10, 10)$ is the perfect computer. The computer ratings are shown in Figure 8.

Figure 8　Rating of computers

Computer	Performance	Affordability
A	6.4	8.5
B	7.3	7.5
C	9.3	3.8
D	5.0	8.2
E	8.8	6.0
F	7.3	6.0
G	3.2	5.1
H	4.4	8.0
I	6.0	3.6
J	5.5	9.7

Which computer comes closest to being the perfect computer? To make sense of this data set, look at the scatter plot in Figure 9.

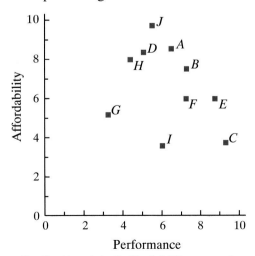

Figure 9　Scatter plot of affordability vs. performance

From the scatter plot, it is clear that a number of computers can be eliminated. The farther to the right a point is located, the better the performance of the computer it represents. If a point is high on the plot, then the computer it represents is more affordable than some other computers. Thus, we want computers that are far to the right and also near the top of the scatter plot. If there were a computer farthest to the right and above all the others, it would be the clear choice. However, high performing computers are generally more expensive and, therefore, less affordable.

The scatter plot in Figure 9 shows that Computer C has the best performance, while Computer J is the most affordable. Computers A, B, and E all have good performance and are reasonably affordable. We seek a method for choosing which computer to purchase.

Source: "The Computer Problem," *Everybody's Problems,* **Consortium**, Number 56, Winter, 1995, COMAP, Inc., Lexington, MA.

Greatest Sum Method

One method of rating involves the greatest total of the performance and affordability measures. That is, Computer A, with a score of 6.4 for performance and 8.5 for affordability, has a total score of 14.9. Computer ratings based on total scores are shown in Figure 10.

Figure 10   **Ratings of computers using the greatest sum method**

Computer	Performances (P)	Affordability (A)	Sum (P + A)
A	6.4	8.5	14.9
B	7.3	7.5	14.8
C	9.3	3.8	13.1
D	5.0	8.2	13.2
E	8.8	6.0	14.8
F	7.3	6.0	13.3
G	3.2	5.1	8.3
H	4.4	8.0	12.4
I	6.0	3.6	9.6
J	5.5	9.7	15.2

With this criterion, we see that Computers J and A are rated first and second, while Computers B and E are tied for third.

Greatest Product Method

Another way to compare the data is to consider the product of the performance and affordability measures. The greater the product, the better the computer. These results are shown in Figure 11.

Figure 11 Ratings of computers using the greatest product method

Computer	Performance (P)	Affordability (A)	Product (PA)
A	6.4	8.5	54.40
B	7.3	7.5	54.75
C	9.3	3.8	35.34
D	5.0	8.2	41.00
E	8.8	6.0	52.80
F	7.3	6.0	43.80
G	3.2	5.1	16.32
H	4.4	8.0	35.20
I	6.0	3.6	21.60
J	5.5	9.7	53.35

Using this criterion, we find that Computers *B*, *A*, and *J* are rated first, second, and third, respectively.

Comparing the Sum and Product Methods

Why would you choose to rate using a product rather than a sum, or vice versa? One way to think about this question is to use the geometric properties of the problem in conjunction with the analytic results. Which computer in Figure 12 is better, the one represented by the point (4, 8) or the one represented by the point (6, 6)?

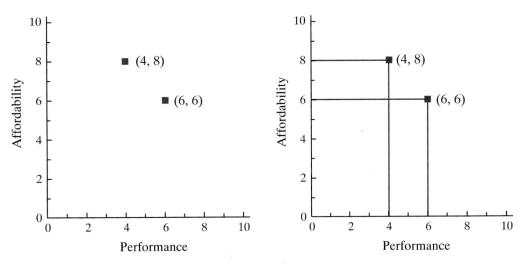

Figure 12   Comparing two computers

Imagine two rectangles, each with one corner at the origin and another corner diagonally opposite the origin at the indicated point. Which rectangle is larger? You may think that the larger rectangle is the one with the larger perimeter. This is equivalent to the greatest sum method since the sum is the semiperimeter of the rectangle. In the example shown in Figure 12, both rectangles have the same semiperimeter, and so we would consider the rectangles to be equally large and the computers they represent equally good.

However, you may think that the larger rectangle is the one with the larger area. This is equivalent to the greatest product model. The rectangle with coordinates (6, 6) would be larger since it has the larger area. Consequently, the computer represented by the point (6, 6) would be considered the better computer.

There is another geometric interpretation of the sum and product measures of "goodness." First, consider the greatest sum as the measure of goodness. Two computers with the same sum, for example, a Computer Q, with performance measure 8 and affordability measure 6, and a Computer R, with performance measure 4 and affordability measure 10, would be equally valued. Any two computers for which $P + A = 14$ would be valued the same as these two. If the sum is greater than 14, then the computer is considered a better computer. The equation $P + A = 14$ is the equation of a line in the plane. Any point in the half-plane above the line represents a computer that is considered better than any computer represented by a point in the half-plane below the line. The line $P + A = 14$ and the ratings are shown in Figure 13.

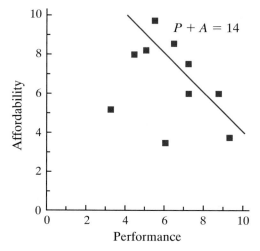

Figure 13   Data and line $P + A = 14$

The best computer, then, will be represented by a point lying on the line $P + A = k$ for the largest possible value of k. From Figure 10, the largest sum is 15.2, and the line $P + A = 15.2$ contains the point representing Computer J, as shown in Figure 14.

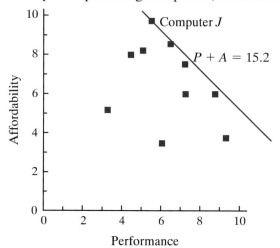

Figure 14   Ratings and line $P + A = 15.2$

The greatest product model works similarly, although the curve representing equally "good" computers is a hyperbola rather than a line. If the product $P \cdot A = k$ is the same for two computers, then the computers are valued equally. Figure 15 shows the graph of $P \cdot A = 54.75$, which contains the point representing Computer B.

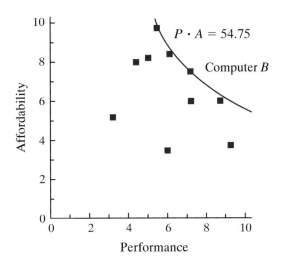

Figure 15   Ratings and hyperbola $P \cdot A = 54.75$

The greatest product method gives the advantage to computers that have a large value for both measures. Notice that computers with one high rating and one low rating have a lower product and fall below the curve $P \cdot A = 54.75$.

Weighted Measures Method

Both the greatest sum method and the greatest product method assume that performance and affordability are equally valued. If the two measures are not equally valued, then one can be weighted, that is, counted more than the other. For example, if affordability is twice as important as performance, then we could consider the sum $P + 2A$ as our measure.

Geometrically, this would change the slope of the boundary line drawn through the plane, as shown in Figure 16. The best computer then will be on the line $P + 2A = k_1$ for the largest possible value of k_1. If you look at values of $P + 2A$, you find that the largest value is 24.9. The line $P + 2A = 24.9$ contains the point representing Computer J, and we would say that Computer J is best. However, if performance was a little more important than affordability, then perhaps we would use the measure $1.25P + A$. The line $1.25P + A = 17$ contains the point representing Computer E, and we would say that Computer E is best. (See Figure 16.)

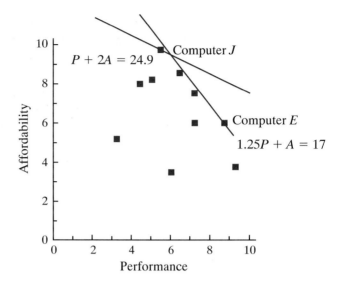

Figure 16   Ratings and lines showing weighted measures

Greatest Ratio Method

Another measure to consider is the ratio of performance to price. Since we are given the measure of affordability rather than price, the simple ratio of performance to affordability is not the measure we want. If the price is high, then the affordability measure is low, and vice versa. However, we do not know how the measure of affordability was determined from the price. Therefore, we need to make an assumption about the relationship. A computer with an affordability rating near 0 would have a price rating near 10, and a computer with an affordability rating near 10 would have a price rating near 0. Therefore, we will assume that a computer with an affordability rating of x has a price rating of $10 - x$. The ratio we want is

$$R = \frac{\text{Performance}}{\text{Price Rating}} = \frac{\text{Performance}}{10 - \text{Affordability}}$$

If this ratio is used, then we have the ratings shown in Figure 17.

Figure 17   **Computers rated by ratios**

Computer	Performance (P)	Affordability (A)	Ratio $\left(\frac{P}{10-A}\right)$
A	6.4	8.5	4.267
B	7.3	7.5	2.920
C	9.3	3.8	1.500
D	5.0	8.2	2.778
E	8.8	6.0	2.200
F	7.3	6.0	1.825
G	3.2	5.1	0.653
H	4.4	8.0	2.200
I	6.0	3.6	0.938
J	5.5	9.7	18.333

Computers for which $\frac{\text{Performance}}{10 - \text{Affordability}} = k$ are equally valued and are represented by points on the line $A = 10 - \frac{P}{k}$. The best computer will be represented by a point on the line $A = 10 - \frac{P}{k}$ for the largest value of k.

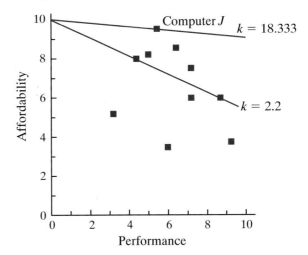

Figure 18 Ratings with lines $A = 10 - \frac{P}{18.333}$ and $A = 10 - \frac{P}{2.2}$

The graph in Figure 18 shows that Computer J is the "best" according to the greatest ratio method. The information given in Figure 18 is based on the assumption that *price rating = 10 − affordability rating*. If this assumption were modified, the outcome of this rating method would also change.

Distance from (10, 10) Method

A perfect computer would be one that had a score of (10, 10). Another way to rate the computers is to determine which computers come "closest" to being a perfect (10, 10) computer. We can determine which computer is closest by measuring the distance between their ratings and the best score, (10, 10). The distance is given by the value of the expression $\sqrt{(10 - P)^2 + (10 - A)^2}$. According to Figure 19, Computer B is the "best" based on the criterion of being closest to the perfect computer. Computers A and E are second and third by this method.

Figure 19 Computers rated using the shortest distance to perfect criterion

Computer	Performance (P)	Affordability (A)	Distance from (10, 10)
A	6.4	8.5	3.900
B	7.3	7.5	3.680
C	9.3	3.8	6.239
D	5.0	8.2	5.314
E	8.8	6.0	4.176
F	7.3	6.0	4.826
G	3.2	5.1	8.382
H	4.4	8.0	5.946
I	6.0	3.6	7.547
J	5.5	9.7	4.510

The distance definition of "best" also has a geometric interpretation. Any two points that are the same distance from the point (10, 10) represent computers that are equally valued. The set of points a given distance from a fixed point is a circle. So the distance method divides the plane into concentric circular regions centered at (10, 10), as shown in Figure 20. A point on the circle with the smallest radius represents the "best" computer.

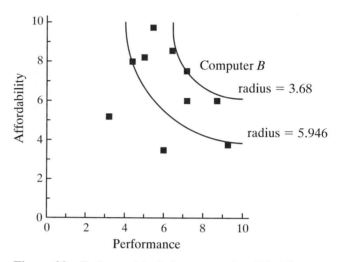

Figure 20   Ratings with circles centered at (10, 10)

Each of the rating methods will be useful when selecting a computer. Comparing the results of several methods may provide greater confidence in the recommendation your class makes to the principal.

Figure 21   Rankings of each computer using four different methods

Computer	A	B	C	D	E	F	G	H	I	J
Sum	2	3	7	6	3	5	10	8	9	1
Product	2	1	7	6	4	5	10	8	9	3
Ratio	2	3	8	4	5	7	10	6	9	1
Distance	2	1	8	6	3	5	10	7	9	4

Figure 21 shows numerical rankings from the greatest sum, greatest product, greatest ratio, and the distance from (10, 10) methods. A rank of 1 means that computer was "best" and a rank of 10 corresponds to "worst" according to that particular method. Figure 21 shows significant variability between ratings; for instance Computer *J* is rated as high as first by some methods and as low as fourth by one method. We see from this final analysis that our rating is highly dependent on the definition of "best."

1. The J.V. basketball team is trying to pick the most valuable player for the season. To be objective, the coach has ranked the five starters according to the number of points scored, the number of rebounds, and the number of fouls. Rankings are from 1 to 5, with 1 being the most and 5 being the least. An ideal player would, therefore, have a ranking of 1 for points and rebounds and 5 for fouls.

Figure 22 Player rankings

Player	Points	Rebounds	Fouls
A	2	3	3
B	1	4	2
C	5	2	1
D	3	1	4
E	4	5	5

Modify the methods discussed in this section to accommodate three measures and then use one or two methods to determine the most valuable player. How does the third measure change the geometric interpretation of your methods?

2. Magazines for runners often test the latest shoes on the market so that the consumer will be able to select a brand and model that meets his or her particular needs. Suppose several years ago a person found the perfect shoe for his knee problems, the Nicke Air Flow. Unfortunately, that shoe has been discontinued. Another shoe with the same name is being produced, but it does not have the same characteristics. Based on the data in Figure 23, what shoe presently being sold comes closest to the structural characteristics of the old Nicke Air Flow?

Figure 23 Ratings for Running Shoes

Shoe	Cushion	Support	Flexibility	Weight	Abrasion Resistance
Old Air Flow	7.5	4.5	3.0	6.5	8.0
Cougar	8.5	3.0	5.0	5.0	6.5
New Air Flow	8.0	6.0	2.5	7.0	8.0
Brokes	5.0	6.5	4.5	6.5	7.0
Asicks	7.5	5.0	4.5	5.0	4.0
Ladidas	8.0	5.0	4.0	4.5	6.0

The Tape Counter Problem

Tape recorders, tape players, and VCRs are common in today's society. Frequently, people will use one of these machines to record several different shows or songs on the same tape. One of the difficulties in using tape players is finding the position of a particular song or show on the tape. Most tape players have a counter to help you determine positions on the tape. The counter reading increases as the tape is being played, but the numbers on the counter can be misleading. If the counter goes from 0 to 500, 250 is not halfway through the tape. When you have 50 counts left, you do not have one tenth of the tape remaining. Just how does the counter work? What is the relationship between the counter reading and the time that the tape has played? Does the counter speed up or slow down as the tape is played? To answer these questions we need a mathematical model that expresses the relationship between the length of time the tape has played and the counter reading.

To gain more understanding of the operation of such a counter, watch a reel-to-reel tape player. Notice that as the tape plays, the supply reel decreases in radius, while the take-up reel increases in radius. Each time the take-up reel rotates, its radius increases, taking up more tape as it goes. (See Figure 24.) The counter is typically turned by either the take-up reel or the supply reel.

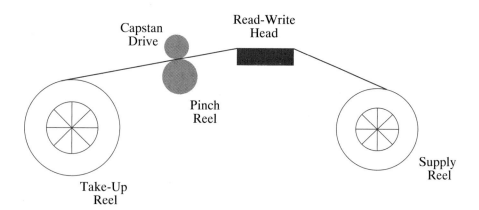

Figure 24 Diagram of a tape player

One way to answer the question about the relationship between the counter reading and the time that the tape has played is to collect data and develop an empirical model. Suppose a tape player was played for 30 minutes and the counter readings were recorded at various times. The data are provided in Figure 25. Since we want to predict the time from the counter reading, we will choose the counter reading, c, as the independent variable and attempt to express time, t, as a function of the counter reading.

Figure 25   Counter reading and elapsed time for a 30-minute tape

Counter	0	22	43	63	100	180	250	311	368	409	420
Time (sec)	0	60	120	180	300	600	900	1200	1500	1740	1800

The scatter plot of the data shown in Figure 26 reveals that the relationship between elapsed time and the counter reading is not linear. Is it quadratic? Since the problem involves the increasing radius of the take-up reel, a quadratic relationship seems likely. After all, as the tape rolls onto the reel, the area defined by the tape is a quadratic function of the radius, namely πr^2.

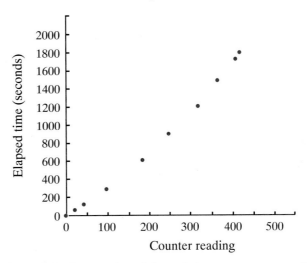

Figure 26   Scatter plot of elapsed time vs. counter reading

If we start the counter at zero, and the time played is initially zero, then we have the point (0, 0). This point is significant in determining a model. If the relationship is indeed quadratic, the point (0, 0) requires that there be no constant term in the model. So, the form of the quadratic function is either $t = Kc^2$, or $t = K_1 c^2 + K_2 c$. If the relationship can be modeled by the power function $t = Kc^2$, re-expressing the data set as $(\ln c, \ln t)$ should linearize it. If $t = K_1 c^2 + K_2 c$ is the correct model, then the re-expression $(c, \frac{t}{c})$ should linearize the data. (We need to delete the point (0, 0) before performing either of these re-expressions.) Figure 27 shows a least squares line fit to each re-expression of the data.

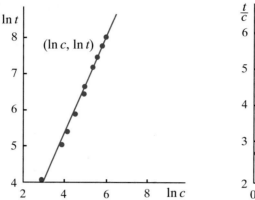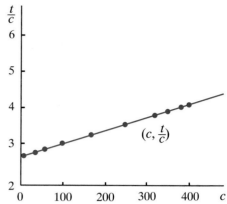

Figure 27 Least squares lines for $(\ln c, \ln t)$ and $(c, \frac{t}{c})$

Analysis of the residual plots corresponding to these fits indicates that the second one is better. The equation of the least squares line is $\frac{t}{c} = 0.004c + 2.62$, so we conclude that the counter reading and time played are related by the quadratic equation $t = 0.004c^2 + 2.62c$. Since the counter reading is the variable whose value we know, we have written the equation so that time is a function of the counter reading. If we are playing a 30-minute tape and the counter reading is 220, then we estimate that the tape has been playing for approximately $0.004(220)^2 + 2.62(220) = 770$ seconds. This means there are about 1030 seconds, or 17 minutes and 10 seconds, that have not yet been played.

empirical model

The *empirical model,* which is a model based on data, was developed using data from a particular 30-minute tape. Is the model the same for a 45-minute or a 60-minute tape? Are those tapes thinner? Will this model be appropriate for all tape players? We could construct an empirical model for cassettes of each size and for several different tape players. However, there is a fundamental unresolved question. What makes the coefficients in our model 0.004 and 2.62? Where do these numbers come from? The empirical model argues for a quadratic relationship, but it does not explain what the components are. It is not an *explicative model,* one which explains what the coefficients represent.

explicative model

An Initial Explicative Model

What are the important features of the tape recorder? Look again at Figure 24. The tape is pulled across the read-write head at a constant speed, *s,* by the capstan drive. (If you have doubts about the constant speed, think about how we can splice tapes together.) The counter is turned by either the take-up reel or the supply reel. For our argument, we will assume that the take-up reel controls the counter. We would like to relate the counter reading, *c,* to the time, *t,* that the tape has been playing.

As usual, several assumptions are necessary to simplify the problem. First, assume the counter reading, *c,* is proportional to the number of turns, *T,* of the take-up reel. Mathematically, $c = kT$, where *k* expresses the number of counts on the counter per turn of the reel. If $k = 0.5$, then the counter increases one-half unit for each turn of the take-up reel. Also, assume that the tape wraps uniformly on the reel. This means that the tape does not stretch significantly as it winds. The thickness of the tape, h, stays the same from the first wrap to the last.

Now let l represent the length of tape that has wrapped around the take-up reel. Since l represents the length of circular wrappings, we know that l is related to circumference. Circumference is given by $2\pi r$, but in our scenario, the radius r is not constant. The radius increases with each rotation. By how much does it increase? Since the tape wraps around the reel each time the take-up reel rotates, the increase in r is one thickness of tape for each rotation. We can describe this mathematically with the expression $r = r_0 + r_T$. The radius of the hub of the take-up reel is denoted r_0. It does not change. However, the outer portion of the radius, denoted r_T, increases by one thickness, h, of tape with each turn. After T turns of the take-up reel, $r_T = hT$.

At this point we should summarize what we know. First, we assumed that $c = kT$. Also,

$$r = r_0 + r_T \quad \text{and} \quad r_T = hT.$$

We can write an expression for the length of the tape wrapped around the take-up reel, l, in terms of the number of rotations of the take-up reel, T. After T rotations, there are T circumference measures contributing to the total length of tape l, so

$$l = T(2\pi r)$$

$$l = T(2\pi(r_0 + hT)).$$

Notice that neither the counter reading c nor the time t is in the equation.

Remember we also assumed the tape moves across the read-write head at a constant speed, s. So another expression for l is $l = st$. Solving for t, we find that

$$t = \frac{l}{s} = \frac{T(2\pi(r_0 + hT))}{s} = \frac{2\pi r_0 T}{s} + \frac{2\pi h T^2}{s}.$$

Based on the assumption that $c = kT$, we have $T = \frac{c}{k}$. This gives the preceding equation the form

$$t = \frac{2\pi r_0 c}{ks} + \frac{2\pi h c^2}{k^2 s}. \qquad \textbf{(8)}$$

This is exactly the quadratic form of our empirical model. And in this model, we know what the coefficients represent. In this sense the model is explicative. The linear coefficient represents the product of: the hub radius r_0, the constant 2π, and $\frac{1}{ks}$ (where s is the tape speed and k is the constant of proportionality representing the number of counts per rotation of the take-up reel). The quadratic coefficient is the product of the tape thickness, h, and the constant 2π, divided by the product of the tape speed, s, and the square of the constant of proportionality.

Checking the Model

Is equation (8) a correct model? An important aspect of developing models is the verification of the model. We can verify this model by measuring the variables s, h, and r_0, finding k, and substituting the values into the equation to see if they agree with our empirical model. On the machine that generated the initial data, the value of s is 4.76 cm/sec, the tape thickness h is 0.0018 cm, and the hub radius of the take-up reel r_0 is 1.1 cm. The value of k can be found by graphing the counter reading c against the number of turns T; k is the slope of that line. For this tape recorder, k is 0.55. Substituting into the equation for t, we find that

$$t = \frac{(2\pi)(1.1)c}{(0.55)(4.76)} + \frac{(2\pi)(0.0018)c^2}{(0.55)^2(4.76)}.$$

Simplifying yields $t = 2.64c + 0.00785c^2$. Recall that our empirical model was $t = 0.004c^2 + 2.62c$. The linear coefficients are very close in both models, but the coefficient of the quadratic term is nearly twice as large in the new model. Which model is correct? Figure 28 gives the values of time calculated using both the empirical and explicative models for the counter readings observed in the example. Notice that both models work reasonably well for small values of c. This is because the linear term in the model dominates for small values of c. Since the linear terms in both models are essentially the same, the results are very close. However, as c increases, the quadratic term becomes increasingly important, and the estimates from the two models begin to differ. Notice that for the largest values of c, the explicative model is a poor predictor.

Figure 28   **Counter reading and elapsed time (in seconds)**

Counter Reading	Observed Time	Predicted Time Empirical Model	Predicted Time Explicative Model
0	0	0	0
22	60	59	62
43	120	120	128
63	180	180	197
100	300	301	343
180	600	599	730
250	900	903	1151
311	1200	1199	1580
368	1500	1502	2035
409	1740	1737	2393
420	1800	1802	2494

It seems that our explicative model is not correct. When a model cannot be verified, the model builder must re-examine his model. Where is the error? The first place to look is the simplifying assumptions used in the beginning. Initially we assumed that the tape wraps uniformly and does not stretch. Perhaps it does, in fact, stretch, and the stretching creates the error. Or perhaps we made a more fundamental error in our analysis.

Improving the Model

Look again at the expression for r_T. We have said that $r_T = hT$ and $r = r_0 + r_T$. This gives us $r = r_0 + hT$, which in turn gives an expression for l, which is $l = T(2\pi(r_0 + hT))$. Look closely at this expression in light of how the tape accumulates on the reel. After one complete rotation, the length of tape that has accumulated on the take-up reel is simply $2\pi r_0$, since on the first turn there was no tape already on the reel. At this point $T = 1$, so according to our expression for l, we have $l = 1(2\pi(r_0 + 1h)) = 2\pi(r_0 + h)$. It appears that we have introduced h in our expression one rotation too soon.

If $T = 2$, then $l = 2(2\pi(r_0 + 2h))$ according to our expression for l. But this is the value we would get if the tape traveled twice around the circumference of a circle with radius $(r_0 + 2h)$. In reality, the tape travels only *once* around a circle with radius r_0 and *once* around a circle of radius $(r_0 + h)$. After three rotations, $T = 3$, the length of the tape given by our model is $l = 3(2\pi(r_0 + 3h))$, indicating that the circumference $2\pi(r_0 + 3h)$ is traversed three times. In reality, the tape winds only once around a circumference $2\pi(r_0)$, once around a circumference of $2\pi(r_0 + h)$, and once around a circumference of $2\pi(r_0 + 2h)$. By the time we have 50 rotations, we are using $50(2\pi(r_0 + 50h))$, which suggests that we have 50 wrappings with circumference $2\pi(r_0 + 50h)$. You can see why the error in our model kept getting larger and larger.

We will recalculate an expression for the length, l, more carefully. When the tape starts initially, it winds approximately the circumference of a circle whose radius is equal to the hub of the take-up reel. That is,

$$\text{if } T = 1, \text{ then } l_1 = 2\pi r_0.$$

The length of tape added on the second rotation, l_2, is approximately the circumference of a circle whose radius is the hub plus one thickness of tape h. So,

$$\text{if } T = 2, \text{ then } l_2 = 2\pi(r_0 + h).$$

Continuing in this fashion, we find

$$\text{if } T = 3, \text{ then } l_3 = 2\pi(r_0 + 2h),$$

$$\text{if } T = 4, \text{ then } l_4 = 2\pi(r_0 + 3h),$$

$$\text{if } T = 5, \text{ then } l_5 = 2\pi(r_0 + 4h),$$

and in general

$$\text{if } T = n, \text{ then } l_n = 2\pi(r_0 + (n - 1)h).$$

The total length of tape accumulated by this process, l, is the sum of all the expressions for l_i. That is,

$$l = 2\pi r_0 T + 2\pi h[1 + 2 + 3 + \cdots + (T - 1)].$$

This can be rewritten as

$$l = 2\pi\left[r_0 T + h[1 + 2 + 3 + \cdots + (T - 1)]\right].$$

Using the formula for the sum of an arithmetic series

$$1 + 2 + 3 + \cdots + (T - 1) = \frac{T(T - 1)}{2}.$$

Therefore, we have

$$l = 2\pi\left[r_0 T + h\frac{T(T - 1)}{2}\right]$$

$$= 2\pi\left[r_0 T + \frac{hT^2}{2} - \frac{hT}{2}\right]$$

$$= 2\pi\left(r_0 - \frac{h}{2}\right)T + \pi h T^2.$$

Substituting $T = \frac{c}{k}$ into this equation gives

$$l = 2\pi\left(r_0 - \frac{h}{2}\right)\frac{c}{k} + \pi h\frac{c^2}{k^2}.$$

Using this new equation and substituting $l = st$ gives the equation

$$t = \left(\frac{2\pi r_0}{ks} - \frac{\pi h}{ks}\right)c + \frac{\pi h}{k^2 s}c^2.$$

Notice that the coefficient of the quadratic term is indeed one-half the quadratic coefficient in the initial erroneous explicative model in equation (8) and that the linear coefficient has been reduced slightly. How does this revised model fit the data? Using $h = 0.0018$ cm, $s = 4.76$ cm/sec, $k = 0.55$, and $r_0 = 1.1$ cm as before, we find that the new equation becomes

$$t = \left(\frac{2\pi(1.1) - 0.0018\pi}{(0.55)(4.76)}\right)c + \frac{0.0018\pi c^2}{(0.55)^2(4.76)}$$

$$= 2.64c + 0.004c^2.$$

If we compare this new result to the empirical model, we see that we now have an explicative model that can be verified. The coefficients are based on the initial conditions of the problem: the hub radius, the speed of the tape across the read-write head, and the constant of proportionality between the number of turns of the take-up reel and the counter. Had we not verified our initial model, we would never have known that we were in error. More importantly, we now have a model that should work for all tape lengths (and thicknesses) and for all tape players. The explicative model is more general and therefore more useful than the empirical model.

This example illustrates one of the most important aspects of developing mathematical models. Once a model has been developed, it must be verified and, if necessary, improved. In this example, we developed an explicative model that was incorrect. By comparing the empirical model to our first explicative model, we were able to focus on the aspect of the model that produced the error. We then proceeded to make the necessary changes and to develop and verify a second explicative model.

SECTION 4.6 Developing a Mathematical Model

The process of developing mathematical models is inventive and creative. The concept is quite simple, though in practice the process can be difficult. Essentially, the goal is to create a mathematical representation of some phenomenon in order to understand it more completely. Whatever the phenomenon, the model is developed by stripping away the extraneous aspects of the problem to allow for a more focused study of its essential aspects. In this way, a model is a caricature of the problem. Certain characteristics are accentuated, while others are de-emphasized. What is essential in one model may be irrelevant in another. In developing a model for projectile motion, for example, friction might be ignored because its role in the phenomenon is minor. However, to the engineers and mathematicians designing the brakes for new aircraft, the model must include the effects of friction as an important feature.

The process of developing a mathematical model can be modeled as well. The model builder begins with a question or curiosity about some perceived phenomenon in the world. As the modeler focuses on the phenomenon, he or she isolates some particular aspect of the phenomenon to model. By isolating a particular aspect to consider, the modeler can develop several submodels that can be combined to produce a larger, more complex model. We did this in the tape counter problem on page 290. We looked at the winding of the tape around the spool as a separate problem and developed a mathematical representation for it. Later, we included this submodel as one piece of the larger model.

Once the phenomenon has been selected, the modeler must idealize the problem or simplify it so that it is manageable and less complex. All simplifying assumptions must be carefully noted so that everyone recognizes the deviations from reality inherent in the model. In the tape counter problem, we assumed uniform thickness of the tape with no thinning as the tape wound around the spool. This assumption is not totally accurate since the tape does in fact stretch as it winds. However, the errors introduced by this assumption do not invalidate the conclusions reached. If, on the other hand, the assumptions are too general, so that the error introduced is significant, the model may not represent reality, and the assumptions will need to be modified so that a more accurate, more useful model can be developed. There is a trade-off between the simplicity and the accuracy of a model.

After the problem has been idealized, a mathematical representation is developed by determining the relationships between the variables remaining in the problem. An empirical relationship was determined between the counter reading and the number of turns of the spool on the tape recorder. If the relationships cannot be determined, stronger assumptions must be made to simplify the problem even more.

Once the model is constructed, it must be analyzed or interpreted. After developing a mathematical function to model the definition of "best" in our problem of buying a computer in Section 4.4, we needed to analyze graphs to find the best computer. The model itself produces no answers or solutions to the original problem. Its power lies in the insight and illumination it offers to help answer questions about the phenomenon being modeled.

Interpretations and conclusions drawn from the model must be checked for accuracy and utility. That is, the model must be verified. Do the conclusions make sense? Does the prediction of the model fit the observations of the phenomenon? In the tape counter problem, we had developed an inaccurate model. We needed to investigate the model to determine the source of the inaccuracy and correct it. If the model cannot be verified, then the model must be inspected step-by-step. Are there mathematical errors? Are the assumptions reasonable? Are they useful? Relax the assumptions a bit. Is the new model of greater utility? Continuing to vary the problem, relaxing assumptions, and adding variables until the model accurately reflects the aspects of the phenomenon under study are important parts of the modeling process.

Never stop too soon when developing a model. What questions does your solution pose? Is it possible to generalize your results? Questioning your results is essential if the model is to be improved. Another important aspect of modeling is analyzing the sensitivity of the model to small changes in the variables. In the tape counter problem, how much will the counter reading be off if our measurements are off by 0.01 cm? Which measurements are most crucial? Can the differences in the predictions of the model be attributed to small errors in measurement, or are the differences too great to be explained this way?

The process of developing a mathematical model is dynamic. As new insight is gained, the model is improved and the process begins again. An existing model can be modified much more easily than a new model can be developed from scratch. If the model developed is too simple, it can be improved by introducing missing components into the existing model. At times, the model will collapse under the weight of these additions and must be constructed again with new initial assumptions. If so, we've learned something. That is the essential ingredient—learning something about the phenomenon under study. If we do, then the process of mathematical modeling is successful.

Some Problems to Model

In previous sections of this chapter, situations were described and models were developed, verified, and modified. In this section, the problems will be described and, in some cases, a few hints given so that you can develop models on your own. Do not limit yourself to those suggestions, however; part of the modeling process is developing your own questions. In each scenario, be sure to identify your assumptions so both you and your reader are aware of premises upon which your model is developed.

Elevators

In some buildings, all elevators go to all floors. In others, certain elevators go only to the lower floors, while the rest go only to the top floors. What is the advantage of having elevators that travel only between certain floors?

Suppose a building has 5 floors (numbered 1 through 5) that are occupied. The ground floor (0) is not used for business purposes. Each floor has 80 people working on it, and there are 4 elevators available. Each elevator can hold 10 people. The elevators take 3 seconds to travel between floors and average 22 seconds on each floor when someone enters or exits. If all of the people arrive at work at about the same time and enter the elevator on the ground floor, how should the elevators be used to get the people to their offices as quickly as possible?

Ideas to Consider

1. On each trip, how much time is spent traveling from floor to floor, and how much time is spent waiting for passengers to exit on each floor?

2. How much time does it take to get everyone to the proper floors if all elevators go to all floors?

3. If a fifth elevator were built, to which floors should it go?

4. Suppose the bottom three floors have 100 people working on each and the top two floors have only 50 people on each. How should the four elevators be used?

5. How sensitive is your solution to the given conditions? Try varying the parameters, such as travel time between floors, exit time, and so forth, and then see if your model needs to be modified.

Source: "The Elevator Problem," *Everybody's Problems*, **Consortium**, Number 57, Spring, 1996, COMAP, Inc., Lexington, MA.

Pollution in the Great Lakes

Most of the water flowing into Lake Erie comes from Lake Huron, and most of the water flowing into Lake Ontario is from Lake Erie. Each year, the percentages of water replaced in Lakes Huron, Erie, and Ontario are 11%, 36%, and 12%, respectively.

For generations, factories on the lakes had been dumping a pollutant into the water. Presently, there are 4000 units of pollutant in Lake Huron, 2000 units in Lake Erie, and 3000 units in Lake Ontario. For the most part, this form of pollution has stopped. Only two such factories remain. One, on Lake Huron, is dumping 25 units of the pollutant into the water each year; the other, on Lake Ontario, is dumping 20 units of the pollutant into the water each year.

How long will it be before the amount of pollutant in the three lakes is reduced to 10% of its present level? What is the long-term level of the pollutant in the lakes?

Source: Intermath: Four Sample Problems, COMAP, Inc., Lexington, MA, 1992.

The Midge Problem

In 1981, two new varieties of a tiny biting insect called a midge were discovered in the jungles of Brazil by biologists W. L. Grogan and W. W. Wirth. They dubbed one kind of midge an *Apf* midge and the other an *Af* midge. The biologists found that the *Apf* midge is a carrier of a debilitating disease that causes swelling of the brain when a human is bitten by an infected midge. Although the disease is rarely fatal, the disability caused by the swelling may be permanent. The other form of the midge, the *Af*, is quite harmless and a valuable pollinator. In an effort to distinguish the two varieties, the biologists took measurements on the midges they caught. The two measurements taken were wing length and antenna length, both measured in centimeters. The data are provided in Figure 29.

Figure 29   Wing and antenna length for *Af* and *Apf* midges

Af midges								
Wing length (cm)	1.72	1.64	1.74	1.70	1.82	1.82	1.90	1.82	2.08
Antenna length (cm)	1.24	1.38	1.36	1.40	1.38	1.48	1.38	1.54	1.56

Apf midges					
Wing length (cm)	1.78	1.86	1.96	2.00	2.00	1.96
Antenna length (cm)	1.14	1.20	1.30	1.26	1.28	1.18

Is it possible to distinguish an *Af* midge from an *Apf* midge on the basis of wing and antenna lengths? Write a report that describes to a naturalist in the field how to classify a just-captured midge.

Sources: Grogan, William L., Jr. and Willis Wirth. 1981. "A new American genus of predaceous midges related to Palpomyia and Bessia (Diptera: Ceratopogonidae)." Proceedings of the Biological Society of Washington 94 (4): 1279–1305.
"The Midge Problem," *Everybody's Problems*, **Consortium**, Number 55, Fall, 1995, COMAP, Inc., Lexington, MA.

Linear Irrigation

Suppose your uncle is designing an irrigation system for his farm in eastern North Carolina. His land is very flat, and the fields, which are rectangular in shape, are roughly 2200 feet long and 1000 feet wide. Your uncle is considering a linear irrigation system, which is essentially one long pipe to water each field. A linear irrigation system is generally set on wheels that keep it above the level of the plants. Nozzles are placed periodically along the pipe, and each nozzle sprays water in a circular region. The entire system moves slowly down the field, watering the plants beneath it as it travels.

There are 20 sprinkler nozzles available to your uncle. He has enough pumps to maintain water pressure so that each nozzle delivers a uniform spray to a circular region 50 feet in radius with a flow rate of 10 gallons per minute.

a. How far apart should the nozzles be placed to produce the most uniform distribution of water on a field 1000 feet wide?

b. Depending on the crop that is planted in the field, different amounts of water are needed. Therefore, your uncle also needs to know how fast the irrigation system must move down each field to supply a particular amount of water to the plants.

Ideas to Consider

1. You do not necessarily need to use all 20 nozzles.

2. What is meant by "uniform distribution of water on a field"?

Sources: Adapted with permission from *Mathematical Modeling in the Secondary School Curriculum: A Resource Guide of Classroom Exercises* by the National Council of Teachers of Mathematics. All rights reserved.

"The Irrigation Problem," *Everybody's Problems*, **Consortium**, Number 63, Fall, 1997, COMAP, Inc., Lexington, MA.

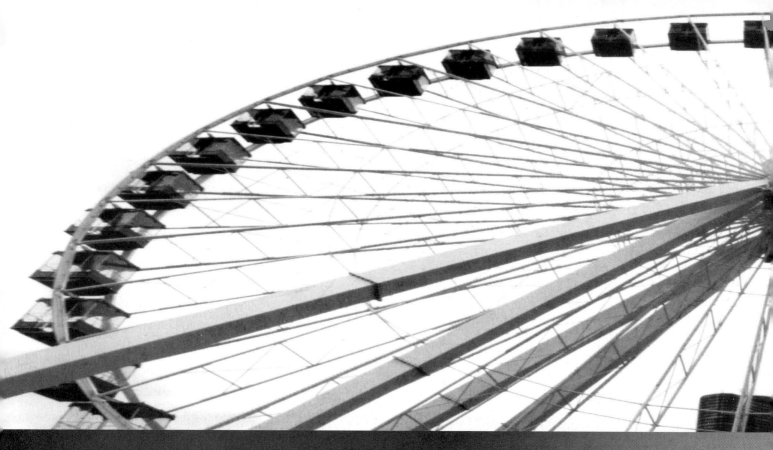

5 Circular Functions and Trigonometry

5.1	The Curves of Trigonometry	302
5.2	Graphing Transformations of Trigonometric Functions	313
5.3	Investigating a Predator-Prey Relationship	321
5.4	Sine and Cosine on the Unit Circle	323
5.5	Getting to Know the Unit Circle	329
5.6	Angles and Radians	337
5.7	Solving Trigonometric Equations	350
5.8	Investigating Trigonometric Identities	358
5.9	Using Trigonometric Identities	360
5.10	Inverse Trigonometric Functions	366
5.11	The Double Ferris Wheel Investigation	373
5.12	Composition with Inverse Trigonometric Fuctions	376
5.13	Solving Triangles with Trigonometry	383
5.14	Investigating Hanging Pictures	396
Chapter 5 Review Exercises		398

The Ferris Wheel

George Washington Gale Ferris was an American engineer whose design of a spectacular wheel, an amusement park ride, won a contest. As a result of the contest win, Ferris was able to erect his wheel for use at the World's Columbian Exposition held in 1893 in Chicago.

A newer version of the wheel is the double Ferris wheel, which offers an exciting ride. A double Ferris wheel at the state fair has a 50-foot bar attached to the main support at the center of the bar. This bar revolves once every 60 seconds. Seats for riders are evenly spaced along two separate wheels, each revolving at the end of the long bar. Each wheel makes one complete revolution every 15 seconds and has a radius of 10 feet. The bar and the two wheels rotate counterclockwise, and the wheels turn independently of the bar. Assume a rider begins at the lowest point, and sketch the path of the rider of the double Ferris wheel.

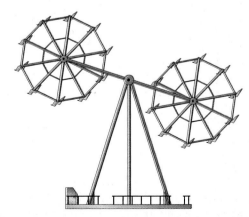

The Curves of Trigonometry

The hands of a clock on a wall move in a predictable way. As time passes, the distance between the ceiling and the tip of the hour hand changes. Suppose that the hour hand is 15 cm long and that at 6 o'clock the distance between the tip of the hour hand and the ceiling is 53 cm. At 9 o'clock the distance will have decreased to 38 cm, and at 12 o'clock the distance will have decreased to 23 cm. This is the minimum possible distance; after 12 o'clock the distance will increase. At 3 o'clock the distance will be 38 cm, and at 6 o'clock the distance will again be 53 cm. Note that 53 cm is the maximum possible distance. The same sequence of values will be repeated every 12 hours. Figure 1 shows several (*time, distance*) ordered pairs. The *t*-coordinate represents the number of hours elapsed since we began recording distances at 6 o'clock.

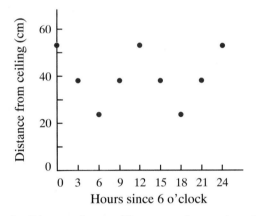

Figure 1 Distance from ceiling versus hours since 6 o'clock

How should the maximum and minimum points in Figure 1 be connected? One possibility is to connect the points with line segments. Linearity implies that the distance from the tip of the hour hand to the ceiling changes at a constant rate. However, the distance changes more quickly around 9 o'clock and 3 o'clock and more slowly around 6 o'clock and 12 o'clock. To display these facts on a graph, the points in Figure 1 should be connected with a curve; the curve is somewhat flatter near its turning points and somewhat steeper away from its turning points. A graph with connected points is shown in Figure 2.

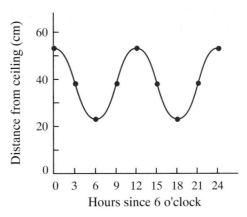

Figure 2 Distance from ceiling versus hours since 6 o'clock

The graph in Figure 2 should remind you of the sine function from your toolkit of functions. It is a continuous curve that oscillates periodically between maximum and minimum values. Such a curve is called a *sinusoid*. The shape of the graph in Figure 2 is characteristic of another trigonometric function called *cosine*.

sinusoid

cosine

amplitude

The cosine function, written $f(x) = \cos x$, has a graph that is very similar to the graph of the sine function. The graph of $y = \cos x$ is shown in Figure 3; the graph is periodic with period 2π. The graph oscillates between a maximum of 1 and a minimum of -1. The *amplitude* of a periodic graph is defined as

$$\text{amplitude} = \frac{\text{maximum value} - \text{minimum value}}{2}.$$

Thus, the amplitude of $y = \cos x$ is $\frac{1 - (-1)}{2} = 1$.

The graph of $y = \cos x$ can be obtained by shifting the graph of $y = \sin x$ to the left $\frac{\pi}{2}$ units; this results in the shifted graph having a maximum on the y-axis. Since the graph of $y = \cos x$ is identical to the graph of $y = \sin x$ shifted to the left $\frac{\pi}{2}$ units, we can write $\cos x = \sin(x + \frac{\pi}{2})$.

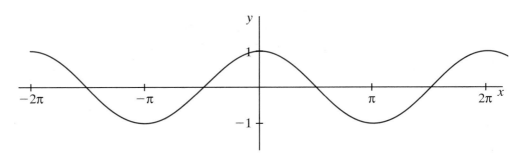

Figure 3 Graph of $y = \cos x$

Note that the graph of $y = \cos x$ is symmetric about the y-axis. The fact that $\cos x = \cos(-x)$ implies that cosine is an even function. In Figure 3, the x-intercepts of $y = \cos x$ occur at $-\frac{3\pi}{2}$, $-\frac{\pi}{2}$, $\frac{\pi}{2}$, and $\frac{3\pi}{2}$, or in general at odd multiples of $\frac{\pi}{2}$. These x-coordinates can be written as $\frac{\pi}{2} + k\pi$ or $\frac{(2k+1)\pi}{2}$, where k is an integer. The maximum values of the function occur at even multiples of π, and the minimum values occur at odd multiples of π. You should add the cosine function to your toolkit and become familiar with the basic shape, x-intercepts, and maxima and minima of its graph.

EXAMPLE 1   Use the periodicity of the sine and cosine functions to evaluate $\cos(7\pi)$ and $\sin\left(\frac{17\pi}{2}\right)$.

Solution   Since the period of the cosine function is 2π, then $\cos(b + 2\pi) = \cos b$ for any real number b. In other words, the y-coordinate on the graph of $f(x) = \cos x$ at $x = b + 2\pi$ is the same as the y-coordinate at $x = b$. This is illustrated in Figure 4.

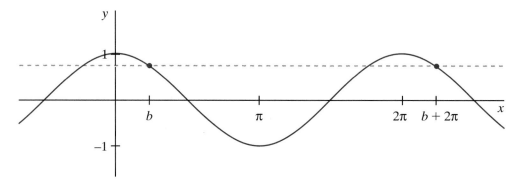

Figure 4 $\cos(b + 2\pi) = \cos(b)$

The key to evaluating $\cos(7\pi)$ is recognizing that the behavior of the graph of $f(x) = \cos x$ on the interval $0 \le x \le 2\pi$ is repeated on additional intervals: $2\pi \le x \le 4\pi$, $4\pi \le x \le 6\pi$, and so forth. Between $x = 0$ and $x = 6\pi$, the cosine graph completes three *cycles*, and $x = 7\pi$ falls within the fourth cycle of the graph. (See Figure 5.)

cycles

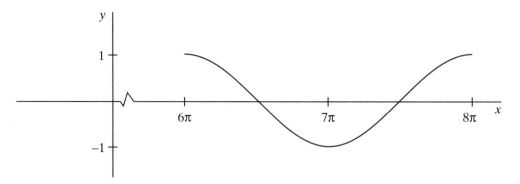

Figure 5 $f(x) = \cos x$ on the interval $6\pi \le x \le 8\pi$

Since the behavior of $f(x) = \cos x$ from 6π to 8π is the same as the behavior from 0 to 2π,

$$\cos(7\pi) = \cos(6\pi + \pi) = \cos \pi = -1.$$

Since the period of the sine function is also 2π, $\sin(b + 2\pi) = \sin b$ is true for all real numbers b. It is also true that the behavior of the graph of $f(x) = \sin x$ on the interval $0 \le x \le 2\pi$ is repeated on the intervals $2\pi \le x \le 4\pi$, $4\pi \le x \le 6\pi$, and so forth. Between $x = 0$ and $x = 8\pi$, the sine graph completes four cycles, and $x = \frac{17\pi}{2}$ falls within the fifth cycle of the graph. (See Figure 6.)

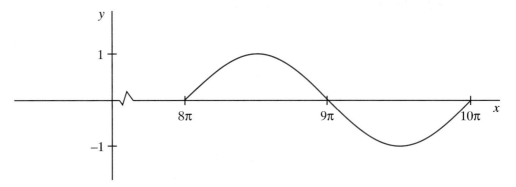

Figure 6 $f(x) = \sin x$ on the interval $8\pi \le x \le 10\pi$

Since the behavior of $f(x) = \sin x$ from 8π to 10π is the same as the behavior from 0 to 2π,

$$\sin\left(\frac{17\pi}{2}\right) = \sin\left(8\pi + \frac{\pi}{2}\right) = \sin\left(\frac{\pi}{2}\right) = 1. \quad \blacksquare$$

EXAMPLE 2   Find all solutions of the inequality $\cos x < 0$.

Solution   The graph of the function $y = \cos x$ in Figure 7 shows intervals where the cosine function has negative values. One interval where the values of $\cos x$ are negative is $\frac{\pi}{2} < x < \frac{3\pi}{2}$.

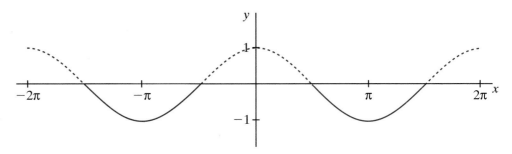

Figure 7   Graph of $y = \cos x$, $-2\pi \le x \le 2\pi$

Since the cosine function has period 2π, values of $\cos x$ will be negative for any x that satisfies $\frac{\pi}{2} + 2k\pi < x < \frac{3\pi}{2} + 2k\pi$, where k is an integer. Substituting an integer value for k in the inequality $\frac{\pi}{2} + 2k\pi < x < \frac{3\pi}{2} + 2k\pi$ produces a particular interval where $\cos x < 0$. For example, if $k = 0$, the interval is $\frac{\pi}{2} < x < \frac{3\pi}{2}$. If $k = 1$, the interval is $\frac{5\pi}{2} < x < \frac{7\pi}{2}$. If $k = -1$, the interval is $-\frac{3\pi}{2} < x < -\frac{\pi}{2}$. Use a graph to verify that these intervals correspond to regions where $\cos x$ has negative values. ■

The graph of $h(x) = \dfrac{\sin x}{\cos x}$

There are many ways in which the sine and cosine functions can be combined to produce new functions. These include addition of functions, such as $y = \sin x + \cos x$; multiplication of functions, such as $y = (\sin x)(\cos x)$; and composition of functions, such as $y = \sin(\cos x)$. Additionally, many important mathematical ideas are based on the quotient of $\sin x$ and $\cos x$. We will use a table of values to investigate the graph of the function $h(x) = \frac{\sin x}{\cos x}$ on the interval $0 \le x \le 2\pi$. The table in Figure 8 shows values of $\sin x$, $\cos x$, and $\frac{\sin x}{\cos x}$ from $x = 0$ to $x = \frac{\pi}{2}$.

Figure 8 Values of $\sin x$, $\cos x$, and $\frac{\sin x}{\cos x}$ from $x = 0$ to $x = \frac{\pi}{2}$

x	$\sin x$	$\cos x$	$y = \frac{\sin x}{\cos x}$
0	0.00000	1.00000	0.00000
$\frac{\pi}{12}$	0.25882	0.96593	0.26795
$\frac{2\pi}{12} = \frac{\pi}{6}$	0.50000	0.86603	0.57735
$\frac{3\pi}{12} = \frac{\pi}{4}$	0.70711	0.70711	1.00000
$\frac{4\pi}{12} = \frac{\pi}{3}$	0.86603	0.50000	1.73205
$\frac{5\pi}{12}$	0.96593	0.25882	3.73205
$\frac{6\pi}{12} = \frac{\pi}{2}$	1.00000	0.00000	undefined

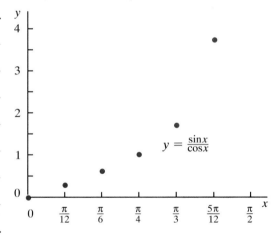

Since values of $\sin x$ and $\cos x$ are nonnegative between $x = 0$ and $x = \frac{\pi}{2}$, the ratio $\frac{\sin x}{\cos x}$ is also nonnegative for all x-values in this interval. The table in Figure 8 shows that the values of $\sin x$ increase from 0 to 1 and the values of $\cos x$ decrease from 1 to 0. The table also shows that the ratio $\frac{\sin x}{\cos x}$ is zero when $x = 0$, and the ratio increases as x increases. When $x = \frac{\pi}{2}$, the function $h(x) = \frac{\sin x}{\cos x}$ will be undefined, since $\cos(\frac{\pi}{2}) = 0$ and division by zero is undefined. However, as x approaches $\frac{\pi}{2}$ from the left, the values of $h(x)$ approach infinity. For example,

$$\frac{\sin\left(\frac{11\pi}{24}\right)}{\cos\left(\frac{11\pi}{24}\right)} = 7.59575, \quad \text{and} \quad \frac{\sin\left(\frac{23\pi}{48}\right)}{\cos\left(\frac{23\pi}{48}\right)} = 15.25705.$$

This occurs because values of $\sin x$ in the numerator approach 1 while values of $\cos x$ in the denominator approach 0. A denominator that is approaching 0 with a numerator approaching 1 causes the ratio $\frac{\sin x}{\cos x}$ to increase without bound.

To further investigate $h(x) = \frac{\sin x}{\cos x}$ on the interval $0 \le x \le 2\pi$, study the table of values of $\sin x$, $\cos x$, and $\frac{\sin x}{\cos x}$ from $x = \frac{\pi}{2}$ to $x = \pi$ in Figure 9.

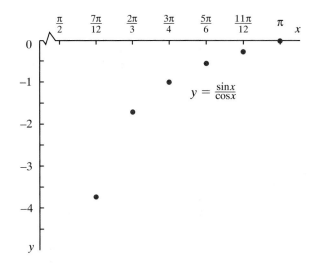

Figure 9 Values of $\sin x$, $\cos x$, and $\frac{\sin x}{\cos x}$ from $x = \frac{\pi}{2}$ to $x = \pi$

x	$\sin x$	$\cos x$	$y = \frac{\sin x}{\cos x}$
$\frac{6\pi}{12} = \frac{\pi}{2}$	1.00000	0.00000	undefined
$\frac{7\pi}{12}$	0.96593	-0.25882	-3.73205
$\frac{8\pi}{12} = \frac{2\pi}{3}$	0.86603	-0.50000	-1.73205
$\frac{9\pi}{12} = \frac{3\pi}{4}$	0.70711	-0.70711	-1.00000
$\frac{10\pi}{12} = \frac{5\pi}{6}$	0.50000	-0.86603	-0.57735
$\frac{11\pi}{12}$	0.25882	-0.96593	-0.26795
$\frac{12\pi}{12} = \pi$	0.00000	-1.00000	0.00000

On the interval from $x = \frac{\pi}{2}$ to $x = \pi$, values of $\sin x$ are nonnegative and values of $\cos x$ are nonpositive, so the ratio $\frac{\sin x}{\cos x}$ is nonpositive on this interval. The table in Figure 9 shows that the values of $\sin x$ decrease from 1 to 0 and the values of $\cos x$ decrease from 0 to -1. The table also shows that the ratio $\frac{\sin x}{\cos x}$ is negative and increases toward zero as x increases.

What have we discovered about $h(x) = \frac{\sin x}{\cos x}$? The sign (positive or negative) of $h(x)$ is determined by the signs of $\sin x$ and $\cos x$. Any x-value that makes $\sin x$ equal zero also makes the value of $h(x)$ equal zero. Any x-value that makes $\cos x$ equal zero is not in the domain of $h(x)$ because division by zero is undefined. Evaluating $h(x)$ at x-values near the zeros of $\cos x$ indicates that $h(x)$ has vertical asymptotes at the zeros of $\cos x$. The graph of $h(x) = \frac{\sin x}{\cos x}$ on the interval $[0, \pi]$ is shown in Figure 10.

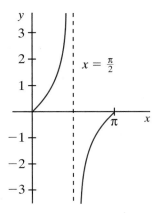

Figure 10 Graph of $h(x) = \frac{\sin x}{\cos x}$, $0 \le x \le \pi$

Figure 11 Values of $\sin x$, $\cos x$, and $\frac{\sin x}{\cos x}$ from $x = \pi$ to $x = \frac{3\pi}{2}$

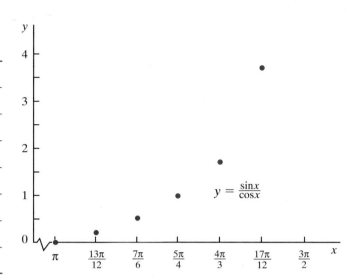

x	$\sin x$	$\cos x$	$y = \frac{\sin x}{\cos x}$
$\frac{12\pi}{12} = \pi$	0.00000	-1.0000	0.00000
$\frac{13\pi}{12}$	-0.25882	-0.96593	0.26795
$\frac{14\pi}{12} = \frac{7\pi}{6}$	-0.50000	-0.86603	0.57735
$\frac{15\pi}{12} = \frac{5\pi}{4}$	-0.70711	-0.70711	1.00000
$\frac{16\pi}{12} = \frac{4\pi}{3}$	-0.86603	-0.50000	1.73205
$\frac{17\pi}{12}$	-0.96593	-0.25882	3.73205
$\frac{18\pi}{12} = \frac{3\pi}{2}$	-1.00000	0.00000	undefined

Figure 11 shows values of $\sin x$, $\cos x$ and $\frac{\sin x}{\cos x}$ from $x = \pi$ to $x = \frac{3\pi}{2}$. On this interval, values of $\sin x$ are nonpositive, values of $\cos x$ are nonpositive, and values of $\frac{\sin x}{\cos x}$ are nonnegative. Compare the values of $\frac{\sin x}{\cos x}$ on the interval $\pi \leq x \leq \frac{3\pi}{2}$ with the values of $\frac{\sin x}{\cos x}$ on the interval $0 \leq x \leq \frac{\pi}{2}$. Although the values of $\sin x$ and $\cos x$ differ on these two intervals, the values of their ratio are identical. You can make a table of values of $\sin x$, $\cos x$, and $\frac{\sin x}{\cos x}$ from $x = \frac{3\pi}{2}$ to $x = 2\pi$ to confirm that the behavior of $h(x) = \frac{\sin x}{\cos x}$ on $\frac{3\pi}{2} \leq x \leq 2\pi$ is identical to the behavior of $h(x)$ on $\frac{\pi}{2} \leq x \leq \pi$.

The repetition of values of $h(x) = \frac{\sin x}{\cos x}$ indicates that this function is periodic. The period of $h(x) = \frac{\sin x}{\cos x}$ is π, since $h(x + \pi) = h(x)$ for all values of x. The fact that $h(x) = \frac{\sin x}{\cos x}$ has period π can be investigated using the graph shown in Figure 12. For all x in the interval $(0, \pi)$, the value of $\sin x$ and the value of $\sin(x + \pi)$ are opposites. Similarly, the value of $\cos x$ and the value of $\cos(x + \pi)$ are opposites. Therefore,

$$\frac{\sin(x + \pi)}{\cos(x + \pi)} = \frac{\sin x}{\cos x}.$$

Thus, the values of $\frac{\sin x}{\cos x}$ for $0 \leq x < \pi$ repeat for $\pi \leq x < 2\pi$.

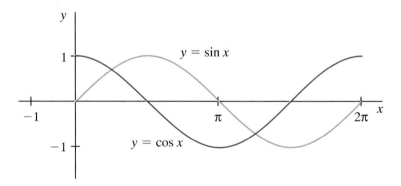

Figure 12 Graphs of $y = \sin x$ and $y = \cos x$ for $0 \leq x \leq 2\pi$

Based on information about zeros, asymptotes, and periodicity, the graph of $h(x) = \frac{\sin x}{\cos x}$ on the interval $0 \leq x \leq 2\pi$ is shown in Figure 13.

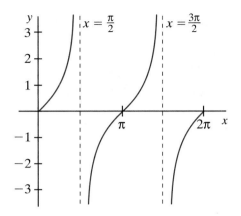

Figure 13 Graph of $h(x) = \frac{\sin x}{\cos x},\ 0 \le x \le 2\pi$

There are four other combinations of the sine and cosine functions that occur frequently and are given special names. These new trigonometric functions are defined as follows:

$$\text{secant}(x) = \sec x = \frac{1}{\cos x}$$

$$\text{cosecant}(x) = \csc x = \frac{1}{\sin x}$$

$$\text{tangent}(x) = \tan x = \frac{\sin x}{\cos x}$$

$$\text{cotangent}(x) = \cot x = \frac{\cos x}{\sin x}.$$

The graph in Figure 13, which was developed as the graph $h(x) = \frac{\sin x}{\cos x}$, is a graph of $y = \tan x$. We can use knowledge of symmetries and periodicity to graph $y = \tan x$ on a larger domain than is shown in Figure 13. We first determine the symmetry of $\tan x$ by simplifying $\tan(-x)$. Recall that a function $f(x)$ is odd if $f(-x) = -f(x)$ for all x-values. Similarly, a function $f(x)$ is even if $f(-x) = f(x)$ for all x-values.

$$\tan(-x) = \frac{\sin(-x)}{\cos(-x)}$$

$$= \frac{-\sin x}{\cos x}$$

$$= -\tan x.$$

This means that $\tan x$ is an odd function, and its graph is symmetric about the origin. A more complete graph of the tangent function is shown in Figure 14 on the next page. The graph shows that the period of the tangent function is π. The concept of amplitude does not apply to the tangent function since values of $\tan x$ get infinitely large. The domain of the tangent function excludes the odd multiples of $\frac{\pi}{2}$, which can be written as $x \ne \frac{(2k+1)\pi}{2}$, where k is any integer. The range is all real numbers. You should add the tangent function to your toolkit and become familiar with the basic shape, intercepts, and asymptotes of its graph.

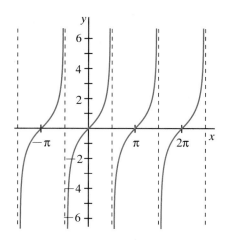

Figure 14 Graph of $y = \tan x$

CLASS PRACTICE

1. Sketch a graph of $y = \sec x$. Use the graph of $f(x) = \cos x$, the fact that $\sec x = \frac{1}{f(x)}$, and your knowledge of composition graphing.

2. Discuss the similarities and differences between the functions $f(x) = \cot x$ and $g(x) = \frac{1}{\tan x}$. Are their y-values equal for all x-values? Are their domains the same?

3. Identify the domain and the range of each of the six trigonometric functions: sine, cosine, tangent, cotangent, secant, and cosecant.

4. Tell whether each of the six trigonometric functions is even, odd, or neither.

Exercise Set 5.1

1. Use the fact that both the sine and cosine functions have period 2π to evaluate the following. Do not use a calculator.

   a. $\cos(-7\pi)$ **b.** $\sin(-\frac{9\pi}{2})$

   c. $\cos(-\frac{\pi}{2})$ **d.** $\sin(\frac{29\pi}{2})$

   e. $\sin(-4\pi)$ **f.** $\cos(\frac{101\pi}{2})$

2. Estimate the value of each of the following and then use your calculator in radian mode to find the decimal value.

   a. $\sin 1$ **b.** $\sin 0.5$

   c. $\cos(-5)$ **d.** $\cos 5$

   e. $\sec 6$ **f.** $\csc(-1)$

   g. $\tan 1.4$ **h.** $\cot(\frac{\pi}{2})$

3. Sketch the graph of $y = \cot x$ for $-7 < x < 7$. Identify the intercepts, the asymptotes, and the period of the function. Use your calculator to check your graph.

4. Sketch the graph of $y = \csc x$ for $-7 < x < 7$. Identify the intercepts, the asymptotes, and the period of the function. Use your calculator to check your graph.

5. How can you transform the graph of $y = \sec x$ to obtain the graph of $y = \csc x$? Write an equation to show the relationship between the graphs.

6. How can you transform the graph of $y = \tan x$ to obtain the graph of $y = \cot x$? Write an equation to show the relationship between the graphs.

7. Find all solutions in the interval $[0, 2\pi]$. Give exact solutions (in terms of π, no decimal approximations).

   a. $\sin x = 1$

   b. $\sin x > 0$

   c. $\tan x = 0$

   d. $\cos x = 1$

8. Sketch a graph to represent each situation described below.

   a. Imagine a diving board just after someone has dived off. Assume there is no friction or air resistance so that the diving board does not slow down over time. Sketch a graph of the motion of the diving board if the diving board vibrates 10 inches from equilibrium each second.

   b. Estimate and then graph the number of hours of daylight per day in your hometown over a two-year time span.

   c. A pendulum that consists of a small weight suspended by a length of string hangs vertically when it is at rest. If the pendulum's bob is displaced by a small amount from its equilibrium position and then released, the bob will swing back and forth in an arc which subtends an angle twice as large as the initial angle of displacement. That is, the pendulum sweeps out equal angles on each side of equilibrium as seen in Figure 15 on the next page. Suppose that the total angle through which the pendulum moves is 10° and that it takes 1 second to complete one swing from left to right and back again. Make a graph showing the pendulum's angular displacement from equilibrium over time. Assume there is no friction or air resistance.

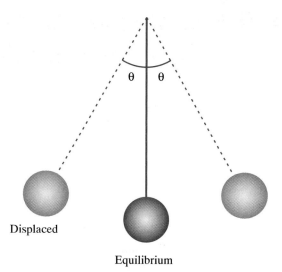

Displaced

Equilibrium

Figure 15   Pendulum

9. Sketch graphs of $y = \sin x$ and $y = x$ on the interval $-\frac{\pi}{2} < x < \frac{\pi}{2}$. Use your calculator (in radian mode) to investigate the relationship between values of $\sin x$ and x for x-values close to zero. Write a few sentences about this relationship.

SECTION 5.2 Graphing Transformations of Trigonometric Functions

Many phenomena are periodic and can be modeled by either the sine or the cosine function. However, most periodic phenomena do not have period 2π and do not oscillate between 1 and -1. The trigonometric functions that model these phenomena are variations of the toolkit functions, which incorporate various horizontal and vertical stretches, compressions, and shifts.

EXAMPLE 1   A weight is suspended from the ceiling on a spring. When the weight is pulled down a small distance from its equilibrium position and then released, it oscillates about its original equilibrium position. Assume that ideal conditions allow the suspended weight to oscillate indefinitely. The weight is pulled down to 1.3 cm below its equilibrium position and then released at time $t = 0$. After one second the weight will be 1.3 cm above its equilibrium position, and when $t = 2$ it will have completed one oscillation and will again be 1.3 cm below equilibrium. (See Figure 16.) Sketch a graph to illustrate the motion of the weight over time. Write an equation for your graph.

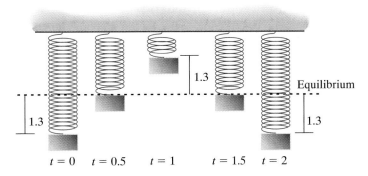

Figure 16   Motion of weight on spring

displacement

Solution   The problem asks for a graph of the motion of the weight over time, which is a graph of displacement over time. *Displacement* is the difference between current position and equilibrium position. We will say that displacement is positive if the spring is above equilibrium and negative if the spring is below equilibrium. Displacement is zero at equilibrium. Figure 17 on the next page lists some ordered pairs (*time, displacement*) and illustrates the graph of spring displacement over time.

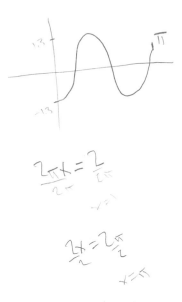

Time (seconds)	0	0.5	1.0	1.5	2.0	2.5	3.0
Displacement (cm)	−1.3	0	1.3	0	−1.3	0	1.3

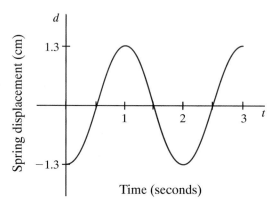

Figure 17   Ordered pairs and graph for spring displacement over time

Note that the graph in Figure 17 has amplitude 1.3 units, period 2 units, and a minimum value when $t = 0$. The function $d = 1.3\cos t$ has a graph with the correct amplitude and includes the point (0, 1.3). The period can be changed from 2π to 2 by doing a horizontal compression. Thus, $d = 1.3\cos(\pi t)$ has the correct amplitude and the correct period. However, its graph has a maximum value when $t = 0$. We can achieve a minimum value when $t = 0$ by reflecting the graph over the t-axis. Thus,

$$d = -1.3\cos(\pi t)$$

is an equation for the graph in Figure 17. Other equations will also produce the desired graph. You will be asked to write other equations for this graph in the exercise set at the end of this section.

EXAMPLE 2   Sketch a graph of $g(x) = \csc(\frac{1}{3}x)$.

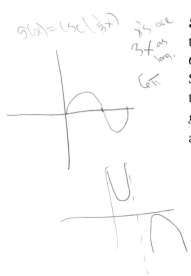

Solution   This is the graph of the cosecant function, $f(x) = \csc x$, stretched horizontally. Since the period of the cosecant is 2π, the period of g will be 3 times as long, or 6π. The graph of $f(x) = \csc x$ has vertical asymptotes at integer multiples of π, or $k\pi$. So, on the graph of $g(x) = \csc(\frac{1}{3}x)$ the vertical asymptotes will be $x = 3k\pi$. Similarly, the turning points on the graph $f(x) = \csc x$ are π units apart at $x = \frac{(2k + 1)\pi}{2}$, and on $g(x) = \csc(\frac{1}{3}x)$ they will be 3π units apart. Therefore, the turning points will occur at $x = \frac{3(2k + 1)\pi}{2}$. The graph of g is shown in Figure 18.

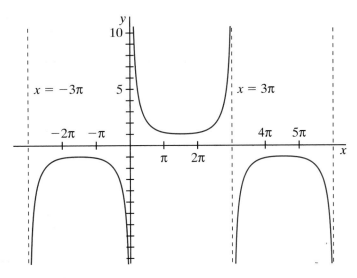

Figure 18   Graph of $g(x) = \csc(\frac{1}{3}x)$ ▪

EXAMPLE 3 ⟩ Sketch a graph of the function $g(x) = 4\cos(x - \frac{\pi}{2}) - 1$.

phase shift

Solution   You should recognize that the cosine function has undergone three transformations. The coefficient 4 causes a vertical stretch; therefore, the amplitude of this transformed function will be 4. The subtraction of $\frac{\pi}{2}$ causes a horizontal shift of $\frac{\pi}{2}$ units to the right. A horizontal shift of a trigonometric curve is called a *phase shift*. The -1 causes a vertical shift one unit down creating new maximum and minimum values. To sketch the graph, it may be helpful to draw an intermediate graph of $y = 4\cos x$ that shows the change of shape and then draw a final graph that shows the change in position. The final graph is shown in Figure 19.

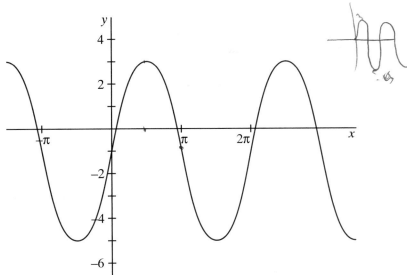

Figure 19   Graph of $g(x) = 4\cos(x - \frac{\pi}{2}) - 1$ ▪

EXAMPLE 4   Graph $h(x) = \sin(2x - \frac{\pi}{4})$.   *(handwritten: right 1/4 & period 2   2x=2π/2, x=π)*

Solution   The graph will exhibit two transformations of the graph of $f(x) = \sin x$: a horizontal compression and a horizontal shift. You know from earlier work that the combination of two horizontal transformations affects graphs in a slightly surprising way. Rewrite the equation for h as $h(x) = \sin(2(x - \frac{\pi}{8}))$ to see that the graph of h can be obtained from the graph of $f(x) = \sin x$ through a horizontal compression and a shift of $\frac{\pi}{8}$ units to the right. Figure 20 shows the graph of $y = \sin(2x)$ and the graph of $y = \sin(2x)$ shifted $\frac{\pi}{8}$ units to the right. This shifted graph is $h(x) = \sin(2x - \frac{\pi}{4})$.

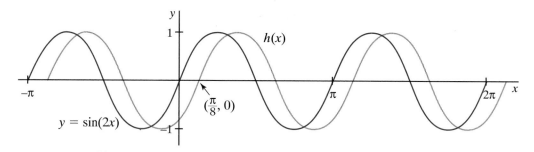

Figure 20   Graphs of $y = \sin(2x)$ and $h(x) = \sin(2x - \frac{\pi}{4})$

Another way to graph $h(x)$ is to study one cycle of $f(x) = \sin x$. Since $f(x)$ completes one cycle when its argument, x, varies from 0 to 2π, $h(x) = \sin(2x - \frac{\pi}{4})$ completes one cycle when its argument, $2x - \frac{\pi}{4}$, varies from 0 to 2π. Thus, h completes one cycle when $0 \le 2x - \frac{\pi}{4} \le 2\pi$. Solve this inequality for x as follows:

$$0 \le 2x - \frac{\pi}{4} \le 2\pi$$

$$\frac{\pi}{4} \le 2x \le \frac{9\pi}{4}$$

$$\frac{\pi}{8} \le x \le \frac{9\pi}{8}.$$

Therefore, $h(x) = \sin(2x - \frac{\pi}{4})$ completes one cycle on the interval $\frac{\pi}{8} \le x \le \frac{9\pi}{8}$. This interval is consistent with the graph shown in Figure 20. The shape of the graph on the interval $\frac{\pi}{8} \le x \le \frac{9\pi}{8}$ is repeated every π units since h has period π. ■

EXAMPLE 5   Solve $\sin x = -0.2$ accurate to four decimal places.

Solution   Use a calculator to graph $y = \sin x$ and $y = -0.2$ on the same axes. (See Figure 21.) Use a trace or intersect feature to find the x-coordinates of the points of intersection of the two curves. These x-values are solutions to the equation $\sin x = -0.2$. One point of intersection occurs at approximately $x = -0.2014$; it is the first point of intersection to the left of the origin. Because the sine curve has a period of 2π, every multiple of 2π added to -0.2014 will produce an x-value whose sine is -0.2. Hence, numbers of the form $x = -0.2014 + k \cdot 2\pi$, where k is an integer, are solutions of $\sin x = -0.2$. However, this set of numbers represents only one-half of the solutions.

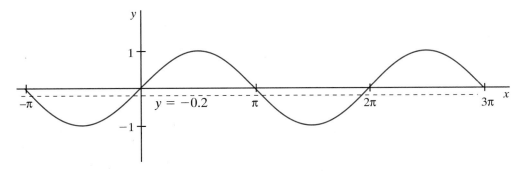

Figure 21 Graphs of $y = \sin x$ and $y = -0.2$

Since the sine curve has symmetry about the vertical line $x = \frac{\pi}{2}$, there is another solution to $\sin x = -0.2$ just to the right of $x = \pi$. This solution is the same distance to the right of π as -0.2014 is to the left of 0 and is $\pi + 0.2014 \approx 3.3430$. Therefore, another solution is $x \approx 3.3430$. Since the sine function has a period of 2π, any number of the form $x = 3.3430 + k \cdot 2\pi$, where k is an integer, is also a solution of the equation $\sin x = -0.2$. Therefore, the complete solution can be written as:

$$x = -0.2014 + k \cdot 2\pi, \text{ or } x = 3.3430 + k \cdot 2\pi, \text{ where } k \text{ is an integer.} \quad \blacksquare$$

Generally in this chapter, we will report values of trigonometric functions and solutions to trigonometric equations accurate to four decimal places.

Exercise Set 5.2

1. A weight suspended from a spring oscillates so that the displacement, d, is related to the elapsed time, t, by the equation $d = -1.3\cos(\pi t)$. Find all t-values at which the displacement is equal to 1.3.

2. Write an equation for the graph in Figure 17 on page 314 that involves a horizontal translation of a cosine function.

3. Write an equation for the graph in Figure 17 that involves a horizontal translation of a sine function.

4. For each of the following functions, sketch at least two complete cycles of the graph. Label the scale on both axes. For each function, write one or two sentences to describe the transformations of the basic toolkit function.

a. $y = 2\cos x$

b. $y = \sin(-2x)$

c. $y = \sin(x - \frac{\pi}{2}) + 1$

d. $y = 1 - \cos(\frac{x}{2})$

e. $y = \sec(\frac{x}{3})$

f. $y = \sin(3\pi x) - 2$

g. $y = \tan(\pi x)$

h. $y = \cos(\pi - x)$

i. $y = \sec(2x) + 1$

j. $y = \csc(-\frac{x}{3})$

k. $y = 3\sin(2x - \frac{\pi}{2})$

l. $y = \tan(\frac{1}{2}x + \frac{\pi}{4})$

m. $y = \cos(2x + \pi)$

n. $y = \sin(\frac{1}{2}x + \pi)$

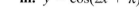

5. Sketch at least two cycles of (x, y). Label the scale on both axes.

a. $x(t) = 3t + 2$
   $y(t) = 2\sin t$

b. $x(t) = \frac{1}{2}t - 1$
   $y(t) = 3\cos t$

6. Use graphing technology to find all the solutions of each equation accurate to four decimal places.

a. $\cos x = 0.4$

b. $\tan x = 10$

c. $\sin x = -0.5$

d. $\cos x = 3$

7. Write an equation of a sine function with the following characteristics.

a. period 2π, amplitude 4, horizontal shift 0

b. period $\frac{\pi}{3}$, amplitude 0.1, horizontal shift $\frac{\pi}{4}$ right

c. period π, amplitude $\frac{1}{3}$, horizontal shift 1 left

d. period 2, amplitude 1, horizontal shift π right

e. period 0.25, amplitude 2, horizontal shift 2 left

f. period 0.0001, amplitude $\frac{1}{2}$, horizontal shift 0

8. Suppose a Ferris wheel has radius 33.2 feet and makes three complete revolutions every minute. For clearance, the bottom of the Ferris wheel is 4 feet above ground. Sketch a graph and write a function that shows how a particular passenger's height above ground varies over time as he or she rides the Ferris wheel. Assume that the passenger is at the bottom of the Ferris wheel when $t = 0$.

9. The rotation of the Ferris wheel in Exercise 8 can be described in two ways. The fact that the wheel takes one-third of a minute to complete a cycle indicates that the period of the motion is one-third minute per cycle. The fact that the wheel completes three revolutions in a minute indicates that the *frequency* of the motion is three cycles per minute. Thus, the same motion can be described either by specifying that the period is one-third or by specifying that the frequency is three. Identify the frequency of each of the sine functions in Exercise 7.

10. In Los Angeles on the first day of summer (June 21) there are 14 hours 26 minutes of daylight, and on the first day of winter (December 21) there are 9 hours 54 minutes of daylight. On average, there are 12 hours 10 minutes of daylight; this average amount occurs on March 20 and September 22.

    a. Sketch a graph that displays this information.

    b. Write an equation that expresses d, the number of minutes of daylight per day, as a function of t, the number of months after June 21.

11. A weight is suspended from a spring that hangs vertically from the ceiling of a room. At time $t = 0$ the weight is pulled down so that its distance from the ceiling is 24.4 cm. When released, it oscillates about its previous equilibrium position. At time $t = \frac{1}{2}$ second it reaches its minimum distance from the ceiling, which is 21.8 cm. Assume that ideal conditions allow the suspended weight to oscillate indefinitely. Identify the period and the frequency of the weight's motion. Write an equation that expresses the weight's distance from the ceiling as a function of time.

12. Use your calculator to graph each function with a viewing window that shows $-2\pi \le x \le 2\pi$ and $-2 \le y \le 2$. How many cycles do you expect to see on the interval $-2\pi \le x \le 2\pi$ for each function? Compare this to the number of cycles you actually see on your calculator screen. Discuss why your calculator seems to make graphing errors.

    a. $f_1(x) = \sin x$

    b. $f_2(x) = \sin(10x)$

    c. $f_3(x) = \sin(100x)$

    d. $f_4(x) = \sin(300x)$

    e. $f_5(x) = \cos(48x)$

frequency

13. A precalculus class went on a field trip to the park. One student started swinging on a swing. Once the student was able to maintain a periodic motion and consistent maximum height, the other class members took the following measurements:

- The height of the swing above ground at equilibrium is 0.33 meter.

- The height of the swing above ground at its highest point is 1.7 meters.

- The horizontal distance from the equilibrium point to the highest position of the swing is 2.26 meters.

- The length of time for five complete swing cycles is 17.1 seconds. (A swing cycle is completed when the swinger returns to the same position and is heading in the same direction.)

Write parametric equations that describe the horizontal and vertical positions of the swinger as a function of time. Graph the equations.

SECTION 5.3 — Investigating a Predator-Prey Relationship

Suppose an isolated island is inhabited by only hawks and doves. The hawks are the predators, and the doves are the prey. Based on observations made over several years, scientists know that the sizes of the hawk and dove populations are interdependent and that both populations exhibit cyclic periods of growth and decline. Based on simplifying assumptions, the hawk and dove populations can be modeled by the following functions, in which t represents the number of years since observations began.

$$hawk\ population = -10\cos\left(\frac{\pi t}{2}\right) + 20$$

$$dove\ population = 100\sin\left(\frac{\pi t}{2}\right) + 150$$

1. Graph the hawk and dove populations over time. Describe the behavior of the populations. Be specific. Include information about maximum and minimum populations and about the period of each function.

2. According to the models, how does the dove population behave when the hawk population reaches a maximum? Is this consistent with your intuition about a predator-prey relationship? How does the dove population behave when the hawk population reaches a minimum? Is this reasonable behavior based on your intuition about a predator-prey relationship?

3. According to the models, how does the hawk population behave when the dove population reaches a maximum? Is this consistent with your intuition about a predator-prey relationship? How does the hawk population behave when the dove population reaches a minimum? Is this reasonable behavior based on your intuition about a predator-prey relationship?

4. With paper and pencil, make a graph of ordered pairs (*hawk population, dove population*) by plotting the values of each population at time t. For instance, at $t = 0$, the hawk population is 10 and the dove population is 150. Therefore, the graph includes the point (10, 150). At $t = 0.5$, the hawk population is 12.9 and the dove population is 220.7. Therefore, the graph of doves versus hawks includes the point (12.9, 220.7). Notice that values of t are invisible on your graph, but they determine the location of each point. Mark on your graph several values of t so that you can see how populations change as t-values change. The graph of doves versus hawks is called a *phase portrait*.

phase portrait

5. Draw arrows on the phase portrait to show the direction that corresponds to increasing t-values. Comment on information represented by the phase portrait and information given by your graphs from Problem 1. What is different about the information presented by the two types of graphs? What is the same? Use the phase portrait to consider the relationships between the hawk and dove

populations. How does the dove population behave when the hawk population is at a maximum? How does the dove population behave when the hawk population is increasing?

6. Using your calculator in parametric mode, you can easily make and modify a phase portrait. Enter the equations for the hawk and dove populations as

$$x(t) = -10\cos\left(\frac{\pi t}{2}\right) + 20$$

$$y(t) = 100\sin\left(\frac{\pi t}{2}\right) + 150.$$

Find a viewing window that will allow you to see the entire phase portrait. What is the smallest set of t-values you can use if you want to see one complete cycle of the phase portrait? Where is the center of the phase portrait? In what way do the values of x and y at the center of the phase portrait relate to the x- and y-values from the graphs of Problem 1?

7. Create a different predator-prey situation by modifying the constants in the equations that model the population sizes so that the phase portrait is circular rather than elliptical.

8. Modify the equations you created in Problem 7. Change the equations so that t-values from 0 to 2π define one complete cycle of the phase portrait.

9. Modify the equations you created in Problem 8 so that the phase portrait is centered at the origin rather than at (20, 150). Note that this modification, even though it makes sense mathematically, results in parametric equations that are no longer realistic population models, since they result in negative population sizes.

10. Create a different phase portrait by modifying the parametric equations so that the phase portrait is a circle of radius 1 centered at the origin and so that t-values from 0 to 2π define one complete cycle of the phase portrait.

11. If necessary, make a final modification to your parametric equations so that the phase portrait goes through the point (1, 0) when $t = 0$ and through the point (0, 1) when $t = \frac{\pi}{2}$.

SECTION 5.4 Sine and Cosine on the Unit Circle

unit circle

In the investigation about hawks and doves, we saw that the graph of the parametric equations $x(t) = \cos t$ and $y(t) = \sin t$ is a circle of radius 1 whose center is at the origin. This circle is called the *unit circle*.

CLASS PRACTICE

Sketch a unit circle. Imagine in each question that you start at the point (1, 0) and walk in a counterclockwise direction along an arc on the circumference of the circle.

1. If you walk 2π units and then stop, where are you? What is your x-coordinate?

2. If you walk π units and then stop, where are you? What is your x-coordinate? What is your y-coordinate?

3. If you walk $\frac{\pi}{2}$ units and then stop, where are you? What are your x- and y-coordinates?

4. If you walk $\frac{\pi}{2}$ units in a clockwise direction and then stop, what are your x- and y-coordinates?

5. Suppose you walk t units, then stop and observe that your x-coordinate is 0. What are some of the possible values for t?

6. Suppose you walk t units, then stop and observe that your y-coordinate is 0. What are some of the possible values for t?

We will define traveling counterclockwise on the unit circle to be the positive direction and clockwise to be the negative direction. Assume that all arcs begin from (1, 0) unless specified otherwise. The table in Figure 22 lists various lengths for arcs, together with the coordinates of the arc's endpoint.

Arc length	Endpoint of arc
0	(1, 0)
$\frac{\pi}{2}$	(0, 1)
π	(−1, 0)
$\frac{3\pi}{2}$	(0, −1)
2π	(1, 0)
$-\frac{\pi}{2}$	(0, −1)

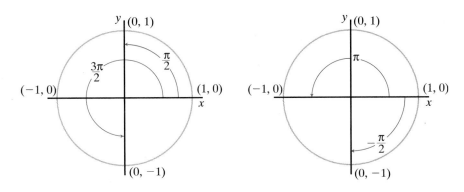

Figure 22 Table of lengths and endpoints of arc lengths and location on unit circle

As you study the entries in the table on the previous page, the ordered pairs (*length of arc, x-coordinate at endpoint of arc*). For instance, an arc length of 0 is paired with an *x*-coordinate of 1, an arc length of π is paired with an *x*-coordinate of −1, and an arc length of $\frac{3\pi}{2}$ is paired with an *x*-coordinate of 0. These ordered pairs are graphed in Figure 23. You should recognize the shape of the cosine curve. The *x*-coordinate of the endpoint of the arc varies from 1 to −1 as the length of the arc changes; this is why the range of the cosine function includes all real numbers from 1 to −1. Whether the arc length is *t* units or *t* + 2π units, the arc ends at the same point. This is why values in the range of the cosine function repeat every 2π units, and the cosine function has period 2π.

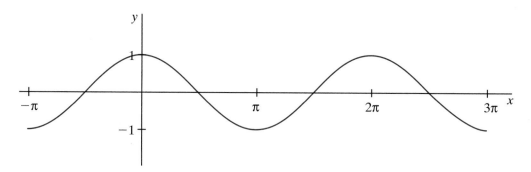

Figure 23   Ordered pairs (*length of arc, x-coordinate of endpoint*)

Now consider the ordered pairs (*length of arc, y-coordinate at endpoint of arc*), which include (0, 0), ($\frac{\pi}{2}$, 1), ($-\frac{\pi}{2}$, −1), and so forth. These ordered pairs are graphed in Figure 24; you should recognize the shape of the sine curve.

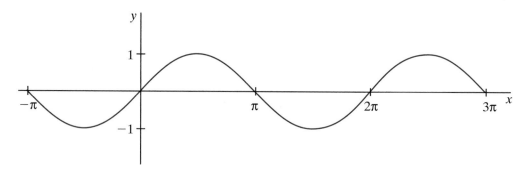

Figure 24   Ordered pairs (*length of arc, y-coordinate of endpoint*)

If we let *t* represent the length of an arc starting at (1, 0) and moving counterclockwise on the unit circle, then

$$\cos t = \text{the } x\text{-coordinate of the arc endpoint,}$$

and

$$\sin t = \text{the } y\text{-coordinate of the arc endpoint.}$$

circular functions

These definitions of $\sin t$ and $\cos t$ are a precise way of describing the sine and cosine functions we have already studied.

The sine and cosine functions are often called *circular functions* since their definitions are based on a unit circle. Figure 25 shows an important geometric interpretation of sine and cosine. If point P is the endpoint of an arc of length t that begins at $(1, 0)$ on the unit circle, then the coordinates of P are $(\cos t, \sin t)$.

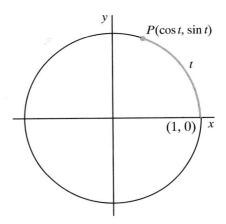

Figure 25   Unit circle interpretation of sine and cosine

Since the coordinates of all points on the unit circle satisfy the equation $x^2 + y^2 = 1$, it follows that

$$(\cos t)^2 + (\sin t)^2 = 1.$$

We usually write $(\cos t)^2$ as $\cos^2 t$, so this equation can be written as

$$\cos^2 t + \sin^2 t = 1.$$

identity

This equation is an *identity* since it is true for all values of the variable t. You can use your calculator to verify that $\cos^2(0.436) + \sin^2(0.436) = 1$ and $\cos^2(-5.278) + \sin^2(-5.278) = 1$. The fact that $\cos^2 t + \sin^2 t = 1$ is an identity means that in any equation or expression, $\cos^2 t + \sin^2 t$ can be replaced by 1, $\cos^2 t$ can be replaced by $1 - \sin^2 t$, and $\sin^2 t$ can be replaced by $1 - \cos^2 t$.

EXAMPLE 1   An ant starts at $(1, 0)$ and walks in a counterclockwise direction 4 units along the circumference of a unit circle. In what quadrant does the ant stop? What are the coordinates of the point where the ant stops?

Solution The ant walked more than halfway around the circle since $4 > \pi$. Since $4 < \frac{3\pi}{2}$, the ant walked less than three-quarters of the way around the circle. Therefore, it ended its trip in the third quadrant. (See Figure 26.)

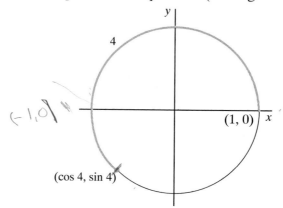

Figure 26 Arc length of 4 units

The coordinates of the point where the ant stops are $(\cos 4, \sin 4)$. Using your calculator in radian mode, you can approximate $\cos 4$ and $\sin 4$. You should find that the ant stops at approximately $(-0.6536, -0.7568)$. Verify that $(-0.6536)^2 + (-0.7568)^2$ is approximately 1. ■

Exercise Set 5.4

1. Sketch a unit circle to evaluate each of the following. Verify your answers using your calculator.

    a. $\sin(-\frac{3\pi}{2})$ **b.** $\cos(-\frac{3\pi}{2})$ **c.** $\sec(\frac{33\pi}{2})$

    d. $\csc(33\pi)$ **e.** $\tan(-49\pi)$ **f.** $\sin(-49\pi)$

    g. $\sin(7\pi)$ **h.** $\cos(7\pi)$ **i.** $\tan(\frac{7\pi}{2})$

2. Determine whether each of the following is approximately equal to 1, approximately equal to 0, or approximately equal to -1. Use your calculator to check your answers.

    a. $\sin 3$ **b.** $\sin(-1.6)$ **c.** $\cos 0.04$

    d. $\cos 6$ **e.** $\cos 1.6$ **f.** $\sin(-0.02)$

3. An ant starts at $(1, 0)$ and walks in a counterclockwise direction 5 units along the circumference of a unit circle. In what quadrant does the ant stop? What are the coordinates of the point where the ant stops?

4. Suppose an ant starts at (1, 0) and walks in a clockwise direction 2 units along the circumference of a unit circle. In what quadrant does the ant stop? What are the coordinates of the point where the ant stops?

5. The Unit Circle Race is to be run on a circular track of radius 1 mile. The race will begin and end at the point (1, 0) and will be run in the counterclockwise direction.

   a. Clark has traveled 1.2 miles. Sketch a unit circle and label his location on the circle. What are his x- and y-coordinates?

   b. Josie has the same y-coordinate as Clark, but she is in a different location. Label her possible locations on the circle. At each location, how many miles has she run?

   c. If a runner could run around the track an unlimited number of times, what are all the possible distances he or she could run and end up at Josie's location?

   d. Crystal runs d miles and is still in the first quadrant. Her x-coordinate is 0.2 mile. Label her location on the circle. Find her y-coordinate without finding d.

   e. Crystal decides to take a shortcut. She runs from her current location (in Part d) directly to a point on the track that is a distance of $d + \frac{3\pi}{2}$ miles from the start of the race. Label her new location on the circle. Use symmetry to find her new x- and y-coordinates.

   f. Michael is a judge for the race. He is standing at the point (0, 1) on the circle's circumference. He walks 0.4 mile in a clockwise direction along the track back toward the race's starting place. How far along the track is he from the starting place?

   g. Maggie walks 0.4 mile from the starting place in a counterclockwise direction. How do her x- and y-coordinates compare to Michael's x- and y-coordinates?

6. We will now define two new functions that are variations of the sine and cosine functions. Replace the unit circle with a square whose center is at the origin. (See Figure 27 on the next page.)

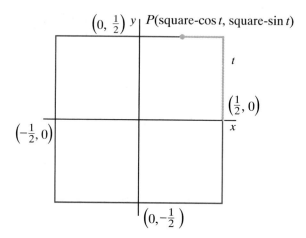

$\left(0, \frac{1}{2}\right)$ y $P(\text{square-cos } t, \text{square-sin } t)$

t

$\left(\frac{1}{2}, 0\right)$

x

$\left(-\frac{1}{2}, 0\right)$

$\left(0, -\frac{1}{2}\right)$

Figure 27 Unit square

A person starts at $\left(\frac{1}{2}, 0\right)$ and walks t units along the perimeter of the square. The two new functions we define will be called square-sine and square-cosine. Let square-sin(t) be equal to the y-coordinate of the point where the person stops on the square, and let square-cos(t) be equal to the x-coordinate of the point where the person stops on the square. For example, square-sin$(1) = \frac{1}{2}$.

a. Find the values of square-sin$\left(\frac{1}{2}\right)$, square-sin$\left(-\frac{7}{2}\right)$, square-sin$(26)$, square-cos$(-1)$, square-cos$\left(-\frac{151}{2}\right)$, and square-cos$(62)$.

b. Make a sketch of the square-sine function. Is this function periodic? If so, what is the period? What are the domain and the range of this function?

c. Sketch the square-cosine function.

d. Can the square-sine function be shifted to obtain the square-cosine function? If so, write an equation to express this relationship.

SECTION 5.5 Getting to Know the Unit Circle

In Section 5.4 the sine and cosine functions were defined using a unit circle. The unit circle has many symmetries that can help you evaluate values of the circular functions. In this section we will explore those symmetries and relate them to the sine and cosine functions.

The circle $x^2 + y^2 = 1$ is symmetric about the x-axis, the y-axis, and the origin. Therefore, if the coordinates of one point on the circle are known, the coordinates of three other points can easily be determined. For example, in Figure 28 point P has coordinates (a, b). Because of symmetry about the y-axis, the coordinates of point Q in the second quadrant are $(-a, b)$. Because of symmetry about the origin, the coordinates of point R in the third quadrant are $(-a, -b)$. Because of symmetry about the x-axis, the coordinates of point S in the fourth quadrant are $(a, -b)$.

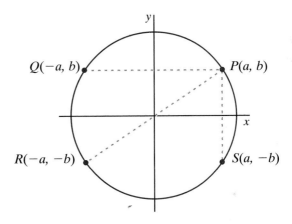

Figure 28 Symmetry of the unit circle

How does this symmetry relate to the circular functions?

How does this symmetry relate to the circular functions? To explore this question, suppose an arc of length t begins at $(1, 0)$ and ends at point P. What length arc would terminate at Q? Since a person starting at $(1, 0)$ can get to Q by walking π units in the positive direction followed by t units in the negative direction, an arc of length $\pi - t$ will end at Q. Similarly, an arc of length $\pi + t$ will end at R. Verify that an arc of length $2\pi - t$ and an arc of length $-t$ will both end at S. (See Figure 29 on the next page.)

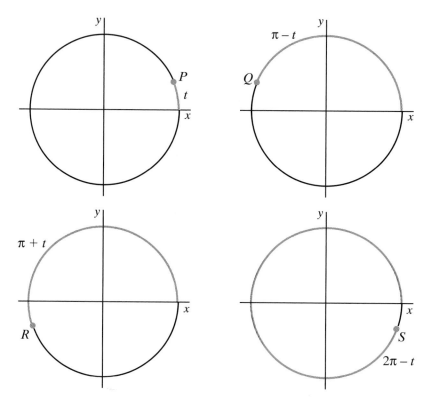

Figure 29 Arc lengths on the unit circle

The information in Figure 28 and Figure 29 can be combined with the definitions of sine and cosine to yield some important relationships. Recall that $\sin t$ is equal to the y-coordinate of P and $\cos t$ is equal to the x-coordinate of P. Since P and Q have the same y-coordinate, $\sin t$ and $\sin(\pi - t)$ are equal. Similarly, since P and S have the same x-coordinate, $\cos t$, $\cos(2\pi - t)$, and $\cos(-t)$ are equal. Since P and R have opposite x-coordinates and opposite y-coordinates, $\sin(\pi + t) = -\sin t$ and $\cos(\pi + t) = -\cos t$. Since each equation is true for any value of t, we refer to each of these as an identity.

EXAMPLE 1 Suppose you know that $\sin t = \frac{5}{13}$. Find the exact value of $\sin(\pi - t)$, $\sin(\pi + t)$, and $\cos t$.

Solution Because of the symmetry of the circle, arcs of length t and $\pi - t$ have endpoints with the same y-coordinates, and therefore have the same sine. Thus, $\sin(\pi - t) = \frac{5}{13}$. The endpoints of arcs of length t and $\pi + t$ have opposite y-coordinates, and therefore have opposite values of sine. Thus, $\sin(\pi + t) = -\frac{5}{13}$.

Since the sine of t is a positive number, t must end in either the first or second quadrant. If t ends in the first quadrant, then $\pi - t$ ends in the second quadrant and $\pi + t$ ends in the third quadrant, which is shown in Figure 30. If t ends in the second quadrant, then $\pi - t$ ends in the first quadrant and $\pi + t$ ends in the fourth quadrant, which is shown in Figure 31. Both of these possibilities are consistent with the fact that $\sin(\pi - t)$ is positive and $\sin(\pi + t)$ is negative.

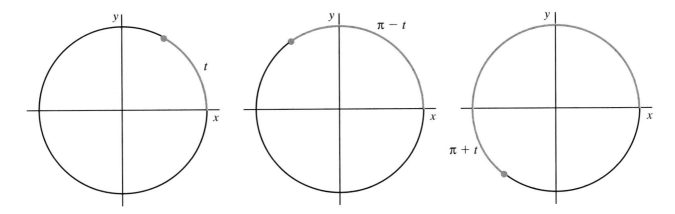

Figure 30   Arcs of length t, $\pi - t$, and $\pi + t$ where $0 < t < \frac{\pi}{2}$

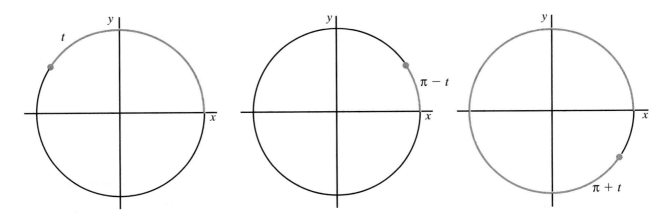

Figure 31   Arcs of length t, $\pi - t$, and $\pi + t$, where $\frac{\pi}{2} < t < \pi$

To find the values of $\cos t$, we will use the fact that $\sin t$ and $\cos t$ are the y- and x-coordinates of a point on the unit circle $\sin^2 t + \cos^2 t = 1$ and thus, $\cos^2 t = 1 - \sin^2 t$. Since $\sin t = \frac{5}{13}$, $\cos^2 t = 1 - \left(\frac{5}{13}\right)^2$, and therefore,

$$\cos t = \pm \sqrt{1 - \frac{25}{169}} = \pm \sqrt{\frac{144}{169}} = \pm \frac{12}{13}.$$

We would need to know whether t ends in the first or the second quadrant to determine whether the cosine of t is positive or negative. Note that $\pm \frac{12}{13}$ are exact values for $\cos t$, and are not decimal approximations. If decimal approximations are used, $\sin t \approx 0.3846$, and $\cos t \approx \pm \sqrt{1 - (0.3846)^2} \approx \pm 0.9231$, which is not exactly equal to $\pm \frac{12}{13}$. Also, $(0.3846)^2 + (\pm 0.9231)^2$ does not exactly equal 1. Rather than use the approximately equals notation ($\approx$) every time we use a decimal representation from a calculator, we will continue to use the convention of writing the values accurate to four decimal places. ■

EXAMPLE 2   Suppose an arc of length θ terminates in the second quadrant and $\sin\theta = 0.62$. (See Figure 32.) Find $\sin(\pi - \theta)$, $\cos\theta$, and $\tan(-\theta)$.

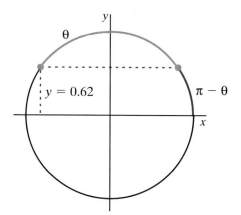

Figure 32   Arc length for Example 2

Solution   Again, we use the symmetry of the circle to recognize that θ and π − θ have the same sine. Thus,

$$\sin(\pi - \theta) = 0.62.$$

We know that $\sin^2\theta + \cos^2\theta = 1$, which implies that $\cos\theta = \pm\sqrt{1 - \sin^2\theta}$. Since θ terminates in the second quadrant, its cosine must be negative, so

$$\cos\theta = -\sqrt{1 - (0.62)^2} = -0.7846.$$

Since the tangent function is defined as $\frac{\sin\theta}{\cos\theta}$,

$$\tan(-\theta) = \frac{\sin(-\theta)}{\cos(-\theta)} = \frac{-\sin\theta}{\cos\theta} = \frac{-0.62}{-0.7846} = 0.7902. \quad ■$$

EXAMPLE 3   On a unit circle, an arc ends at a point whose x-coordinate is 0.45. Find the length of the arc.

inverse cosine

Solution   There are many different arcs that will terminate at a point whose x-coordinate is 0.45. We need to find the values of t that satisfy $\cos t = 0.45$, where t is the length of the arc. One way to solve this equation is to find the points of intersection of the graphs of $y = \cos t$ and $y = 0.45$. Another way to solve the equation involves using the notation $t = \cos^{-1}0.45$. This notation means that t is a number whose cosine is 0.45; t is called the *inverse cosine* or *arccosine* of 0.45. Thus, the notation $\cos^{-1}0.45$, which is read "the inverse cosine of 0.45," represents a number whose cosine is 0.45. Use your calculator (in radian mode) to find a number whose cosine is 0.45. You should find that $\cos^{-1}0.45 = 1.104031$. Thus, an arc about 1.104 units long will terminate at a point whose x-coordinate is 0.45. Remember that arcs of length 1.104 units and $1.104 + 2\pi$ units and $1.104 + 4\pi$ units will all end at the same point. In fact, all numbers of the form $1.104 + 2k\pi$, where k is an integer, are solutions of $\cos t = 0.45$ accurate to three decimal places.

You can find other values of t that satisfy $\cos t = 0.45$ by using the symmetry of the circle. Based on symmetry, $\cos(-t) = \cos t$. Since $\cos(1.104) = 0.45$, it is also true that $\cos(-1.104) = 0.45$. Because the circumference of the circle is 2π, the numbers -1.104, $-1.104 + 2\pi$, and so on, all have cosine equal to 0.45. The complete set of possible arc lengths is $1.104 + 2k\pi$ or $-1.104 + 2k\pi$, where k is an integer. ▧

CLASS PRACTICE

1. If θ ends in the first quadrant and $\cos\theta = \frac{2}{3}$, find the exact value of $\sin\theta$ (without finding θ).

2. If θ ends in the first quadrant and $\cos\theta = \frac{2}{3}$, find the value of θ. Then find the value of $\sin\theta$.

3. Compare the exact value of $\sin\theta$ from Problem 1 to the decimal approximation you obtained in Problem 2 above.

We can evaluate the sine and cosine of some numbers without using technology. For instance, we know $\sin\frac{\pi}{2} = 1$ and $\cos\pi = -1$ from the definitions of sine and cosine as circular functions and from the graphs of these functions. There are several other special numbers whose sine and cosine can be determined without technology.

EXAMPLE 4   Evaluate $\sin\frac{\pi}{4}$ and $\cos\frac{\pi}{4}$.

Solution   Let point P with coordinates (c, s) be the endpoint of the arc that begins at $(1, 0)$ and has length $\frac{\pi}{4}$. Note that $c = \cos\frac{\pi}{4}$ and $s = \sin\frac{\pi}{4}$. Figure 33 shows that P lies on the line $y = x$, which implies that $c = s$.

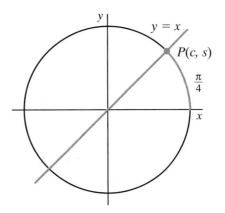

Figure 33   Arc of length $\frac{\pi}{4}$ on the unit circle

Since P is on the unit circle, its coordinates must satisfy the equation of the circle, which is $x^2 + y^2 = 1$. Therefore,

$$c^2 + s^2 = 1.$$

Substituting $c = s$ gives

$$c^2 + c^2 = 1$$
$$2c^2 = 1$$
$$c^2 = \frac{1}{2}$$
$$c = \sqrt{\frac{1}{2}} = \frac{\sqrt{2}}{2}.$$

We chose the positive square root for the value of c because P is in the first quadrant. Since $c = s$, $\sin\frac{\pi}{4} = \frac{\sqrt{2}}{2}$ and $\cos\frac{\pi}{4} = \frac{\sqrt{2}}{2}$. Using the values of $\sin\frac{\pi}{4}$ and $\cos\frac{\pi}{4}$, you can evaluate sine and cosine of $\frac{3\pi}{4}$, $\frac{5\pi}{4}$, and $\frac{7\pi}{4}$ based on the circle's symmetries. ■

EXAMPLE 5   Evaluate $\sin\frac{\pi}{3}$ and $\cos\frac{\pi}{3}$.

Solution   Let point B be the endpoint of the arc that begins at $(1, 0)$ and has length $\frac{\pi}{3}$. (See Figure 34.) We can use some geometric facts to find the coordinates of B.

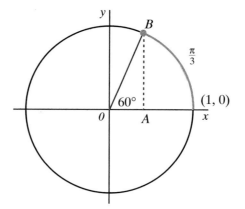

Figure 34   Arc of length $\frac{\pi}{3}$ on the unit circle

The length of $\overline{BO}$ is 1 since $\overline{BO}$ is a radius of the circle. From B draw a line segment that is perpendicular to the x-axis at A. The length of this line segment is equal to the y-coordinate of B, which is equal to $\sin\frac{\pi}{3}$. The length AO represents the x-coordinate at B, which is equal to $\cos\frac{\pi}{3}$.

If we can find the lengths of $\overline{AB}$ and $\overline{AO}$, then we will know $\sin\frac{\pi}{3}$ and $\cos\frac{\pi}{3}$. Since $\frac{\pi}{3}$ is one-sixth of 2π, the measure of $\angle BOA$ is one-sixth of $360°$, or $60°$. This means that triangle AOB is a 30-60-90 triangle. The side opposite the $30°$ angle is one-half as long as the hypotenuse, and the side opposite the $60°$ angle is $\frac{\sqrt{3}}{2}$ times as long as the hypotenuse. Therefore, $AO = \frac{1}{2}$ and $AB = \frac{\sqrt{3}}{2}$. We can conclude that $\sin\frac{\pi}{3} = \frac{\sqrt{3}}{2}$ and $\cos\frac{\pi}{3} = \frac{1}{2}$.

The symmetry of the unit circle and the values of $\sin\frac{\pi}{3}$ and $\cos\frac{\pi}{3}$ can be used to evaluate the sine and cosine of $\frac{2\pi}{3}$, $\frac{4\pi}{3}$, and $\frac{5\pi}{3}$. ■

Exercise Set 5.5

1. Use the symmetry of the unit circle to explain why cosine is an even function.

2. Use the symmetry of the unit circle to explain why sine is an odd function.

3. Evaluate $\sin\frac{\pi}{6}$ and $\cos\frac{\pi}{6}$ using geometric arguments like those used in Example 5 on page 334. You should determine exact values and then compare them to the decimal approximations you can obtain with your calculator.

4. Label a unit circle with the endpoints of the arcs which begin at $(1, 0)$ and have length $\frac{\pi}{6}, \frac{\pi}{4}, \frac{\pi}{3}, \frac{\pi}{2}, \frac{2\pi}{3}, \frac{3\pi}{4}, \frac{5\pi}{6}, \pi, \frac{7\pi}{6}, \frac{5\pi}{4}, \frac{4\pi}{3}, \frac{3\pi}{2}, \frac{5\pi}{3}, \frac{7\pi}{4}$, and $\frac{11\pi}{6}$. Use exact values you know and symmetries of the unit circle to label the coordinates of each endpoint.

5. Use the symmetry of the circle and your knowledge of the exact values determined in Example 4, Example 5, and Exercise 3 to evaluate the following without using your calculator.

   a. $\sin(-\frac{5\pi}{4})$ b. $\sin(\frac{71\pi}{6})$ c. $\sin(\frac{5\pi}{3})$

   d. $\cot(\frac{23\pi}{3})$ e. $\cos(\frac{33\pi}{4})$ f. $\csc(\frac{57\pi}{3})$

   g. $\tan(-\frac{17\pi}{6})$ h. $\tan(-\frac{33\pi}{4})$ i. $\sec(-\frac{11\pi}{4})$

6. If $\cos t = -\frac{1}{4}$ and t ends in the second quadrant, find the exact values of $\sin t$ and $\tan t$ without finding t.

7. If $\sin\theta = -\frac{9}{25}$ and θ ends in the fourth quadrant, find the exact value of each of the following. (Hint: First finding θ on a calculator will not give an exact value.)

   a. $\sin(-\theta)$ b. $\cos\theta$

   c. $\tan(\pi - \theta)$ d. $\tan(\pi + \theta)$

8. Suppose $\sin t = 0.14$ and one value of t is approximately 3.00, accurate to two decimal places. Use symmetry to find all other possible values for t, accurate to two decimal places.

9. Suppose $\cos t = 0.85$ and one value of t is approximately 0.5548, accurate to four decimal places. Use symmetry to find all other possible values for t, accurate to four decimal places.

10. Suppose $\sin t = 0.3$ and t ends in the second quadrant. Find a value of t accurate to four decimal places.

11. Suppose $\cos t = -0.548$ and t ends in the third quadrant. Find a value of t accurate to four decimal places.

12. Suppose that an arc of length a ends in the first quadrant at a point whose x-coordinate is 0.44. Find $\sin a$, $\cos(a + \pi)$, and $\sin(2\pi - a)$ without first finding the value of a.

13. Suppose $\cos \theta = w$ and θ ends in the fourth quadrant. Express each of the following in terms of w.

    a. $\sin \theta$ b. $\tan \theta$

    c. $\cos(\frac{\pi}{2} - \theta)$ d. $\sec(\pi - \theta)$

    e. $\sin(\frac{\pi}{2} + \theta)$ f. $\cot(-\theta)$

14. Suppose $\cos \theta = q$ and θ ends in the third quadrant. Express each of the following in terms of q.

    a. $\sin \theta$ b. $\tan \theta$

    c. $\cos(\frac{\pi}{2} - \theta)$ d. $\csc(\frac{\pi}{2} - \theta)$

    e. $\sin(\pi + \theta)$ f. $\cot(-\theta)$

Angles and Radians

We have defined $\cos t$ and $\sin t$ as the x-and y-coordinates of the endpoint of an arc of length t on the unit circle. There is another way to think about cosine and sine. In many situations we use cosine and sine to represent ratios of lengths in a right triangle. You may have studied this approach when you studied right triangle trigonometry in a previous mathematics class. For instance, $\sin 60°$ and $\cos 60°$ represent ratios of lengths of sides, and 60 represents the degree measure of an angle in a right triangle. (See Figure 35.)

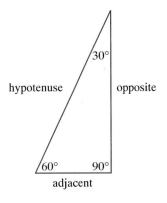

Figure 35   30-60-90 triangle

If one of the angles of a right triangle is $60°$, as shown in Figure 35, then $\cos 60°$ is equal to the length of the adjacent side divided by the length of the hypotenuse, and $\sin 60°$ is equal to the length of the opposite side divided by the length of the hypotenuse.

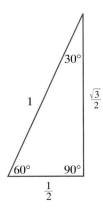

Figure 36   30-60-90 triangle with unit hypotenuse

If a $60°$ angle is in a right triangle whose hypotenuse is 1 unit long, then the geometric relationships among the parts of the triangle determine that the adjacent leg is 0.5 unit long and the opposite leg is $\frac{\sqrt{3}}{2} \approx 0.8660$ unit long. (See Figure 36.) If the hypotenuse were twice as long, then each of the legs would be twice as long. The ratios of the lengths of the sides would not change. Therefore, the values of cos 60° and sin 60° are the same in any right triangle, no matter how large or small. The value of cos 60° is

$$\frac{\frac{1}{2}}{1} = \frac{1}{2},$$

and the value of sin 60° is

$$\frac{\frac{\sqrt{3}}{2}}{1} = \frac{\sqrt{3}}{2}.$$

These ideas can be generalized using the right triangle illustrated in Figure 37.

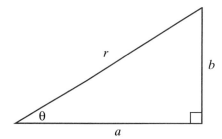

Figure 37   Right triangle with angle θ

The relationships are summarized below:

$$\sin \theta = \frac{b}{r} = \frac{\text{length of side opposite } \theta}{\text{length of hypotenuse}} = \frac{\text{opp}}{\text{hyp}},$$

$$\cos \theta = \frac{a}{r} = \frac{\text{length of side adjacent } \theta}{\text{length of hypotenuse}} = \frac{\text{adj}}{\text{hyp}},$$

and

$$\tan \theta = \frac{b}{a} = \frac{\text{length of side opposite } \theta}{\text{length of side adjacent to } \theta} = \frac{\text{opp}}{\text{adj}}.$$

These ratios have many applications.

EXAMPLE 1   A plane takes off from the airport in Chicago and flies on a heading of 35°. If the pilot maintains the same course, how far north of the airport is the plane after flying 100 miles?

Solution   The situation is illustrated in Figure 38; the airport is shown at the origin of a coordinate system.

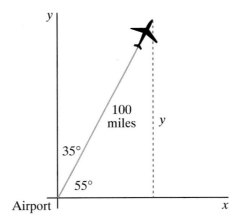

Figure 38 Plane on 35° heading

By definition, when a plane is flying on a heading of 35°, the angle between due north (the positive *y*-axis) and the plane's path is 35°. Therefore, the angle between the plane's path and the positive *x*-axis is 55°. Using the ratio $\sin\theta = \frac{\text{opp}}{\text{hyp}}$ in the right triangle,

$$\frac{y}{100} = \sin 55°$$

$$y = 100\sin 55°$$

$$y = 81.915.$$

The plane is approximately 81.9 miles north of Chicago after flying a distance of 100 miles.

EXAMPLE 2   In the right triangle *ABC*, *a* = 3, *b* = 4, and *c* is the length of the hypotenuse. (See Figure 39.) Find the measure of ∠*A*. (Notice that it is conventional to label the length of a side of a triangle with the lowercase letter that is used to label the opposite angle.)

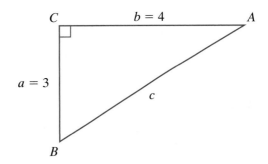

Figure 39   Right triangle *ABC* for Example 2

Solution This problem can be solved in several different ways. The sides whose lengths are known are opposite and adjacent to $\angle A$. This suggests we use the tangent function to determine the measure of $\angle A$. We know that $\tan A = \frac{\text{opp}}{\text{adj}} = \frac{3}{4} = 0.75$. Therefore, $A = \tan^{-1} 0.75$ and we can use a calculator in degree mode to show that the measure of $\angle A$ is about $36.9°$. (Generally in this textbook, degree measures of angles will be given to the nearest tenth of a degree.)

Another way to solve this problem is to first use the Pythagorean Theorem to find that the hypotenuse is 5 units long. We know that $\sin A = \frac{\text{opp}}{\text{hyp}} = \frac{3}{5} = 0.60$. Thus, $A = \sin^{-1} 0.60$ and $A = 36.9°$. Finally, we know that $\cos A = \frac{\text{adj}}{\text{hyp}} = \frac{4}{5} = 0.80$. This means that $A = \cos^{-1} 0.80$ and $A = 36.9°$. ■

How do the right triangle interpretations of sine and cosine relate to the unit circle interpretation of the trigonometric functions?

How do the right triangle interpretations of sine and cosine relate to the unit circle interpretation of the trigonometric functions? Figure 40 shows a right triangle with a $20°$ angle embedded in a unit circle. The angle is positioned in the coordinate system so that its vertex is at the origin and one leg lies along the positive x-axis. The length of the hypotenuse of the right triangle is 1. The $20°$ angle subtends (marks off) the arc from $(1, 0)$ to point P along the circumference. The x-coordinate of point P is equal to the length of the leg adjacent to the $20°$ angle, and the y-coordinate of point P is equal to the length of the leg opposite the $20°$ angle. The right triangle interpretation tells us that the leg opposite the $20°$ angle has length $\sin 20° = 0.3420$ unit. Similarly, the leg adjacent to the $20°$ angle is $\cos 20° = 0.9397$ unit long. Thus, based on right triangles, we know that the coordinates of point P are approximately $(0.9397, 0.3420)$. Since $20°$ is $\frac{1}{18}$ of $360°$, the length of the arc from $(1, 0)$ to P is $\frac{1}{18}$ of the circumference of the circle. The arc is therefore $\frac{1}{18} \cdot 2\pi = \frac{\pi}{9}$ units long.

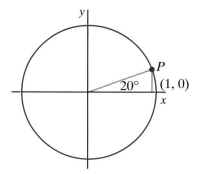

Figure 40 Right triangle embedded in a unit circle

According to the unit circle definitions of cosine and sine stated in Section 5.4,

$$\cos\left(\frac{\pi}{9}\right) = x\text{-coordinate of } P,$$

$$\sin\left(\frac{\pi}{9}\right) = y\text{-coordinate of } P.$$

Use a calculator in the radian mode to confirm that $\cos(\frac{\pi}{9}) = 0.9397$ and $\sin(\frac{\pi}{9}) = 0.3420$. This shows that the right triangle interpretations of $\cos 20°$ and $\sin 20°$ are consistent with the unit circle interpretations of $\cos(\frac{\pi}{9})$ and $\sin(\frac{\pi}{9})$.

CLASS PRACTICE

1. Sketch a 90° angle with its vertex at the center of a unit circle. What arc length does this angle subtend? Now sketch a 90° angle with its vertex at the center of a circle of radius 5. What arc length does this angle subtend? What arc length will the angle subtend in a circle of radius $\frac{1}{2}$?

2. Draw four concentric circles with centers at (0, 0) and radii of lengths 1, 2, 3, and 6 units, respectively. On the circle of radius 1, start at (1, 0) and draw an arc of length 1, and label its endpoint. On the circle of radius 2, start at (2, 0) and draw an arc of length 2, and label its endpoint. On the circle of radius 3, start at (3, 0) and draw an arc of length 3. On the circle of radius 6, start at (6, 0) and draw an arc of length 6. For each circle, compute the value of the ratio $\frac{\text{arc length}}{\text{radius}}$.

3. How are the values of the ratios $\frac{\text{arc length}}{\text{radius}}$ related to each other in Problem 2? How are the angles that the arc lengths subtend related to each other? How are the endpoints of the different arc lengths related?

For each of the arc lengths you drew in Class Practice Problem 2, you should have found that the ratio $\frac{\text{arc length}}{\text{radius}}$ was constant. In each circle the ratio $\frac{\text{arc length}}{\text{radius}}$ had a value of 1. Even though the lengths of the arc changed and the radii changed, the ratio of arc length to radius remained constant. The fact that this ratio is constant is the basis for a unit of angle measure called the *radian*. The radian measure of an angle is defined to be the ratio of the subtended arc length to the radius of the circle.

radian

$$\text{radian measure} = \frac{\text{arc length}}{\text{radius}}$$

An angle whose radian measure is 1 subtends an arc that is the same length as the radius of the circle.

One radian is the measure of an angle if the length of the subtended arc is equal to the radius. Thus, in a unit circle a 1 radian angle subtends an arc of length 1. In a circle of radius 3, a 1 radian angle subtends an arc of length 3, an angle of 2 radians subtends an arc of length 6. (See Figure 41 on the next page.)

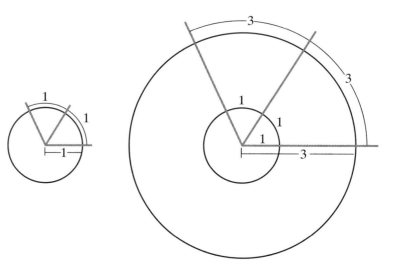

Figure 41 Radian measure

In a unit circle, an angle of 2π radians subtends an arc length of 2π. This arc length is equal to the circumference of the circle, so the angle must be equal to 360°. An angle of π radians in a unit circle subtends half the circumference, so the angle is equal to 180°. In general, the ratio of an angle's degree measure to 360 is equal to the ratio of an angle's radian measure to 2π. This relationship can be expressed with the proportion

$$\frac{d}{360} = \frac{r}{2\pi},$$

where d is an angle's degree measure and r is an angle's radian measure. Letting r equal one radian and solving for d reveals that one radian is approximately 57.3°.

Familiar angles measured in radians have the following degree measures:

$$\pi \text{ radians} = 180°$$

$$\frac{\pi}{6} \text{ radians} = 30°,$$

$$\frac{\pi}{4} \text{ radians} = 45°,$$

$$\frac{\pi}{3} \text{ radians} = 60°,$$

$$\frac{\pi}{2} \text{ radians} = 90°.$$

You should learn these measures and multiples of each, such as $\frac{2\pi}{3}$, $\frac{5\pi}{4}$, and $\frac{11\pi}{6}$.

EXAMPLE 3 In Figure 42, the arc length is $\frac{5\pi}{4}$ inches and the radius of the circle is 2 inches. Find the radian measure of the angle.

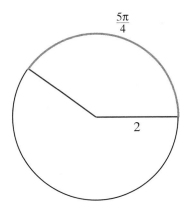

$$\frac{5\pi}{4}$$

$$2$$

Figure 42   Arc length in circle of radius 2

Solution   Since the radian measure is equal to the ratio of arc length to radius, the radian measure of the angle is

$$\frac{\frac{5\pi}{4} \text{ inches}}{2 \text{ inches}}, \quad \text{or} \quad \frac{5\pi}{8} \text{ radians.}$$

Notice that the arc length and the radius were both measured in inches; the units cancel in their ratio, leaving a number with no units. This means that a radian must be regarded as a quantity without units, often called a dimensionless unit. Hence, when we write $\sin\left(\frac{5\pi}{8}\right)$, we assume $\frac{5\pi}{8}$ is measured in radians, since no unit of measure is shown. ▨

At this point we know two definitions of sine and cosine. One definition is related to the coordinates of the endpoint of an arc on the unit circle. The other definition is based on the ratio of the lengths of the sides in a right triangle. Though these definitions may seem different, we can show that they are equivalent in the first quadrant of a unit circle. When we work with $\sin t$ in a unit circle, we can think of t either as the length of arc or as the radian measure of the central angle that subtends that arc. (See Figure 43.) This results from the fact that

$$\text{radian measure} = \frac{\text{arc length}}{\text{radius}} = \frac{\text{arc length}}{1} = \text{arc length}$$

in a unit circle. In Figure 43 we show $\sin t$ as the y-coordinate of point P, which is the endpoint of the arc that has length t. Now examine the embedded right triangle with the acute angle measuring t radians. In this triangle,

$$\sin t = \frac{\text{length of side opposite } t}{\text{length of hypotenuse}} = \frac{y\text{-coordinate of } P}{1} = y\text{-coordinate of } P.$$

Thus, the endpoint of the arc and the embedded right triangle result in the same value for $\sin t$.

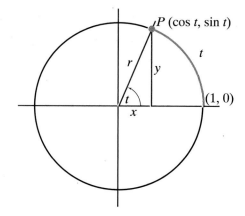

Figure 43   Radian measure and arc length

In Figure 43, we see that for the angle t,

$$\sin t = \frac{y}{r} = \frac{y}{1} = y,$$

$$\cos t = \frac{x}{r} = \frac{x}{1} = x,$$

and

$$\tan t = \frac{y}{x}.$$

In these equations x and y are the lengths of the legs of the embedded triangle. They are also coordinates of point P. The hypotenuse of the embedded triangle, r, is the radius of the unit circle. These relationships are true, even if the angle is not in the first quadrant.

If the angle t is not in the first quadrant, we can still make a connection between the value of the circular functions and the ratio of the lengths of the sides of a right triangle as shown in Figures 44−46.

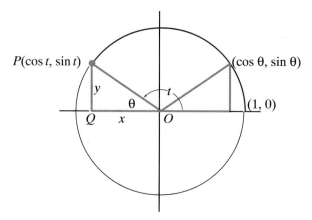

Figure 44   Angle t in the second quadrant

If $\frac{\pi}{2} < t < \pi$, then triangle OPQ defined by t is shown in Figure 44. Notice that the x-coordinate of P is negative, while the y-coordinate of P is positive, so $\sin t$ is positive and $\cos t$ is negative. For the angle θ in triangle OPQ, $\theta = \pi - t$, so $t = \pi - \theta$. Therefore, $\cos t = \cos(\pi - \theta) = -\cos\theta$, while $\sin t = \sin(\pi - \theta) = \sin\theta$.

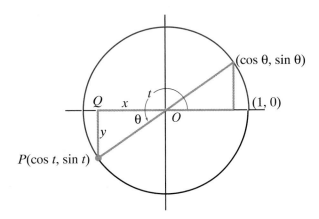

Figure 45 Angle t in the third quadrant

If $\pi < t < \frac{3\pi}{2}$, then triangle OPQ defined by t is shown in Figure 45. Notice that both the x-coordinate and the y-coordinate of P are negative. For the angle θ in triangle OPQ, $\theta = t - \pi$, so $t = \pi + \theta$. Therefore, $\cos t = \cos(\pi + \theta) = -\cos\theta$, while $\sin t = \sin(\pi + \theta) = -\sin\theta$.

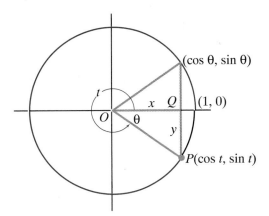

Figure 46 Angle t in the fourth quadrant

If $\frac{3\pi}{2} < t < 2\pi$, then triangle OPQ defined by t is shown in Figure 46. Notice that the x-coordinate of P is positive, while the y-coordinate of P is negative. For the angle θ in triangle OPQ, $\theta = 2\pi - t$, so $t = 2\pi - \theta$. Thus, we know that $\cos t = \cos(2\pi - \theta) = \cos\theta$, while $\sin t = \sin(2\pi - \theta) = -\sin\theta$.

1. Label a unit circle with the following radian measures: $\frac{\pi}{6}, \frac{\pi}{4}, 1, \frac{\pi}{3}, \frac{\pi}{2}, 2, \frac{2\pi}{3}, \frac{3\pi}{4}$, $\frac{5\pi}{6}, 3$, and π. Note the positions of 1, 2, and 3 relative to the other measures.

2. Express each angle measure in degrees.

   a. $\frac{2\pi}{3}$ radians

   b. $\frac{9\pi}{2}$ radians

   c. 3π radians

   d. $\frac{5\pi}{12}$ radians

   e. 2 radians

   f. 0.5 radian

3. Express each angle measure in radians.

   a. 270°

   b. 135°

   c. 240°

   d. 72°

   e. 100°

   f. −25°

   g. 1°

4. Draw four concentric circles with centers at (0, 0) and radii of lengths 1, 2, 3, and 6 units, respectively. On the circle of radius 1, start at (1, 0) and draw an arc of length $\frac{\pi}{3}$. On the circle of radius 2, start at (2, 0) and draw an arc of length $\frac{2\pi}{3}$. On the circle of radius 3, start at (3, 0) and draw an arc of length π. On the circle of radius 6, start at (6, 0) and draw an arc of length 2π. For each circle, compute the value of the ratio $\frac{\text{arc length}}{\text{radius}}$.

5. In right triangle ABC, $a = 35$, $c = 62$, and $\angle C = 90°$. Find the length of side b and the measures of $\angle A$ and $\angle B$.

6. In right triangle PQR, $\angle R = 90°$, $\angle P = 34°$, and $q = 15$. Find the lengths of sides p and r and the measure of $\angle Q$.

7. A ladder 4.2 meters long is leaning against a vertical building and the bottom of the ladder is 1.5 meters from the base of the building. How high is the top of the ladder? What angle does the top of the ladder make with the wall?

8. An observer on the ground finds that the angle between the horizontal and the line of sight to a balloon is 51°. The point directly under the balloon is 235 feet horizontally from the observer. How high is the balloon?

9. A ski run extends in a straight line down a mountain that is 1.1 km high. The run is 5 km long. Assuming the ground is horizontal, what angle does the ski run make with the ground?

10. A 27° angle is positioned in the coordinate system so that its vertex is at the origin and one leg lies along the positive x-axis. The angle intersects the circle $x^2 + y^2 = 20$ at the point (a, b). Find a and b.

11. Suppose $\angle T$, which has its vertex at the origin and one leg along the positive x-axis, intersects the circle $x^2 + y^2 = 12$ at the point $(2, 2\sqrt{2})$. Find the measure of $\angle T$.

12. Make a sketch to illustrate the fact that $\sin 10 = -0.5440$. Make a different sketch to illustrate the fact that $\sin 10° = 0.1736$.

13. Your younger brother started playing around with your calculator after he had finished his homework. He noticed that sometimes your calculator reported that $\cos 1 = 0.5403$, and other times it reported that $\cos 1 = 0.9998$. Explain to your brother what was going on.

14. An arc of length 6.5 begins at the point $(3, 0)$ on the circle $x^2 + y^2 = 9$. What are the possible coordinates of the endpoints of the arc?

15. A unit circle and a circle of radius $\sqrt{5}$ are drawn on the same axes. $\angle T$ intersects the larger circle at approximately $(1.208, 1.882)$ as shown in Figure 47.

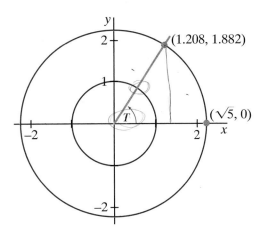

Figure 47   Circles $x^2 + y^2 = 1$ and $x^2 + y^2 = 5$

    a. What are the coordinates of the points where $\angle T$ intersects the unit circle? How can you interpret these coordinates?

    b. Find the measure of $\angle T$.

16. In this exercise, we will provide motivation for using the words "tangent" and "secant" for two of the trigonometric functions. Figure 48 shows a unit circle with center O. Segment OA intersects the circle at B, and segment AP is tangent to the circle at P.

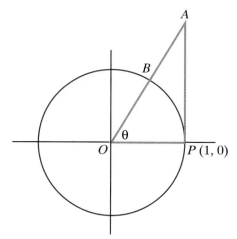

Figure 48 Tangent and secant segments

    a. Express the coordinates of B in terms of θ.

    b. What are the coordinates of A? How long is segment AP? (Notice that segment AP is a tangent to the circle.)

    c. How long is segment OA? (Notice that line OA is a secant of the circle.)

17. A southbound road turns southeast as shown in Figure 49. If the radius of curvature is 100 feet, what is the length of the curved portion of the road?

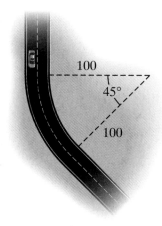

Figure 49 Turn in southbound road

18. In a circle of radius 3.2 cm, what arc length is subtended by an angle that measures 1.4 radians?

19. Find the measure of the angle that subtends an arc of length 8 cm in a circle of radius 5 cm.

20. A winch with a radius of 6 inches is to lift a bucket from a well that is 40 feet deep. How many turns of the winch are required to lift the bucket. (See Figure 50.) Ignore the thickness of the rope.

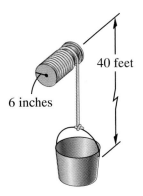

Figure 50   Winch lifting bucket

21. A circle of radius 2 is centered at the origin of the coordinate plane. Suppose a person starts at the point (2, 0) and walks counterclockwise along the circumference of the circle, which is 4π units long. Define the function $f(t)$ = the x-coordinate of the person's location after walking t units.

   a. Evaluate $f(0), f(\pi)$, and $f(2\pi)$.

   b. Identify the domain and the range of f.

   c. Explain why f is a periodic function. Identify the period.

   d. Graph f.

   e. Write an equation for f.

22. Using the information in the preceding exercise, define the function $g(t)$ = the y-coordinate of the person's location after walking t units along the circumference. Complete Parts a through e above for function g.

23. Generalize the results of the two preceding exercises for a circle of radius r. Write one or two sentences to explain why the value of r appears where it does in the equations for f and g.

Solving Trigonometric Equations

Periodic phenomena can be modeled with trigonometric functions. After the model is determined, questions about the situation can be answered by solving an equation based on the model. The examples in this section will illustrate several techniques for solving trigonometric equations.

CLASS PRACTICE

1. Suppose θ is a number whose cosine is 0.3. Illustrate what this means using a triangle.

2. Suppose θ is a number whose cosine is 0.3. Illustrate what this means using a unit circle.

3. Suppose θ is a number whose cosine is 0.3. Illustrate what this means using a graph of the cosine function.

4. Write one or two sentences to describe the relationships among the three different representations of a number whose cosine is 0.3.

EXAMPLE 1   Solve $\cos x = 0.3$.

Solution   The solutions of this equation are numbers whose cosine is 0.3. If you think of this problem in terms of the unit circle, solutions are arc lengths that begin at the point $(1, 0)$ and that terminate at points whose x-coordinate is 0.3. You can also think of this problem in terms of the cosine curve, where solutions are the x-coordinates of the points of intersection of $y = \cos x$ and $y = 0.3$. (See Figure 51.) Either way you visualize this problem, it should be clear that there are infinitely many numbers whose cosine is 0.3. If $x = a$ is one solution, then by the symmetry of the circle or symmetry of the graph $f(x) = \cos x$, $x = -a$ is another solution.

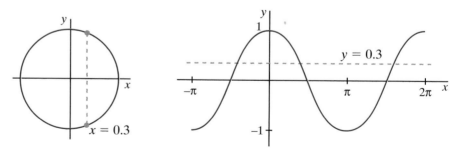

Figure 51   Solutions of $\cos x = 0.3$

One solution of $\cos x = 0.3$ can be found using your calculator in radian mode in which $\cos^{-1} 0.3$ gives approximately 1.2661. Another number whose cosine is 0.3 is -1.2661, which we can determine using the symmetry of either the graph of the cosine function or the unit circle. We can add multiples of 2π to ± 1.2661 to yield more solutions. The complete set of solutions is $x = \pm 1.2661 + 2k\pi$, where k is any integer. One of these solutions, $x = 1.2661$, is also the radian measure of an angle in a right triangle. This is the angle for which the ratio

$$\frac{\text{length of adjacent leg}}{\text{length of hypotenuse}} = 0.3.$$

Once we know that $\cos^{-1} 0.3 = 1.2661$ from the calculator, there is another method to find the second value that produces a cosine value of 0.3. Studying the graph of $y = \cos x$ or the unit circle, we see there are two values in the interval $[0, 2\pi]$ that produce a cosine value of 0.3. There is one value at $x = 1.2661$ and another at $x = 2\pi - 1.2661 = 5.0171$. The solutions represented by $x = -1.2661 + 2k\pi$ are the same set of numbers as those represented by $x = 5.0171 + 2k\pi$. For example, if we let $k = 1$ in $-1.2661 + 2k\pi$ we obtain 5.0171. ■

Suppose the equation in Example 1 is modified to $\cos\left(\frac{5x-1}{3}\right) = 0.3$. The technique for solving this equation is similar to the technique used for $\cos x = 0.3$. However, in the new equation the argument of cosine is $\frac{5x-1}{3}$ instead of x. Therefore, instead of solutions $x = \pm 1.2661 + 2k\pi$, the new equation has solutions $\frac{5x-1}{3} = \pm 1.2661 + 2k\pi$, where k is any integer. The equation $\frac{5x-1}{3} = \pm 1.2661 + 2k\pi$ must be solved for x.

$$5x - 1 = 3(\pm 1.2661 + 2k\pi)$$
$$5x = 1 + 3(\pm 1.2661 + 2k\pi)$$
$$x = \frac{1 + 3(\pm 1.2661 + 2k\pi)}{5}$$
$$x = \frac{1 \pm 3.7983 + 6k\pi}{5}$$
$$x = 0.9597 + \frac{6k\pi}{5}$$

or

$$x = -0.5597 + \frac{6k\pi}{5}$$

Notice that the function $y = \cos\left(\frac{5x-1}{3}\right)$ in the modified equation has period $\frac{6\pi}{5}$, which corresponds to the $\frac{6k\pi}{5}$ in the solution of $\cos\left(\frac{5x-1}{3}\right) = 0.3$.

CLASS PRACTICE

1. Use the results of Example 1 to help you solve $\cos\left(\frac{1}{2}x - 6\right) = 0.3$. How are the solutions of this equation similar to those of $\cos x = 0.3$?

EXAMPLE 2   Find all solutions of $\sin(2x) = -0.5$ between 0 and 2π.

Solution   Solutions of $\sin(2x) = -0.5$ correspond to points where $y = \sin(2x)$ intersects $y = -0.5$. The graph in Figure 52 shows that there are two intersections in each cycle of the sine curve. The period of $y = \sin(2x)$ is π, so one cycle is completed every π units. This means there are four solutions to the equation between 0 and 2π.

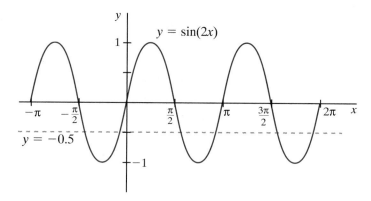

Figure 52   Graphs of $y = \sin(2x)$ and $y = -0.5$

With your calculator in radian mode, you should find that -0.5236 is a number whose sine is -0.5. By the symmetry of the unit circle, $\pi - (-0.5236) = 3.6652$ is another number whose sine is -0.5. Therefore,

$$2x = -0.5236 + 2k\pi \quad \text{or} \quad 3.6652 + 2k\pi$$

$$x = -0.2618 + k\pi \quad \text{or} \quad 1.8326 + k\pi.$$

Since we want only the solutions between 0 and 2π, we can substitute integer values of k that produce x-values between 0 and 2π. The x-values 1.8326, 2.8798, 4.9742, and 6.0214 are all solutions of $\sin(2x) = -0.5$ between 0 and 2π. Verify that all four of these numbers do indeed satisfy the original equation. ■

As an alternative method of obtaining a solution for Example 2, you may recognize that -0.5 is one of the special values of the sine function that can be determined without a calculator: $\sin(\frac{7\pi}{6}) = -0.5$. This means that $\frac{7\pi}{6}$ is one possible value for $2x$. By the symmetry of the unit circle another value for $2x$ is $\frac{11\pi}{6}$. Thus, $2x = \frac{7\pi}{6}$ or $\frac{11\pi}{6}$, so $x = \frac{7\pi}{12}$ or $\frac{11\pi}{12}$. Other solutions can be obtained by adding or subtracting π (the period of $\sin(2x)$) to these two solutions, resulting in $\frac{7\pi}{12}, \frac{11\pi}{12}, \frac{19\pi}{12}$, and $\frac{23\pi}{12}$. This technique leads to exact solutions to the equation $\sin(2x) = -0.5$. Notice that exact solutions are possible only because -0.5 is a special value of the sine function.

EXAMPLE 3   Solve $\sin x > 0.8$.

Solution   Solutions of this inequality are numbers whose sine is greater than 0.8. Solutions correspond to *x*-values on the sine curve whose *y*-coordinate is greater than 0.8. (See Figure 53.) Because the sine curve oscillates, it is useful to find where $\sin x$ is equal to 0.8 and then study the graph to see where $\sin x$ is greater than 0.8. Using a calculator in radian mode to solve $\sin x = 0.8$ gives $x = \sin^{-1}(0.8) = 0.9273$. By symmetry another solution is $x = \pi - 0.9273 = 2.2143$. Therefore, the solution set of $\sin x = 0.8$ is $x = 0.9273 + 2k\pi$ or $2.2143 + 2k\pi$, where *k* is any integer. These values are marked on the graphs in Figure 53. Note that $\sin x > 0.8$ for *x*-values between 0.9273 and 2.2143. Because of periodicity, intervals containing solutions of the inequality are repeated every 2π units. Therefore, the complete solution set of $\sin x > 0.8$ is $0.9273 + 2k\pi < x < 2.2143 + 2k\pi$, where *k* is any integer.

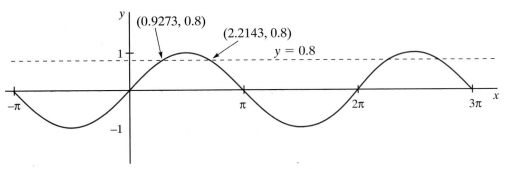

Figure 53   Solutions of $\sin x > 0.8$

EXAMPLE 4   The graph in Figure 54 shows how the distance between the ceiling and the tip of the hour hand on a clock changes over time, which was investigated in Section 5.1 on page 302. At what times is this distance less than 30 cm?

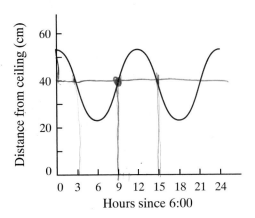

Figure 54   Distance from ceiling versus hours since 6 o'clock

Solution We first need to write an equation for the graph in Figure 54. The period is 12, the amplitude is 15, and the graph starts at a maximum and has been shifted up 38 units. If t represents the number of hours elapsed since 6 o'clock, then an equation for the graph is

$$d = 15 \cos\left(\frac{\pi t}{6}\right) + 38.$$

To determine when the distance is less than 30 cm, we need to solve the inequality

$$15 \cos\left(\frac{\pi t}{6}\right) + 38 < 30.$$

Since the graph oscillates, it is useful to first find solutions of $15 \cos(\frac{\pi t}{6}) + 38 = 30$. Subtracting 38 from both sides and then dividing by 15 yields

$$\cos\left(\frac{\pi t}{6}\right) = -\frac{8}{15}.$$

Two solutions of this equation are given by

$$\frac{\pi t}{6} = \cos^{-1}\left(-\frac{8}{15}\right) = 2.1333 \quad \text{or} \quad \frac{\pi t}{6} = 2\pi - \cos^{-1}\left(-\frac{8}{15}\right) = 4.1499.$$

Since $\cos(\frac{\pi t}{6}) = -\frac{8}{15}$ has infinitely many solutions, we must write $\frac{\pi t}{6} = 2.1333 + 2k\pi$ or $\frac{\pi t}{6} = 4.1499 + 2k\pi$. Solving for t gives $t = 4.0744 + 12k$ or $t = 7.9256 + 12k$.

These are t-values at which the distance is equal to 30 cm. The graph in Figure 55 shows that the distance is less than 30 cm when t is between 4.0744 and 7.9256. Since t represents the number of hours that have elapsed since 6 o'clock, $t = 4.0744$ corresponds to about 10:04, and $t = 7.9256$ corresponds to about 1:56. This means that the distance between the ceiling and the tip of the hour hand is less than 30 cm between approximately 10:04 and 1:56.

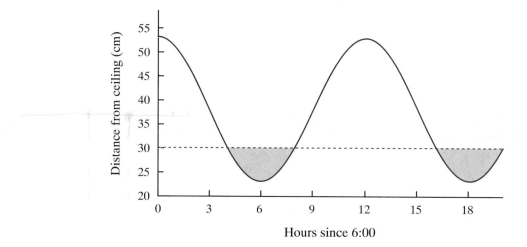

Figure 55 Solutions of $15 \cos(\frac{\pi t}{6}) + 38 < 30$ ■

We can solve equations of the form $\sin(ax + b) = k$ using methods described in the previous examples. We can combine this new knowledge with algebra and previously learned techniques to solve a variety of other equations and inequalities. Some examples of these equations are:

$$(\sin x - 1)\left(\cos x + \frac{1}{2}\right) = 0,$$
$$\cos^2 x - 4 \cos x = 0,$$
$$10^{\sin x} = 5,$$
$$\frac{\cos(4x)}{\sin(3x)} > 0.$$

Knowledge of solving quadratic equations enables us to solve equations such as $\tan^2 x + \tan x - 1 = 0$. This quadratic equation is not easily factored, but with the substitution of $a = \tan x$, the equation can be written as

$$a^2 + a - 1 = 0.$$

Using the quadratic formula we find $a = \frac{-1 \pm \sqrt{5}}{2}$. Therefore, $\tan x = \frac{-1 + \sqrt{5}}{2}$ or $\tan x = \frac{-1 - \sqrt{5}}{2}$, which will lead to an infinite number of solutions for x.

However, there are some equations that cannot be solved using these techniques; an example is the equation $x = \cos x$. For an equation of this type, we must rely on graphs to find the x-coordinates of points for which $x = \cos x$.

CLASS PRACTICE   Solve the following.

1. $10^{\sin x} = 5$

2. $2\cos^2(4x) = \cos(4x)$

3. $x = \cos x$

Exercise Set 5.7

1. Suppose that the solution set of an equation includes numbers of the form $x = 1.6 + 3k\pi$. Write the decimal approximations for six values in the solution set.

2. Suppose an inequality has solutions $-0.2 + k\pi \le x \le 0.3 + k\pi$. Sketch a number line and shade solutions of the inequality (between 0 and 10) on the number line.

3. Find all the solutions (accurate to four decimal places) of each equation or inequality. Use your calculator and your knowledge of symmetry and periodicity.

   a. $\cos x = 0.75$ **b.** $\sin(x - 1) = 0.4$ **c.** $\cos\left(\frac{1}{2x}\right) + 1 = 3.4$

   d. $3\sin(2x) = 0.6$ **e.** $\cos\left(2x - \frac{\pi}{3}\right) = 0.8$ **f.** $\sin\left(x - \frac{\pi}{4}\right) < 0.4$

   g. $\tan(x + 2) > 0.6$

4. At what times is the distance between the ceiling and the tip of the hour hand of the clock in Example 4 on page 353 greater than 40 cm?

5. When an object is suspended from a spring, the equation $x = 0.1\cos(4\pi t)$ describes the displacement, x (in meters), of the object from its equilibrium position at time t seconds. The positive direction is downward.

   a. What is the maximum vertical distance through which the object moves?

   b. What is the minimum time that it takes the object to move through the distance described in Part a?

   c. When is the second time (after it is released at $t = 0$) that the object is at its equilibrium position?

   d. When does the object reach its equilibrium position for the third time after release?

   e. Draw a graph of the object's displacement from its equilibrium position as a function of time. Does your graph reflect each of the answers to the previous questions?

6. A Ferris wheel has radius 33.2 feet and makes a complete revolution every 80 seconds. The bottom of the wheel is 4 feet above ground. Margo is riding the Ferris wheel and starts her stopwatch when she is at the highest point. During the first 2 minutes of her ride, what time(s) will be displayed on her stopwatch when she is 35 feet high?

7. The distance y inches between a yo-yo and Susan's hand at time t seconds is given by the equation $y = -15\cos\left(\frac{2\pi t}{2.8}\right) + 15$. Steven needs to take a picture when the yo-yo is 7 inches from the hand and moving away from the hand. At what times can he take the picture?

8. A population of lynx oscillates in a four-year cycle. Kate is a biologist who keeps records of the lynx population in a small area. She has counted lynx on January 1 of each year. Her first count in 1994 was 40; in 1995, 60; in 1996, back to 40; and in 1997, 20. The population was back to 40 in 1998. Assuming the population continues as a sinusoidal function, during which months of the next four years should Kate expect to find fewer than 35 lynx?

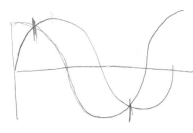

9. Find all solutions of each equation or inequality. Find exact solutions whenever possible. Find approximate solutions accurate to four decimal places.

a. $4\sin^2 x = 1$

b. $2\cos^2 x + \cos x = 1$

c. $3\sin^2(2x) = \sin(2x)$

d. $2\tan^2 x + 7\tan x + 4 = 0$

e. $4\cos^2 x - \cos x = 0$

f. $\cos x = x^2 - 1$

g. $\cos x < \sin x$

h. $2^{\sin x} = 1$

i. $\dfrac{\sin(2x)}{x} = 1$

j. $\tan x = \sin x$

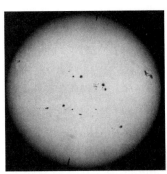

Sunspots

10. A sunspot is a dark area on the surface of the sun that is about $4000°$ cooler than the surrounding surface. Over the years astronomers have counted the number of sunspots that occur within a given year. The graph of these data can be modeled by the function $spots = 50\cos(\frac{2\pi \cdot years}{11}) + 60$, where *years* is the number of years since 1750. Find two years after 2000 in which the number of sunspots will exceed 40.

11. In a tidal river, the time between high tide and low tide is approximately 6.2 hours. The average depth of the water in a port on the river is 4 meters; at high tide the depth is 5 meters.

a. Sketch a graph of the depth of the water in the port over time if the relationship between time and depth is sinusoidal and there is a high tide at 12:00 noon. Write an equation for your curve. Let *t* represent the number of hours after 12:00 noon.

b. If a boat requires a depth of at least 4 meters of water to sail, how many minutes before noon can it enter the port, and by what time must it leave to avoid being stranded?

c. If a boat requires a depth of at least 3.5 meters of water to sail, at what time before noon can it enter the port, and by what time must it leave to avoid being stranded?

d. A boat that requires a depth of 4 meters of water to sail is at a dock in the port. As some of the cargo is unloaded, the depth of water required to sail decreases. Suppose at noon the crew begins to unload cargo. The unloading of the cargo decreases the draft of the boat at a rate of 0.1 meters per hour. At what time must the ship stop unloading and leave the port to ensure that it sails before 6 P.M.?

 **Investigating Trigonometric Identities**

To solve the equation $\sin(x) + \sin(-x) = 0$, we can use graphing technology to locate the zeros of the function $f(x) = \sin(x) + \sin(-x)$. We will find that the graph of $f(x) = \sin(x) + \sin(-x)$ is identical to the graph of $g(x) = 0$ over the entire domain of f. This implies that $f(x) = 0$ for all x-values. Therefore, the equation $\sin(x) + \sin(-x) = 0$ is true for all values of x. Recall that an equation that is true for all values of the variable is called an identity. The terms in the identity $\sin(x) + \sin(-x) = 0$ can be easily rewritten as $\sin(-x) = -\sin(x)$. This identity should be a familiar one. It is the algebraic statement that $h(x) = \sin(x)$ is an odd function.

In this investigation you will use graphs, the unit circle, right triangles, and algebra to show that various equations are identities.

1. Graph each function given below. Based on what you know about transformations and trigonometry, write a second equation whose graph would be identical to the one you have graphed. For example, if the given function were $f(x) = \sin(-x)$, then the equation $g(x) = -\sin(x)$ produces an identical graph. Therefore, you know that $\sin(-x) = -\sin(x)$ is an identity. If you find several answers for a particular part, record them all.

   a. $h(x) = \sin(x) \cos(2x) + \cos(x) \sin(2x)$

   b. $m(x) = \sin(\pi x) \cos(\frac{\pi}{2}x) + \cos(\pi x) \sin(\frac{\pi}{2}x)$

   c. $p(x) = \cos(x) \cos(\frac{1}{2}x) - \sin(x) \sin(\frac{1}{2}x)$

   d. $r(x) = \cos(\pi x) \cos(\frac{\pi}{2}x) - \sin(\pi x) \sin(\frac{\pi}{2}x)$

   e. $g(x) = \cos^2 x - \sin^2 x$ f. $f(x) = 2\sin(x)\cos(x)$

   g. $k(x) = 1 - 2\sin^2(x)$ h. $t(x) = \sin^2 x$

2. The coordinates of points labeled on the unit circle in Figure 56 suggest several identities. For example, point P has coordinates (a, b) and point Q has coordinates (b, a) because it is the reflection of P across the line $y = x$.

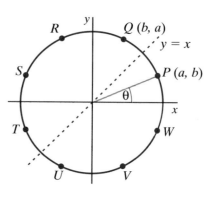

Figure 56   Points on unit circle

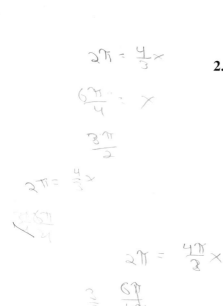

Point P is the endpoint of an arc of length θ that begins at $(1, 0)$, and point Q is the endpoint of an arc of length $\frac{\pi}{2} - \theta$. The x-coordinate of P is the same as the y-coordinate of Q. This yields the identity $\cos\theta = \sin(\frac{\pi}{2} - \theta)$.

Determine the coordinates of points R, S, T, U, V, and W in terms of a and b. Then write at least two other identities suggested by the coordinates you have labeled. Your identities should involve the sine, the cosine, or the tangent functions.

3. Refer to the diagram of a right triangle in Figure 57. Use right triangle ratios to show that each of the following equations is an identity, for $0° \leq \theta \leq 90°$.

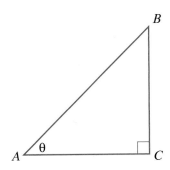

Figure 57   Right triangle

a. $\cos(90° - \theta) = \sin\theta$

b. $\sin(90° - \theta) = \cos\theta$

c. $\sin^2\theta + \cos^2\theta = 1$

4. It is often possible to write new identities based on your existing knowledge of identities. Use algebra and known identities to establish each new identity.

a. Divide both sides of $\sin^2\theta + \cos^2\theta = 1$ by $\cos^2\theta$ to write a new identity involving $\sec^2\theta$.

b. Divide both sides of $\sin^2\theta + \cos^2\theta = 1$ by $\sin^2\theta$ to write a new identity involving $\csc^2\theta$.

SECTION 5.9 Using Trigonometric Identities

In pre-calculator days identities were very useful for finding particular values of trigonometric functions. For instance, you can find an exact value for $\sin\left(\frac{\pi}{12}\right)$ by writing it as $\sin\left(\frac{\pi}{3} - \frac{\pi}{4}\right)$ and then using the identity for $\sin(a - b)$. With access to a calculator the importance of identities has changed. Even if you have access to technology to evaluate $\sin\left(\frac{\pi}{12}\right)$, it is important that you be sufficiently familiar with identities to recognize situations in which they may be useful. The following identities are ones that should be familiar. Several are generalizations of results found in the previous investigation. Several of these identities will be verified in the exercises of this section.

$$\sin^2\theta + \cos^2\theta = 1$$

$$\sin(-\theta) = -\sin\theta$$

$$\cos(-\theta) = \cos\theta$$

$$\sin\left(\frac{\pi}{2} - \theta\right) = \cos\theta$$

$$\sin(A + B) = \sin A \cos B + \cos A \sin B$$

$$\cos(A + B) = \cos A \cos B - \sin A \sin B$$

$$\sin(2\theta) = 2\sin\theta\cos\theta$$

$$\cos(2\theta) = \cos^2\theta - \sin^2\theta$$

There are many other identities that may prove useful in some situations. A large number of these other identities can easily be derived from familiar identities. For example, dividing both sides of $\sin^2\theta + \cos^2\theta = 1$ by $\cos^2\theta$ yields the new identity

$$\frac{\sin^2\theta}{\cos^2\theta} + 1 = \frac{1}{\cos^2\theta}$$

$$\tan^2\theta + 1 = \sec^2\theta.$$

It is probably more efficient to derive this new identity if and when you need it, rather than to memorize it.

EXAMPLE 1   Show that $\sin(2\theta) = 2\sin\theta\cos\theta$ is an identity.

Solution   The expression $\sin(2\theta)$ represents the sine of double an arc length θ. Since $\sin(2\theta) = \sin(\theta + \theta)$, we can use the identity for the sine of the sum of two numbers:

$$\sin(A + B) = \sin A \cos B + \cos A \sin B.$$

Substituting θ for the value of A and for the value of B in the above identity yields

$$\sin(\theta + \theta) = \sin\theta\cos\theta + \cos\theta\sin\theta,$$

which becomes

$$\sin(2\theta) = 2\sin\theta\cos\theta. \quad \blacksquare$$

There are many situations in which you will use trigonometric identities. The following examples illustrate some of these situations.

EXAMPLE 2   Find the exact value of $\sin(2\theta)$, if $\sin\theta = \frac{1}{3}$.

Solution   To determine $\sin(2\theta)$, we can use the identity established in Example 1, $\sin(2\theta) = 2\sin\theta\cos\theta$. Since we know that $\sin\theta = \frac{1}{3}$, we need to find $\cos\theta$. The identity $\sin^2\theta + \cos^2\theta = 1$ allows us to find $\cos\theta$. The value of $\cos\theta$ will be either positive or negative, depending on whether the endpoint of arc θ is in the first or second quadrant of the unit circle.

$$\cos\theta = \pm\sqrt{1 - \sin^2\theta}$$

$$\cos\theta = \pm\sqrt{1 - \frac{1}{9}} = \pm\frac{2\sqrt{2}}{3}.$$

Therefore,

$$\sin(2\theta) = 2\sin\theta\cos\theta = 2 \cdot \frac{1}{3} \cdot \left(\pm\frac{2\sqrt{2}}{3}\right) = \pm\frac{4\sqrt{2}}{9}. \quad \blacksquare$$

EXAMPLE 3   Solve $3\sin(2x) + \sin(-2x) + 1 = 0.7$.

Solution   The identity $\sin(-\theta) = -\sin\theta$ allows us to replace $\sin(-2x)$ with $-\sin(2x)$. Thus,

$$3\sin(2x) + \sin(-2x) + 1 = 0.7$$

becomes

$$3\sin(2x) - \sin(2x) + 1 = 0.7.$$

We now can solve for $\sin(2x)$ as follows.

$$3\sin(2x) - \sin(2x) + 1 = 0.7$$

$$2\sin(2x) + 1 = 0.7$$

$$2\sin(2x) = -0.3$$

$$\sin(2x) = -0.15$$

Since $\sin^{-1}(-0.15) = -0.1506$ and $\pi - (-0.1506) = 3.2922$, we know that $2x = -0.1506 + 2k\pi$ or $2x = 3.2922 + 2k\pi$. Therefore, the solutions of the equation are $x = -0.0753 + k\pi$ or $x = 1.6461 + k\pi$, where k is any integer. ■

EXAMPLE 4   Find all solutions of $2\cos x + \sin^2 x = 0$.

Solution   First rewrite this equation in terms of a single trigonometric function. The identity $\sin^2 x + \cos^2 x = 1$ can be written as $\sin^2 x = 1 - \cos^2 x$, so replace $\sin^2 x$ with $1 - \cos^2 x$ to obtain

$$2\cos x + 1 - \cos^2 x = 0$$

$$\cos^2 x - 2\cos x - 1 = 0.$$

Now we have a quadratic equation in $\cos x$. The quadratic formula yields

$$\cos x = 1 \pm \sqrt{2}.$$

The number $1 + \sqrt{2}$ is greater than 1. Since $\cos x$ is never greater than 1, we must reject this value. Therefore, $\cos x = 1 - \sqrt{2}$, and

$$x = \cos^{-1}(1 - \sqrt{2}) = 1.9979.$$

Thus, $x = 1.9979$ is a number whose cosine is $1 - \sqrt{2}$. By the symmetry of the unit circle, -1.9979 is another number whose cosine is $1 - \sqrt{2}$. The complete solution set of the equation is $x = 1.9979 + 2k\pi$ or $x = -1.9979 + 2k\pi$, k is any integer. ■

EXAMPLE 5   Solve $5\cos^3 t - \cos t \cos(2t) + 4\cos^2 t = 0$.

Solution   This equation involves only the cosine function, but the arguments of the cosines are not all the same. Every term contains $\cos t$, which is a common factor. The equation can be rewritten as

$$\cos t(5\cos^2 t - \cos(2t) + 4\cos t) = 0$$

The identity $\cos(2t) = \cos^2 t - \sin^2 t = 2\cos^2(t) - 1$ can be used to change all arguments to t. We can substitute $2\cos^2(t) - 1$ for $\cos(2t)$ in our equation, which becomes

$$\cos t \left[5\cos^2 t - (2\cos^2 t - 1) + 4\cos t \right] = 0,$$

or

$$\cos t \left[3\cos^2 t + 4\cos t + 1 \right] = 0.$$

The expression $3\cos^2 t + 4\cos t + 1$ can be factored to yield the equation

$$\cos t(3\cos t + 1)(\cos t + 1) = 0.$$

Either $\cos t = 0$, $3\cos t + 1 = 0$, or $\cos t + 1 = 0$. The solutions of $\cos t = 0$ are $t = \frac{\pi}{2} + k\pi$. The solutions of $3\cos t + 1 = 0$ can be determined by evaluating $\cos^{-1}(-\frac{1}{3})$. These solutions are $t = 1.9106 + 2k\pi$ or $t = -1.9106 + 2k\pi$. The solutions of $\cos t + 1 = 0$ are $t = \pi + 2k\pi$. The complete solution set of the equation is $t = 1.9106 + 2k\pi$ or $t = -1.9106 + 2k\pi$ or $t = \pi + 2k\pi$ or $t = \frac{\pi}{2} + k\pi$, where k is any integer. ▧

Exercise Set 5.9

1. Suppose you know that $\sin\theta = -\frac{1}{4}$ and $\frac{3\pi}{2} < \theta < 2\pi$. Use identities to find the exact value of the following.

   a. $\sin(\theta - \frac{\pi}{4})$ b. $\cos(2\theta)$

   c. $\sin(2\theta)$ d. $\sin(\frac{\pi}{4} + \theta)$

2. Find all solutions of each equation in the interval $[0, 2\pi]$. Find exact values where possible.

   a. $\cos x = \cos(2x)$

   b. $\sin(2x) - \sin(4x) = 0$

   c. $\sec^2 x - \tan x = 1$

   d. $(\sin x + \cos x)^2 = 1$

   e. $\sin(4x)\cos x - \sin x \cos(4x) = 0$

   f. $\cos x \sin x = 0.84$

   g. $\sin(x + 0.3) + \sin(x - 0.3) = 2\cos 0.3$

3. Solve each equation. Find all real solutions.

   a. $2\cos x - \sin^2 x = 2$

   b. $5\sec^2 x + 2\tan x - 8 = 0$

   c. $\sin^2 x - \cos^2 x = -\frac{1}{8}$

   d. $\sin(5x)\cos(2x) + \cos(5x)\sin(2x) = 0.72$

   e. $\sin x + \cos(2x) = 1$

4. Decide if each equation is an identity. If a particular equation is an identity, support that fact analytically; if it is not an identity, find its solutions.

 a. $4\sin^2 x \cos x - \sin(2x) = 0$

 b. $\cos(2x) - \cos^2 x + \sin^2 x = 0$

 c. $\cos(3x) = 4\cos^3 x - 3\cos x$ (Hint: $3x = 2x + x$.)

 d. $\cos(2x) - 2\cos x = 0$

 e. $\cos^4 x - \sin^4 x = \cos^2 x - \sin^2 x$

 f. $\cos(x + y) + \cos(x - y) = 2\cos x \cos y$

5. Suppose you know that $\cos A = \frac{1}{3}$ and $0 \le A \le \frac{\pi}{2}$. Use identities to find the exact value of the following.

 a. $\sin\left(\frac{A}{2}\right)$ b. $\cos\left(\frac{A}{2}\right)$

6. Assume that the birthrate for an animal species is given by $B = \cos^2 t + 2\cos t$ and the death rate is $D = -\cos(2t)$, where t is the time in months. Find the smallest positive value of t for which the birth and death rates are equal.

7. The graphs of $y = \sin\left(\frac{3\pi}{2} - x\right)$ and $y = -\cos x$ are identical. Therefore, $\sin\left(\frac{3\pi}{2} - x\right) = -\cos x$ must be an identity. Verify this identity using the unit circle and/or other known identities.

8. Write a paragraph that includes three justifications that $\sin^2 \theta + \cos^2 \theta = 1$ is an identity. Your arguments should be based on graphs, a unit circle, and a right triangle.

9. Follow these steps to prove the identity for $\cos(A - B)$.

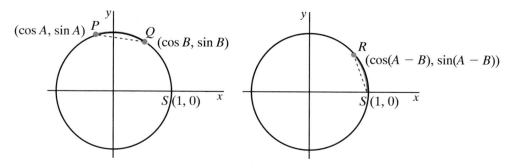

Figure 58 Unit circles for derivation of identity for $\cos(A - B)$

The unit circle on the left in Figure 58 shows an arc of length A beginning at $(1, 0)$ and ending at P. An arc of length B also begins at $(1, 0)$; it ends at point Q. (Assume $0 < B < A < 2\pi$.) The coordinates of P are $(\cos A, \sin A)$ and the coordinates of Q are $(\cos B, \sin B)$. The arc from Q to P has length $A - B$. The unit circle on the right in Figure 58 shows an arc of length $A - B$ beginning at $(1, 0)$ and ending at R. Thus the coordinates of R are $(\cos(A - B), \sin(A - B))$. Chord QP in one circle has the same length as chord SR in the other.

   a. Use the distance formula to express the length of chords QP and SR in terms of A and B.

   b. Set the two expressions equal to each other and simplify to find the identity for $\cos(A - B)$.

10. Use the identity $\cos(A - B) = \cos A \cos B + \sin A \sin B$ to derive each of the following identities.

   a. $\cos(A + B) = \cos A \cos B - \sin A \sin B$
     Hint: $A + B = A - (-B)$.

   b. $\cos(2A) = \cos^2 A - \sin^2 A$
     Hint: $2A = A + A$.

   c. $\sin(A - B) = \sin A \cos B - \cos A \sin B$
     Hint: $\sin x = \cos(x - \frac{\pi}{2})$
     so, $\sin(A - B) = \cos((A - B) - \frac{\pi}{2}) = \cos(A - (B + \frac{\pi}{2}))$.

   d. $\sin(A + B) = \sin A \cos B + \cos A \sin B$

11. Use the results of the previous exercise to show that $\cos(2A) = 1 - 2\sin^2 A$ and that $\cos(2A) = 2\cos^2 A - 1$.

12. In the investigation in Section 5.8, the following identities were shown to be true in a right triangle, where $0° < \theta < 90°$:

$$\cos(90° - \theta) = \sin\theta \qquad \sin(90° - \theta) = \cos\theta$$

Show that these identities are true for all values of θ.

SECTION 5.10 Inverse Trigonometric Functions

We have already used the $\sin^{-1}$ key on a calculator to solve equations like $\sin x = 0.8$. When we evaluate $\sin^{-1}(0.8)$ in radian mode, the display is 0.927295218. This tells us that $\sin(0.927295218) = 0.8$. Another way to think about $\sin^{-1}$ is as a function: $g(x) = \sin^{-1}x$. The fact that we evaluate $\sin^{-1}(0.8)$ and get 0.927295218 means that the function $g(x) = \sin^{-1}x$ pairs an x-value of 0.8 with a y-value of 0.927295218. Since the function $f(x) = \sin x$ pairs an x-value of 0.927295218 with a y-value of 0.8, the functions f and g have an important characteristic of inverse functions.

The sine function is not a one-to-one function, since many different x-values are paired with the same y-value. One consequence of many x-values being paired with the same y-value is that equations such as $\sin x = 0.8$ have infinitely many solutions. Another consequence is that the ordered pairs $(\sin x, x)$ do not produce a function. If you reflect the graph of $f(x) = \sin x$ about the line $y = x$, you will see a graph that is not a function. (See Figure 59.)

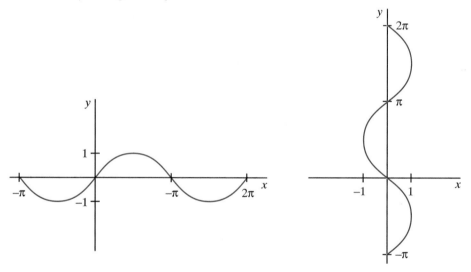

Figure 59 Graphs of $y = \sin x$ and its inverse

How can we can restrict the domain of f to an interval of x-values on which f is one-to-one?

Suppose f is a function whose inverse is not a function. How can we restrict the domain of f to an interval of x-values on which f is one-to-one? For $f(x) = \sin x$, there are many different ways to choose this domain restriction. The interval $\frac{\pi}{2} \le x \le \frac{3\pi}{2}$ would be satisfactory; so would the union of $-\pi \le x \le -\frac{\pi}{2}$ with $\frac{\pi}{2} \le x \le \pi$. Both of these choices include the entire range of f without repeating range values.

However, mathematicians have agreed to restrict the domain of $f(x) = \sin x$ to the interval $-\frac{\pi}{2} \le x \le \frac{\pi}{2}$. This interval has the advantages that it is continuous and it contains $x = 0$. Figure 60 shows $f(x) = \sin x$ with this restricted domain. The domain shown for f is $-\frac{\pi}{2} \le x \le \frac{\pi}{2}$, and the range is $-1 \le y \le 1$. Figure 60 also shows the graph of $f^{-1}(x) = \sin^{-1}x$.

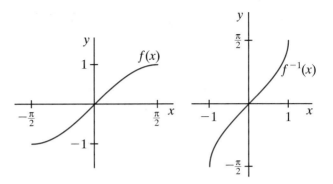

Figure 60 Graphs of $f(x) = \sin x$ with restricted domain and $f^{-1}(x) = \sin^{-1}x$

The choice of $-\frac{\pi}{2} \le x \le \frac{\pi}{2}$ for the restricted domain of $f(x) = \sin x$ determines the range of the inverse function. The inverse function, $f^{-1}(x) = \sin^{-1}x$, has domain $-1 \le x \le 1$ and range $-\frac{\pi}{2} \le y \le \frac{\pi}{2}$. The interval $-\frac{\pi}{2} \le y \le \frac{\pi}{2}$ is universally accepted as the range of the inverse sine function.

The fact that the range of f^{-1} is the interval $-\frac{\pi}{2} \le x \le \frac{\pi}{2}$ means that values of $\sin^{-1}x$ lie between $-\frac{\pi}{2}$ and $\frac{\pi}{2}$ for all x-values. You can use a calculator in radian mode to confirm that the inverse sine of any number is always between $-\frac{\pi}{2}$ and $\frac{\pi}{2}$. For example, when a calculator reports the value of $\sin^{-1}(0.8)$ to be 0.9272952, you can conclude that 0.9272952 is the unique number between $-\frac{\pi}{2}$ and $\frac{\pi}{2}$ whose sine is 0.8.

It is important to understand the difference between the solutions of the two equations $\sin x = 0.8$ and $x = \sin^{-1}(0.8)$. The equation $\sin x = 0.8$ has infinitely many solutions, and $\sin^{-1}(0.8)$ represents only one of the solutions. The symbol $\sin^{-1}(0.8)$ represents the single solution of $\sin x = 0.8$ that is between $-\frac{\pi}{2}$ and $\frac{\pi}{2}$, inclusive.

Like the sine function, the cosine function is not one-to-one. It is necessary to restrict the domain so that the inverse of the cosine function will be a function. As with the sine function, there are many possible choices for this domain restriction. However, the standard restriction is the interval $0 \le x \le \pi$. Figure 61 shows the graphs of $g(x) = \cos x$ with restricted domain and $g^{-1}(x) = \cos^{-1}x$. Note that $g^{-1}(x) = \cos^{-1}x$ has domain $-1 \le x \le 1$; this is also the range of $g(x)$. The range of $g^{-1}(x) = \cos^{-1}x$ is $0 \le y \le \pi$; this is the same as the domain of $g(x)$.

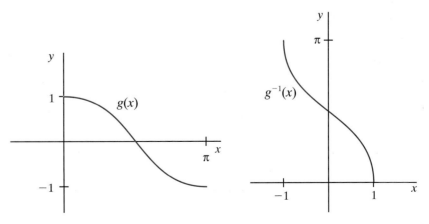

Figure 61 Graphs of $g(x) = \cos x$ with restricted domain and $g^{-1}(x) = \cos^{-1}x$

1. Explain the difference between the solutions to the equation $\cos x = 0.3$ and the equation $x = \cos^{-1} 0.3$.

2. Find the value of $\tan^{-1}(-12)$, $\tan^{-1}(-3.4)$, $\tan^{-1}(-1.81)$, $\tan^{-1}(0.058)$, $\tan^{-1}(1.65)$, $\tan^{-1}(2.54)$, and $\tan^{-1}(23.7)$. Use the numerical results and the graph of the tangent function to help you guess the range of the inverse tangent function. Make a sketch of $f(x) = \tan^{-1}(x)$.

Like the sine and the cosine functions, the tangent function is not one-to-one. The inverse of the tangent function will be a function only if we restrict its domain. There are many ways to restrict the domain of the tangent function to make it one-to-one. The results of Class Practice Problem 2 show that all the values of inverse tangent produced by a calculator in radian mode are between $-\frac{\pi}{2}$ and $\frac{\pi}{2}$ (between approximately -1.6 and 1.6). This indicates that the range of the inverse tangent function is $-\frac{\pi}{2} < x < \frac{\pi}{2}$. Therefore, we can conclude that the standard domain chosen to make tangent a one-to-one function is $-\frac{\pi}{2} < x < \frac{\pi}{2}$.

Figure 62 shows the graphs of $h(x) = \tan x$ and $h^{-1}(x) = \tan^{-1}x$. The restricted domain of tangent and the range of inverse tangent is the interval $(-\frac{\pi}{2}, \frac{\pi}{2})$. The range of the tangent function and the domain of the inverse tangent function are both the set of all real numbers. Note that the graph of $h^{-1}(x) = \tan^{-1}x$ has horizontal asymptotes at $y = -\frac{\pi}{2}$ and at $y = \frac{\pi}{2}$.

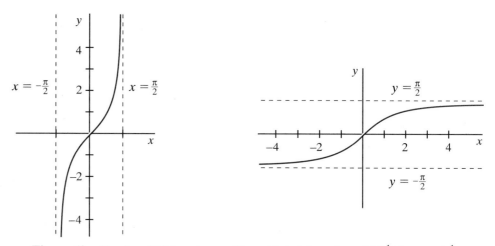

Figure 62 Graphs of $h(x) = \tan x$ with restricted domain and $h^{-1}(x) = \tan^{-1}x$

EXAMPLE 1 Evaluate $\tan^{-1}(-1)$ and $\cos^{-1}(-0.5)$. Explain what these values mean.

Solution Since the range of the inverse tangent function includes numbers from $-\frac{\pi}{2}$ to $\frac{\pi}{2}$, $\tan^{-1}(-1)$ represents the number in this interval whose tangent is -1. In radian mode your calculator reports that $\tan^{-1}(-1) = -0.785$. The number -0.785 answers the question "What number between $-\frac{\pi}{2}$ and $\frac{\pi}{2}$ has a tangent of -1?"

Since the range of the inverse cosine function includes numbers from 0 to π, $\cos^{-1}(-0.5)$ represents a number between 0 and π whose cosine is -0.5. In radian mode your calculator reports that $\cos^{-1}(-0.5) = 2.0944$. The number 2.0944 answers the question "What number between 0 and π has a cosine of -0.5?"

You may recognize that $\tan^{-1}(-1)$ and $\cos^{-1}(-0.5)$ involve special values of the trigonometric functions and can be evaluated without a calculator. The value of $\tan^{-1}(-1)$ is a number between $-\frac{\pi}{2}$ and $\frac{\pi}{2}$ whose tangent is -1. The number that meets these criteria is $-\frac{\pi}{4}$. While it is true that $\tan\left(\frac{3\pi}{4}\right) = -1$ and $\tan\left(\frac{7\pi}{4}\right) = -1$, neither $\frac{3\pi}{4}$ nor $\frac{7\pi}{4}$ is in the range of the inverse tangent function. The value of $\cos^{-1}(-0.5)$ is a number between 0 and π whose cosine is -0.5. The number that meets these criteria is $\frac{2\pi}{3}$. Again, even though it is true that $\cos\left(\frac{4\pi}{3}\right) = -0.5$, $\frac{4\pi}{3}$ is not in the range of the inverse cosine function, so $\cos^{-1}(-0.5) \neq \frac{4\pi}{3}$. ■

Most calculators do not have keys for the remaining three trigonometric functions: secant, cosecant, and cotangent. Nor do they have keys for the inverses of these three functions. Therefore, we cannot directly determine values of the inverse secant, inverse cosecant, and inverse cotangent functions on a calculator. Since cotangent is the reciprocal of tangent, you may suspect that $\cot^{-1}$ is the reciprocal of $\tan^{-1}$. This is generally not the case:

$$\cot^{-1}x \neq \frac{1}{\tan^{-1}x}.$$

The symbol $\cot^{-1}x$ represents a number whose cotangent is x, and $\tan^{-1}x$ represents a number whose tangent is x. These numbers are not reciprocals of each other.

For example, in the equation $x = \cot^{-1}(2)$, x represents the number whose cotangent is 2. That is, $\cot x = 2$. Since the cotangent of x is 2, x is a number whose tangent is $\frac{1}{2}$. Therefore, $\cot x = 2$ can be rewritten as $\tan x = \frac{1}{2}$ or $x = \tan^{-1}\left(\frac{1}{2}\right)$. Therefore,

$$\cot^{-1}(2) = \tan^{-1}\left(\frac{1}{2}\right).$$

Similar arguments will show that the following identities are true under certain conditions:

$$\csc^{-1}(b) = \sin^{-1}\left(\frac{1}{b}\right) \quad \text{and} \quad \sec^{-1}(b) = \cos^{-1}\left(\frac{1}{b}\right).$$

In Exercise Set 5.10, you will investigate these conditions.

EXAMPLE 2   Solve $\cot x = 3$.

Solution   Rewrite the equation as follows:

$$\cot x = 3$$
$$\tan x = \frac{1}{3}$$
$$x = \tan^{-1}\left(\frac{1}{3}\right).$$

We can use a calculator in radian mode to find that $\tan^{-1}(\frac{1}{3}) = 0.3218$. Verify that 0.3218 is a number whose cotangent is 3. Since the cotangent function has period π, the set of all solutions of the equation $\cot x = 3$ is given by $x = 0.3218 + k\pi$, where k is any integer. ▩

Exercise Set 5.10

1. Evaluate the following with a calculator. Write a sentence to explain what the value of each expression means.

   a. $\sin^{-1} 0.95$

   b. $\cos^{-1}(-0.67)$

   c. $\tan^{-1} 100$

   d. $\sin^{-1} 1.95$

   e. $\sec^{-1} 10$

   f. $\csc^{-1}(-4.9)$

2. Use your calculator to evaluate the following.

   a. $\sin(\sin^{-1} 0.9)$

   b. $\sin^{-1}(\sin 2)$

   c. $\cos[\cos^{-1}(-0.39)]$

   d. $\cos^{-1}(\cos 1)$

3. The symbol $\cos^{-1}(\frac{1}{2})$ represents a number between 0 and π whose cosine is $\frac{1}{2}$.

   a. Use a right triangle to illustrate/explain the meaning of $\cos^{-1}(\frac{1}{2})$.

   b. Use the unit circle to illustrate/explain the meaning of $\cos^{-1}(\frac{1}{2})$.

   c. Use the graph of the cosine function to illustrate/explain the meaning of $\cos^{-1}(\frac{1}{2})$.

4. You and a classmate are doing homework together. Your classmate writes the following steps to solve $\cos x = 0.83$:

   $$\cos x = 0.83$$
   $$x = \cos^{-1}(0.83)$$
   $$x = 0.5917.$$

   Write several sentences to explain what is right and what is wrong with your classmate's work. Explain how to get the correct and complete solution set.

5. a. Sketch a graph of the function $y = \csc^{-1}x$. Choose the domain and range so that $y = \csc^{-1}x$ is a function.

   b. Compare the domain and range of $y = \csc^{-1}x$ that you chose in Part a and the domain and range of $y = \sin^{-1}x$. How does this information apply to the identity $\csc^{-1}(b) = \sin^{-1}(\frac{1}{b})$?

6. a. Sketch a graph of the function $y = \sec^{-1}x$. Choose the domain and range so that $y = \sec^{-1}x$ is a function.

   b. Compare the domain and range of $y = \sec^{-1}x$ that you chose in Part a and the domain and range of $y = \cos^{-1}x$. How does this information apply to the identity $\sec^{-1}(b) = \cos^{-1}(\frac{1}{b})$?

7. Sketch a graph of each function using your knowledge of transformations of functions.

   a. $y = \sin^{-1}(2x)$ **b.** $y = 2\tan^{-1}x$

   c. $y = 1 + \cos^{-1}x$ **d.** $y = \sin^{-1}(x + 1)$

8. Find the inverse of each function. State the domain and range of each inverse function.

   a. $f(x) = \sin^{-1}(2x)$ **b.** $g(x) = \cos^{-1}(x - \pi)$

   c. $h(x) = 2\tan^{-1}x$ **d.** $f(x) = a\sin^{-1}(bx + c)$

9. Use your calculator to graph $y = \cos^{-1}x + \sin^{-1}x$. Find its domain. Explain how you could have anticipated the results of the graph.

10. Tell whether the following statement is true. Support your answer.

$$\sin^{-1}x = \frac{1}{\sin x}.$$

11. Sketch a graph of $y = \cot x$. The period and the position of the vertical asymptotes suggest restricting the domain to $0 < x < \pi$ to make the function one-to-one. This is the domain restriction that is commonly used. With this domain restriction, the range of the inverse cotangent function will be $0 < y < \pi$. Sketch a graph of $y = \cot^{-1}x$ that is consistent with this domain restriction on $y = \cot x$.

12. When the inverse tangent key on a calculator is used to find values of the inverse cotangent function, the values of inverse cotangent will be in the same range as those of the inverse tangent, that is, from $-\frac{\pi}{2}$ to $\frac{\pi}{2}$. Therefore, the graph of $y = \cot^{-1}x$ would be as shown in Figure 63 if it were based on values obtained from a calculator.

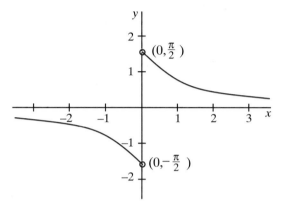

Figure 63 Graph of $y = \cot^{-1}x$ based on calculator values of $\tan^{-1}(\frac{1}{x})$

a. Compare the graph of $y = \cot^{-1}x$ in Figure 63 to the graph you sketched for Exercise 11. What disadvantages does the graph in Figure 63 have?

b. Note that whenever x is negative your calculator will give a negative value for $\tan^{-1}(\frac{1}{x})$, but we would like the range of $\cot^{-1}$ to include only numbers from 0 to π. How can we modify the calculator procedure for evaluating $\cot^{-1}x$ when x is negative to achieve this range?

SECTION 5.11 The Double Ferris Wheel Investigation

1. The radius of a Ferris wheel is 25 feet and the center of the wheel is located 30 feet off the ground. The wheel rotates counterclockwise and completes one revolution every 60 seconds. Assume the rider begins at the lowest point on the wheel. Using parametric equations describe the vertical and horizontal position of a person riding the Ferris wheel over time. Sketch a graph showing the position of the rider in the first two minutes of the ride. In parametric mode, plot the path of the rider on your graphing calculator.

Figure 64   A Ferris wheel

2. Modify the parametric equations you wrote in Problem 1 to establish (0, 0) as the center of the Ferris wheel. In parametric mode, plot the path of the rider on your graphing calculator.

3. Change the parametric equations in Problem 2 so that the Ferris wheel rotates clockwise.

4. A double Ferris wheel is an exciting ride. As shown in Figure 65 on the next page, a 50-foot bar is attached to the main support at the center of the bar, and the bar revolves once every 60 seconds. Seats for riders are evenly spaced along two separate wheels, each revolving at the end of the long bar. Each wheel makes one complete revolution every 15 seconds and has a radius of 10 feet. Assume both the bar and the wheels are rotating counterclockwise, the rider begins at the lowest point, and the wheels turn independently of the bar.

    a. Write equations that describe the vertical and horizontal position of the rider over time. Establish the point (0, 0) of the coordinate system as the center of the bar. In parametric mode, plot the path of the rider on your graphing calculator. In function mode, plot the vertical position of the rider over time, and plot the horizontal position of the rider over time.

    b. Predict the number of loops that will occur in 60 seconds. Is the graph consistent with your prediction?

    Source: Presentation by Bill Rousseau at Contemporary Precalculus through Applications Summer Workshop, NCSSM, Durham, NC, July, 1992.

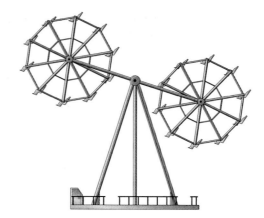

Figure 65 A double Ferris wheel

5. When looking at a double Ferris wheel or creating an animation or a physical model, we discover that the two wheels do not turn independently of the bar. Each wheel turns some on its own; it also turns a small amount because the bar is turning too. Therefore, the angle the wheel turns is the sum of the rotation of the wheel and the rotation of the bar.

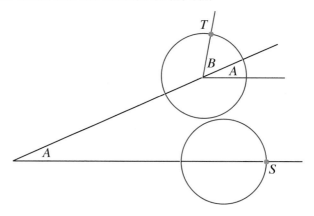

Figure 66 Changes in the position of a wheel in one second

Figure 66 allows us to follow the motion of the bar and the wheel of the double Ferris wheel for one second. The figure is not drawn to scale so the positions of the bar and wheel may be more easily seen. Assume the rider begins at point S in Figure 66. One second later the rider is at point T. In that second, the bar has completed one-sixtieth of a revolution, so it has turned counterclockwise $\frac{1}{60}(2\pi) = \frac{2\pi}{60}$, or 6°. During the same second, the wheel has completed one-fifteenth of a revolution and has turned counterclockwise $\frac{1}{15}(2\pi) = \frac{2\pi}{15}$, or 24°. Therefore, the rider has rotated through angle B and, because the bar turns, also through angle A. Therefore, the rider on the small wheel has rotated $\frac{2\pi}{60} + \frac{2\pi}{15}$ radians, or 30°.

With this refinement of the model, the equations for vertical and horizontal position are the following:

$$x(t) = 25 \sin\left(\frac{2\pi t}{60}\right) + 10 \sin\left(\frac{2\pi t}{15} + \frac{2\pi t}{60}\right)$$

$$y(t) = -25 \cos\left(\frac{2\pi t}{60}\right) - 10 \cos\left(\frac{2\pi t}{15} + \frac{2\pi t}{60}\right).$$

Sketch a graph of the path of the rider with these modifications. How many loops does the rider complete in 60 seconds?

6. One way to study the loops that the rider moves through is to determine the distance of the rider from the center of the bar as a function of time t. Since the rider's position is described by the ordered pair (x, y), the distance between the center $(0, 0)$ and (x, y) can be measured using the distance formula. Write a function to represent the rider's distance from the center in terms of time. Graph the function to determine the number of loops per minute.

7. What equations describe the vertical and horizontal position of the rider if the bar completes one clockwise revolution every 60 seconds and the wheel completes one counterclockwise revolution every 15 seconds? When graphed in parametric mode, what features of the graph change? Does the ride seem more exciting if both wheels are turning counterclockwise or if the wheels are going in opposite directions? Graph the function that represents the distance of the rider from the center of the Ferris wheel over time. Determine the number of loops per minute.

8. Experiment with changes in the time for the small wheel to complete one revolution. How do these changes affect the path of the rider? Which rides seem most exciting?

9. What is the maximum height of a rider on the double Ferris wheel as described in Problem 5? Can you determine this height algebraically? Support your answer.

SECTION 5.12 Composition with Inverse Trigonometric Functions

In Exercise Set 5.10 you investigated the composition of trigonometric functions and their inverses. For instance, you found that

$$\sin(\sin^{-1} 0.9) = 0.9 \quad \text{and} \quad \sin^{-1}(\sin 2) = 1.1416.$$

For $f(x) = \sin x$ and $g(x) = \sin^{-1} x$ these examples show that the compositions $f \circ g$ and $g \circ f$ may or may not produce the identity function.

EXAMPLE 1   Let $f(x) = \sin x$ and let $g(x) = \sin^{-1} x$. For what values of x is it true that $f(g(x)) = \sin(\sin^{-1} x) = x$? For what values of x is it true that $g(f(x)) = \sin^{-1}(\sin x) = x$? Explain your answers.

Solution   The graphs in Figure 67 show $y = \sin(\sin^{-1} x)$ on the left and $y = \sin^{-1}(\sin x)$ on the right. The graph of $f(g(x)) = \sin(\sin^{-1} x)$ is identical to the graph of $y = x$ on the interval $[-1, 1]$. The graph of $g(f(x)) = \sin^{-1}(\sin x)$ is identical to the graph of $y = x$ on the interval $[-\frac{\pi}{2}, \frac{\pi}{2}]$. Therefore, $\sin(\sin^{-1} x) = x$ for all x between -1 and 1, and $\sin^{-1}(\sin x) = x$ for all x between $-\frac{\pi}{2}$ and $\frac{\pi}{2}$.

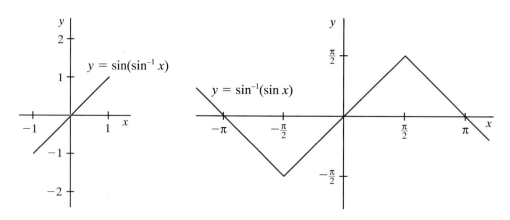

Figure 67   Graphs of $y = \sin(\sin^{-1} x)$ and $y = \sin^{-1}(\sin x)$

To understand why $\sin(\sin^{-1} x) = x$ only on the interval $[-1, 1]$, consider the domain of the composition $f(g(x)) = \sin(\sin^{-1} x)$. The domain of $f \circ g$ is the subset of the domain of $g(x) = \sin^{-1} x$ where the values of $g(x)$ are in the domain of $f(x) = \sin x$. The domain of g is $-1 \le x \le 1$. The values of $g(x)$ are between $-\frac{\pi}{2}$ and $\frac{\pi}{2}$; since the domain of f is all real numbers, all values of $g(x)$ are in the domain of f. Therefore, the domain of $f \circ g$ is equal to the entire domain of g, which is $-1 \le x \le 1$. Therefore, $\sin(\sin^{-1}(x)) = x$ for all values of x in the domain of the composition, that is, for all x between -1 and 1, inclusive.

To understand why $\sin^{-1}(\sin x) = x$ only on the interval $[-\frac{\pi}{2}, \frac{\pi}{2}]$, consider the range of $g(f(x)) = \sin^{-1}(\sin x)$. The values of $g(f(x))$ are in the range of $g(x) = \sin^{-1}x$, which includes values between $-\frac{\pi}{2}$ and $\frac{\pi}{2}$. For $\sin^{-1}(\sin x)$ to be equal to x, the value of x must also be in the interval from $-\frac{\pi}{2}$ to $\frac{\pi}{2}$, inclusive. Therefore, $\sin^{-1}(\sin(x)) = x$ for all x between $-\frac{\pi}{2}$ and $\frac{\pi}{2}$.

Note that $\sin(\sin^{-1}x)$ is equal to x on its entire domain, which is $-1 \le x \le 1$. In contrast, $\sin^{-1}(\sin x)$ is equal to x only when $-\frac{\pi}{2} \le x \le \frac{\pi}{2}$, even though its domain is all real numbers. ■

CLASS PRACTICE

1. Determine the values of a for which $\cos(\cos^{-1}a) = a$. Explain your results.

2. Determine the values of a for which $\cos^{-1}(\cos a) = a$. Explain your results.

3. Determine the values of a for which $\tan(\tan^{-1}a) = a$. Explain your results.

4. Determine the values of a for which $\tan^{-1}(\tan a) = a$. Explain your results.

5. Comment on the statement: "The composition of a function and its inverse yields the identity function." Is this statement always true? Under what conditions, if any, might it be false?

As you know, functions that are not inverses of each other can also be composed. For these kinds of compositions, we need to recall that inverse trigonometric functions produce numbers that are arc lengths or angle measures. Therefore, it will sometimes be convenient to use our understanding of right triangle trigonometry to evaluate these compositions.

EXAMPLE 2   Find the exact value of $\sin(\cos^{-1}(\frac{3}{8}))$.

Solution   We want to find the sine of a number whose cosine is $\frac{3}{8}$. If we use a calculator to approximate $\cos^{-1}(\frac{3}{8})$ and then take the sine of that value, our result will be an approximation rather than an exact answer. Instead, you can use right triangle trigonometry and let θ represent $\cos^{-1}(\frac{3}{8})$. This means that θ is the number between 0 and π whose cosine is $\frac{3}{8}$. The value we want to find can be written as $\sin \theta$. The triangle in Figure 68 on the next page will help us find the sine of θ using the fact that $\cos \theta = \frac{3}{8}$. The cosine of θ is defined as the ratio of the length of the adjacent side to the length of the hypotenuse, which is 3:8.

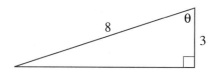

Figure 68 $\theta = \cos^{-1}(\tfrac{3}{8})$ in a right triangle

The Pythagorean Theorem can be used to show that the side opposite θ has length $\sqrt{8^2 - 3^2} = \sqrt{55}$. Therefore, the sine of θ, defined as the ratio of the length of the opposite side to the length of the hypotenuse, is equal to $\frac{\sqrt{55}}{8}$. Since θ is between 0 and π, we know that θ has a positive sine.

The value of $\sin(\cos^{-1}(\tfrac{3}{8}))$ can also be determined using the identity $\sin^2\theta + \cos^2\theta = 1$, which can be written $\sin\theta = \pm\sqrt{1 - \cos^2\theta}$. Since $\cos\theta = \tfrac{3}{8}$,

$$\sin\theta = \pm\sqrt{1 - \left(\frac{3}{8}\right)^2}$$

$$= \pm\sqrt{\frac{64}{64} - \frac{9}{64}} = \pm\frac{\sqrt{55}}{8}.$$

Since θ must be between 0 and π, the sine of θ is positive. Thus, $\sin\theta = \frac{\sqrt{55}}{8}$ and $\sin(\cos^{-1}(\tfrac{3}{8})) = \frac{\sqrt{55}}{8}$.

If, in Example 2, we had been asked to find the value of $\sin(\cos^{-1}(-\tfrac{3}{8}))$, we would have again found $\sin\theta = \pm\frac{\sqrt{55}}{8}$. Since $\cos\theta = -\tfrac{3}{8}$ and the range of the inverse cosine function is $[0, \pi]$, θ is in the second quadrant. In the second quadrant $\sin\theta$ is positive, and $\frac{\sqrt{55}}{8}$ is again the answer.

EXAMPLE 3   Express $\tan(\cos^{-1}(x + 1))$ in terms of x without trigonometric functions.

Solution   The expression $\tan(\cos^{-1}(x + 1))$ represents the tangent of an angle θ whose cosine is $x + 1$. The technique of Example 2 can be used here with slight adaptation. Let $\theta = \cos^{-1}(x + 1)$. Since $x + 1$ could be either positive or negative and the range of inverse cosine is $[0, \pi]$, we must consider two possible positions for angle θ. (See Figure 69.)

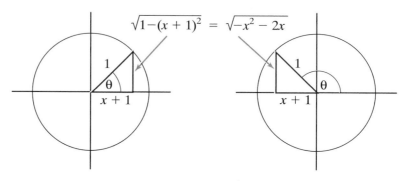

Figure 69 $\tan(\cos^{-1}(x + 1))$

Since $\cos\theta = x + 1$ and the expression involves $\tan\theta$, first find $\sin\theta$. Since $0 \le \theta \le \pi$, $\sin\theta$ will be nonnegative. The Pythagorean Theorem or the identity $\sin^2\theta + \cos^2\theta = 1$ can be used to show that the value of $\sin\theta$ is $\sqrt{1^2 - (x + 1)^2}$, which can be written as $\sqrt{-x^2 - 2x}$. The tangent of θ is the ratio of $\sin\theta$ to $\cos\theta$, so

$$\tan(\cos^{-1}(x + 1)) = \frac{\sin(\cos^{-1}(x + 1))}{\cos(\cos^{-1}(x + 1))} = \frac{\sqrt{-x^2 - 2x}}{x + 1}.$$

Note that the tangent will be positive if $x + 1$ is positive (θ in the first quadrant) and will be negative if $x + 1$ is negative (θ in the second quadrant). ■

CLASS PRACTICE

These questions refer to Example 3.

1. Substitute $x = -\frac{1}{2}$ in $\tan(\cos^{-1}(x + 1))$ and in $\frac{\sqrt{-x^2 - 2x}}{x + 1}$. Compare the results.

2. Substitute $x = \frac{1}{2}$ in $\tan(\cos^{-1}(x + 1))$ and in $\frac{\sqrt{-x^2 - 2x}}{x + 1}$. Compare the results.

3. For what values of x is the following equation true?

$$\tan(\cos^{-1}(x + 1)) = \frac{\sqrt{-x^2 - 2x}}{x + 1}$$

EXAMPLE 4 Solve $\sin^{-1}(2x) = \cos^{-1}x$.

Solution The expression $\sin^{-1}(2x)$ represents an angle whose sine is $2x$. Similarly, the expression $\cos^{-1}x$ represents an angle whose cosine is x. The equation $\sin^{-1}(2x) = \cos^{-1}x$ means that these two angles are equal. Therefore, there is an angle θ whose sine is $2x$ and whose cosine is x. The angle θ must be in the interval $[0, \frac{\pi}{2}]$ because it is in the range of the inverse sine function, which is $[-\frac{\pi}{2}, \frac{\pi}{2}]$, and also in the range of the inverse cosine function, which is $[0, \pi]$. Since $\sin\theta = 2x$, $\cos\theta = x$, and $\sin^2\theta + \cos^2\theta = 1$, we have

$$(2x)^2 + x^2 = 1.$$

We can now solve for x.

$$4x^2 + x^2 = 1$$

$$5x^2 = 1$$

$$x^2 = \frac{1}{5}$$

$$x = \pm\sqrt{\frac{1}{5}}$$

We know that the angle θ must be in the first quadrant, so only $x = \sqrt{\frac{1}{5}}$ is a solution of the original equation. ▪

An alternate way to solve $\sin^{-1}(2x) = \cos^{-1}x$ is by "undoing" one of the inverse trigonometric functions to isolate one of the variables. This "undoing" can be accomplished by composing a function with its inverse. We choose to take the sine of both sides of the equation. This "undoes" the inverse sine function as follows:

$$\sin(\sin^{-1}(2x)) = \sin(\cos^{-1}x)$$

$$2x = \sin(\cos^{-1}x).$$

Note that the expression $\sin(\sin^{-1}(2x))$ is undefined for some values of x, in particular $x > \frac{1}{2}$ or $x < -\frac{1}{2}$. However, if we assume that x is a number for which $\sin^{-1}(2x)$ is defined, then we can be sure that $\sin(\sin^{-1}(2x)) = 2x$.

We can use the technique of Example 3 on page 378 to evaluate $\sin(\cos^{-1}x)$. Figure 70 shows the sine of an angle θ whose cosine is x.

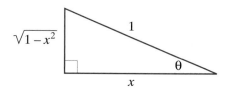

Figure 70   Right triangle illustrating $\theta = \cos^{-1}x$

Since the range of $\cos^{-1}x$ is $[0, \pi]$, the sine of $\cos^{-1}x$ is positive. The triangle in Figure 70 shows that $\sin(\cos^{-1}x) = \sqrt{1 - x^2}$.

By taking the sine of both sides of $\sin^{-1}(2x) = \cos^{-1}x$, we have produced the equation $2x = \sqrt{1 - x^2}$. We now need to solve $2x = \sqrt{1 - x^2}$ for x as follows:

$$2x = \sqrt{1 - x^2}$$

$$4x^2 = 1 - x^2$$

$$5x^2 = 1$$

$$x = \pm\sqrt{\frac{1}{5}}.$$

The two possible values for x are $\sqrt{\frac{1}{5}}$ and $-\sqrt{\frac{1}{5}}$. Checking these x-values in the original equation shows that only $x = \sqrt{\frac{1}{5}}$ is a solution. The value $x = -\sqrt{\frac{1}{5}}$ is an extraneous solution that was introduced when we squared both sides of our equation. The negative value for x is not a solution because the inverse sine of a negative number is between $-\frac{\pi}{2}$ and 0, whereas the inverse cosine of a negative number is between $\frac{\pi}{2}$ and π.

Exercise Set 5.12

1. Find the exact value of each of the following.

   a. $\sin(\cos^{-1}(\frac{1}{4}))$ **b.** $\tan(\cos^{-1}(-\frac{1}{2}))$ **c.** $\cos(\sin^{-1}(-\frac{7}{9}))$

2. Find the domain and range of the function $f(x) = \sin(\cos^{-1}x)$.

3. Sketch and label a unit circle to illustrate the fact that $\sin^{-1}(\cos 2.3) = -0.7292$.

4. A picture of height 3 feet is placed on a wall with the bottom of the picture 4 feet above the eye level of an observer. Write a function that expresses the angle θ subtended at the observer's eye in terms of the distance x between the observer and the wall. (See Figure 71.) Find the distance x that maximizes the subtended angle. This distance will provide the best viewing angle for the observer.

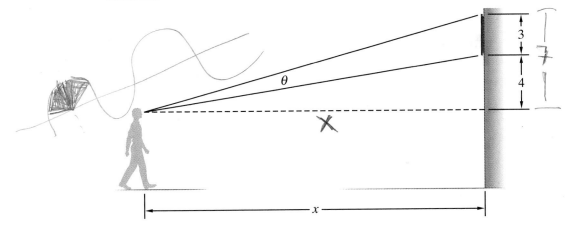

Figure 71 Angle subtended by a picture

5. Sketch a graph of each function, giving special attention to the domain and range.

   a. $y = \cos(\cos^{-1}x)$ **b.** $y = \cos^{-1}(\cos x)$

   c. $y = \tan(\tan^{-1}x)$ **d.** $y = \tan^{-1}(\tan x)$

6. For some values of x, $\sin^{-1}(\sin x) \neq x$. For these x-values, what is the value of $\sin^{-1}(\sin x)$? Explain why your answer is reasonable.

7. For some values of x, $\cos^{-1}(\cos x) \neq x$. For these x-values, what is the value of $\cos^{-1}(\cos x)$? Explain why your answer is reasonable.

8. Rewrite each of the following expressions without trigonometric functions. State any necessary restrictions on the value of the variable.

a. $\sin(\cos^{-1}x)$

b. $\cos(\tan^{-1}x)$

c. $\sin(\cos^{-1}x + \sin^{-1}x)$

d. $\tan(\cos^{-1}(-\frac{1}{x-1}))$

9. A solar receptor is placed at ground level between two buildings whose heights are 115 meters and 70 meters, respectively. (See Figure 72.) The buildings are 100 meters apart. The angle θ is the angle swept out by the sun at the receptor between the time it appears over the top of the taller building and the time it disappears behind the smaller building. For the greatest efficiency, the receptor should be located where it receives the most sunlight, in other words, where θ is a maximum. Write a function that expresses the angle θ in terms of the distance x between the receptor and the taller building. Then determine the most effective location for the receptor.

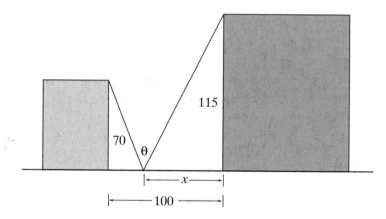

Figure 72 Location of a solar receptor

10. Rashid was riding on a single Ferris wheel that is situated adjacent to a carnival tent. His view from the Ferris wheel was obstructed by the tent until his seat reached the same height as the top of the tent. During his second trip around, Rashid began to time the ride. Four seconds after starting his watch he reached the height of the tent. Six seconds later he was below the level of the top of the tent. The diameter of the Ferris wheel is 40 feet, and the center is 25 feet from ground level. The tent is 35 feet high. Determine an equation for Rashid's height as a function of the elapsed time since he started timing the ride. How long does it take for the Ferris wheel to make one complete revolution?

11. Solve each equation.

a. $\cos^{-1}(2x) = \sin^{-1}x$

b. $\cos^{-1}(3x - 1) = \sin^{-1}x$

c. $\sin^{-1}(2x) = \pi + \cos^{-1}x$

SECTION 5.13 Solving Triangles with Trigonometry

If θ is an acute angle in a right triangle, the sine, cosine, and tangent of θ are given by the following ratios:

$$\sin\theta = \frac{\text{length of side opposite }\theta}{\text{length of hypotenuse}}$$

$$\cos\theta = \frac{\text{length of side adjacent to }\theta}{\text{length of hypotenuse}}$$

$$\tan\theta = \frac{\text{length of side opposite }\theta}{\text{length of side adjacent to }\theta}.$$

These ratios enable us to determine the measures of all parts of a right triangle if we know the measure of at least three parts, including the right angle. Recall from your study of geometry that at least one of the known parts must be the measure of a side, since three angles do not determine a unique triangle. The process of determining the measures of all unknown angles and sides is called solving a triangle.

EXAMPLE 1
angle of elevation

During an air show, a stunt pilot flies over a crowd of spectators at an altitude of 2500 feet. One minute later the *angle of elevation* of the plane is 15°. The angle of elevation is formed by two rays originating at an observer's eye. One ray extends horizontally, and the other is the line of sight from the observer to an elevated object. Assuming the pilot maintains an altitude of 2500 feet and a constant speed, how fast is the plane flying? Ignore the height of the observer for this problem.

Solution Figure 73 shows a sketch of the given information. The angle of elevation is ∠CAD and length BC represents the distance traveled by the plane during the minute it was observed.

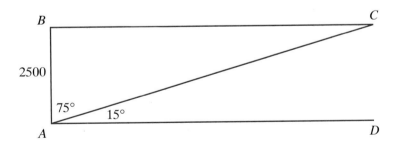

Figure 73 Diagram for Example 1

If we find the distance the plane travels in one minute, we can convert feet per minute to miles per hour to obtain the speed of the plane. Since ∠BAC and ∠CAD are complementary angles, we know that ∠BAC = 90° − 15° = 75°. Using right triangle ABC and the tangent ratio, we can now solve for side BC as follows:

$$\tan\angle BAC = \frac{BC}{AB}$$

$$\tan 75° = \frac{BC}{2500 \text{ ft}}$$

$$BC = (2500 \text{ ft})\tan 75° = 9330 \text{ ft.}$$

The plane travels about 9330 feet in one minute. We can convert to miles per hour as follows:

$$\frac{9330 \text{ ft}}{1 \text{ min}} = \frac{9330 \text{ ft}}{1 \text{ min}} \cdot \frac{60 \text{ min}}{1 \text{ hr}} \cdot \frac{1 \text{ mile}}{5280 \text{ ft}} \approx 106 \text{ mph.} \quad \blacksquare$$

EXAMPLE 2 During the same airshow described in Example 1, another plane, flying at a constant speed, passes over the crowd at an altitude of 2500 feet. When it is directly above the crowd it begins ascending. One minute later the angle of elevation to the plane is 40°, and the plane's altitude is 7200 feet. Find the speed of the plane during its ascent.

Solution The given information is illustrated in Figure 74.

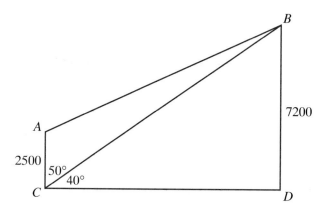

Figure 74 Diagram for Example 2

The plane ascends along side AB of $\triangle ABC$, so this is the side whose measure we need. Since $\angle BCD$ and $\angle ACB$ are complementary, we know that $\angle ACB = 50°$. We need at least three parts of a triangle to solve it, but we have only one angle and one side of $\triangle ABC$. Using right $\triangle BCD$, we can solve $\sin 40° = \frac{7200}{BC}$ to determine that side $BC \approx 11{,}201$ feet. Now we know two sides and an included angle of $\triangle ABC$. We cannot use right triangle ratios to solve for side AB, however, since $\triangle ABC$ is not a right triangle.

     We will complete the solution of this problem after we develop methods for solving triangles that are not right triangles. These are called *oblique triangles*.

oblique triangles

We will develop the Law of Cosines to solve some oblique triangles, such as $\triangle ABC$ in Figure 75. We will follow the convention that the side opposite $\angle A$ has length a, the side opposite $\angle B$ has length b, and so forth.

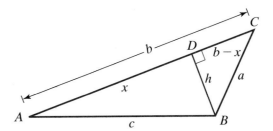

Figure 75   Oblique triangle ABC

Assume that the lengths b and c and the measure of angle A are known. To find the length a, we draw the altitude DB with length h from $\angle B$ to side AC. This will divide side AC into two segments whose lengths are x and $(b - x)$.

In right triangle BCD the Pythagorean Theorem gives

$$(b - x)^2 + h^2 = a^2. \tag{1}$$

In right triangle ABD, we have

$$x^2 + h^2 = c^2. \tag{2}$$

We can rewrite equation (2) as $h^2 = c^2 - x^2$ and substitute for h^2 in equation (1) to obtain

$$(b - x)^2 + c^2 - x^2 = a^2 \tag{3}$$

Squaring $(b - x)$ on the left side of equation (3) gives

$$b^2 - 2bx + x^2 + c^2 - x^2 = a^2,$$

which can be written as

$$a^2 = b^2 + c^2 - 2bx. \tag{4}$$

We would like to express x in terms of sides or angles in $\triangle ABC$. In right triangle ABD, $\cos A = \frac{x}{c}$ so $x = c \cos A$. Substituting in equation (4) yields

$$a^2 = b^2 + c^2 - 2bc \cos A. \tag{5}$$

Law of Cosines

The relationship expressed in equation (5) is the *Law of Cosines*. Three versions of this law are stated below:

$$a^2 = b^2 + c^2 - 2bc \cos A$$
$$b^2 = a^2 + c^2 - 2ac \cos B$$
$$c^2 = a^2 + b^2 - 2ab \cos C.$$

The Law of Cosines contains four variables. Three represent lengths of sides and one represents an angle measure. There are several different circumstances under which this law allows us to find unknown sides or angles in a triangle.

If we know the lengths of three sides, we can find the measure of any angle. For instance, if we know a, b, and c, we can use $a^2 = b^2 + c^2 - 2bc \cos A$ to find angle A, $b^2 = a^2 + c^2 - 2ac \cos B$ to find angle B, and so forth.

If we know the lengths of two sides and the measure of the included angle, we can find the measure of the third side. For instance, if we know b, c, and A, we can use $a^2 = b^2 + c^2 - 2bc \cos A$ to find a.

If we know the lengths of two sides and the measure of the nonincluded angle, we can find the measure of the third side. For instance, if we know a, b, and B, we can use $b^2 = a^2 + c^2 - 2ac \cos B$ to find c. Once the known values for a, b, and B are substituted in $b^2 = a^2 + c^2 - 2ac \cos B$, the resulting equation is quadratic in c and can have 0, 1, or 2 real solutions. We will explore this situation in Example 4. Now we have the tools to answer the question posed in Example 2.

CLASS PRACTICE

1. Complete the work begun in Example 2 on page 384 to determine the distance traveled during the first minute of the plane's ascent. (See Figure 76, side AB.) After finding the distance traveled in one minute, convert the speed from feet per minute to miles per hour.

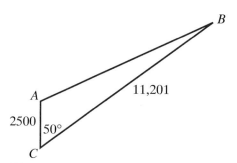

Figure 76 Diagram for Class Practice

The Law of Cosines is useful for solving many oblique triangles, but it does not help in all situations. Suppose, for example, that you know two angles and one side of an oblique triangle. You simply do not know the appropriate information to use the Law of Cosines. In this situation, the Law of Sines can be used to solve the triangle.

Figure 77 shows an oblique triangle *ABC* in which an altitude of length *h* has been drawn from $\angle B$ to side *AC*. In $\triangle ABD$, $\sin A = \frac{h}{c}$, and in $\triangle BCD$, $\sin C = \frac{h}{a}$. Rewriting these ratios shows that $c \sin A = h$ and $a \sin C = h$. Therefore,

$$c \sin A = a \sin C, \text{ or } \frac{\sin A}{a} = \frac{\sin C}{c}.$$

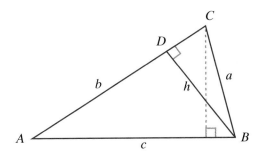

Figure 77 Oblique triangle *ABC*

Law of Sines

Similarly, if an altitude is drawn from *C*, it can be shown that $\frac{\sin B}{b} = \frac{\sin A}{a}$. These results can be combined to produce the *Law of Sines*:

$$\frac{\sin A}{a} = \frac{\sin B}{b} = \frac{\sin C}{c}.$$

There is a variety of circumstances in which the Law of Sines allows us to find unknown sides or angles in a triangle, as illustrated in the following examples.

EXAMPLE 3 Margaret and John want to sail their boat from a marina to an island 15 miles east of the marina. Along the path there are several small islands that they must avoid. When charting the course, they decide to sail first on a heading of 70° and then turn and sail on a 120° heading. What is the total distance they will travel before reaching their destination?

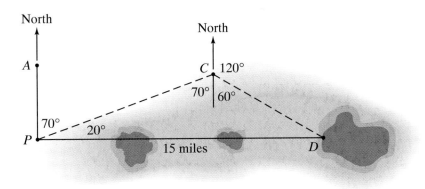

Figure 78 Diagram for Example 3

Solution Figure 78 shows the islands and the course that Margaret and John follow. To sail on a heading of 70° mean that the angle between their path and due north is 70°. Since $\angle APC = 70°$ and is complementary to $\angle CPD$, we know that $\angle CPD = 20°$. The measure of $\angle PCD$ is 130°, because it is the sum of 70° (obtained from an angle that forms an alternate interior pair with $\angle APC$) and 60° (obtained as the supplement of the 120° angle that gives the second heading). The sum of the angles of the triangle must be 180°, so we can use the fact that $\angle D = 30°$ in our calculations. We want to find the lengths of sides PC and CD in $\triangle PCD$.

We know the length of only one side, so we use the Law of Sines rather than the Law of Cosines to find the length of side CD.

$$\frac{\sin 20°}{CD} = \frac{\sin 130°}{15}$$

$$CD = \frac{15}{\sin 130°} \cdot \sin 20° \approx 6.7 \text{ miles}$$

Similarly, we can find the length of side PC.

$$\frac{\sin 30°}{PC} = \frac{\sin 130°}{15}$$

$$PC = \frac{15}{\sin 130°} \cdot \sin 30° \approx 9.8 \text{ miles}$$

Adding the lengths of sides PC and CD, we conclude that they traveled approximately 16.5 miles to reach their destination.

EXAMPLE 4 A triangular piece of land in a park is to be made into a flower bed. Stakes have previously been driven into the ground at the vertices of the triangle, which we will call B, E, and D, but the gardener can locate only the two stakes at B and E. The length of side BE is 6.2 meters. The gardener recalls that the angle at B is 60° and that the side opposite the 60° angle is to be 5.5 meters in length. Based on this information, where should the gardener search for the missing stake?

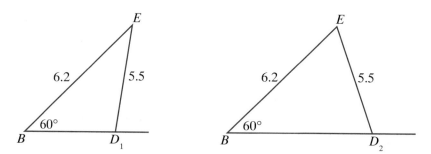

Figure 79 Diagram for Example 4

Solution There are two possible placements for point D as shown in Figure 79. Both $\triangle BED_1$ and $\triangle BED_2$ satisfy the description of the flower bed. The lengths and angle measures that are given in the problem allow us to use either the Law of Sines or the Law of Cosines to solve the triangle. When we use the Law of Cosines to locate D with the variable e representing the length of side BD, we have

$$5.5^2 = 6.2^2 + e^2 - 2(6.2)(e)\cos 60°.$$

This is a quadratic equation in e, which can be solved using the quadratic formula.

$$0 = e^2 - 2(6.2)(\cos 60°)e + (6.2^2 - 5.5^2)$$

$$0 = e^2 - 6.2e + 8.19$$

$$e = \frac{6.2 \pm \sqrt{6.2^2 - 4(8.19)}}{2}$$

$$e = 4.292 \text{ or } 1.908$$

The fact that the quadratic equation $0 = e^2 - 6.2e + 8.19$ has two real solutions implies that there are two possible lengths for e. This means that the information given in the problem does not determine a unique triangle. That is, there are two triangles that satisfy all of the conditions stated in the problem. This situation can occur when you know two sides and a nonincluded angle of a triangle.

We can also use the Law of Sines to locate D, since we know the measure of $\angle B$ and the length of the opposite side. Based on the Law of Sines, we have

$$\frac{\sin D}{6.2} = \frac{\sin 60°}{5.5}$$

$$\sin D = 6.2\left(\frac{\sin 60°}{5.5}\right)$$

$$\sin D = 0.9762.$$

There are two angles between $0°$ and $180°$ with sine that equals 0.9762. One is approximately $77.5°$, and the other is approximately $180° - 77.5° = 102.5°$. This means that the information given in the problem does not determine a unique triangle, and there are two triangles that satisfy all of the conditions stated in the problem. Figure 80 on the next page shows these two triangles; $\triangle BED_1$ has $\angle D_1 = 102.5°$ and $\triangle BED_2$ has $\angle D_2 = 77.5°$. We can finish solving each of these two triangles to find the possible locations of the third stake.

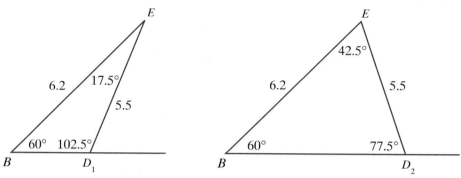

Figure 80 Two triangles for Example 4

In $\triangle BED_1$, $\angle E = 180° - (60° + 102.5°) = 17.5°$ and $\frac{BD_1}{\sin 17.5°} = \frac{5.5}{\sin 60°}$, so $BD_1 = 1.910$ meters. In $\triangle BED_2$, $\angle E = 180° - (60° + 77.5°) = 42.5°$ and $\frac{BD_2}{\sin 42.5°} = \frac{5.5}{\sin 60°}$, so $BD_2 = 4.291$ meters.

Notice that these answers vary slightly from the values we obtained using the Law of Cosines. This is a result of round-off error associated with approximating the angle at D. ▨

the ambiguous case

In Example 4, we solved a triangle for which we were given two sides and a nonincluded angle. The given information did not describe a unique triangle, but rather corresponded to two distinct triangles. This situation is often referred to as *the ambiguous case*. This ambiguity can occur when two sides and a nonincluded angle are known (SSA). If you are given sides with lengths a and b and acute angle A, there are four distinct possibilities for the relationship among a, b, and $\angle A$. The first three cases are illustrated in Figure 81.

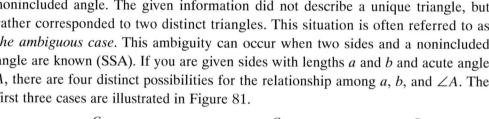

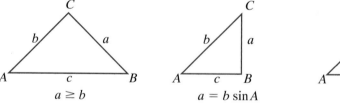

Figure 81 SSA triangles

Case 1: If a is equal to or greater than b, then only one triangle exists. To swing length CB back to the left would not produce a triangle. So, if $a \geq b$, there is exactly one triangle.

Case 2: If the side with length a is exactly long enough to touch the opposite side, then a right triangle is formed. In this case $\sin A = \frac{a}{b}$ or $a = b \sin A$.

Case 3: The side with length a is not long enough to form a triangle. If $a < b \sin A$, then the side with length a is not long enough to form a right triangle (since the shortest distance between a point and a line is the perpendicular distance), so no triangle is formed.

Case 4: Suppose $b \sin A < a < b$. There are two ways to form a triangle. Figure 82 shows the two triangles that contain sides of lengths a and b and acute angle A: triangle ABC and triangle $AB'C$.

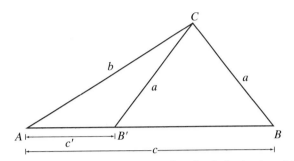

Figure 82 Ambiguous case triangle: $b \sin A < a < b$

In Example 4, we knew the measures of two sides and a nonincluded angle. Since $6.2 \sin 60° < 5.5 < 6.2$, we could have anticipated that there were two possible triangles.

When you are given two sides and a nonincluded angle and asked to solve a triangle, your first task is to determine the number of triangles that can be formed. In Example 4, the algebra that resulted from the use of the Law of Cosines alerts you to the two possible solutions. The algebra resulting from the use of the Law of Sines does not. If you use the Law of Sines, you must remember to consider two angles.

Notice that when the nonincluded angle A is $90°$ or greater, there will be at most one triangle formed. Since the longest side must be opposite the largest angle, a must be greater than b.

EXAMPLE 5   In Figure 83, $AC = 503$ cm, $DC = 101$ cm, $AC \perp CD$, $\angle BAD = 20°$, and $\angle BCA = 47°$. Find BC.

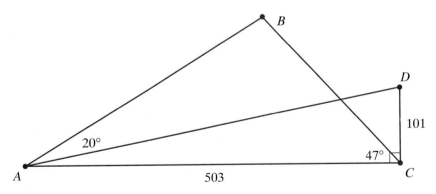

Figure 83 Diagram for Example 5

Solution To use the Law of Sines to find the length of side *BC*, we first need the measure of $\angle DAC$ in right triangle *ACD*. Since the lengths of sides *AC* and *DC* are known, we can use the tangent ratio.

$$\tan(\angle DAC) = \frac{101}{503}$$

$$\angle DAC = \tan^{-1}\left(\frac{101}{503}\right) = 11.35°$$

The measure of $\angle BAC$ is 20° plus the measure of $\angle DAC$. Therefore, the measure of $\angle BAC$ is 31.35°, and the measure of $\angle ABC$ is 101.65° Using triangle *ABC* and the Law of Sines,

$$\frac{BC}{\sin 31.35°} = \frac{503 \text{ cm}}{\sin 101.65°}$$

$$BC = 267.2 \text{ cm.} \quad \blacksquare$$

In the solution of Example 5, the angles were calculated to the nearest hundredth of a degree. If the angles had been rounded to the nearest degree, the Law of Sines would have produced a different result:

$$\frac{BC}{\sin 31°} = \frac{503 \text{ cm}}{\sin 102°}$$

$$BC = 264.9 \text{ cm.}$$

As you would expect, the angle measures to the nearest hundredth of a degree produce a more accurate answer for the length of *BC*. In solving problems, it is best to retain more digits in the intermediate steps and to round in the final answer.

When using technology, a good strategy is to solve the equation(s) using algebra without making any numerical substitutions. Using this method in Example 5 leads to the following steps:

$$\tan(\angle DAC) = \frac{101}{503}$$

$$\angle DAC = \tan^{-1}\left(\frac{101}{503}\right).$$

Substituting in the Law of Sines,

$$\frac{BC}{\sin\left(20° + \tan^{-1}\left(\frac{101}{503}\right)\right)} = \frac{503}{\sin\left(180° - 47° - 20° - \tan^{-1}\left(\frac{101}{503}\right)\right)} \qquad \textbf{(6)}$$

$$BC = 267.2.$$

Equation (6) is exactly true because no intermediate values have been estimated. Once the value of *BC* is calculated, using equation (6), the reported answer can be rounded to be consistent with the accuracy in the original measurements.

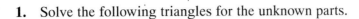
1. Solve the following triangles for the unknown parts.

   a. $c = 8, B = 50°, a = 10$ b. $a = 6, b = 12, A = 30°$

   c. $B = 30°, c = 12, b = 7$ d. $a = 3, b = 5, c = 7$

2. A ladder 28 feet long leans against the side of a building. If the angle between the ladder and the building is 20°, how far does the top of the ladder move down the building if the distance between the building and the ladder increases by 2 feet?

3. The Great Pyramid of Cheops in Egypt has a square base with dimensions 230 meters by 230 meters. The face of the pyramid makes an angle of 51.83° with the horizontal base. How tall is the pyramid?

4. A statue that is 70 meters tall stands on top of a hill. From a point at the base of the hill, the angle of elevation to the base of the statue is 20.75°, and the angle of elevation to the top of the statue is 28.30°. What is the height of the hill?

5. An art gallery has a painting of a tomato soup can. The painting has a height of 18 feet. If you stand 6 feet from the wall where the painting is hung, you must look up at an 18° angle to focus on the artist's name, which is located along the bottom edge of the painting. What is the distance between your eye and the top edge of the painting?

6. Two hikers come to a fork in a road. One walks down one straight branch at 3 mph and the other walks down the other straight branch at 3.5 mph. If the angle between the branches is 30°, find the distance between the two hikers after one hour.

7. A carpenter wants to build a ramp 30 feet long that rises to a height of 6 feet above level ground. What angle should the ramp make with the horizontal?

8. A triangular plot of land has sides of lengths 400 feet, 350 feet, and 100 feet. Find the smallest angle between the sides of the plot.

9. From the top of a mountain 3000 feet above sea level you can see two ships traveling along the same line toward you. The angles of depression to the two ships are 18° and 13°, respectively. Find the distance between the ships, to the nearest 10 feet.

10. A jogger running at a constant speed of one mile every ten minutes runs on a heading of 150° for 30 minutes and then on a heading of 40° for the next 15 minutes. Find the distance between the jogger and her starting point, to the nearest tenth of a mile.

11. Suppose you want to install a vertical TV tower on the side of a hill which makes a 25° angle with the horizon. The tower is 75 meters high, and guy wires are to be attached two-thirds of the way up the tower. If the guy wires are anchored 30 meters up the hill and 35 meters down the hill from the bottom of the tower, find the length of the guy wires. (See Figure 84.)

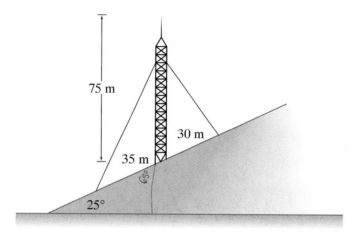

Figure 84 TV tower with guy wires

12. A balloon is sighted from two points on level ground. From point *A* the angle of elevation is 18°. From point *B* the angle of elevation is 12°.

   a. If points *A* and *B* are 8.4 miles apart and on the same side of the balloon, what is the height of the balloon?

   b. If points *A* and *B* are 8.4 miles apart and the balloon is between them, what is the height of the balloon?

13. Engineers want to measure the distance from *P* to *Q*, but the span from *P* to *Q* is across the tip of a lake. So they select a point *R* on land and find that the distance from *R* to *Q* is 100 feet and from *R* to *P* is 120 feet. Angle *QPR* measures 47°. How far is it from *P* to *Q*?

14. A golfer takes two putts to get the golf ball into the hole. The first putt rolls the ball 10.2 feet in the northwest direction, and the second putt sends the ball due north 3.7 feet into the hole. How far, and in what direction, should the golfer have aimed the first putt to get the ball into the hole with one stroke? (Assume that the green is level.)

15. Suppose an airplane is traveling toward an observer at 200 mph (293 ft/sec) at an altitude of 3000 feet. Eventually, the plane will pass directly over an observer on the ground.

    a. If sound traveled infinitely fast, then the sound that the plane made when its angle of elevation was 20° would be heard immediately. However, since sound travels at 1100 ft/sec, if the observer looks in the direction of 20° upon hearing the sound, the plane will no longer be there. At what angle should the observer look to see the plane?

    b. Suppose the observer views the plane at an angle of 25° above the horizon and simultaneously hears the sound of its engines. At what angle was the plane above the horizon when it made the sound the observer hears?

16. The pilot of a plane heading directly north finds it necessary to detour around a group of thunderstorms. To get around the storms, the pilot follows a heading of 339° for some distance and then changes heading to 35°. She heads north again when this path intersects her original flight path heading directly north. The distance between the exit point and the point of re-entry on the original flight path is 70 kilometers. How many extra miles did the pilot fly to avoid the thunderstorms?

17. Lines of latitude run east to west around the earth. The equator is the 0° line of latitude. Longitudinal lines run north to south around the earth through the North and South Poles. (See Figure 85.) Assume that the earth is a sphere with radius 3960 miles.

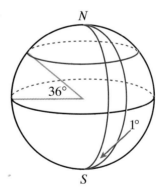

Figure 85 Latitude and longitude

    a. What is the length, in miles, of one degree of longitude at the equator?
    b. What is the length, in miles, of one degree of longitude at 36° N?
    c. What is the length, in miles, of one degree of longitude at 40° S?
    d. At what latitude does one degree of longitude encompass 30 miles?

SECTION 5.14 Investigating Hanging Pictures

Henry wants to hang a heavy picture on one wall in his living room. Since the picture is quite heavy, he has only two options for hanging the picture. Henry can install a bracket to hang the picture from the molding along the ceiling or to hang it at a reinforced area on the wall. If he uses the molding along the ceiling, he will need to use a long wire to attach the picture. If he uses the reinforced area on the wall, he can use a much shorter wire and hide the wire behind the picture. Both positions will support the picture. Henry wants to know whether having the long or short wire will make it easier to insure that the picture is at the proper height and hangs straight on the wall.

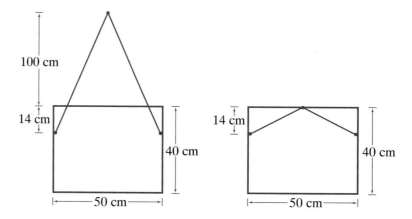

Figure 86   Two options for hanging a picture

As shown in Figure 86, the picture is 50 cm wide and 40 cm tall. On each side of the back of the frame, a small hook is attached to the frame in order to attach the wire that will be used to hang the picture. Each hook is placed 14 centimeters from the top edge of the frame and is very close to the outer edge of the frame. (Assume the hooks are 50 cm apart.) Complete the following tasks. Give solutions to the nearest hundredth of a centimeter.

1. How long should the piece of wire be to hang the picture 100 centimeters from the ceiling? Calculate the length of the wire for both options. Assume that the midpoint of the wire is at the bracket.

2. Suppose the length of wire is not measured correctly. How will an error of 2 cm affect the height of the picture using each method? Is the height affected more by the wire being 2 cm too long or 2 cm too short?

3. Suppose the error in the length of the wire is x centimeters. Write a function to describe the relationship between x and the resulting error in the height of the picture for both methods of hanging the picture. Graph the functions or make tables of values for the functions.

4. Determine which method is more sensitive to incorrectly measuring the length of the wire.

5. Suppose Henry hangs the picture so that the bracket is not at the midpoint of the wire. For which of the two options will this have the greatest effect on the position of the picture? Explain why.

6. Write a summary to compare the two methods Henry can use to hang the picture.

Source: Presentation by Henry Pollak at Teaching Contemporary Mathematics Conference, NCSSM, Durham, NC, February, 1996.

Chapter 5 Review Exercises

1. Consider the equation $2\sin^2\theta - \sin\theta - 3 = 0$.

   a. Find all solutions between 0 and 2π.

   b. Find all the solutions.

2. Solve $2\cos(\frac{\pi t}{2}) + 1 = 0$ for t in the interval $-2\pi \le t \le 2\pi$.

3. Write an equation for the graph of the sine function shown below.

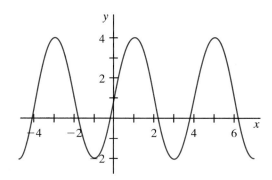

Figure 87

4. The student with whom you are working finishes a problem and announces his answer is $\cos(5.7000)$. You get an answer of the form $\sin(7.2708)$. Explain why these answers have the same numerical value.

5. Express $\theta = 47°$ in radians.

6. Two ships, A and B, are 60 miles apart at sea. Ship A is directly north of ship B. They both receive a distress signal from a third ship at the same time. Ship A receives the signal from a direction of 35° south of east. Ship B receives the signal from a direction of 50° north of east. How far is the third ship from ship A?

7. Find the exact value of each of the expressions below. Assume $\sin\theta = \frac{1}{3}$ and $\frac{\pi}{2} < \theta < \pi$.

   a. $\cos\theta$

   b. $\cos(\frac{\pi}{2} - \theta)$

   c. $\sin(2\theta)$

   d. $\cos(2\theta)$

   e. $\sin(\frac{\pi}{4} - \theta)$

8. Solve for all solutions. Find exact values whenever possible.

   a. $-4 \sin x = 6 \cos x$

   b. $2 \sin(2x - \frac{\pi}{2}) = 1$

   c. $\sin^{-1} x = \cos^{-1}(3x)$

   d. $\cos(2x) = \sin^2 x$

9. Sketch a graph of each function. You should be able to complete these "by hand." Label important points.

   a. $y = 2 \sin(\pi x)$

   b. $y = 1 - \cos(x - \frac{\pi}{4})$

   c. $y = \sec(x + \frac{\pi}{4})$

   d. $y = \sin(3x - \pi)$

   e. $y = \tan^{-1}(x + 2)$

10. Find the inverse function for each of the following functions f. State the domain and range of f^{-1}.

    a. $f(x) = 2 \cos^{-1}(x + 1)$

    b. $f(x) = \sin(x - \frac{\pi}{4}) + 1$

11. Solve each triangle.

    a. $\angle A = 90°, \angle B = 22°, a = 10$

    b. $\angle A = 40°, a = 12, b = 15$

12. Explain how the right triangle definition of sine is related to the unit circle definition of sine.

13. Describe what the expression $\tan^{-1}(0.7)$ represents.

14. The temperature on a normal spring day in Durham is a sinusoidal function of time. The temperature begins at 55° at 6 A.M. and reaches a maximum at noon of 72°. The temperature falls back to 55° by 6 P.M. and drops to 38° by midnight. This cycle continues for a number of days. A high school biology class wants to put some potted seeds outside during the times of the day when the temperature is warm enough for the seeds to germinate. If the temperature needs to be at least 60° for the seeds to germinate, during what times of the day can the class leave the pots of seeds outside?

6 Combinations of Functions

6.1	Introduction to Combinations of Functions	402
6.2	Investigating Sums and Products of Functions	404
6.3	Sums and Products of Functions	408
6.4	Investigating CO_2 Concentration	421
6.5	Investigating Beats	424
6.6	Introduction to Polynomial Functions	426
6.7	Investigating Polynomial Functions	431
6.8	Polynomial Functions	433
6.9	Rational Functions	442
6.10	Applications of Rational Functions	455
6.11	Investigating a Traffic Flow Model	461
Chapter 6 Review Exercises		464

Maintaining Traffic Flow

The North Carolina Department of Transportation is adding a third lane to the belt-way around Raleigh. Unfortunately, one of the existing lanes must be closed for several hours each day, creating a one-lane road for approximately 10 miles. Naturally, the Department of Transportation would like to maximize the flow of traffic through this region. What speed limit should be set for this stretch of road to ensure the greatest traffic flow?ˆ

Source: Adapted from Stone, Alan and Ian Huntley. "Easing the Traffic Jam" *Solving Real Problems with Mathematics*, Volume 2, The Spode Group, Cranfield Press, Cranfield, UK, 1982.

Introduction to Combinations of Functions

Throughout this course you have used functions to model phenomena in the world around you. You have found mathematical models for data using linear functions, quadratic functions, exponential functions, power functions, and trigonometric functions. In this section you will combine several of these toolkit functions to model more complicated data sets.

Students monitored the temperature of a liquid as it cooled in a refrigerator. The data they gathered are shown in the table with Figure 1.

Time (min)	0	1	2	3	4	5	6	7	8	9	10	11
Temperature (°F)	55.0	53.2	51.8	50.9	50.5	50.1	49.7	48.9	48.0	47.2	46.6	46.5
Time (min)	12	13	14	15	16	17	18	19	20	21	22	23
Temperature (°F)	46.7	46.8	46.8	46.5	45.9	45.4	45.1	45.2	45.5	45.8	46.0	45.7

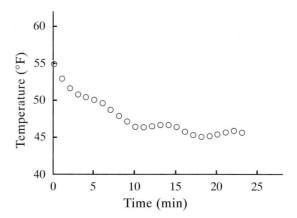

Figure 1 Cooling data and scatter plot

The students believe that the internal air temperature of the refrigerator is 45° and that the temperature of the liquid decreases exponentially with the graph approaching a horizontal asymptote at 45. Therefore, they re-expressed the data by taking the natural logarithm of the difference between the temperatures and 45. The least squares line fit to the ordered pairs $(x, \ln(y - 45))$ has slope -0.14542 and intercept 2.21439. The students believe that the function $T(x) = 9.15579e^{-0.14542x} + 45$ models their original data. Figure 2 shows this model superimposed on the data.

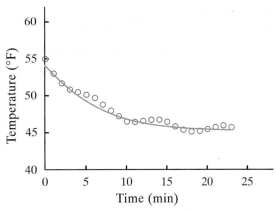

Figure 2　Cooling data modeled by $T(x) = 9.15579e^{-0.14542x} + 45$

Based on the scatter plot, it appears that the liquid's temperature oscillates around the exponential model. This oscillation is more easily observed by looking at a residual plot, which you will do in the exercises that follow.

Exercise Set 6.1

1. Use your calculator to create the residual plot associated with fitting $T(x) = 9.15579e^{-0.14542x} + 45$ to the original data.

2. Consider the residuals as a set of ordered pairs (*time, residual*), and fit a transformation of a sine wave, $F(x)$, to these ordered pairs. The fit will not be perfect, but you should be able to model the amplitude and the period of the residuals.

3. By definition, the residuals r_i are equal to the data values y_i minus the values of the model $T(x_i)$. Since $r_i = y_i - T(x_i)$, it follows that $y_i = T(x_i) + r_i$. Therefore, if the residuals can be modeled by a function $F(x)$, then the data values can be more accurately modeled by the sum $T(x) + F(x)$. Superimpose the graph of $y = T(x) + F(x)$ on the original data. Describe the fit.

4. Calculate the residuals associated with fitting $y = T(x) + F(x)$ to the original data. Examine the residual plot and comment on the goodness of the fit.

Investigating Sums and Products of Functions

In this investigation you will graph functions that are sums or products of familiar functions. Your goal will be to understand how the graphs of f and g influence the graphs of $h(x) = f(x) + g(x)$ and $k(x) = f(x) \cdot g(x)$.

INVESTIGATION 1   Sums of Functions

1.   The function $h(x) = x + \cos x$ can be written as $h(x) = f(x) + g(x)$, where $f(x) = x$ and $g(x) = \cos x$.

   a.   Use a calculator or computer to graph $h(x) = x + \cos x$ and $f(x) = x$ on the same axes. Experiment with the viewing window until you can see a graph of h that shows both global and local behavior. Sketch the graphs of h and f.

   To understand the graph of h and how it is related to the graphs of f and g, answer the following questions.

   b.   What is the domain of h? What is the range of h?

   c.   Find the values of $f(\pi)$, $g(\pi)$, and $h(\pi)$. How are these values related to the graphs of f, g, and h?

   d.   How does the graph of h behave when $\cos x = 0$?

   e.   For what x-values is the graph of h above the graph of $f(x) = x$? below the graph of $f(x) = x$?

   f.   What general relationships do you notice among the graphs of h, f, and g that would help you graph another function, such as $y = x + \sin x$?

2.   The function $h(x) = \sqrt{x} + \sin x$ can be written as $h(x) = f(x) + g(x)$, where $f(x) = \sqrt{x}$ and $g(x) = \sin x$.

   a.   Use a calculator or computer to graph $h(x) = \sqrt{x} + \sin x$ and $f(x) = \sqrt{x}$ on the same axes. Experiment with the viewing window so that you can see both global and local behavior of the function h. Sketch the graphs of h and f.

   To understand the graph of h and how it is related to the graphs of f and g, answer the following questions.

   b.   What is the domain of h? What is the range of h?

   c.   How does the graph of h behave when $\sin x = 0$?

d. For what x-values is the graph of h above the graph of $f(x) = \sqrt{x}$? below the graph of $f(x) = \sqrt{x}$?

e. Based on the graphs of $h(x) = x + \cos x$ and $h(x) = \sqrt{x} + \sin x$, what predictions can you make about the graph of $y = f(x) + \sin x$ for a different function f?

3. The function $h(x) = -x + e^x$ is the sum of $f(x) = -x$ and $g(x) = e^x$.

a. Use a calculator or computer to graph $h(x) = -x + e^x$, $f(x) = -x$, and $g(x) = e^x$ on the same axes. Experiment with the viewing window as needed. If possible, graph h with a different graphing style or color to help you distinguish among the three graphs. Sketch the graphs of h, f, and g.

b. To understand the graph of h, ask yourself (and then answer) questions similar to those you answered in Problems 1 and 2.

4. Write a short paragraph to explain in general how the graphs of f and g influence the graph of $h(x) = f(x) + g(x)$. Your paragraph should include generalizations about domain, points of intersection, positive and negative function values, and so forth. ▨

INVESTIGATION 2   Products of Functions

1. The function $k(x) = x \cos x$ can be written as $k(x) = f(x) \cdot g(x)$, where $f(x) = x$ and $g(x) = \cos x$.

a. Use a calculator or computer to graph $k(x) = x \cos x$, $f(x) = x$, and $g(x) = \cos x$ on the same axes. You may need to experiment with the viewing window so that you can see a "complete" graph of k. If possible, graph k with a different graphing style or color to help you distinguish among the three graphs. Sketch the graphs of k, f, and g.

To understand the graph of k and how it is related to the graphs of f and g, answer the following questions.

b. For what x-values does $f(x) = 0$ or $g(x) = 0$? How does the graph of k behave at these x-values?

c. For what x-values does $f(x) = \pm 1$ or $g(x) = \pm 1$? How does the graph of k behave at these x-values?

d. How does the graph of k behave when f and g both have positive y-values? when both have negative y-values? How does the graph of k behave when one function has positive y-values and the other function has negative y-values?

e. Graph only $y = x$ and $y = -x$ on the same axes as $k(x) = x \cos x$. Describe how the graph of k is related to the graphs of these two lines.

f. What type of symmetry is exhibited in the graph of k?

g. What general relationships do you notice among the graphs of k, f, and g that would help you graph another function, such as $y = x \sin x$?

2. The function $k(x) = 2^x \sin x$ is the product of $f(x) = 2^x$ and $g(x) = \sin x$.

a. Use a calculator or computer to graph $k(x) = 2^x \sin x$, $f(x) = 2^x$, and $g(x) = \sin x$ on the same axes. You may need to experiment with the viewing window. Sketch the graphs.

To understand the graph of k and how it is related to the graphs of f and g, answer the following questions.

b. For what x-values does $f(x) = 0$ or $g(x) = 0$? How does the graph of k behave at these x-values?

c. For what x-values does $f(x) = \pm 1$ or $g(x) = \pm 1$? How does the graph of k behave at these x-values?

d. Graph only $y = 2^x$ and $y = -(2^x)$ on the same axes as $k(x) = 2^x \sin x$. Describe how the graph of k is related to the graphs of these two curves.

e. For what values of x does the graph of k have positive y-values? For what values of x does the graph of k have negative y-values?

f. Describe the behavior of the graph of k as $x \to \infty$ and as $x \to -\infty$.

g. Based on the graphs of $k(x) = x \cos x$ and $k(x) = 2^x \sin x$, what predictions can you make about the graph of $y = f(x) \sin x$ for a different function f?

3. The function $k(x) = (5 - x) \ln x$ is the product of $f(x) = 5 - x$ and $g(x) = \ln x$.

a. Use a calculator or computer to graph $k(x) = (5 - x) \ln x$, $f(x) = 5 - x$, and $g(x) = \ln x$ on the same axes. You may need to experiment with the viewing window so that you can see "complete" graphs of k, f, and g. Sketch the graphs of k, f, and g.

b. To understand the graph of k, ask yourself (and then answer) questions similar to those you answered in Problems 1 and 2.

4. The function $k(x) = (x + 3) e^{-x}$ is the product of $f(x) = x + 3$ and $g(x) = e^{-x}$.

    a. Use a calculator or computer to graph $k(x) = (x + 3) e^{-x}$. Graph $f(x) = x + 3$ and $g(x) = e^{-x}$ on the same axes. You may need to experiment with the viewing window so that you can see "complete" graphs of k, f, and g. Sketch the graphs of k, f, and g.

    b. To understand the graph of k, ask yourself (and then answer) questions similar to those you answered in Problems 1, 2, and 3.

5. Write a short paragraph to explain in general how the graphs of f and g influence the graph of $k(x) = f(x) \cdot g(x)$. Your paragraph should include generalizations about domain, x-intercepts, positive and negative function values, and so forth.

SECTION 6.3 Sums and Products of Functions

Based on your work in the previous section, you should have some general ideas about how the graphs of $f(x) + g(x)$ and $f(x) \cdot g(x)$ behave. In this section we will look at some more examples in detail.

EXAMPLE 1 Explain how the graphs of $f(x) = x$ and $g(x) = \sin x$ influence the graph of $h(x) = x + \sin x$.

Solution Figure 3 shows the graph of $h(x) = x + \sin x$ and the graph of $f(x) = x$ on the same coordinate axes.

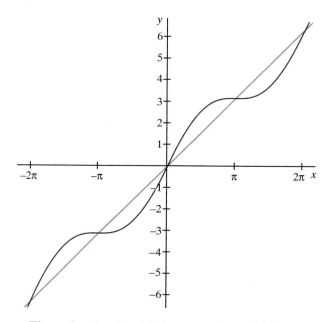

Figure 3 Graphs of $h(x) = x + \sin x$ and $f(x) = x$

At any particular x-value, the y-coordinate on h is the sum of the y-coordinate on $f(x) = x$ and the y-coordinate on $g(x) = \sin x$. Since y-coordinates of points simply measure vertical distances from the x-axis, the vertical distance between the graph of h and the x-axis is the sum of two vertical distances, one from f and one from g. We can understand the shape of the graph of h by analyzing the sums of these distances.

Whenever $\sin x$ is equal to zero, the y-value on the graph of $g(x) = \sin x$ is zero, so the graph of h has the same y-coordinate as the graph of f. This means that the graph of h intersects the graph of $f(x) = x$ when $\sin x = 0$. Whenever $\sin x$ is equal to 1, the y-value of $h(x)$ is one more than the y-value of $f(x) = x$, so the graph of h is one unit above the graph of $f(x) = x$. Whenever $\sin x$ is equal to -1, the graph of $h(x)$ is one unit below the graph of $f(x) = x$. Thus, the graph of $h(x)$ oscillates between the graphs of $y = x + 1$ and $y = x - 1$ as shown in Figure 4. Notice that the oscillation is in the vertical direction, perpendicular to the x-axis. The oscillation is not perpendicular to the graph of $f(x) = x$.

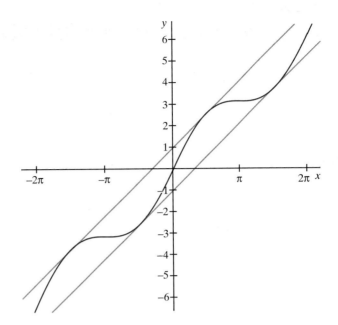

Figure 4 Graphs of $h(x) = x + \sin x$, $y = x + 1$, and $y = x - 1$

Another way to think of the process of graphing $h(x) = x + \sin x$ is to graph $g(x) = \sin x$ relative to the graph of $f(x) = x$. That is, instead of using the x-axis as the "baseline" from which vertical distances are measured, you can use the graph of $f(x) = x$ as a baseline and graph points whose vertical distance from this diagonal line is $\sin x$. When $\sin x > 0$ the graph of h is above the baseline, and when $\sin x < 0$ the graph of h is below the baseline. ▨

EXAMPLE 2   Sketch a graph of the function $h(x) = x + \ln x$.

Solution   We can write $h(x) = x + \ln x$ as $h(x) = f(x) + g(x)$, where $f(x) = x$ and $g(x) = \ln x$. We can then apply ideas about sums of functions that we learned in Section 6.2 and in Example 1.

The value of $h(x)$ at any particular x-value is obtained by adding the value of $f(x) = x$ and the value of $g(x) = \ln x$. This means that any x-value is in the domain of h only if it is in the domains of both f and g. Since the domain of $g(x) = \ln x$ includes only positive numbers and the domain of $f(x) = x$ is all real numbers, the domain of h is the set of positive real numbers.

The graph of h can be created by using $f(x) = x$ as a baseline and graphing $g(x) = \ln x$ relative to this baseline. The graph of h will be below $f(x) = x$ when $\ln x$ has negative values; this occurs on the interval $0 < x < 1$. The graph of h will intersect the graph of $f(x) = x$ when $\ln x = 0$; this occurs when $x = 1$. The graph of h will be above $f(x) = x$ when $\ln x$ has positive values; this occurs when $x > 1$.

The graph of $g(x) = \ln x$ has a vertical asymptote at $x = 0$ since $\ln x \to -\infty$ as $x \to 0^+$. The sum of x and $\ln x$ will also approach $-\infty$ as $x \to 0^+$, so the graph of h will also have a vertical asymptote at the y-axis. Since $\ln x$ increases very slowly as *x*-values increase, the vertical distance between $h(x) = x + \ln x$ and $f(x) = x$ will increase slowly as *x* increases. The graphs of $f(x) = x$, $g(x) = \ln x$, and $h(x) = x + \ln x$ are shown in Figure 5.

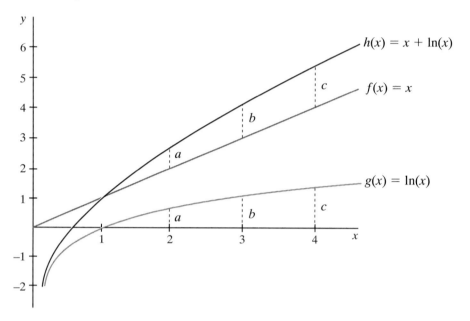

Figure 5 Graphs of $f(x) = x$, $g(x) = \ln x$, and $h(x) = x + \ln x$

The graphs we have investigated in these examples should enable you to make some general conclusions about the graph of $h(x) = f(x) + g(x)$ if *f* and *g* are functions whose graphs you know well. You may need to supplement these general conclusions with details you obtain from your calculator. However, an informative calculator graph requires a good choice of viewing window, so it is important to know some general features of the graph of $h(x) = f(x) + g(x)$ without relying on your calculator. In general, the graph of *h* will intersect the graph of *f* whenever $g(x)$ is equal to zero and will intersect the graph of *g* whenever $f(x)$ is equal to zero. The graph of *h* will be above the graph of *f* when $g(x)$ has positive values, and the graph of *h* will be below the graph of *f* when $g(x)$ has negative values. In particular, if *g* is a sine or cosine function, then the graph of $h(x) = f(x) + g(x)$ will look like the graph of *g* oscillating around the graph of *f*. This idea can be applied to a function such as $h(x) = 2 + \sin x$, which can be thought of either as the graph of $y = \sin x$ shifted up two units or as a sine wave oscillating around the graph of $y = 2$.

EXAMPLE 3   Explain how the graphs of $f(x) = x$ and $g(x) = \sin x$ contribute to the behavior of the graph of $k(x) = x \sin x$.

Solution   The graph of $k(x) = x \sin x$ is shown in Figure 6.

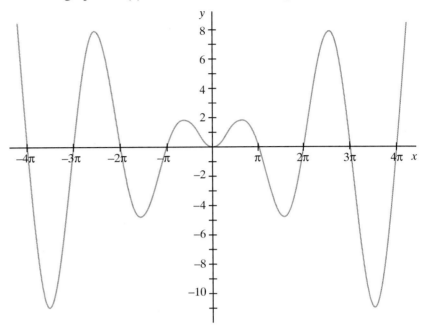

Figure 6   Graph of $k(x) = x \sin x$

To understand why the graph of k looks the way it does, we will use three simple facts about multiplication of real numbers: If a is any real number, then $0 \cdot a = 0$, $1 \cdot a = a$, and $-1 \cdot a = -a$.

We know that the product of two real numbers is zero if at least one of the numbers is zero. Therefore, the y-coordinate on the graph of k will be zero when either $f(x) = 0$ or $g(x) = 0$. This implies that the graph of k has x-intercepts at $x = 0$ (where $f(x) = 0$) and at $x = n\pi$, where n is an integer (where $g(x) = 0$). Whenever $g(x) = 1$, the product of $f(x)$ and $g(x)$ will be equal to $f(x)$. This means that the graph of k intersects the graph of $y = x$ when $\sin x = 1$, or $x = \frac{\pi}{2} + 2n\pi$, n any integer. Similarly, whenever $g(x) = -1$, the product of $f(x)$ and $g(x)$ will be equal to $-f(x)$. This means that the graph of k intersects the graph of $y = -x$ when $\sin x = -1$, or $x = \frac{3\pi}{2} + 2n\pi$, again where n is an integer.

Note that the graph of k has an x-intercept anywhere that either $g(x) = \sin x$ or $f(x) = x$ has an x-intercept. Figure 6 also shows that the graph of k is symmetric about the y-axis. This symmetry occurs because $k(-x) = (-x)\sin(-x) = (-x)(-\sin x) = x \sin x = k(x)$, which implies that k is an even function.

Superimposing the graphs of $y = x$ and $y = -x$ on the graph of $k(x) = x \sin x$ helps show that the graph of k oscillates between the graphs of $y = x$ and $y = -x$, as shown in Figure 7 on the next page. The two curves between which the graph of k oscillates are often called the *envelopes* of the graph. In this example, the lines $y = x$ and $y = -x$ are envelopes for the graph of k.

envelopes of the graph

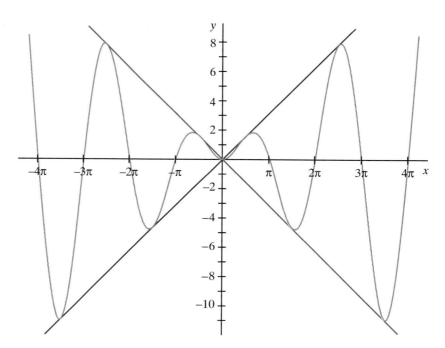

Figure 7 Graphs of $k(x) = x \sin x$, $y = x$, and $y = -x$

EXAMPLE 4 Explain how the graphs of $f(x) = x^2 - 4$ and $g(x) = e^{x-1}$ influence the graph of $k(x) = (x^2 - 4)e^{x-1}$.

Solution Graphs of $f(x) = x^2 - 4$ and $g(x) = e^{x-1}$ are shown in Figure 8.

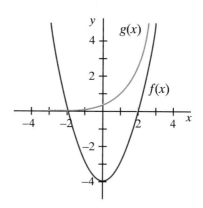

Figure 8 Graphs of $f(x) = x^2 - 4$ and $g(x) = e^{x-1}$

The domain of k includes all real numbers that are in the domains of both f and g. This means that the domain of k is all real numbers.

The product $k(x) = (x^2 - 4)e^{x-1}$ has an x-intercept where either f or g has an x-intercept. The graph of g has no x-intercepts, and the graph of f has x-intercepts at $x = -2$ and $x = 2$. Thus, k has x-intercepts at $x = -2$ and $x = 2$.

Since $g(x) = e^{x-1}$ is never negative, values of $k(x)$ are positive only when $f(x) = x^2 - 4$ is positive. This occurs when $x < -2$ and when $x > 2$.

It is also important to understand the behavior of the graph of k as x increases or decreases without bound. As $x \to \infty$, $f(x) \to \infty$ and $g(x) \to \infty$. The y-values of k are thus the product of two very large positive numbers, so k has large positive y-values when x is large. Since larger x-values produce larger y-values on both f and g, we can see that $k(x) \to \infty$ as $x \to \infty$.

As $x \to -\infty$, $f(x) \to \infty$ and $g(x) \to 0$. We need to understand how the tendencies in f and g combine to influence the graph of k. For x-values that are negative numbers far from zero, f has y-values that are positive and far from zero, while g has y-values that are positive and close to zero. Thus, the y-values of k are positive and are the product of one number which is far from zero and one number which is close to zero. The y-values from f tend to make the product k far from zero and the y-values from g tend to make the product k close to zero. In effect, the functions f and g pull the product k in opposite directions. The winner of this tug-of-war is the function that "pulls harder." The table in Figure 9 shows values of $f(x) = x^2 - 4$, $g(x) = e^{x-1}$, and $k(x) = (x^2 - 4)e^{x-1}$ as $x \to -\infty$.

Figure 9   Values of x, $f(x)$, $g(x)$, and $k(x)$

x	$f(x) = x^2 - 4$	$g(x) = e^{x-1}$	$k(x) = (x^2 - 4)e^{x-1}$
-2	0	0.04979	0
-8	60	$1.2341 \cdot 10^{-4}$	0.00740
-14	192	$3.0590 \cdot 10^{-7}$	$5.8733 \cdot 10^{-5}$
-20	396	$7.5826 \cdot 10^{-10}$	$3.0027 \cdot 10^{-7}$
-100	9996	$1.3685 \cdot 10^{-44}$	$1.3680 \cdot 10^{-40}$

These values show that the function g dominates in this tug-of-war; that is, in this case the pull of g toward zero is stronger than the pull of f away from zero, so that $k(x) \to 0$ as $x \to -\infty$.

The characteristics of the graph of $k(x) = (x^2 - 4)e^{x-1}$ that we have discussed are shown in Figure 10.

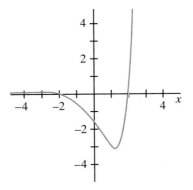

Figure 10   Graph of $k(x) = (x^2 - 4)e^{x-1}$

The graphs of $f(x) = x^2 - 4$, $g(x) = e^{x-1}$, and $k(x) = (x^2 - 4)e^{x-1}$ are shown together in Figure 11.

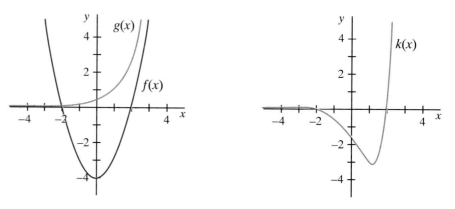

Figure 11   Graphs of $f(x) = x^2 - 4$, $g(x) = e^{x-1}$, and $k(x) = (x^2 - 4)e^{x-1}$

EXAMPLE 5   Use the graphs of $f(x) = x + 6$ and $g(x) = (x - 1)^2$ to explain the behavior of the graph of $k(x) = (x + 6)(x - 1)^2$.

Solution   Graphs of $f(x) = x + 6$ and $g(x) = (x - 1)^2$ are shown in Figure 12.

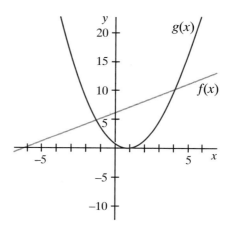

Figure 12   Graphs of $f(x) = x + 6$ and $g(x) = (x - 1)^2$

The domain of k includes all real numbers, since all real numbers are in the domains of both f and g.

The graph of g has an x-intercept at $x = 1$, and the graph of f has an x-intercept at $x = -6$. Thus, k has x-intercepts at $x = 1$ and $x = -6$.

The graph of k has positive y-values whenever $f(x)$ and $g(x)$ are both positive or both negative. Since $g(x) = (x - 1)^2$ is nonnegative for all values of x, the sign of $f(x) = x + 6$ determines whether k has positive or negative values. The graph of $f(x) = x + 6$ in Figure 12 implies that $k(x) < 0$ only when $x < -6$.

As $x \to \infty$, $f(x) \to \infty$ and $g(x) \to \infty$; therefore, $k(x) \to \infty$. As $x \to -\infty$, $f(x) \to -\infty$ and $g(x) \to \infty$. Negative $f(x)$-values and positive $g(x)$-values result in negative values for $k(x)$. As x-values approach $-\infty$, the resulting y-values become large in magnitude on both f and g. This implies that $k(x) \to -\infty$ as $x \to -\infty$.

The characteristics of the graph of $k(x) = (x + 6)(x - 1)^2$ that we have discussed are shown in Figure 13.

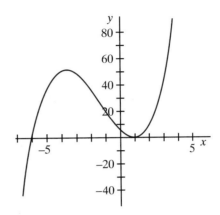

Figure 13 Graph of $k(x) = (x + 6)(x - 1)^2$

The graphs of $f(x) = x + 6$, $g(x) = (x - 1)^2$, and $k(x) = (x + 6)(x - 1)^2$ are shown together in Figure 14.

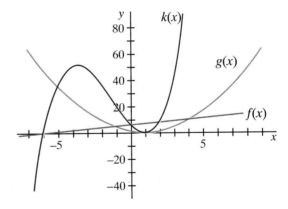

Figure 14 Graphs of $f(x) = x + 6$, $g(x) = (x - 1)^2$, and $k(x) = (x + 6)(x - 1)^2$

1. Each function graphed below is the sum of two simpler functions. Use the functions that are added to explain the shape of the graph of *h*.

   a. $h(x) = (x - 1) + \frac{1}{x + 2}$

   b. $h(x) = (1 - x) + e^x$

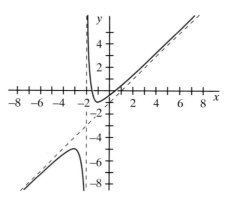

Figure 15

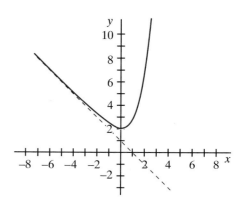

Figure 16

   c. $h(x) = |x| + \ln(x + 4)$

   d. $h(x) = \ln x + \sin x$

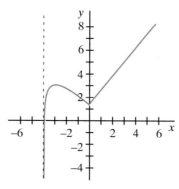

Figure 17

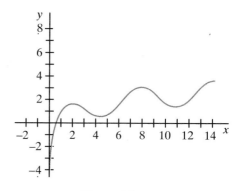

Figure 18

2. Sketch a graph of each function. Label important points and features. Identify the domain.

   a. $y = \sqrt{x} + \cos x$

   b. $y = x + \sin(2x)$

   c. $y = \frac{1}{x - 1} + x$

   d. $y = \frac{1}{x} + 2\sin x$

   e. $y = x^2 + \cos(2x)$

   f. $y = e^{-x} + \tan x$

3. Consider the function $k(x) = e^{x+1}\left(\frac{1}{x - 5}\right)$.

   a. Identify the domain of *k*.

b. For which x-values is $k(x) > 0$? $k(x) < 0$? $k(x) = 0$?

c. How do the y-values of k behave as $x \to \pm\infty$?

d. Explain why the calculator graph of k shown in Figure 19 looks the way it does. The viewing window is $-7 \le x \le 7$, $-5 \le y \le 5$.

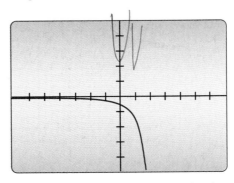

Figure 19 Graph of $k(x) = e^{x+1}\left(\frac{1}{x-5}\right)$

e. Suggest a better viewing window for the graph of k. Sketch a graph of k that shows the function's global behavior.

4. Consider the function $k(x) = \ln(x+3)\,\frac{1}{x+3}$.

a. Identify the domain of k.

b. For which x-values is $k(x) > 0$? $k(x) < 0$? $k(x) = 0$?

c. How do the y-values of k behave as $x \to \pm\infty$?

d. Explain why the calculator graph of k shown in Figure 20 looks the way it does. The viewing window is $-7 \le x \le 7$, $-5 \le y \le 5$.

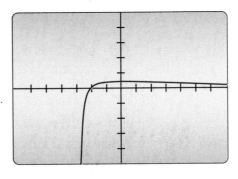

Figure 20 Graph of $k(x) = \ln(x+3)\,\frac{1}{x+3}$

e. Suggest a better viewing window for the graph of k. Sketch a graph of k that shows the function's local behavior near $x = 0$.

5. Each function graphed is the product of two simple factors. Use the factors to explain the graph's features, including domain, x-intercepts, asymptotes, enveloping curves, positive or negative y-values, and any other significant features.

a. $y = \cos(2x) \cdot \frac{1}{x}$

b. $y = x^2 \ln(x + 4)$

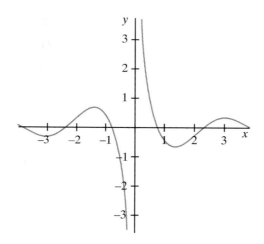

Figure 21

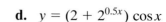

Figure 22

c. $y = \frac{1}{x} \cdot e^{x-1}$

d. $y = (2 + 2^{0.5x}) \cos x$

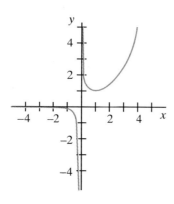

Figure 23

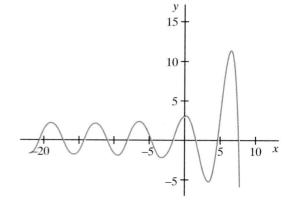

Figure 24

6. After studying several graphs, a student concludes that the product of two odd functions is always an even function. A member of her study group thinks that multiplying two odd functions may result in an even function or may result in a function that is neither even nor odd. Which student is correct? Support your answer algebraically.

7. After studying more graphs, the same student concludes that the product of two even functions is always an even function. Her friend thinks that multiplying two even functions may result in an even function or may result in a function that is neither even nor odd. Which student is correct? Support your answer algebraically.

8. The graph of $k(x) = f(x) \cdot g(x) = (x^2 - 4)e^{x-1}$ intersects the graph of each of its factors, as shown in Figure 25. When $g(x) = 1$, $k(x) = f(x)$ and the graph of k intersects the graph of f. When $f(x) = 1$, $k(x) = g(x)$ and the graph of k intersects the graph of g.

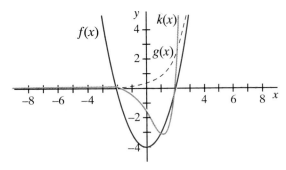

Figure 25 Graphs of $f(x) = x^2 - 4$, $g(x) = e^{x-1}$, and $k(x) = (x^2 - 4)e^{x-1}$

a. Find the x-value(s) where $f(x) = 1$. What are the coordinates of the point(s) on the graph of k for these x-value(s)?

b. Find the x-value(s) where $g(x) = 1$. What are the coordinates of the point(s) on the graph of k for these x-value(s)?

9. Answer the following questions about each function and then make a sketch of its graph. Label important points and features on the graph.

  i. What is the domain of k?

  ii. Where are the zeros of k located?

  iii. For what x-values is $k(x) > 0$?

  iv. What happens to the y-values of k as $x \to \infty$? as $x \to -\infty$?

a. $k(x) = (x + 4)\ln(x - 1)$

b. $k(x) = \ln(x - 3)\frac{1}{x - 6}$

c. $k(x) = \sin(3x)e^{-x}$

d. $k(x) = \sqrt{x} + \pi \cos(0.5x)$

e. $k(x) = xe^x$

f. $k(x) = \frac{\ln(x + 1)}{x}$

g. $k(x) = (x - 1)(x + 4)^2$

h. $k(x) = (x + 2)^2 \frac{1}{x - 4}$

10. Each function graphed below is the product of two trigonometric functions. Use the factors to explain the graph's x-intercepts and envelopes. Determine the period of each function.

a. $h(x) = \sin(5x)\sin(0.5x)$

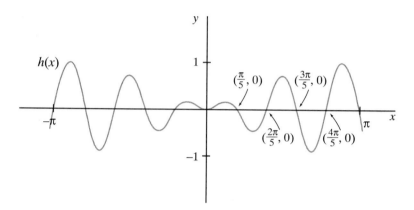

Figure 26

b. $k(x) = \sin(8x)\cos x$

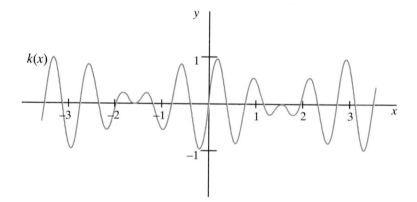

Figure 27

SECTION 6.4 Investigating CO_2 Concentration

The graph in Figure 28 shows the level of CO_2 (measured in parts per million) in the air at the Mauna Loa Observatory in Hawaii. The scatter plot exhibits an overall increase in CO_2 level together with short term declines in CO_2 level. This phenomenon can be modeled by a combination of functions with which you are familiar.

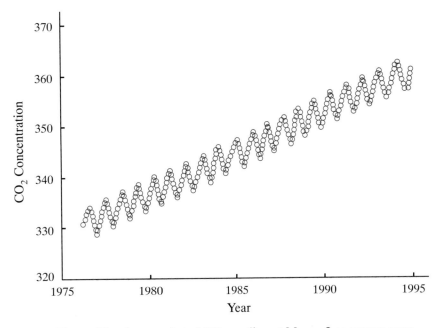

Figure 28   Scatter plot of CO_2 reading at Mauna Loa versus year

The scatter plot shows that from year to year, the concentration of CO_2 in the air has increased. However, within each individual year the concentration of CO_2 oscillates. On some time intervals the concentration is increasing, and on some intervals it is decreasing. The average CO_2 level in any year is between the high and low of that year. If a function f were used to model the average CO_2 level from year to year, the graph of f would run through the middle of the scatter plot between the individual peaks and valleys.

The table in Figure 29 shows the actual concentration for each month and the average for each year from 1976−1995.

Figure 29 Concentration of CO$_2$ for each month from 1976–1995

Year	Jan	Feb	Mar	Apr	May	June	July	Aug	Sept	Oct	Nov	Dec	Avg
1976	331.6	332.4	333.3	334.4	334.7	334.2	332.9	330.8	329.1	328.8	330.1	331.5	332.0
1977	332.8	333.3	334.5	335.9	336.6	336.1	334.8	332.6	331.4	331.0	332.2	333.7	333.7
1978	334.8	335.2	336.5	337.6	337.8	337.7	336.4	334.5	332.6	332.4	333.8	334.8	335.3
1979	336.1	336.6	337.8	338.7	339.3	339.1	337.6	335.9	333.7	333.7	335.1	336.6	336.7
1980	337.8	338.2	339.9	340.6	341.3	341.0	339.4	337.4	335.7	335.8	336.9	338.0	338.5
1981	339.1	340.3	341.2	342.3	342.7	342.1	340.3	338.3	336.5	336.7	338.2	339.4	339.8
1982	340.6	341.4	342.5	343.4	344.0	343.2	341.9	339.7	337.8	337.7	339.1	340.3	341.0
1983	341.2	342.4	342.9	344.8	345.6	345.1	343.8	342.2	339.7	339.8	341.0	342.8	342.6
1984	343.5	344.3	345.1	346.9	347.3	346.6	345.2	343.1	340.9	341.2	342.8	344.0	344.2
1985	344.8	345.8	347.3	348.2	348.8	348.1	346.4	344.5	342.9	342.6	344.1	345.4	345.7
1986	346.1	346.8	347.7	349.4	350.0	349.4	347.8	345.7	344.7	344.0	345.5	346.7	347.0
1987	347.8	348.3	349.2	350.8	351.7	351.1	349.3	347.9	346.3	346.2	347.6	348.8	348.8
1988	350.3	351.5	352.0	353.4	354.0	353.6	352.2	350.3	348.5	348.7	349.9	351.2	351.3
1989	352.6	352.9	353.5	355.2	355.5	354.9	353.7	351.5	349.6	349.8	351.1	352.3	352.7
1990	353.5	354.5	355.2	356.0	357.0	356.0	354.6	352.7	350.8	351.0	352.7	354.0	354.0
1991	354.6	355.6	357.0	358.4	359.2	358.1	356.0	353.9	352.0	352.1	353.6	354.9	355.4
1992	355.9	356.6	357.7	359.1	359.6	359.2	356.9	354.9	352.9	353.2	354.1	355.3	356.3
1993	356.6	357.1	358.3	359.4	360.2	359.6	357.5	355.5	353.6	353.9	355.3	356.7	357.0
1994	358.3	358.8	359.9	361.2	361.6	360.8	359.4	357.4	355.7	355.9	357.5	358.9	358.8
1995	359.8	360.9	361.5	363.3	363.6	363.1	361.7	359.3	357.9	357.6	359.4	360.5	360.7

Source: Keeling, Dr. Charles D., Scripps Institution of Oceanography, La Jolla, CA, LDEO/IRI Data Library and http://ingrid.ldgo.columbia.edu/SOURCES/.KEELING/.MAUNA_LOA.cdf/

Notice that in each year, the January concentration is approximately equal to the annual average. This is true again in the middle of the year (June–August) and again at the end of the year in December. In the spring of each year the concentration is greater than the annual average, and in the fall the concentration is less than the annual average.

If we use the function f to model the average CO$_2$ level as a function of time t in years, then we can examine the difference between each data value plotted in Figure 28 and the corresponding $f(t)$-value. These differences will oscillate in a periodic way between maximum and minimum values. In January of each year the difference will be approximately zero, in the spring the differences will be positive and will reach a maximum, in the summer the differences will again be approximately zero, in the fall the differences will be negative and will reach a minimum, and in the winter the differences will again be near zero. Thus, the differences between the actual CO$_2$ level and the model $f(t)$ can be represented by a sinusoidal function.

If we use the function g to model the difference between the actual CO_2 level and the level predicted by the model f, then $g(t)$ = actual CO_2 level $-f(t)$ and $g(t)$ will be a sinusoidal function. Therefore, the sum of the functions f and g will model the actual monthly CO_2 levels as a function of time. Thus, the function $C(t) = f(t) + g(t)$ will model both the year-to-year trend in the CO_2 level and the oscillation within each year.

1. Create the function $C(t) = f(t) + g(t)$ described above and compare it to the CO_2 data.

2. The CO_2 concentration for 1970 is given in Figure 30. How well does your model fit these data?

Figure 30 CO_2 concentration for 1970

Year	Jan	Feb	Mar	Apr	May	June	July	Aug	Sept	Oct	Nov	Dec	Avg
1970	324.3	325.3	326.3	327.5	327.5	327.2	326.0	324.4	322.9	322.9	323.9	325.0	325.3

3. If the CO_2 concentration continues to increase in a manner consistent with the growth shown in Figure 28, what CO_2 concentration would be expected in January of the year 2010?

4. Use the Internet to find the most recent data available for CO_2 concentration level. Does your model still fit the data well? Would some other model fit better?

5. Use the Internet to find data for the CO_2 concentration level from 1958–1975. Verify that for these years an exponential function is a better model for the average CO_2 level as a function of time than a linear function. What does the change from exponential growth to linear growth indicate about the long-term concentration of CO_2 in the atmosphere?

Investigating Beats

Sound travels in waves that can be modeled by sinusoidal functions. The pitch of the sound is associated with the frequency of the sine wave, and the intensity of the sound is associated with the amplitude of the wave. Physicists often write the equation of a sine wave in the form $y = \sin(2\pi f t)$, where t represents time and f represents the frequency of the sine wave. Recall that the frequency of a sine wave is the reciprocal of the period of the sine wave. If the period of a sine wave is the number of seconds required to complete one cycle, then the frequency is the number of cycles completed in one second. Frequency is often measured in units called hertz, abbreviated Hz. One hertz corresponds to a frequency of one cycle per second. The period of $y = \sin(2\pi f t)$ is $\frac{2\pi}{2\pi f}$, or $\frac{1}{f}$ seconds per cycle, so its frequency is f cycles per second.

Suppose two tuning forks are struck at the same time. One fork produces a sound whose frequency is f_1. The sound produced by the other fork has frequency f_2. What kind of sound will be produced by these two tuning forks vibrating simultaneously? You can simulate the effect of combining two sounds in this way by using your calculator to graph the functions that represent these sound waves.

Two tuning forks

1. The effect of two notes being played at the same time can be simulated by graphing the sum of the functions representing the individual notes. Choose values for f_1 and f_2 that are between 200 Hz and 500 Hz. Make your choices so that $f_1 \neq f_2$ and $|f_1 - f_2| \leq 50$. Let the amplitude of both sound waves be equal to 1. Use your calculator to graph the functions that model each sound wave. Since the periods of these waves are very small, $\frac{1}{f_1}$ and $\frac{1}{f_2}$ respectively, your viewing window should include only a small part of the x-axis. Use a calculator or computer to graph the sum of these two functions. Modify your viewing window as necessary. Sketch the graph on paper.

   The graph of the sum should resemble a periodic function oscillating inside an envelope that is a sine wave. If this is not obvious from your graph, adjust your viewing window so that the difference between the maximum and minimum x-values is approximately equal to 2 divided by the difference between the frequencies f_1 and f_2.

2. The sound associated with the two notes being played together has alternating intervals of intensity and silence. At times when the envelope is far from the x-axis, we hear sound. At times when the envelope is on or near the x-axis, we "hear silence." The alternating pattern of sound and silence creates an audio effect called *beats*. When beats occur, we hear a "throbbing" sound in which the sound intensity rapidly changes from high to low and back to high.

   a. Use a transformation of a cosine function to write the equation of the function that acts as the envelope for the graph you sketched in Problem 1.

b. Write the equation of a transformed sine function that has an amplitude of 1 and has the same frequency as the wave oscillating inside the envelope for the graph you sketched in Problem 1.

c. Combine the equations from Parts a and b above to create a second equation that models the graph you sketched in Problem 1.

3. At this point, you know two different equations that produce the graph in Problem 1. One equation involves the sum of two sine functions, and the other equation involves the product of a sine function and a cosine function. Since you know two different equations that have identical graphs, you have discovered a new trigonometric identity. We will now try to understand why the graph of a sum has the same appearance as the graph of a product.

From our work in Chapter 5, we know that

$$\sin(A + B) = \sin A \cos B + \cos A \sin B$$

and that

$$\sin(A - B) = \sin A \cos B - \cos A \sin B.$$

Adding these two equations produces a new identity:

$$\sin(A + B) + \sin(A - B) = 2 \sin A \cos B. \tag{1}$$

To apply this identity to our problem about beats, let $A + B = x$ and $A - B = y$. Solve the equations $A + B = x$ and $A - B = y$ simultaneously to find A and B in terms of x and y. Substitute expressions for $A + B, A - B, A$, and B in terms of x and y in equation (1) to obtain an identity of the form

$$\sin(\underline{}) + \sin(\underline{}) = 2 \sin(\underline{}) \cos(\underline{}).$$

This identity expresses a sum in terms of a product.

4. Use the identity you wrote in Problem 3 to rewrite the sum you wrote in Problem 1 as a product. How does the product obtained from the identity compare to the one you wrote in Problem 2? Graph the sum and the product on the same axes.

SECTION 6.6 Introduction to Polynomial Functions

In the preceding sections we have investigated sums and products of functions. If f and g are familiar functions, we can determine how the functions $y = f(x) + g(x)$ and $y = f(x) \cdot g(x)$ behave. There are some functions whose equations can be written either as a sum or as a product. For instance, $y = x^3 - 4x^2$ can be considered as the sum of the functions $f(x) = x^3$ and $g(x) = -4x^2$. Alternatively, $y = x^3 - 4x^2$ can be written as $y = x^2(x - 4)$ and can thus be considered the product of the functions $p(x) = x^2$ and $q(x) = x - 4$. The choice of whether to view $y = x^3 - 4x^2$ as a sum or as a product is influenced by which problems we want to solve using the function. For instance, if we need to know the function's x-intercepts it is more useful to view $y = x^3 - 4x^2$ as the product of $p(x) = x^2$ and $q(x) = x - 4$.

To further investigate the relationship between sums and products, we will study a situation in which an object is shot upward from ground level. If we shoot an object that has a small surface area upward at a low speed, the effect of air resistance is negligible. This means that only two things influence the height of the object over time—the acceleration due to gravity, g, and the object's initial velocity, v_0. Thus, there are two components that contribute to the object's height: the expression $-\frac{1}{2}gt^2$ represents the influence of gravity on the height at time t seconds, and the expression $v_0 t$ represents the influence of initial velocity on the height at time t seconds. The model for height at time t is obtained by adding $-\frac{1}{2}gt^2$ and $v_0 t$, so the model $h(t) = -\frac{1}{2}gt^2 + v_0 t$ represents the object's height at time t. If we assume that $g = 32$ ft/sec^2 and $v_0 = 64$ ft/sec, then $h(t) = -16t^2 + 64t$. A graph of h is shown in Figure 31.

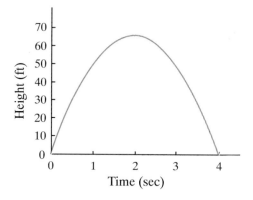

Figure 31 Graph of $h(t) = -16t^2 + 64t$

The graph includes the points $(0, 0)$ and $(4, 0)$. The value of h at $t = 0$ is zero because $h(0) = -16(0)^2 + 64(0) = 0$. The value of h at $t = 4$ is zero because $h(4) = -16(4)^2 + 64(4) = 0$. These values tell us that the object is at ground level at times $t = 0$ and $t = 4$, which means that the object stays in the air for 4 seconds.

By factoring the expression $-16t^2 + 64t$, we can rewrite the function h as a product: $h(t) = -16t(t - 4)$. Thus, we can think of h as the product of the two functions $f(t) = -16t$ and $g(t) = t - 4$, and we can apply what we have learned about products to understand the behavior of $h(t) = -16t(t - 4)$. Even though the domain of the function $h(t) = -16t(t - 4)$ is all real numbers, t-values greater than 4 or less than 0 produce heights that are negative, which is not realistic. In the context of this problem, the domain of h is $0 \le t \le 4$.

CLASS PRACTICE

1. The height of an object that is shot into the air with an initial velocity of 29.4 m/sec is represented by $h(t) = -4.9t^2 + 29.4t$. In this situation, the acceleration due to gravity is 9.8 m/sec^2.

   a. Use what you know about sums of functions to sketch the graph of $h(t) = -4.9t^2 + 29.4t$ as the sum of the functions $f(t) = -4.9t^2$ and $g(t) = 29.4t$.

   b. Use what you know about products of functions to sketch the graph of $h(t) = -4.9t^2 + 29.4t$ as the product of the functions $f(t) = -4.9t$ and $g(t) = t - 6$.

In Chapter 2 we analyzed a function that models the volume of a box made by cutting congruent squares from the corners of a piece of cardboard measuring 18 inches by 24 inches. The equation for the volume was $V(x) = x(18 - 2x)(24 - 2x)$, where x is the length of the side of the square cut from each corner. Notice that this function is the product of three linear factors: x, $18 - 2x$, and $24 - 2x$. The x-intercepts of V occur at $x = 0$, $x = 9$, and $x = 12$. The function V will have positive values when either none or two of its three factors are negative. The number line in Figure 32 shows that V is positive when $0 < x < 9$, or when $x > 12$.

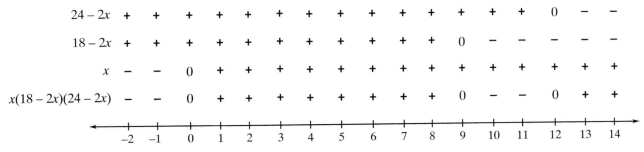

Figure 32 Signs of factors of $V(x) = x(18 - 2x)(24 - 2x)$

As x increases without bound, the values of the factor x are positive and increase without bound, while the values of the factors $18 - 2x$ and $24 - 2x$ are both negative and decrease without bound. Therefore, $V(x) \to \infty$ as $x \to \infty$. Similarly, as $x \to -\infty$, $18 - 2x \to \infty$ and $24 - 2x \to \infty$. Since two factors are positive and one factor is negative, the product of the three factors is negative. So, $V(x) \to -\infty$ as $x \to -\infty$.

A graph of $V(x) = x(18 - 2x)(24 - 2x)$ is shown in Figure 33.

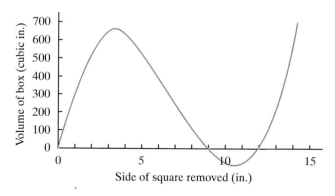

Figure 33 Graph of $V(x) = x(18 - 2x)(24 - 2x)$

The domain of V within the problem context is $0 < x < 9$. This is because x represents the length of a side of a square cut from a rectangle whose dimensions are 18 by 24. The graph of $V(x) = x(18 - 2x)(24 - 2x)$ on the interval $0 \le x \le 9$ is shown in Figure 34.

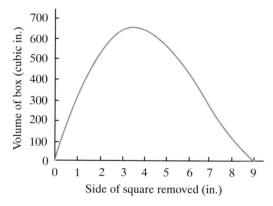

Figure 34 Graph of $V(x) = x(18 - 2x)(24 - 2x)$ on the interval $0 \le x \le 9$

By expanding the product $x(18 - 2x)(24 - 2x)$, we can also write the function V as a sum $V(x) = 4x^3 - 84x^2 + 432x$. The equation $V(x) = 4x^3 - 84x^2 + 432x$ does not provide the same information that we can obtain from $V(x) = x(18 - 2x)(24 - 2x)$. But it is important to realize that the expression for V can be written as either a sum or a product.

polynomial functions

The functions $h(t) = -16t^2 + 64t$ and $V(x) = 4x^3 - 84x^2 + 432x$ are both examples of *polynomial functions*. A polynomial function is a function that can be written in the form

$$p(x) = a_n x^n + a_{n-1} x^{n-1} + a_{n-2} x^{n-2} + \cdots + a_2 x^2 + a_1 x + a_0,$$

degree of the polynomial

where the coefficients a_i are real numbers and the exponents are nonnegative integers. The *degree of the polynomial* is n, the largest exponent. For example, $f(x) = 7x^2 - 3x + 6$ has degree 2, $g(x) = -2x^3 - x^2 + 4x - 3$ has degree 3, and $h(x) = x^4 + 2x$ has degree 4. The functions in the examples, $h(t) = -16t^2 + 64t$ and $V(x) = 4x^3 - 84x^2 + 432x$, have degree 2 and degree 3, respectively.

As we have seen, the two example polynomials, $h(t) = -16t^2 + 64t$ and $V(x) = 4x^3 - 84x^2 + 432x$, can be written as products of linear factors with real-number coefficients. Many polynomials cannot be written in factored form as products of linear factors with real-number coefficients. For example, the polynomial function $y = 5x^4 - 2x + 1$ cannot be factored into factors with real-number coefficients. For many other polynomials, the factors are difficult to determine. For example, the fourth degree polynomial $f(x) = x^4 + x^3 - 8x^2 - 2x + 12$ factors into:

$$f(x) = (x - \sqrt{2})(x + \sqrt{2})(x + 3)(x - 2).$$

Often it is a difficult task to determine whether or not a particular polynomial can be written in factored form.

Exercise Set 6.6

1. Write each polynomial as a product of factors.

    a. $y = x^2 - 6x - 27$

    b. $y = x^3 + 6x^2 - 27x$

    c. $y = x^4 - 6x^2 - 27$

    d. $y = 2x^3 - 54$

2. Write each polynomial as a sum of terms.

    a. $y = (2x - 3)(x + 14)$

    b. $y = -2(x^2 + 3)(x - 7)$

    c. $y = (x - 4)^2(3x + 7)$

    d. $y = (x + 3)(x - 1)(x + 1)$

3. For each polynomial in Exercise 2, identify the interval(s) of x-values for which $y \geq 0$.

4. Identify the degree of each polynomial.

    a. $y = (x - 3)^2(x + 1)$

    b. $y = x^3(x - 1)(x + 5)$

    c. $y = 4x(x - 4)^2(7 - x)$

    d. $y = (x + 3)^2(x - 1)^2(2x + 1)^2$

5. For each polynomial in Exercise 4, identify the coefficient of the term that determines the degree of the polynomial.

6. On page 426 we discussed an object whose height is modeled by the equation

$$h(t) = -16t^2 + 64t.$$

Rewrite the function by completing the square to find the maximum height the object reaches.

7. Explain how you can use the bisection algorithm to approximate the zeros of a polynomial.

SECTION 6.7 Investigating Polynomial Functions

In this investigation you will graph a variety of polynomial functions. You will also answer questions about the relationships between the equations of functions and the features of their graphs. As you graph, you should think about the following features of the graphs and/or the equations: degree, factors, x-intercept(s), y-intercept, and behavior as $x \to \pm\infty$.

1. Graph each polynomial function using a calculator or computer. You may need to experiment with different viewing windows so that you can see global behavior as well as local behavior. Carefully sketch each graph.

   a. $f(x) = x^3 - x$

   b. $f(x) = \frac{1}{2}(x - 6)^2 - 4$

   c. $f(x) = -\frac{1}{3}(x + 5)(x - 4)$

   d. $f(x) = (x - 1)^3 + 2$

   e. $f(x) = -5x^3 - 2x + 1$

   f. $f(x) = (x + 4)^2(x - 1)$

   g. $f(x) = (x + 5)^2$

   h. $f(x) = x^2 + 4$

2. Based on the graphs you sketched in Problem 1, if a polynomial f has degree 2, how do the y-values of f behave as $x \to \pm\infty$? How many x-intercepts could the graph of f have? Answer the same questions about a polynomial of degree 3.

3. Graph each polynomial function using a calculator or computer. You may need to experiment with different viewing windows so that you can see global behavior as well as local behavior. Carefully sketch each graph.

   a. $f(x) = -x(x + 2)(x - 3)(x - 4)$

   b. $f(x) = x^4 - 13x^2 + 36$

   c. $f(x) = (x + 2)(x - 3)(x + 1)(5 - x)$

   d. $f(x) = (x - 3)(x + 2)^2(x - 6)$

   e. $f(x) = (x - 1)^2(x + 3)^2$

4. Based on the graphs you sketched in Problem 3, what is the relationship between the factors of f and the zeros of f? If a polynomial f has degree 4, how do the y-values of f behave as $x \to \pm\infty$? How many x-intercepts could the graph of f have?

5. Graph each polynomial function using a calculator or computer. You may need to experiment with different viewing windows so that you can see global behavior as well as local behavior. Sketch each graph.

   a. $f(x) = x^4 + 5x^3 + 3x^2 - 6x - 4$

   b. $f(x) = (x - 4)^2(x + 2)(x - 1)$

   c. $f(x) = x(x + 5)(x - 1)^2(x + 1)$

   d. $f(x) = (x + 2)^2(x - 3)^3(x - 4)$

   e. $f(x) = x^5 - 6x + 4$

   f. $f(x) = x^6 - 3x^3 + x^2 - 7$

6. Based on the graphs you sketched in Problems 1, 3, and 5, how do the y-values of f behave as $x \to \pm\infty$ if f is a polynomial function whose degree is even? whose degree is odd? Based on the graphs you sketched in Problems 5b, 5c, and 5d, describe the behavior of f near $x = a$ if $(x - a)^2$ is a factor of f. Describe the behavior of f near $x = a$ if $x - a$ is a factor of f.

7. Look back over the functions you have graphed and identify patterns in the relationships between the equation of the polynomial and the shape of the graph. Write a paragraph that summarizes these relationships. You should include information about the relationship between the degree and the number of x-intercepts, the relationship between factors and x-intercepts, the relationship between the degree and the global shape of the graph, and the graph's behavior as $x \to \infty$ and as $x \to -\infty$. ▪

SECTION 6.8 Polynomial Functions

Based on the work you did in the preceding investigation, you should now be aware of many general features that are shared by the graphs of *n*th degree polynomial functions. A polynomial of degree *n* has at most *n* *x*-intercepts. The global shape of an *n*th degree polynomial depends on whether *n* is an odd or even integer. If *n* is odd, then globally the graph resembles a cubic curve; the *y*-values have opposite signs as $x \to \infty$ and $x \to -\infty$. If *n* is even, then globally the graph resembles a parabola; the signs of the *y*-values are the same as $x \to \infty$ and $x \to -\infty$. Illustrative examples shown in Figure 35 include $f(x) = (x + 3)^2(x - 1)$, a polynomial of odd degree, and $h(x) = 0.1x(2x + 5)(4 - x)(x + 1)$, a polynomial of even degree. Two graphs of each function are shown to illustrate global and local behavior.

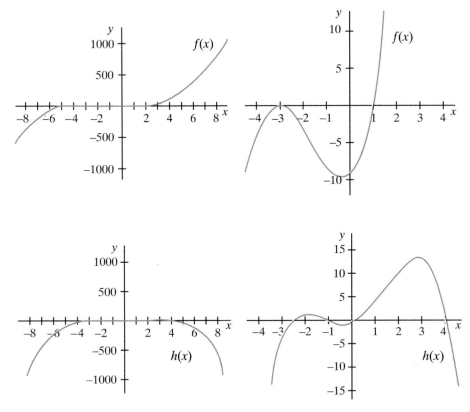

Figure 35 Graphs of $f(x) = (x + 3)^2(x - 1)$ and $h(x) = 0.1x(2x + 5)(4 - x)(x + 1)$ showing global and local behavior

The polynomial $f(x) = (x + 3)^2(x - 1)$ shown in Figure 35 has degree 3 but has only two x-intercepts. This occurs because the polynomial has only two unique factors, $x + 3$ and $x - 1$. The factor $x + 3$ is raised to the second power, and the x-intercept $x = -3$ is called a *zero of multiplicity two*. Notice that the graph of $f(x) = (x + 3)^2(x - 1)$ touches the x-axis at $(-3, 0)$ but does not cross the x-axis at this point. You are already familiar with other functions whose graphs behave this way. For example, the quadratic function $y = (x - 5)^2$ has $x = 5$ as a zero of multiplicity two. The graph of $y = (x - 5)^2$ contains the point $(5, 0)$ but does not cross the x-axis at this point.

zero of multiplicity two

The graph of $h(x) = 0.1x(2x + 5)(4 - x)(x + 1)$ shown in Figure 35 has 4 distinct zeros, $x = 0$, $x = -\frac{5}{2}$, $x = 4$, and $x = -1$. Each of these zeros has multiplicity one, and the graph of h crosses the x-axis at each of them. The graphs of f and h illustrate an important relationship between zeros and factors known as the *Factor Theorem*. The Factor Theorem states that a polynomial f has a factor $x - a$ if and only if $f(a) = 0$.

Factor Theorem

EXAMPLE 1 Find the zeros of each polynomial function. State the multiplicity of each zero.

  a. $f(x) = (x + 6)^2(x - 1)(x + 11)$

  b. $p(x) = x^3 - 7x^2 + 13x - 3$

  c. $g(x) = (4 - x)(x^2 + 1)$

Solution

  a. Since the equation for f is written in factored form, we know that the zeros are $x = -6$, $x = 1$, and $x = -11$. The zero $x = -6$ has multiplicity two; the graph of f does not cross the x-axis at the point $(-6, 0)$. Each of the other zeros has multiplicity one, so the graph of f crosses the x-axis at $(1, 0)$ and at $(-11, 0)$.

  b. To find the zeros of $p(x) = x^3 - 7x^2 + 13x - 3$ we need to solve the equation $x^3 - 7x^2 + 13x - 3 = 0$. Since $x^3 - 7x^2 + 13x - 3$ is not written in factored form and the factorization of this expression is not apparent, there is no straightforward algebraic technique that will allow us to solve $x^3 - 7x^2 + 13x - 3 = 0$. However, a graph of $p(x) = x^3 - 7x^2 + 13x - 3$ in Figure 36 on the next page shows three distinct zeros, one of which appears to be $x = 3$. Since $p(3) = 0$ we can conclude that $x - 3$ is a factor of p. Using long division, we find that $p(x) = (x - 3)(x^2 - 4x + 1)$. We can then apply the quadratic formula to the equation $x^2 - 4x + 1 = 0$ to find the remaining zeros of $p(x)$, which are $2 + \sqrt{3}$ and $2 - \sqrt{3}$. Each zero has a multiplicity of one. Note that p can be factored as

  $$p(x) = (x - 3)(x - (2 + \sqrt{3}))(x - (2 - \sqrt{3})).$$

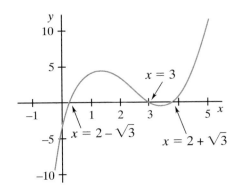

Figure 36   Graph of $p(x) = x^3 - 7x^2 + 13x - 3$
$$= (x - 3)(x - (2 + \sqrt{3}))(x - (2 - \sqrt{3}))$$

c. To find the zeros of g, we need to solve the equations $4 - x = 0$ and $x^2 + 1 = 0$. The solution of $4 - x = 0$ corresponds to the zero $x = 4$, which has multiplicity one. The equation $x^2 + 1 = 0$ has no real solutions, which means that the quadratic factor $x^2 + 1$ does not correspond to any x-intercept on the graph of g. The complex number solutions of $x^2 + 1 = 0$ cannot be graphed in the real coordinate plane. The graph is shown in Figure 37. Because of the quadratic factor $x^2 + 1$, the graph has two turning points, but has only one x-intercept.

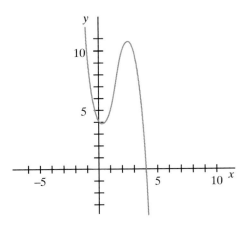

Figure 37   Graph of $g(x) = (4 - x)(x^2 + 1)$ ◼

EXAMPLE 2   Sketch the graph of $p(x) = x(x - 1)(2x + 1)^3(x - 3)^4$ by hand.

Solution   The polynomial $p(x) = x(x - 1)(2x + 1)^3(x - 3)^4$ is a ninth degree polynomial, so its graph will have a shape that is characteristic of odd degree polynomials. If the factors are multiplied so that the polynomial is written as a sum, the coefficient of x^9 is 8. Since 8 is a positive number, the y-values of p will increase without bound as $x \to \infty$ and will decrease without bound as $x \to -\infty$. The zeros of p are $x = 0, x = 1, x = -\frac{1}{2}$, and $x = 3$. The zeros $x = 0, x = 1$, and $x = -\frac{1}{2}$ have odd multiplicity, and the graph of p crosses the x-axis at $(0, 0)$, $(1, 0)$, and $(-\frac{1}{2}, 0)$. The zero $x = 3$ has even multiplicity, so the graph of p does not cross the x-axis at $(3, 0)$. Although $p(3) = 0$, p does not change sign at $x = 3$ because $(x - 3)^4$ is never negative. A sketch of p is shown in Figure 38.

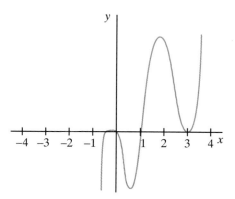

Figure 38 Sketch of $p(x) = x(x - 1)(2x + 1)^3(x - 3)^4$

The sketch shown in Figure 38 has four x-intercepts and four turning points. We can be sure that there are turning points between each pair of consecutive zeros, but we cannot determine the coordinates of these maximum and minimum points without using a calculator or computer graphing software.

A calculator-generated graph of p is shown in Figure 39. The viewing window for this graph extends over $-2 \leq x \leq 6$, $-100 \leq y \leq 300$. To see the maximum between $x = 1$ and $x = 3$, we need a window that shows y-values as large as 300. Notice that this choice for a viewing window prevents us from seeing clearly that p has positive y-values between $x = -\frac{1}{2}$ and $x = 0$.

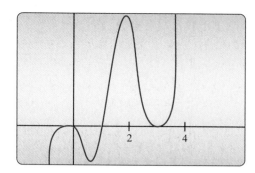

Figure 39 Graph of $p(x) = x(x - 1)(2x + 1)^3(x - 3)^4$

Figure 40 shows that if we choose a window that allows us to see the behavior near the x-axis, then we cannot see all the turning points of the graph. The viewing window for this graph is $-1 \leq x \leq 0.5$, $-6 \leq y \leq 6$.

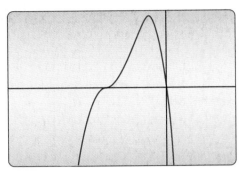

Figure 40 Behavior of $p(x) = x(x - 1)(2x + 1)^3(x - 3)^4$ near the origin

The information your calculator gives you about the graph of a polynomial function is strongly influenced by your choice of viewing window. It is therefore very important that you understand polynomial graphs well enough to be able to make good predictions about what your calculator "should" be showing you.

The Factor Theorem tells us that a polynomial f has a factor $x - a$ if and only if $f(a) = 0$. This relationship between factors and zeros will allow us to write the equation of a polynomial function if we know its zeros. For instance, suppose we know that a polynomial function f has exactly two zeros that are located at $x = -2$ and $x = 5$. Since $f(-2) = 0$ and $f(5) = 0$, we know that both $x + 2$ and $x - 5$ are factors of f. If $(-2, 0)$ and $(5, 0)$ are the only x-intercepts that f has, then $x + 2$ and $x - 5$ are the only linear factors f has. The simplest equation for f is $f(x) = (x + 2)(x - 5)$, but since we don't know the multiplicities of the zeros, this is not the only possible equation for f. Each of the zeros of f could have any multiplicity, so the equation for f could be any equation of the form $f(x) = (x + 2)^m(x - 5)^n$, where m and n are positive integers. Furthermore, the graph of f could be stretched vertically by any amount without changing its x-intercepts, so we can conclude that the equation of f may have the form $f(x) = k(x + 2)^m(x - 5)^n$, where m and n are positive integers and k is any real number. Several possible graphs for f are shown in Figure 41 on the next page. Notice that each graph contains the points $(-2, 0)$ and $(5, 0)$.

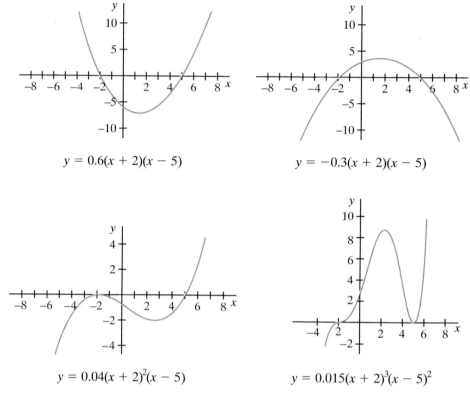

$$y = 0.6(x + 2)(x - 5)$$

$$y = -0.3(x + 2)(x - 5)$$

$$y = 0.04(x + 2)^2(x - 5)$$

$$y = 0.015(x + 2)^3(x - 5)^2$$

Figure 41 Possible graphs of $f(x) = k(x + 2)^m(x - 5)^n$

EXAMPLE 3   Write the equation of the polynomial function of least degree whose graph passes through $(-6, 0)$, $(\frac{1}{2}, 0)$, $(4, 0)$, and $(2, 8)$.

Solution   The polynomial must have three linear factors, $(x + 6)$, $(x - \frac{1}{2})$, and $(x - 4)$. Since we want the polynomial of least degree, each zero must have multiplicity one and the polynomial must be of degree 3. The equation of the polynomial has the form $p(x) = k(x + 6)(x - \frac{1}{2})(x - 4)$. Since the graph of the polynomial must pass through the point $(2, 8)$, it must be true that $p(2) = 8$. We can use this information to solve for k as follows.

$$p(x) = k(x + 6)\left(x - \frac{1}{2}\right)(x - 4)$$

$$p(2) = k(2 + 6)\left(2 - \frac{1}{2}\right)(2 - 4)$$

$$p(2) = -24k$$

Since $p(2) = 8$, we know that $-24k = 8$, so $k = -\frac{1}{3}$. Thus, the equation for the unique third degree polynomial passing through $(-6, 0)$, $(\frac{1}{2}, 0)$, $(4, 0)$, and $(2, 8)$ is $p(x) = -\frac{1}{3}(x + 6)(x - \frac{1}{2})(x - 4)$. A graph of p is shown in Figure 42.

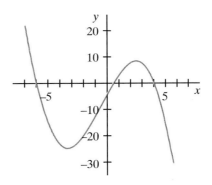

Figure 42   Graph of $p(x) = -\frac{1}{3}(x + 6)(x - \frac{1}{2})(x - 4)$ ▩

In Example 3, the task of writing the equation of the polynomial through four points was fairly straightforward because three of the points were x-intercepts. Because of the Factor Theorem, information about x-intercepts is equivalent to information about factors. The task of writing the equation of the polynomial through four arbitrary points is not as simple. You will tackle a problem of that type in the exercises that follow.

Exercise Set 6.8

1. Explain why the polynomial of least degree that goes through exactly four distinct zeros must have degree 4.

2. Sketch a graph of each function by hand. Label coordinates of the zeros on your graphs. Check each graph using your calculator.

   a. $y = x^2(x - 1)(x + 1)^2(x - 3)$

   b. $y = (2x - 1)(x - 4)(2 - x)$

   c. $y = \frac{1}{5}(x - 10)(x + 2)^2$

   d. $y = -3(x + 1)(x - 2)(x + 3)(x - 4)$

   e. $y = (x - 1)^3(2x + 4)(x + 6)$

   f. $y = -2(x - 3)^5 + 4$

   g. $y = 2(x - 1)^3 - 18$

   h. $y = 2x^3 - 2x$

   i. $y = x^3 + 6x^2 + 9x$

3. Write the equations of three different polynomial functions whose graphs pass through the zeros $x = -1$, $x = 3$, and $x = 0$. Sketch a graph of each polynomial.

4. Write the equations of two fourth degree polynomial functions whose graphs pass through the points $(-4, 0)$, $(-1, 0)$, $(0, 5)$, and $(5, 0)$. Sketch the graph of each function.

5. Write the equations of two cubic functions whose only x-intercepts are $(-2, 0)$ and $(5, 0)$ and whose y-intercept is $(0, 20)$. Sketch a graph of each function.

6. For a certain polynomial function, $x = 2$ is a zero with multiplicity three, $x = 1$ is a zero with multiplicity two, and $x = -2$ is a zero with multiplicity one. Write a possible equation of this function and sketch its graph.

7. Write the equation of the polynomial function of least degree whose graph passes through the points $(-3, 0)$, $(2, 0)$, $(1, 0)$, $(0, -24)$, and $(-4, 0)$.

8. Write the equation of a cubic function whose zeros are $x = 3$, $x = 2 - \sqrt{5}$, and $x = 2 + \sqrt{5}$. Write your final answer in the form $y = ax^3 + bx^2 + cx + d$, rather than in factored form.

9. Write an equation for the polynomial function associated with each graph in Figures 43–46.

a.
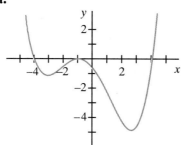

Figure 43

b.
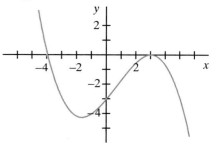

Figure 44

c.
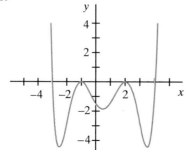

Figure 45

d.
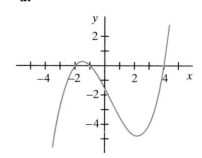

Figure 46

10. To understand the Factor Theorem, recall that whenever a polynomial function f is divided by $(x - a)$, the result of the division is a quotient $q(x)$ and a remainder r, so we can write $f(x) = q(x)(x - a) + r$, where $q(x)$ is a polynomial function and r is a real number.

   a. Assume that $(x - a)$ is a factor of f, which means that f is divisible by $(x - a)$. Explain why it must be true that $f(a) = 0$.

   b. Assume we know that $f(a) = 0$. Explain why it must be true that $(x - a)$ is a factor of f.

11. Write an equation for the polynomial function of least degree whose graph passes through $(-2, 5)$, $(0, 4)$, $(2, 7)$, and $(3, 5)$. Since two points determine a unique line and three points determine a unique parabola, four points determine a unique polynomial of degree three, which has an equation of the form $y = ax^3 + bx^2 + cx + d$. Substituting each of the four values for x and y into the equation $y = ax^3 + bx^2 + cx + d$ produces a system of four equations in four variables.

$$5 = -8a + 4b - 2c + d$$
$$4 = 0a + 0b + 0c + d$$
$$7 = 8a + 4b + 2c + d$$
$$5 = 27a + 9b + 3c + d$$

Solve this system to find the equation of the polynomial through the four given points.

12. Write a cubic polynomial function that approximates the sine function on the interval $-\pi \le x \le \pi$ and has the same x-intercepts as $y = \sin x$ on this interval.

Rational Functions

We have seen that the graphs of polynomial functions have characteristic shapes that are influenced by the degree of the polynomial. Many polynomials can be written as the product of simple factors, and these factors determine the shape of the polynomial's graph. Another class of functions are those that can be written as the ratio of two polynomials. Functions of this type are called *rational functions* and include functions such as

rational functions

$$y = \frac{x+2}{x^2+6}, \qquad y = \frac{(x-6)(2x-1)}{x-7}, \qquad \text{and} \qquad y = \frac{x^2+3x-8}{x^4-6x^3+5x-2}.$$

The graphs of rational functions also have characteristic shapes that are influenced by the degrees of the numerator and denominator and by the factors of each.

Since the rational function $h(x) = \frac{x+1}{x-3}$ can be written as the product of $f(x) = x + 1$ and $g(x) = \frac{1}{x-3}$, our initial study of rational functions will make use of what we learned about products earlier in this chapter.

EXAMPLE 1   Predict how the graph of $k(x) = (2x+1) \cdot \frac{1}{x-1}$ will behave.

Solution   Graphs of $f(x) = 2x + 1$ and $g(x) = \frac{1}{x-1}$ are shown in Figure 47.

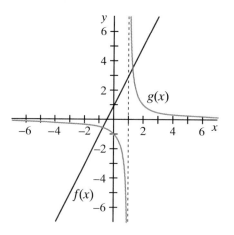

Figure 47   Graphs of $f(x) = 2x + 1$ and $g(x) = \frac{1}{x-1}$

We know that the product k will have a zero wherever either f or g has a zero. Since $f\left(-\frac{1}{2}\right) = 0$ and $g(x)$ is never equal to zero, the graph of k will have a single zero at $x = -\frac{1}{2}$. The y-values on the graph of k will be positive whenever f and g are both positive or both negative. Thus, $k(x) > 0$ when $x < -\frac{1}{2}$ and when $x > 1$. For x-values in the interval $-\frac{1}{2} < x < 1$, the y-values on the graph of k will be negative. Since the domain of k includes only x-values that are in the domains of both f and g, the domain of k includes all real numbers except 1. This information is displayed in Figure 48.

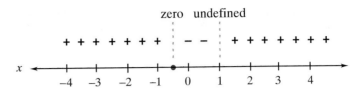

zero   undefined

x

$$+ + + + + +\ \ |\ \ -\ -\ \ |\ + + + + + +$$

−4 −3 −2 −1   0   1   2   3   4

Figure 48   Information about graph of $k(x) = (2x + 1) \cdot \dfrac{1}{x - 1}$

We can analyze the behavior of f and g to determine the behavior of $k(x)$ as x increases or decreases without bound. As $x \to \infty$, $f(x) \to \infty$ and $g(x) \to 0^+$. For large positive x-values, we know that the y-values of k are positive. We also know that the y-values from f tend to make the product far from zero, and the y-values from g tend to make the product close to zero. The table in Figure 49 shows that this "tug-of-war" results in y-values for k that approach 2 as $x \to \infty$. This implies that the graph of k is asymptotic to the horizontal line $y = 2$.

Figure 49   Values of $2x + 1$, $\dfrac{1}{x - 1}$, and $(2x + 1) \cdot \dfrac{1}{x - 1}$ as $x \to \infty$

x	$f(x) = 2x + 1$	$g(x) = \dfrac{1}{x - 1}$	$k(x) = (2x + 1) \cdot \dfrac{1}{x - 1}$
5	11	$\dfrac{1}{4}$	$\dfrac{11}{4} = 2.7500$
10	21	$\dfrac{1}{9}$	$\dfrac{21}{9} \approx 2.3333$
50	101	$\dfrac{1}{49}$	$\dfrac{101}{49} \approx 2.0612$
100	201	$\dfrac{1}{99}$	$\dfrac{201}{99} \approx 2.0303$
500	1001	$\dfrac{1}{499}$	$\dfrac{1001}{499} \approx 2.0060$

A similar table in Figure 50 shows values of $2x + 1$, $\dfrac{1}{x - 1}$, and $(2x + 1) \cdot \dfrac{1}{x - 1}$ as $x \to -\infty$. For negative x-values that are large in magnitude, the values of $f(x)$ and $g(x)$ are both negative, so their product is positive. The table also shows that $k(x) \to 2^-$ as $x \to -\infty$.

Figure 50   Values of $(2x + 1)$, $\dfrac{1}{x - 1}$, and $(2x + 1) \cdot \dfrac{1}{x - 1}$ as $x \to -\infty$

x	$f(x) = 2x + 1$	$g(x) = \dfrac{1}{x - 1}$	$k(x) = (2x + 1) \cdot \dfrac{1}{x - 1}$
−5	−9	$-\dfrac{1}{6}$	$\dfrac{-9}{-6} = 1.5000$
−10	−19	$-\dfrac{1}{11}$	$\dfrac{-19}{-11} \approx 1.7273$
−50	−99	$-\dfrac{1}{51}$	$\dfrac{-99}{-51} \approx 1.9412$
−100	−199	$-\dfrac{1}{101}$	$\dfrac{-199}{-101} \approx 1.9703$
−500	−999	$-\dfrac{1}{501}$	$\dfrac{-999}{-501} \approx 1.9940$

The fact that $k(1)$ is undefined suggests that we should investigate the behavior of $k(x)$ for x-values near 1. As $x \to 1^+$, $f(x) \to 3$ and $g(x) \to \infty$. The product of a number near 3 and a number that approaches infinity is a number that approaches infinity. Therefore $k(x) \to \infty$. As $x \to 1^-$, $f(x) \to 3$ and $g(x) \to -\infty$. The product of a number near 3 and a number that approaches negative infinity is a number that approaches negative infinity, so $k(x) \to -\infty$. The behavior of k near $x = 1$ suggests that the graph of k is asymptotic to the line $x = 1$.

A graph is shown in Figure 51. This graph incorporates all the predicted features of the graph of k.

- $k\left(-\frac{1}{2}\right) = 0$

- $k(x) > 0$ when $x > 1$ and when $x < -\frac{1}{2}$

- $k(x) < 0$ when $-\frac{1}{2} < x < 1$

- $k(1)$ is undefined; vertical asymptote at $x = 1$

- horizontal asymptote at $y = 2$

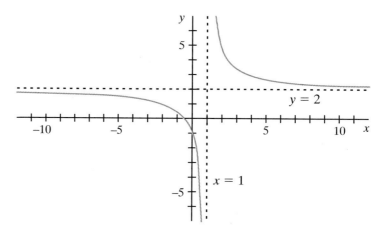

Figure 51   Graph of $k(x) = (2x + 1) \cdot \dfrac{1}{x - 1}$

The graph in Figure 51 should remind you of the toolkit function $y = \frac{1}{x}$ with some shifts and stretches. Using long division, we can rewrite the equation for $k(x)$ as $k(x) = 2 + \frac{3}{x - 1}$. This way of rewriting the equation for $k(x)$ shows that the graph of k is a transformation of $y = \frac{1}{x}$. The toolkit graph of $y = \frac{1}{x}$ has been stretched vertically by a factor of 3, shifted 1 unit to the right, and shifted 2 units up. These shifts should help you understand why the graph of k has a vertical asymptote at $x = 1$ and a horizontal asymptote at $y = 2$. ▨

EXAMPLE 2   Describe the behavior of the graph of $k(x) = \frac{(x-3)(x+2)}{x-2}$.

Solution   Graphs of $f(x) = (x-3)(x+2)$ and $g(x) = \frac{1}{x-2}$ are shown in Figure 52.

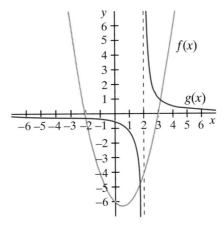

Figure 52   Graphs of $f(x) = (x-3)(x+2)$ and $g(x) = \frac{1}{x-2}$

The graph of k has zeros whenever either f or g has a zero, so the zeros of k are located at $x = 3$ and at $x = -2$. Positive y-values on the graph of k occur when $f(x)$ and $g(x)$ are either both positive or both negative. Therefore, $k(x)$ is positive when $x > 3$ and when $-2 < x < 2$. The domain of k includes only those real numbers that are in the domain of both f and g. This means that $x = 2$ is not in the domain of k. This information is displayed in Figure 53.

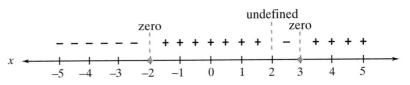

Figure 53   Information about graph of $k(x) = \frac{(x-3)(x+2)}{x-2}$

Since $k(2)$ is undefined, we should investigate the behavior of k for x-values near 2. As $x \to 2^+, f(x) \to -4$ and $g(x) \to \infty$. The product of a number near -4 and a number that approaches infinity is a number that approaches negative infinity, so $k(x) \to -\infty$. As $x \to 2^-, f(x) \to -4$ and $g(x) \to -\infty$. The product of a number near -4 and a number that approaches negative infinity is a number that approaches positive infinity, so $k(x) \to \infty$. The behavior of k near $x = 2$ indicates that the graph of k is asymptotic to the vertical line $x = 2$.

We can also use our knowledge of the graphs of f and g to predict the behavior of $k(x)$ as x increases or decreases without bound. As $x \to \infty$, $f(x) \to \infty$ and $g(x) \to 0^+$. If we were to make a table of values similar to those in Figures 49 and 50, we would see that f dominates in this "tug-of-war": $k(x) \to \infty$ as $x \to \infty$. As $x \to -\infty$, $f(x) \to \infty$ and $g(x) \to 0^-$. The result of this "tug-of-war" is that $k(x) \to -\infty$ as $x \to -\infty$.

A graph of $k(x) = \frac{(x-3)(x+2)}{x-2}$ is shown in Figure 54. This graph is consistent with what we predicted about zeros, domain, positive function values, and behavior as x increases or decreases without bound.

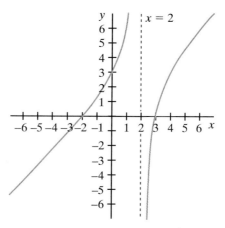

Figure 54   Graph of $k(x) = \frac{(x-3)(x+2)}{x-2}$

We can get some additional information about the graph of k by rewriting its equation as follows:

$$k(x) = \frac{(x-3)(x+2)}{x-2}$$

$$= \frac{x^2 - x - 6}{x - 2}$$

$$= x + 1 + \frac{-4}{x-2}.$$

When we write the equation in this form, we can think of k as the sum of two familiar functions: $y = x + 1$ and $y = \frac{-4}{x-2}$. When x increases without bound, $\frac{-4}{x-2}$ is a negative number close to zero, so y-values on the graph of k will be slightly less than $x + 1$. That is, as $x \to \infty$, $k(x) \to (x + 1)^-$. When x decreases without bound, $\frac{-4}{x-2}$ is a positive number close to zero, so y-values on the graph of k will be slightly more than $x + 1$. That is, as $x \to -\infty$, $k(x) \to (x + 1)^+$. This analysis shows that as x increases or decreases without bound, the graph of k approaches the graph of *oblique asymptote* | $y = x + 1$. The line $y = x + 1$ is called an *oblique asymptote* or a *slant asymptote* of the graph. Figure 55 on the next page shows the graph of k with the line $y = x + 1$. Notice that the degree of the numerator of k is one more than the degree of the denominator. As $x \to \pm\infty$, the ratio $\frac{(x-3)(x+2)}{x-2}$ approaches $x + 1$, which is a polynomial of degree 1.

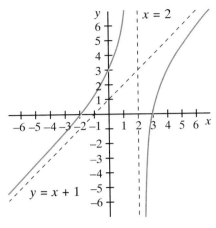

Figure 55 Graph of $k(x) = \dfrac{(x-3)(x+2)}{(x-2)}$ with $y = x + 1$ ▨

CLASS PRACTICE

1. Sketch the graph of $k(x) = \dfrac{(x-3)(x+2)(x-2)}{x-4}$. Focus on domain, zeros, positive and negative y-values, asymptotes, and behavior as x increases or decreases without bound.

2. Look back over the three rational functions you have studied in Examples 1 and 2 and Class Practice Problem 1. Write a paragraph describing what you have discovered. You should address the following questions:

    *   What features of a function's equation influence the location of the graph's x-intercepts?

    *   What features of a function's equation influence the location of vertical asymptotes?

    *   What features of a function's equation influence the behavior of the graph as x increases or decreases without bound?

In the Class Practice you were asked to make generalizations about the graphs of rational functions. Based on the examples you have studied, it appears that the rational function $k(x) = \dfrac{p(x)}{q(x)}$ has zeros where the numerator p has zeros and has vertical asymptotes where the denominator q has zeros. It also appears that the degrees of p and q determine the behavior of k as $x \to \pm\infty$. If the degree of p is equal to the degree of q, then the graph of k approaches a horizontal asymptote (degree 0) as $x \to \pm\infty$. If the degree of p is greater than the degree of q, then the difference between the degrees tells the degree of the oblique asymptote. For example, if p has degree 5 and q has degree 3, then $k(x) = \dfrac{p(x)}{q(x)}$ behaves like a function of degree 2 as x increases or decreases without bound. These generalizations about rational functions have been true for all the examples we have studied so far. After looking at additional examples, we will need to modify some of these generalizations.

EXAMPLE 3   Predict how the graph of $k(x) = \frac{x-2}{x^2+x-12}$ will behave.

Solution   The function k has $p(x) = x - 2$ in the numerator and $q(x) = x^2 + x - 12$ in the denominator. Since $x = 2$ is a zero of p, you should know that the graph of k will have an x-intercept at $(2, 0)$. Since $q(x) = x^2 + x - 12$ can be factored as $q(x) = (x + 4)(x - 3)$, you know that the graph of k will have vertical asymptotes at $x = -4$ and $x = 3$. Since $k(0) = \frac{1}{6}$, the graph of k crosses the y-axis at $(0, \frac{1}{6})$. You should also be able to predict the behavior of k as x increases or decreases without bound. As $x \to \infty$, $p(x) \to \infty$ and $q(x) \to \infty$. Since the degree of p is less than the degree of q, for large enough x-values we can be sure that $p(x) < q(x)$ and that values of $q(x)$ are increasing faster than values of $p(x)$. This implies that the value of $\frac{p(x)}{q(x)}$ approaches 0 as $x \to \infty$. Similar analysis shows that the value of $\frac{p(x)}{q(x)}$ approaches 0 as $x \to -\infty$. The graph of k in Figure 56 suggests that $k(x) \to 0^+$ as $x \to \infty$ and $k(x) \to 0^-$ as $x \to -\infty$. The locations of zeros and vertical asymptotes are consistent with our predictions.

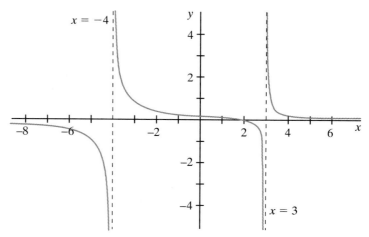

Figure 56   Graph of $k(x) = \frac{x-2}{x^2+x-12}$

Notice that the middle section of the graph of $k(x) = \frac{x-2}{x^2+x-12}$ crosses the x-axis, which is the graph's horizontal asymptote. This occurs because $k(2) = 0$. The graph of k is asymptotic to the x-axis as x-values increase or decrease without bound. The graph is not asymptotic to the x-axis when x-values are near zero. ■

EXAMPLE 4   Sketch a graph of $k(x) = \frac{x^2+3x-10}{x-2}$ by hand.

Solution   The rational function k has $p(x) = x^2 + 3x - 10$ in the numerator and $q(x) = x - 2$ in the denominator. The degree of p is one greater than the degree of q, so the graph of k will behave like a polynomial of degree 1 as $x \to \pm\infty$. Since $p(x) = x^2 + 3x - 10$ can be factored as $p(x) = (x + 5)(x - 2)$, you should predict that the graph of k will have x-intercepts at $x = -5$ and $x = 2$. Since $x = 2$ is a zero of q, you should anticipate that the graph of k will have a vertical asymptote at $x = 2$. It is clear that these predictions need to be modified, because a graph cannot have a zero and a vertical asymptote at the same x-value. How does the graph of k behave at $x = 2$? We can understand the behavior of k near $x = 2$ if we rewrite its equation as $k(x) = \frac{(x + 5)(x - 2)}{x - 2}$. Since $k(2) = \frac{0}{0}$, which is undefined, it is clear that $x = 2$ is not in the domain of k, so $x = 2$ cannot be an x-intercept. Could k have a vertical asymptote at $x = 2$? Since $\frac{x - 2}{x - 2} = 1$ for all $x \neq 2$, these factors have the effect of canceling each other out at all x-values other than $x = 2$. This means that when $x \neq 2$, $k(x) = \frac{(x + 5)(x - 2)}{x - 2}$ is equal to $f(x) = x + 5$. When $x = 2$, $k(x)$ is undefined. The behavior of k as $x \to 2^{+}$ and as $x \to 2^{-}$ is illustrated in the table of values in Figure 57.

Figure 57 Values of $k(x) = \frac{x^2 + 3x - 10}{x - 2}$ as $x \to 2$

x	$\dfrac{x^2 + 3x - 10}{x - 2}$
1.95	6.95
1.96	6.96
1.97	6.97
1.98	6.98
1.99	6.99
2.00	undefined
2.01	7.01
2.02	7.02
2.03	7.03

The entries in the table show that values of $k(x)$ do not "shoot up" or "shoot down" near $x = 2$ as they would if $x = 2$ were a vertical asymptote. Rather, the table shows that as $x \to 2$, $k(x) \to 7$. The graph of $k(x) = \frac{x^2 + 3x - 10}{x - 2}$ does not contain a point whose x-coordinate is 2, but the graph gets arbitrarily close to the point $(2, 7)$. This behavior is indicated with an open circle on the graph of k located at $(2, 7)$.

In conclusion, we have determined that the graph of k contains the point $(-5, 0)$ and has an open circle at $(2, 7)$. We also know that when $x \neq 2$, the graph of k is identical to the graph of $f(x) = x + 5$. The fact that $f(2) = 7$ should help you understand why $k(x) \to 7$ as $x \to 2$. A graph of k is shown in Figure 58 on the next page.

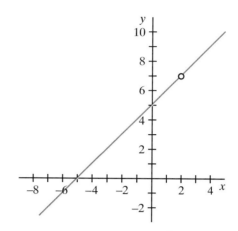

Figure 58   Graph of $k(x) = \frac{x^2 + 3x - 10}{x - 2}$ ▨

Example 4 illustrates that the rational function $k(x) = \frac{p(x)}{q(x)}$ will not always have a vertical asymptote where $q(x) = 0$. Rather, the graph will have a vertical asymptote at any x-value for which $q(x) = 0$ and $p(x) \neq 0$. Similarly, the graph will not always have an x-intercept where $p(x) = 0$, but will have an x-intercept where $p(x) = 0$ and $q(x) \neq 0$. At any x-value for which $p(x) = 0$ and $q(x) = 0$, k will be undefined and the graph of k may either have an open circle or a vertical asymptote at this x-value. Canceling common factors from $p(x)$ and $q(x)$ can help to show how $k(x)$ behaves near x-values where $p(x) = 0$ and $q(x) = 0$.

EXAMPLE 5   Sketch a graph of $k(x) = \frac{3x^2 - 3}{x^2 + 1}$ by hand. Use your calculator to check your graph.

Solution   The function k has $p(x) = 3x^2 - 3$ in the numerator and $q(x) = x^2 + 1$ in the denominator. Since $p(x) = 3x^2 - 3$ can be factored as $p(x) = 3(x + 1)(x - 1)$, the graph of k will have x-intercepts $x = -1$ and $x = 1$. The function $q(x) = x^2 + 1$ in the denominator has no zeros, so the function k has no vertical asymptotes. The sketch of $k(x) = \frac{3x^2 - 3}{x^2 + 1}$ in Figure 59 also shows that $k(0) = -3$ and that the graph of k is symmetric about the y-axis. The degree of p is equal to the degree of q, so the graph of k will approach a horizontal asymptote. This asymptote is the line $y = 3$.

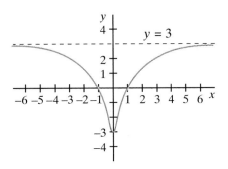

Figure 59 Graph of $k(x) = \dfrac{3x^2 - 3}{x^2 + 1}$

Another way to see why $k(x) = \dfrac{3x^2 - 3}{x^2 + 1}$ approaches $y = 3$ as $x \to \pm \infty$ is to rewrite the equation of k as shown:

$$k(x) = \frac{3x^2 - 3}{x^2 + 1}$$

$$= \frac{3x^2 - 3}{x^2 + 1} \cdot \frac{\dfrac{1}{x^2}}{\dfrac{1}{x^2}}$$

$$= \frac{3 - \dfrac{3}{x^2}}{1 + \dfrac{1}{x^2}}.$$

Both $-\dfrac{3}{x^2}$ and $\dfrac{1}{x^2}$ approach zero as x increases or decreases without bound, so $k(x) = \dfrac{3x^2 - 3}{x^2 + 1}$ approaches $y = \dfrac{3}{1} = 3$ as $x \to \pm \infty$. ▨

Examples 1 and 5 both involve rational functions in which the degrees of the numerator and the denominator are equal. In Example 1 the coefficients of the highest degree terms in the numerator and denominator were 2 and 1, respectively, and the horizontal asymptote was $y = \dfrac{2}{1}$. Similarly, in Example 5 the coefficients of the highest degree terms were 3 and 1, and the horizontal asymptote was $y = \dfrac{3}{1}$. In both cases the highest degree terms in the numerator and denominator determine the behavior of the function as x-values increase or decrease without bound.

Rewriting $k(x) = \frac{3x^2 - 3}{x^2 + 1}$ as

$$k(x) = \frac{3 - \dfrac{3}{x^2}}{1 + \dfrac{1}{x^2}}$$

shows why the graph of k approaches $y = 3$ as a horizontal asymptote. The algebraic manipulation we used to rewrite k can be used to determine horizontal asymptotes only when the degrees of the numerator and the denominator are equal.

In summary, we know that if the degree of the numerator is equal to the degree of the denominator, then the graph of the rational function has a horizontal asymptote whose equation is determined by the coefficients of the highest degree terms. If the degree of the numerator is less than the degree of the denominator, the graph has the line $y = 0$ as a horizontal asymptote. If the degree of the numerator is greater than the degree of the denominator, then the difference between the degrees tells the degree of the asymptote, and y-values will increase or decrease without bound as x-values increase and decrease without bound.

CLASS PRACTICE

The graph of $k(x) = \frac{x^2 - 4}{x^2 + 2x + 1}$ is shown in Figure 60.

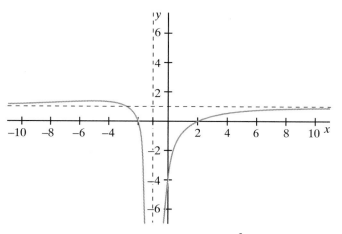

Figure 60   Graph of $k(x) = \frac{x^2 - 4}{x^2 + 2x + 1}$

1. Explain the behavior of the graph as $x \to \infty$ and as $x \to -\infty$.

2. Analyze the behavior of the graph around the vertical asymptote at $x = -1$. Why does the graph not change sign around $x = -1$?

3. Identify the x- and y-intercepts of the graph.

4. Find the coordinates of the point where the graph crosses its horizontal asymptote.

1. Identify the intercepts, asymptotes, and "holes" on the graph of each rational function. Sketch each graph.

   a. $f(x) = \dfrac{x-1}{(x-2)^2}$

   b. $f(x) = \dfrac{x+2}{x}$

   c. $f(x) = \dfrac{x}{x^2 - x - 6}$

   d. $f(x) = \dfrac{3x+2}{x-1}$

   e. $f(x) = \dfrac{x-2}{x^2 - 9}$

   f. $f(x) = \dfrac{x}{x^2 - 3x}$

   g. $f(x) = \dfrac{x^2 - 2x - 8}{x^2 - x - 6}$

   h. $f(x) = \dfrac{x+3}{x^2 + 2}$

   i. $f(x) = \dfrac{(x+5)(x-4)(x-5)}{(x+1)^2(x-2)}$

   j. $f(x) = \dfrac{x^3 + 4x^2 - 5x}{x^2 - 1}$

   k. $f(x) = \dfrac{x^2 - 3x + 4}{x-2}$

   l. $f(x) = \dfrac{-x^2 - 2x + 5}{x^2 + 3}$

2. Neither the numerator nor the denominator of the function $f(x) = \dfrac{x^3 - 3x + 1}{x^3 - x^2 - 3}$ can be factored. Explain how you would use your calculator to help you find the locations of the x-intercepts and the vertical asymptotes on the graph of f.

3. Determine the x-coordinate of the point where each function crosses its horizontal asymptote.

   a. $f(x) = \dfrac{x^2 + 1}{x^2 - 2x + 1}$

   b. $f(x) = \dfrac{x^3 - 4x}{(x-1)^2(x+1)}$

4. Suppose the graph of the rational function $k(x)$ has the lines $x = -2$ and $x = 3$ as vertical asymptotes, $x = 1$ and $x = 4$ as x-intercepts, and a horizontal asymptote at $y = \frac{1}{2}$. Sketch a possible graph of k. Write an equation for your graph.

5. Suppose the graph of the rational function $k(x)$ has the lines $x = 2$ and $y = -x + 3$ as asymptotes. Sketch a possible graph of k. Write an equation for your graph.

6. Sketch graphs of $f(x) = \dfrac{1}{x+1}$ and $g(x) = \dfrac{2x}{3x-1}$ on the same coordinate axes. Use the graphs to help you solve the inequality $\dfrac{1}{x+1} \le \dfrac{2x}{3x-1}$.

7. The principles you used to graph rational functions can be applied to other functions of the form $y = \frac{p(x)}{q(x)}$ even if p and q are not polynomial functions. Use these principles to identify the intercepts, asymptotes, positive and negative y-values, and behavior as $x \to \pm\infty$ for the graph of each function below.

a. $k(x) = \dfrac{e^x}{x + 1}$

b. $k(x) = \dfrac{\ln(x - 2)}{x^2}$

c. $k(x) = \dfrac{\sin(2x)}{(x - \pi)^2}$

d. $k(x) = \dfrac{e^{x+1}}{\ln(x + 2)}$

SECTION 6.10 Applications of Rational Functions

Rational functions play an important role in many mathematical models. They are often used in combination with other toolkit functions. In this section we will consider applications in which rational functions occur.

EXAMPLE 1   A food packaging company wants to make cylindrical cans that hold 1 liter (which is 1000 cm^3) of its product. What dimensions will enable the company to use a minimum amount of material?

Solution   Figure 61 shows that the total surface area of a cylindrical can is made up of a rectangular section and two circular sections. For the purpose of simplifying the problem, we will ignore the material needed to construct the seams of the can.

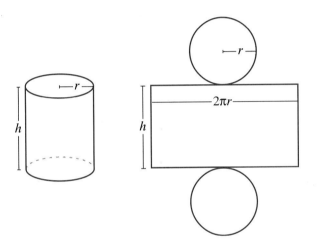

Figure 61   Surface area of cylindrical can

The total surface area is a function of two variables, the radius and the height of the can. This function is $f(r, h) = 2\pi r^2 + 2\pi rh$. The notation $f(r, h)$ implies that the function f depends on two variables, in this case r and h. If we could write the surface area as a function of one variable, our work would be much simpler. Since the volume is given by $\pi r^2 h$ and the volume of the can is 1000 cm^3, we know that $\pi r^2 h = 1000$, so $h = \frac{1000}{\pi r^2}$. We can now express the surface area as a function of r alone:

$$S(r) = 2\pi r^2 + \frac{2000}{r}.$$

The domain of the function S is all real numbers except zero. In the context of the problem, the domain is $r > 0$. Since the range that we are interested in consists only of positive values, we are concerned with the characteristics of S in the first quadrant. What does the graph of S look like in the first quadrant?

The function S is the sum of two simple functions, $f(r) = 2\pi r^2$ and $g(r) = \frac{2000}{r}$. If $0 < r < 1$, then r^2 is very small, and $f(r)$ is nearly zero, so the graph of S should resemble that of function g. As r increases, $g(r)$ approaches zero, and the graph of S should resemble that of function f. The relationship between the graph of S and the two component functions f and g is shown in Figure 62. In addition to thinking of S as the sum of two functions, we can rewrite the equation for S as the ratio of two polynomials:

$$S(r) = 2\pi r^2 + \frac{2000}{r}$$

$$= \frac{2\pi r^3}{r} + \frac{2000}{r}$$

$$= \frac{2\pi r^3 + 2000}{r}.$$

The graph of S has a vertical asymptote at $r = 0$ and is asymptotic to a curve of degree 2 as $r \to \infty$.

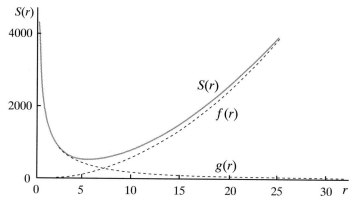

Figure 62 Graphs of $S(r) = 2\pi r^2 + \frac{2000}{r}$, $f(r) = 2\pi r^2$, and $g(r) = \frac{2000}{r}$

Using technology to find the coordinates of the turning point on the graph of S as shown in Figure 63, we see that when the radius r is approximately 5.42 cm, the surface area has a minimum value of approximately 553.58 cm².

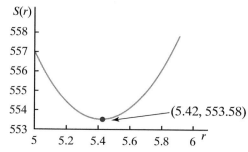

Figure 63 Minimum value for $S(r) = 2\pi r^2 + \frac{2000}{r}$

Substituting $r = 5.42$ into $h = \frac{1000}{\pi r^2}$, we find that the height that minimizes surface area is approximately 10.84 cm. ▩

EXAMPLE 2   After watching a TV commercial, students were interested in investigating the pH of their saliva after eating a piece of candy. The pH level is used to measure the acidity or alkalinity of liquids. On a scale from 0 to 14, a pH of 0 represents the highest level of acid and a pH of 14, the most basic or alkaline. Pure water has a pH of 7. The commercial had suggested that the candy decreased the pH, indicating the mouth had become more acidic, with resulting damage to the teeth. In the commercial, the graph of pH over time was shown to have the shape given in Figure 64.

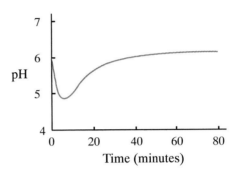

Figure 64   Graph of pH of saliva after chewing candy

This graph does not resemble any familiar toolkit functions. By doing some research in science textbooks, the students found that the model for pH level is

$$\text{pH} = A - \frac{Bt}{t^2 + C},$$

where t is measured in minutes. Does the function $f(t) = A - \frac{Bt}{t^2 + C}$ have a graph like the one shown in Figure 64 when A, B, and C are positive and $t \geq 0$?

Source: Presentation by Ron Lancaster at the Seventh Annual Conference on Secondary School Technology and Mathematics, Exeter, NH, June, 1991.

Solution   The graph of $y = \frac{Bt}{t^2 + C}$ has a zero at $t = 0$ and is asymptotic to the t-axis, since the degree of the denominator is larger than the degree of the numerator. Since $y = 0$ when $t = 0$ and $y \to 0$ as $t \to \infty$, there must be a "hump" in the graph where y has a maximum value, as shown in Figure 65. Also, since C is positive, the denominator is never zero, so the graph of $y = \frac{Bt}{t^2 + C}$ has no vertical asymptotes.

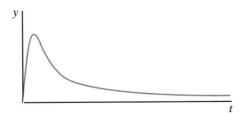

Figure 65   Graph of $y = \frac{Bt}{t^2 + C}$

The graph of $y = -\frac{Bt}{t^2 + C}$ is the graph in Figure 65 reflected across the t-axis as shown in Figure 66.

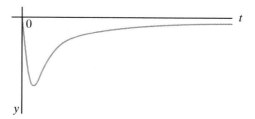

Figure 66   Graph of $y = -\frac{Bt}{t^2 + C}$

Finally, add the constant A to the function graphed in Figure 66. This shifts the graph A units vertically, resulting in the graph shown in Figure 67.

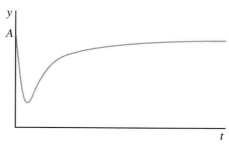

Figure 67   Graph of $f(t) = A - \frac{Bt}{t^2 + C}$

The graph of $f(t) = A - \frac{Bt}{t^2 + C}$ has the desired shape. We do not know where the minimum value of the function will be, but we do know the basic shape of the graph.

Note that the function $f(t) = A - \frac{Bt}{t^2 + C}$ can be rewritten as

$$f(t) = \frac{A(t^2 + C) - Bt}{t^2 + C} = \frac{At^2 - Bt + AC}{t^2 + C}.$$

The rational function f has numerator and denominator of degree 2. This way of writing the function shows that the pH level approaches A as $t \to \infty$. It also shows that the pH is equal to A when $t = 0$. Whether or not the graph of f crosses the horizontal axis is determined by whether the numerator $At^2 - Bt + AC$ has any zeros. Using the quadratic formula to solve for zeros leads to a discriminant of $B^2 - 4A^2C$. If the constants A, B, and C satisfy the inequality $B^2 - 4A^2C < 0$, then the graph of $f(t) = \frac{At^2 - Bt + AC}{t^2 + C}$ does not cross the t-axis. Thus, the equation $f(t) = \frac{At^2 - Bt + AC}{t^2 + C}$ is consistent with the graph in Figure 64 provided $B^2 - 4A^2C < 0$. ▨

1. The Durham Tin Can Company minimizes costs by constructing cans from the least possible amount of material. This company supplies many different sizes of cans to packing firms. A designer with the company needs to find the radius that gives the minimum surface area for cylindrical cans with volumes between 100 cc and 1500 cc.

   a. Use the technique from Example 1 to find the radius that minimizes surface area for a volume of 100 cc, 300 cc, 500 cc, 700 cc, 900 cc, 1100 cc, 1300 cc, and 1500 cc. Make a scatterplot of the ordered pairs (*volume*, *radius*). What type of relationship appears to exist between these variables?

   b. Use techniques of data analysis to find a function that expresses the relationship between volume and radius for a can of minimal surface area.

   c. Use your function to predict the radius that gives the minimum surface area for a cylindrical can with volume of 2000 cc.

   d. What is the ratio of diameter to height for the can of minimal surface area that holds a given volume? Do manufacturers generally produce cans with diameter and height in this ratio? Why or why not?

2. The weight of an object above the earth's surface is given by the function $w(h) = \left(\frac{r}{r+h}\right)^2 w_0$, where r is the radius of the earth, w_0 is the weight of the object at sea level, and h is the height of the object above the earth's surface. The radius of the earth is approximately 3960 miles.

   a. The function $w(h) = \left(\frac{r}{r+h}\right)^2 w_0$ is also used to determine the heights of mountain peaks, such as Mt. Everest. Suppose an object that weighs 10 pounds on the earth's surface weighs 9.992 pounds when it is at the top of a mountain. How high is the mountain?

   b. Mt. Everest is 29,028 feet high. How much would a 180-pound hiker weigh at the top of the mountain?

3. The volume of an open box with a square base and rectangular sides is 250 in^3. If sides are double thickness and the bottom is triple thickness, what size box will use the least amount of material?

4. The volume of a cylindrical can is 500 cm^3. The material used to make the top and bottom costs 0.012 cent/cm^2, the material used for the sides costs 0.01 cent/cm^2, and the seam joining the top and bottom to the sides costs 0.015 cent/cm. What size can would cost the least to produce?

5. Some students decided to measure the pH of their saliva and use the data to determine the values of A, B, and C in the model in Example 2. After chewing a piece of candy, they measured the pH of their saliva every thirty seconds for a period of ten minutes. Some students measured the pH by spitting in a cup and dipping litmus paper into the spit; others wiped a wooden stick around their mouths and used the litmus paper on the stick. One student's results are given in Figure 68. The initial pH before eating the candy was 6.5.

Figure 68   **Data for pH of student's saliva and time**

time (sec)	30	60	90	120	150	180	210	240	270	300
pH	6.2	6.0	5.7	5.5	5.3	5.1	5.1	4.9	4.9	4.9
time (sec)	330	360	390	420	450	480	510	540	570	600
pH	4.8	4.8	4.8	4.8	4.9	4.9	4.9	4.9	5.0	5.0

a. Determine the value of A by considering the initial pH.

b. Do the following to find the values of B and C. As you proceed, remember that t should be measured in minutes. Re-express the given data as $\left(t^2, \frac{t}{A - \text{pH}}\right)$ and examine a scatterplot to observe that the re-expression linearizes the data.

c. Show how to derive the linearization $\left(t^2, \frac{t}{A - \text{pH}}\right)$ from the model

$$f(t) = A - \frac{Bt}{t^2 + C}.$$

d. Fit a linear model to the re-expressed data and use it to find the values of B and C.

e. Compare this function f to the original data. Does the model fit the data well?

f. According to your model, how long will it take for the pH level to reach 6.0?

g. Measure the pH in a person's mouth to create your own data set. How well does this model work for your pH data?

6. To help understand the graph of $y = \frac{Bt}{t^2 + C}$ from Example 2, make the following graphs using $B = 8$ and $C = 2$.

a. Graph $y = \frac{8t}{t^2 + 2}$ and $y = 4t$ on the closed interval $\left[0, \frac{1}{2}\right]$. Explain why the graphs resemble each other.

b. Graph $y = \frac{8t}{t^2 + 2}$ and $y = \frac{8}{t}$ on the closed interval $[5, 20]$. Explain why the graphs resemble each other.

SECTION 6.11 Investigating a Traffic Flow Model

The North Carolina Department of Transportation is adding a third lane to the beltway around Raleigh. Unfortunately, one of the existing lanes must be closed for several hours each day, creating a one-lane road for approximately 10 miles. Naturally, the Department of Transportation would like to maximize the flow of traffic through this region. What speed limit should be set for this stretch of road to ensure the greatest traffic flow?

When developing a mathematical model of a real-world situation, it is often necessary to make some simplifying assumptions. In this model, we assume that all the cars are the same length L, measured in feet, and that the cars follow each other at a distance, d, also measured in feet. (See Figure 69.) We know from experience that the faster we are driving, the more distance we should leave between our car and the car in front of us. Therefore, we assume that the following distance depends upon the assigned speed limit, s, measured in miles per hour.

Figure 69 Diagram of the flow of cars on the highway

Source: Adapted from Stone, Alan and Ian Huntley. "Easing the Traffic Jam" *Solving Real Problems with Mathematics*, Volume 2, The Spode Group, Cranfield Press, Cranfield, UK, 1982.

What is meant by traffic flow? It is the measure of the number of cars that pass a given point in a given unit of time. We will use cars per second to measure the traffic flow. Imagine the situation described in the paragraph above. Cars are flowing uniformly down the road, each traveling s miles per hour and following d feet behind the car in front.

How can we develop a model of the flow of highway traffic from these basic assumptions? Imagine that you are standing beside the road watching the cars pass you by. Consider the situation in which a car has just gone past. When will the next car pass? The time t required for the next car to pass you is the ratio of the distance it has to travel and the speed at which it is moving, that is,

$$t = \frac{L + d}{s}. \tag{2}$$

The distance the car must travel is the sum of the length of the car and the distance between cars. However, there is a problem with the units in equation (2). The time is measured in seconds, the distances in feet, and the speed in miles per hour. We need to express the speed in feet per second. Since there are 5280 feet in a mile and 3600 seconds in an hour, we know that

$$\frac{1 \text{ mile}}{1 \text{ hour}} = \frac{5280 \text{ feet}}{3600 \text{ seconds}} \approx 1.467 \text{ feet per second.}$$

Equation (2) can be rewritten as

$$t = \frac{L + d}{1.467s}.$$

The units for time are seconds per car. The traffic flow is measured in cars per second. The model for traffic flow is simply the reciprocal of t. That is,

$$F(s) = \frac{1.467s}{L + d}.$$

This relationship between time and traffic flow is the same as the relationship between period and frequency of sine waves.

We have already noted that the distance between the cars, d, also depends upon the speed at which the cars are traveling, s. The table in Figure 70 from a driver's handbook provides some approximate figures on the components of stopping distance.

Figure 70 Table of speed and stopping distances

Speed (mph)	Thinking distance (ft)	Braking distance (ft)
25	27	35
35	38	68
45	49	113
55	60	168
65	71	235

Notice that the stopping distance has two components. Thinking distance, T, is the distance the car travels while the driver is reacting and putting his foot on the brake. Braking distance, B, is the distance the car travels as the brake slows the car to a stop. Assume that $d = T + B$. The traffic flow function can be written more specifically as

$$F(s) = \frac{1.467s}{L + T(s) + B(s)}. \tag{3}$$

1. Fit a linear model $T(s) = m \cdot s$ to the data for thinking distance and a power model $B(s) = as^n$ to the data for braking distance. Substitute these function models into equation (3) to write F as a function of s. Determine the speed that maximizes the flow of traffic as defined by F when the average car length is 12 feet.

2. Suppose the length of the car was not specified, but is given as the parameter L.

    a. For $L = 8$, 15, 20, 40, 80, and 100, determine the optimum speed limit. Use data analysis to determine the relationship between L and the speed that maximizes traffic flow.

    b. The New Jersey Turnpike has lanes that are restricted to large trucks. Should the speed limit be set higher or lower for these lanes? Explain your answer.

3. Our experience suggests that drivers do not leave as much space between cars as they should. Let p be the fraction of the required stopping distance that the drivers actually leave between cars. Assume $0.1 \leq p < 1$. Then the model can be written in terms of both L and p as

$$F(s) = \frac{1.467s}{L + p(T(s) + B(s))} \, .$$

   a. Let $L = 12$ as before, and determine the speed that maximizes the flow of traffic for at least six different values of p.

   b. Use data analysis to determine the relationship between the value of p and the speed that maximizes traffic flow.

   c. Often the speed limit on one-lane roads is set at 45 mph. Assuming our model is correct, what does this speed limit imply about the perceived value of p?

4. Some might argue that drivers would never drive so close that they would not have sufficient time to react. These drivers might leave all the thinking distance but might leave only a fraction of the braking distance. In this case,

$$F(s) = \frac{1.467s}{L + T(s) + p \cdot B(s)}.$$

   How different from the previous solution is the optimal speed if this model is used?

5. Which seems to be more important in determining the best speed, the length of the car or the fraction of desired stopping distance allowed? Explain your answer.

Source: Stone, Alan and Ian Huntley, "Easing the Traffic Jam," *Solving Real Problems with Mathematics, Volume 2*, The Spode Group, Cranfield Press, Cranfield, UK, 1982.

Chapter 6 Review Exercises

1. The function $h(x) = \frac{1}{x-1} + \sin(4x)$ is graphed below. Use the sum of two functions to explain the shape of the graph of h.

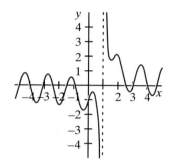

Figure 71

2. Consider the function $k(x) = 2^{x+1}\left(\frac{1}{x-4}\right)$.

    a. Identify the domain of k.

    b. For what x-values is $k(x) > 0$? $k(x) < 0$? $k(x) = 0$?

    c. How do the y-values of k behave as $x \to \pm\infty$?

    d. Sketch a graph of k.

3. The function $f(x) = \sin(2x) \cdot (2^x + 1)$ is the product of two simple factors and is graphed below. Use the factors to explain the graph's features, including domain, x-intercepts, asymptotes, enveloping curves, positive or negative y-values, and any other significant features.

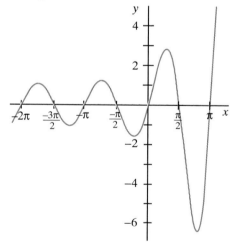

Figure 72

4. Answer the following questions about the function $k(x) = e^{-x}\ln(x + 3)$ and then make a sketch of its graph. Label important points and features on the graph.

    a. What is the domain of k?

    b. Where are the zeros of k located?

    c. For what x-values is $k(x) > 0$?

    d. What happens to the y-values of k as $x \to \infty$? as $x \to -\infty$?

5. Use the identity for $\sin A + \sin B$ to write $f(t) = \sin(20\pi t) + \sin(30\pi t)$ as a product of two factors. Graph f and the product function on the same coordinate axes.

6. Write the equations of two fourth degree polynomial functions that pass through the points $(-3, 0)$, $(1, 0)$, $(4, 0)$, and $(0, 2)$. Sketch the graph of each function.

7. Write the equations of two cubic functions whose only x-intercepts are $(-3, 0)$ and $(4, 0)$ and whose y-intercept is $(0, 6)$. Sketch a graph of each function.

8. Write an equation for the polynomial function associated with the graph in Figure 73.

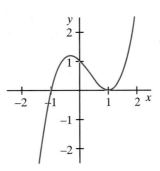

Figure 73

9. Identify the intercepts, asymptotes, and "holes" on the graph of each rational function. Sketch each graph.

    a. $f(x) = \dfrac{x - 4}{(x + 2)^2}$ **b.** $f(x) = \dfrac{x + 3}{x^2 + 3x}$

10. **a.** Under what conditions will a rational function have a "hole?"

    b. Under what conditions will a rational function have a horizontal asymptote at $y = 2$?

    c. Under what conditions will a rational function have an oblique asymptote?

7 Matrices

7.1	Introduction to Matrices	468
7.2	Matrix Addition and Scalar Multiplication	470
7.3	A Common-Sense Approach to Matrix Multiplication	474
7.4	Computer Graphics	485
7.5	The Leontief Input-Output Model and the Inverse of a Matrix	495
7.6	Additional Applications of the Inverse of a Matrix	509
7.7	The Leslie Matrix Model	512
7.8	Markov Chains	523
	Chapter 7 Review Exercises	541

Age-Specific Population Growth

The population of the United States is growing at an annual rate of about 2.5%. For purposes of planning for schools, preschool childcare, and other needs of children, knowing the overall growth rate is insufficient. Planners need to know the predicted population of children. The following table gives some of the information needed to predict age-specific population patterns. The table concerns only the population of females. But since females make up approximately half the population, the total population can be predicted by doubling the final figures.

Age (yr)	0–9	10–19	20–29	30–39	40–49	50–59	60–69	70–79	80–89	90+
Female pop. (thousands)	19,940	19,302	18,331	21,870	19,307	12,111	9213	6639	2388	312
Birthrate (10 yr)	0	0.2795	0.6025	1.1250	0.5995	0.0355	0	0	0	0
Survival rate (10 yr)	0.9918	0.9976	0.9940	0.9914	0.9833	0.9607	0.8990	0.7789	0.5352	0

Based on these data, how many people will be under the age of 20 in 30 years?

SECTION 7.1 Introduction to Matrices

Three major countries that produce cars for sale in the United States are Japan, Germany, and the United States itself. When it is time to buy a new car, people will choose a car based in part on the satisfaction they received from the car they presently own. Suppose that of car buyers who presently own an American car, 55% will purchase another American car, 25% will buy a Japanese car, 10% will buy a German car, and 10% will buy a car made in none of these three countries. Of those who presently own a Japanese car, 60% will buy another Japanese car, 25% will buy an American car, 10% a German car, and 5% some other car. Of those car buyers who own a German car, 40% will again purchase a German car, 35% will switch to American cars, 15% will switch to Japanese cars, and 10% will buy a car made in another country. Of those who presently own a car from a country other than the three major producers, 20% will switch to American, 25% will switch to Japanese, 15% will switch to German, and 40% will continue to buy from another country.

The details of this information are difficult to grasp all at once; however, the following display of the data offers distinct advantages over the verbal description given above.

$$
T = \begin{array}{c} \\ \text{\textit{United States}} \\ \text{\textit{Japan}} \\ \text{\textit{Germany}} \\ \text{\textit{Other}} \end{array}
\begin{array}{c} \text{\textit{United}} \\ \text{\textit{States \ \ Japan \ \ Germany \ \ Other}} \\ \begin{bmatrix} 0.55 & 0.25 & 0.10 & 0.10 \\ 0.25 & 0.60 & 0.10 & 0.05 \\ 0.35 & 0.15 & 0.40 & 0.10 \\ 0.20 & 0.25 & 0.15 & 0.40 \end{bmatrix} \end{array}
$$

matrix

entries

dimension of a matrix

square matrix

transition matrix

The table of numbers, *T*, shown above is an example of a *matrix*, in which the numbers are known as *entries*. The *dimension of a matrix* is given by the number of rows and the number of columns, so *T* is considered a 4×4 matrix. If a matrix has *m* rows and *n* columns, then it is said to be an $m \times n$ matrix. If $m = n$, as in *T*, then the matrix is called a *square matrix*. Individual entries in a matrix are identified by using a subscript with row number and column number, in that order. For example, the entry 0.05 in row 2 and column 4 of *T* is denoted T_{24}. We also have $T_{32} = 0.15$ and $T_{13} = 0.10$ as two other entries in the matrix *T*. A single, unspecified entry from matrix *T* is represented T_{ij}, where *i* is the row number and *j* is the column number.

Each entry in *T* has a specific, unique meaning; therefore, the dimension of a matrix cannot be reduced without losing essential information. In *T*, the rows represent the country of origin of the presently-owned car, whereas the columns represent the origin of the next car. *T* is an example of a *transition matrix*, because it contains information concerning the owner's transition from the present car to the new car. The concept of a transition matrix is essential to Section 7.8 on Markov chains.

When mathematics is used to analyze real-world phenomena, the interaction of mathematical concepts and the real world is often based upon data. We apply mathematics to information that is gathered through measurement and observation. The discipline of mathematics includes many concepts that aid us in analyzing and interpreting data so that data can be used and expressed in summary form. Various data representations exist that allow us to discern trends, to form generalizations, and to make predictions, all on the basis of data. For example, an economist may use a linear function to forecast the growth of an industry. A demographer may use an exponential function to predict when the world population will exceed 10 billion people. These two examples are characterized by using mathematics to describe the behavior of a given situation, which is the essence of mathematical modeling. Most of the mathematical models you have studied in this course involve real-world phenomena approximated by continuous functions.

A matrix, on the other hand, is used with a collection of data that does not lend itself to the use of a continuous model. *Matrices* (the plural of matrix) are part of a branch of mathematics called *discrete mathematics*. The term *discrete* refers to the fact that these techniques of mathematical modeling deal with finite sets of noncontinuous data rather than continuous functions or continuous sets of data. The advantage of using a matrix to organize a set of data is illustrated by the preceding example of purchasing cars. Just as a rule like $f(t) = -9.8t^2 + 100$ is a good continuous model for the height of a falling object, a matrix is a mathematical tool used to handle noncontinuous data sets in summary form. Because matrices operate with discrete data, they possess dimension; that is, a matrix cannot be reduced to a single number such as the value of a function at a point. A matrix can be thought of as a single entity for the sake of simplicity; however, it is a single entity that contains many data values. In addition, just as special algebraic rules exist for functions, matrices have a special algebra associated with them. Although matrix algebra seems rather peculiar at first, it involves operations defined in ways that allow matrices to be used in many mathematical models. The following sections introduce methods for using matrices to model discrete data.

discrete mathematics

SECTION 7.2 Matrix Addition and Scalar Multiplication

We begin this section with a problem that illustrates the operations of matrix addition and scalar multiplication.

EXAMPLE 1 **The Hobby Shop Problem**

Suppose you have a small woodworking shop in your garage and you make toys for children as a hobby. Lately you have begun selling your toys at the local flea market. You make four different kinds of toys, namely a train (t), an airplane (a), a dragon (d), and a nameplate (n). Each of these can be made very plainly out of pine (p) or with greater detail and ornamentation out of oak (o). Let the following matrices O, N, and D represent your sales for October, November, and December:

$$O = \begin{array}{c} \\ p \\ o \end{array} \begin{array}{cccc} t & a & d & n \\ \left[\begin{array}{cccc} 3 & 5 & 0 & 2 \\ 1 & 2 & 1 & 3 \end{array}\right] \end{array} \qquad N = \begin{array}{c} \\ p \\ o \end{array} \begin{array}{cccc} t & a & d & n \\ \left[\begin{array}{cccc} 4 & 2 & 1 & 3 \\ 1 & 0 & 2 & 4 \end{array}\right] \end{array} \qquad D = \begin{array}{c} \\ p \\ o \end{array} \begin{array}{cccc} t & a & d & n \\ \left[\begin{array}{cccc} 4 & 8 & 5 & 3 \\ 4 & 2 & 1 & 6 \end{array}\right] \end{array}$$

How many of each item were sold for the entire three-month period?

Solution Notice that, as with the transition matrix in the previous section, each entry in matrices O, N, and D has a specific meaning. For example, the number of pine dragons made in December is given by entry D_{13}, which is 5. How many oak trains were made in November? This is given by entry $N_{21} = 1$. What does O_{24} represent? It represents the number of oak nameplates made in October, which is 3.

The total number of pine trains sold is $3 + 4 + 4 = 11$, which is the sum of the numbers in the upper-left corner of each matrix. The total number of oak dragons is $1 + 2 + 1 = 4$, which is the sum of the entries in row 2, column 3, or

$$O_{23} + N_{23} + D_{23} = 4.$$

We can construct a matrix S that has entries representing the total number of each item sold during the three months. From the pattern established above, we see that each entry of S is the sum of the corresponding entries of O, N, and D. Symbolically, we have

$$S_{ij} = O_{ij} + N_{ij} + D_{ij},$$

and S contains the entries

$$S = \begin{array}{c} \\ p \\ o \end{array} \begin{array}{cccc} t & a & d & n \\ \left[\begin{array}{cccc} 11 & 15 & 6 & 8 \\ 6 & 4 & 4 & 13 \end{array}\right] \end{array}.$$

The fact that each entry of S equals the sum of the corresponding entries in O, N, and D is represented by the matrix sum

$$S = O + N + D.$$ ■

Addition of matrices is a common-sense operation. Since each entry in a matrix has a meaning based on its position within the matrix, matrix addition is performed by adding the corresponding entries in each matrix. Matrix addition provides a concise structure for organizing an otherwise complicated operation. For example, in the Hobby Shop Problem, the single matrix sum $S = O + N + D$ represents 8 sums of 3 numbers each.

EXAMPLE 2 Using the data from Example 1, suppose that in the following year you sell the same number of each item in October, double the number sold of each item in November, and triple the number sold in December. How many of each item do you sell during the three months?

Solution The matrices O', N', and D' that we want to add are

$$O' = \begin{matrix} & t & a & d & n \\ p & 3 & 5 & 0 & 2 \\ o & 1 & 2 & 1 & 3 \end{matrix} \qquad N' = \begin{matrix} & t & a & d & n \\ p & 8 & 4 & 2 & 6 \\ o & 2 & 0 & 4 & 8 \end{matrix} \qquad D' = \begin{matrix} & t & a & d & n \\ p & 12 & 24 & 15 & 9 \\ o & 12 & 6 & 3 & 18 \end{matrix}.$$

The entries of O' are identical to those in O. Each entry of N' is twice the corresponding entry of N, which we can represent by the notation

$$N' = 2N.$$

scalar

scalar multiplication

The number 2, which is not a matrix, is a dimensionless quantity called a *scalar*. The result of a matrix multiplied by a scalar, called *scalar multiplication*, is a matrix derived by multiplying each entry of the original matrix by the scalar. Thus, we also have

$$D' = 3D.$$

Using addition and scalar multiplication, we can represent S', the number sold during the three months, by

$$S' = O' + N' + D',$$

or, equivalently,

$$S' = O + 2N + 3D,$$

which gives

$$S' = \begin{matrix} & t & a & d & n \\ p & 23 & 33 & 17 & 17 \\ o & 15 & 8 & 8 & 29 \end{matrix}.$$ ■

Matrix addition and scalar multiplication have reasonable definitions. Likewise, subtraction of matrices is defined just as you would expect: finding the difference of corresponding entries.

EXAMPLE 3 Suppose that in the original Hobby Shop Problem, sales for the entire year are given by the matrix Y, which is

$$Y = \begin{array}{c} \\ p \\ o \end{array} \begin{array}{cccc} t & a & d & n \\ \left[\begin{array}{cccc} 26 & 25 & 14 & 16 \\ 13 & 7 & 8 & 28 \end{array} \right] \end{array}.$$

How many of each item did you sell during the months other than October, November, and December? The answer is given by the matrix R defined as

$$R = Y - (O + N + D)$$
$$= Y - S.$$

The result is

$$R = \begin{array}{c} p \\ o \end{array} \begin{array}{cccc} t & a & d & n \\ \left[\begin{array}{cccc} 26 & 25 & 14 & 16 \\ 13 & 7 & 8 & 28 \end{array} \right] \end{array} - \begin{array}{c} p \\ o \end{array} \begin{array}{cccc} t & a & d & n \\ \left[\begin{array}{cccc} 11 & 15 & 6 & 8 \\ 6 & 4 & 4 & 13 \end{array} \right] \end{array} = \begin{array}{c} p \\ o \end{array} \begin{array}{cccc} t & a & d & n \\ \left[\begin{array}{cccc} 15 & 10 & 8 & 8 \\ 7 & 3 & 4 & 15 \end{array} \right] \end{array}. \quad \blacksquare$$

Addition and subtraction of matrices require that the matrices have the same dimension, meaning that they must have the same number of rows and the same number of columns. Furthermore, the corresponding rows and columns must have identical interpretations. Each row and column of a matrix has a specific meaning; therefore, trying to add or subtract matrices with different row or column labels is an attempt to combine incompatible quantities.

Exercise Set 7.2

1. The Campus Bookstore's inventory of books consists of the following quantities. Hardcover: textbooks—5280; fiction—1680; nonfiction—2320; reference—1890.
    Paperback: textbooks—1940; fiction—2810; nonfiction—1490; reference—2070.
    The College Bookstore's inventory of books consists of the following quantities. Hardcover: textbooks—6340; fiction—2220; nonfiction—1790; reference—1980.
    Paperback: textbooks—2050; fiction—3100; nonfiction—1720; reference—2710.

    a. Represent the inventory of the Campus Bookstore as a matrix.

    b. Represent the inventory of the College Bookstore as a matrix.

c. Use matrix addition to determine the total inventory of a new company formed by the merger of the Campus Bookstore and the College Bookstore.

2. The Lucrative Bank has three branches in Durham: Northgate (N), Downtown (D), and South Square (S). Matrix A shows the number of accounts of each type, checking (c), savings (s), and money market (m), at each branch office on January 1.

$$A = \begin{array}{c} \\ N \\ D \\ S \end{array} \begin{array}{ccc} c & s & m \\ \left[\begin{array}{ccc} 40039 & 10135 & 512 \\ 15231 & 8751 & 105 \\ 25612 & 12187 & 97 \end{array}\right] \end{array}$$

Matrix B shows the number of accounts of each type at each branch that were opened during the first quarter, and matrix C shows the number of accounts closed during the first quarter.

$$B = \begin{array}{c} \\ N \\ D \\ S \end{array} \begin{array}{ccc} c & s & m \\ \left[\begin{array}{ccc} 5209 & 2506 & 48 \\ 1224 & 405 & 17 \\ 2055 & 771 & 21 \end{array}\right] \end{array} \qquad C = \begin{array}{c} \\ N \\ D \\ S \end{array} \begin{array}{ccc} c & s & m \\ \left[\begin{array}{ccc} 2780 & 1100 & 32 \\ 565 & 189 & 25 \\ 824 & 235 & 14 \end{array}\right] \end{array}$$

a. Calculate the matrix representing the number of accounts of each type at each location at the end of the first quarter.

b. The sudden closing of a large textile plant has led bank analysts to estimate that all accounts will decline in number by 7% during the second quarter. Calculate a matrix that represents the anticipated number of each type of account at each branch at the end of the second quarter. Assume that fractions of accounts are rounded to integer values.

c. The bank president announces that the Lucrative Bank will merge with the Me. D. Okra Bank, which has branches at the same locations as those of the Lucrative Bank. The accounts at each branch of the Me. D. Okra Bank on January 1 are given by matrix E.

$$E = \begin{array}{c} \\ N \\ D \\ S \end{array} \begin{array}{ccc} c & s & m \\ \left[\begin{array}{ccc} 1345 & 2531 & 52 \\ 783 & 1987 & 137 \\ 2106 & 3765 & 813 \end{array}\right] \end{array}$$

Find the total number of accounts of each type at each branch of the bank formed by the merger of the two banks. Use the January 1 figures and assume that the accounts stay at their current branch offices.

A Common-Sense Approach to Matrix Multiplication

In the previous section, we saw that matrices can be used to organize information that is otherwise more difficult to grasp. We also saw that matrix addition and scalar multiplication are defined just as one would expect, given the meaning attached to the data in a matrix. Is it likewise possible to define matrix multiplication in a common-sense way that relates to real-life situations? The multiplication of integers can be thought of as repeated additions. For example, 5 times 3 can be thought of as $5 + 5 + 5$. This interpretation can be applied to scalar multiplication, since $3A$ is equal to the sum $A + A + A$. The product AB of two matrices A and B does not fit the same interpretation. That is, A cannot be added to itself B times, since B is a matrix, not a number. The following example provides motivation for matrix multiplication.

EXAMPLE 1 **The Cutting Board Problem**

Sam and a friend decide to go into a partnership making cutting boards and selling them at the local flea market. Suppose that each of them makes three different types of cutting boards.

> *Style 1*: made of alternating oak and walnut strips
> *Style 2*: made of oak, walnut, and cherry strips
> *Style 3*: made in a checkerboard pattern of walnut and cherry

Sam and his partner plan to make the number of cutting boards of each style shown in matrix A.

$$A = \begin{array}{c} \\ Sam \\ Partner \end{array} \begin{array}{ccc} 1 & 2 & 3 \\ \left[\begin{array}{ccc} 8 & 4 & 6 \\ 6 & 6 & 8 \end{array}\right] \end{array}$$

Each cutting board is made by gluing together one-inch strips of wood of the appropriate type in the desired pattern. Matrix B describes the number of strips of oak (o), walnut (w), and cherry (c) needed for each style.

$$B = \begin{array}{c} \\ 1 \\ 2 \\ 3 \end{array} \begin{array}{ccc} o & w & c \\ \left[\begin{array}{ccc} 10 & 10 & 0 \\ 8 & 6 & 6 \\ 0 & 10 & 10 \end{array}\right] \end{array}$$

To determine how much of each type of wood is needed to produce the cutting boards listed in matrix A, we will examine the following questions.

a. How much oak will Sam use to make the cutting boards?

Solution Sam will make 8 boards of Style 1, each of which uses 10 oak strips; 4 boards of Style 2, each of which uses 8 oak strips; and 6 boards of Style 3, which uses no oak. The total number of oak strips Sam will use is expressed by the sum of the products

$$8(10) + 4(8) + 6(0) = 112,$$

so a total of 112 oak strips is required.

b. How much oak will Sam's partner use?

Solution Sam's partner will use an amount of oak given by the sum

$$6(10) + 6(8) + 8(0) = 108,$$

so his partner requires a total of 108 oak strips.

c. How much cherry will Sam use?

Solution The amount of cherry Sam will use is

$$8(0) + 4(6) + 6(10) = 84,$$

for a total of 84 cherry strips. ■

For each of the questions above, we found the amount needed of a particular type of wood through addition of the products obtained from multiplying the number of each style by the corresponding amount of wood needed to make one cutting board. In each of these products, the first factor is a number from a row of matrix A, whereas the second factor is from a column of matrix B. We can summarize the amount of wood Sam and his partner will use with the following matrix C, in which the entries are the numbers of wood strips needed:

$$C = \begin{array}{c} \\ Sam \\ Partner \end{array} \begin{array}{c} o \quad\; w \quad\; c \\ \begin{bmatrix} 112 & 164 & 84 \\ 108 & 176 & 116 \end{bmatrix} \end{array}.$$

Observe that the entry in the first row, first column of C is obtained by lining up the first row of A and the first column of B, and then multiplying the corresponding entries and adding the products. Row 1 of A and column 1 of B are

$$\begin{array}{c} \\ Sam \end{array} \begin{array}{c} 1 \;\; 2 \;\; 3 \\ \begin{bmatrix} 8 & 4 & 6 \end{bmatrix} \end{array} \times \begin{array}{c} \\ 1 \\ 2 \\ 3 \end{array} \begin{array}{c} o \\ \begin{bmatrix} 10 \\ 8 \\ 0 \end{bmatrix} \end{array}.$$

Multiplying corresponding entries and adding gives

$$8(10) + 4(8) + 6(0) = 112.$$

This sum is entry C_{11}. Likewise, entry C_{23} is obtained by multiplying corresponding entries in the second row of A by the third column of B, which gives

$$\textit{Partner} \begin{array}{c} 1 \;\; 2 \;\; 3 \\ \begin{bmatrix} 6 & 6 & 8 \end{bmatrix} \end{array} \times \begin{array}{c} o \\ 1 \\ 2 \\ 3 \end{array}\begin{bmatrix} 0 \\ 6 \\ 10 \end{bmatrix} = 6(0) + 6(6) + 8(10) = 116.$$

Which row of A multiplied by which column of B gives entry C_{12}? The entry in the first row, second column of C is found by multiplying the first row of A by the second column of B.

matrix multiplication

All the entries in C can be found using the method illustrated above. This way of combining entries in two matrices to yield a third matrix is called *matrix multiplication*, and the matrix C is defined as the product of matrices A and B. The operation can be written in the form shown below.

$$C = AB$$

$$= \begin{array}{c} \textit{Sam} \\ \textit{Partner} \end{array}\begin{array}{c} 1 \;\; 2 \;\; 3 \\ \begin{bmatrix} 8 & 4 & 6 \\ 6 & 6 & 8 \end{bmatrix} \end{array} \times \begin{array}{c} 1 \\ 2 \\ 3 \end{array}\begin{array}{c} o \;\;\; w \;\;\; c \\ \begin{bmatrix} 10 & 10 & 0 \\ 8 & 6 & 6 \\ 0 & 10 & 10 \end{bmatrix} \end{array}$$

$$= \begin{array}{c} \textit{Sam} \\ \textit{Partner} \end{array}\begin{array}{c} o \;\;\;\; w \;\;\;\; c \\ \begin{bmatrix} 112 & 164 & 84 \\ 108 & 176 & 116 \end{bmatrix} \end{array}$$

In general, the matrix multiplication $C = AB$ is defined as follows: Each entry C_{ij} is obtained by multiplying corresponding entries in the *i*th row of the left-hand matrix A by the *j*th column of the right-hand matrix B. In symbols, this definition means that

$$C_{ij} = A_{i1}B_{1j} + A_{i2}B_{2j} + A_{i3}B_{3j} + \ldots + A_{in}B_{nj}.$$

EXAMPLE 2   Let matrix D represent the cost in dollars per strip for each type of wood in the Cutting Board Problem.

$$D = \begin{array}{c} o \\ w \\ c \end{array}\begin{array}{c} \textit{Cost} \\ \begin{bmatrix} 0.18 \\ 0.22 \\ 0.20 \end{bmatrix} \end{array}$$

a.   Sam would like to determine the cost of the wood for the cutting boards. What is the total cost of the wood for one cutting board of each style?

Solution The cost of the wood for a cutting board of Style 1 is equal to the number of strips of each type of wood multiplied by the cost per strip, or

$$10(0.18) + 10(0.22) + 0(0.20) = 4.00 \text{ dollars.}$$

The cost of the wood for a cutting board of Style 1 is $4.00. This number is calculated by multiplying corresponding entries in the first row of B by the column in D. Using similar reasoning for Styles 2 and 3, we see that the matrix product BD gives the required information for each style.

$$BD = \begin{array}{c} \\ 1 \\ 2 \\ 3 \end{array} \begin{array}{ccc} o & w & c \\ \left[\begin{array}{ccc} 10 & 10 & 0 \\ 8 & 6 & 6 \\ 0 & 10 & 10 \end{array} \right] \end{array} \times \begin{array}{c} o \\ w \\ c \end{array} \begin{array}{c} Cost \\ \left[\begin{array}{c} 0.18 \\ 0.22 \\ 0.20 \end{array} \right] \end{array}$$

$$= \begin{array}{c} 1 \\ 2 \\ 3 \end{array} \begin{array}{c} Cost \\ \left[\begin{array}{c} 4.00 \\ 3.96 \\ 4.20 \end{array} \right] \end{array}$$

The cost of the wood for a cutting board of Style 1 is $4.00, the cost for Style 2 is $3.96, and the cost for Style 3 is $4.20.

b. Sam and his partner would like to know how much money to budget for purchasing wood for the cutting boards. What are the total costs for Sam and his partner to produce the number of cutting boards listed in matrix A of Example 1?

Solution The product AB from Example 1 gives the number of strips of wood used by Sam and his partner. Multiplying AB by D results in a matrix containing the costs for Sam and his partner, as shown below.

$$AB(D) = \begin{array}{c} Sam \\ Partner \end{array} \begin{array}{ccc} o & w & c \\ \left[\begin{array}{ccc} 112 & 164 & 84 \\ 108 & 176 & 116 \end{array} \right] \end{array} \times \begin{array}{c} o \\ w \\ c \end{array} \begin{array}{c} Cost \\ \left[\begin{array}{c} 0.18 \\ 0.22 \\ 0.20 \end{array} \right] \end{array}$$

$$= \begin{array}{c} Sam \\ Partner \end{array} \begin{array}{c} Cost \\ \left[\begin{array}{c} 73.04 \\ 81.36 \end{array} \right] \end{array}$$

The total cost for Sam is $73.04; the total cost for his partner is $81.36. ▨

The rows and columns of data in the matrices in Examples 1 and 2 are described by labels. In matrix A, the row labels are names (Sam and his partner) and the column labels are styles of cutting boards (1, 2, and 3), so matrix A classifies data according to name and style. We refer to A as a name-by-style matrix. Consistent with this notation, matrix B is a style-by-wood matrix. The row and column labels of matrices are especially helpful in interpreting the results of matrix multiplication.

In Example 1 we multiplied a name-by-style matrix (A) by a style-by-wood matrix (B) to get a name-by-wood matrix (C). In Example 2, we multiplied a style-by-wood matrix (B) by a wood-by-cost matrix (D) to get a style-by-cost matrix. We also found that the product of a name-by-wood matrix (C) and a wood-by-cost matrix (D) is a name-by-cost matrix. In each example, matrix multiplication eliminated the labels of the first factor's columns and the second factor's rows, leaving a product matrix with exactly the row and column labels we desired in our answer. Matrix multiplication, which at first glance may seem very strange, actually is designed to give us the information we want in a straightforward manner.

The product of two matrices can be found by multiplying the elements of a row of the left-hand matrix by the corresponding elements of a column of the right-hand matrix and then adding the products. If matrix S is multiplied by matrix T, the number of columns of S must equal the number of rows of T. If $ST = U$, the product matrix U has the same number of rows as S and the same number of columns as T. In symbols,

$$S_{m \times n} T_{n \times p} = U_{m \times p}.$$

The column labels of S must be the same as the row labels of T if the product ST is to be meaningful. If $ST = U$, then U has the row labels of S and the column labels of T.

vector

column vector

row vector

identity matrix

main diagonal

Some of the matrices in our discussion consisted of only one row or one column. This special type of matrix is called a *vector*. A matrix that consists of one column is called a *column vector*, and a matrix that consists of one row is called a *row vector*.

The *identity matrix* $I_{n \times n}$ is a square matrix with the property that multiplying a matrix A by I returns A as the product. If A is a 2×3 matrix and $I \cdot A = A$, then I must be a 2×2 matrix. Can you explain why? If A is a 2×3 matrix and $A \cdot I = A$, then I must be a 3×3 matrix. The elements of an identity matrix are ones on the *main diagonal* (going from upper left to lower right) and zeros elsewhere. Several examples of identity matrices are shown below.

$$I_{2 \times 2} = \begin{bmatrix} 1 & 0 \\ 0 & 1 \end{bmatrix} \quad I_{3 \times 3} = \begin{bmatrix} 1 & 0 & 0 \\ 0 & 1 & 0 \\ 0 & 0 & 1 \end{bmatrix} \quad I_{n \times n} = \begin{bmatrix} 1 & 0 & \cdots & 0 & 0 \\ 0 & 1 & \cdots & 0 & 0 \\ \vdots & \vdots & \vdots & \vdots & \vdots \\ 0 & 0 & \cdots & 1 & 0 \\ 0 & 0 & \cdots & 0 & 1 \end{bmatrix}$$

1. The following is a set of abstract matrices (without row and column labels).

$$M = \begin{bmatrix} 1 & -1 \\ 2 & 0 \end{bmatrix} \quad N = \begin{bmatrix} 2 & 4 & 1 \\ 0 & -1 & 3 \\ 1 & 0 & 2 \end{bmatrix} \quad O = \begin{bmatrix} 6 \\ -1 \end{bmatrix} \quad P = \begin{bmatrix} 1 & \frac{1}{2} \\ -1 & \frac{1}{2} \end{bmatrix} \quad Q = \begin{bmatrix} 4 \\ 1 \\ 3 \end{bmatrix}$$

$$R = \begin{bmatrix} 3 & 1 \\ -1 & 0 \end{bmatrix} \quad S = \begin{bmatrix} 3 & 1 \\ 1 & 0 \\ 0 & 2 \\ -1 & 1 \end{bmatrix} \quad T = \begin{bmatrix} 1 \\ 2 \\ -3 \\ 4 \end{bmatrix} \quad U = \begin{bmatrix} 4 & 2 & 6 & -1 \\ 5 & 3 & 1 & 0 \\ 0 & 2 & -1 & 1 \end{bmatrix}$$

List all pairs of matrices from this set for which the product is defined. State the dimension of each product. List both products, such as *MR* and *RM,* if both products are defined.

2. Using the matrices *M* and *P* from Exercise 1, find the matrix products *MP* and *PM*. What property do you notice about these matrices?

Is matrix multiplication associative?

3. Is matrix multiplication associative? In other words, is it always true that *A(BC) = (AB)C,* assuming these matrix products are defined? Use some of the matrices from Exercise 1 to test your conjecture.

4. What is the result of multiplying a vector times a vector, if the multiplication is defined?

Is matrix multiplication commutative?

5. Is matrix multiplication commutative? In other words, is it always true that *AB = BA*? Is multiplication by the identity matrix commutative? Use some matrices from Exercise 1 to test your conjectures. Explain why it is necessary to use the terms *left-multiply* and *right-multiply* when referring to matrix multiplication. (For example, in the product *AB*, matrix *A* is right-multiplied by *B*, and *B* is left-multiplied by *A*.)

6. A company has investments in three states—North Carolina, North Dakota, and New Mexico. Its deposits in each state are divided among bonds, mortgages, and consumer loans. The amount of money (in millions of dollars) invested in each category is displayed in Figure 1.

Figure 1 Investments

Categories	NC	ND	NM
Bonds	13	25	22
Mortgages	6	9	4
Loans	29	17	13

The current annual yields on these investments are 7.5% for bonds, 11.25% for mortgages, and 6% for consumer loans. Use matrix multiplication to find the total annual earnings for each state.

7. Several years ago an investor purchased growth stocks, with the expectation that they would increase in value over time. The purchase included 100 shares of stock *A*, 200 shares of stock *B*, and 150 shares of stock *C*. At the end of each year the value of each stock was recorded. Figure 2 shows the price per share (in dollars) of stocks *A*, *B*, and *C* at the end of the years 1994, 1995, and 1996.

Figure 2 Shares of stock

Categories	1994	1995	1996
Stock A	68.00	72.00	75.00
Stock B	55.00	60.00	67.50
Stock C	82.50	84.00	87.00

Calculate the total value of the stock purchases at the end of each year.

8. A virus hits campus. Students are either sick, well, or carriers of the virus. As shown in Figure 3, the following percentages of people are in each category, depending on whether they are juniors or seniors.

Figure 3 Percentages of sick students

Categories	Junior	Senior
Well	15%	25%
Sick	35%	40%
Carrier	50%	35%

The student population is distributed by class and gender as shown in Figure 4.

Figure 4 Student population

Categories	Male	Female
Junior	104	80
Senior	107	103

How many sick males are there? How many well females? How many female carriers?

9. The Sound Company produces stereos. Its inventory includes four models—the Budget, the Economy, the Executive, and the President models. The Budget model needs 50 transistors, 30 capacitors, 7 connectors, and 3 dials. The Economy model needs 65 transistors, 50 capacitors, 9 connectors, and 4 dials. The Executive model needs 85 transistors, 42 capacitors, 10 connectors, and 6 dials. The President model needs 85 transistors, 42 capacitors, 10 connectors, and 12 dials. The daily manufacturing goal in a normal quarter is 10 Budget, 12 Economy, 11 Executive, and 7 President stereos.

a. How many transistors are needed each day? capacitors? connectors? dials?

b. During August and September, production is increased by 40%. How many Budget, Economy, Executive, and President models are produced daily during these months?

c. It takes 5 person-hours to produce the Budget model, 7 person-hours to produce the Economy model, 6 person-hours for the Executive model, and 7 person-hours for the President model. Determine the number of employees needed to maintain the normal production schedule, assuming everyone works an average of 7 hours each day. How many employees are needed in August and September?

10. The parabola $y = -4x^2 + 16x - 12$ contains the points (1, 0), (2, 4), and (3, 0). Find the new ordered pair (x', y') that is produced from each pair (x, y) using the matrix multiplication below, where $\theta = 30°$.

$$\begin{bmatrix} \cos\theta & -\sin\theta \\ \sin\theta & \cos\theta \end{bmatrix} \begin{bmatrix} x \\ y \end{bmatrix} = \begin{bmatrix} x' \\ y' \end{bmatrix}$$

Carefully plot the new points in the same coordinate system as the three given points of the original parabola. Sketch the remainder of the parabola as you think it would appear if every point on the original curve were transformed by the matrix multiplication above. Compare this sketch with a sketch of the original parabola. What happened? Try $\theta = 90°$ to test your conjecture.

11. The president of the Lucrative Bank is hoping for a 21% increase in checking accounts (c), a 35% increase in savings accounts (s), and a 52% increase in money market accounts (m). The current statistics on the number of accounts at each branch are as follows:

	Checking	Savings	Money Market
Northgate	40039	10135	512
Downtown	15231	8751	105
South Square	25612	12187	97

What is the goal at each branch for each type of account? (Hint: Multiply by a 3×3 matrix with certain nonzero entries on the main diagonal and zero entries elsewhere.) What will be the total number of accounts at each branch?

12. Winners at a science fair are determined by a scoring system based on five items with different weights attached to each item. The items and associated weights are: summary of background research—weight 3; experimental procedure—weight 5; research paper—weight 6; project display—weight 8; and creativity of idea—weight 4. Each project is judged by grading each of the five items on a scale from 0 to 10, with 10 being highest. The total score for a project is calculated by adding the products of the corresponding weights and points for each item.

a. What is the maximum total score possible for a project?

b. Calculate the score for a student who earns 8 points on background research, 9 points on experimental procedure, 7 points on research paper, 8 points on project display, and 6 points on creativity.

c. The table in Figure 5 contains the points for the finalists in the biology division. Calculate the total scores to determine the first-, second-, and third-place entries.

Figure 5 Science fair scores

Categories	Peter	Sheila	Arvind	Kathy	Nia	Chris	Maurice
Background research	9	8	10	7	8	9	10
Experimental procedure	10	9	9	10	10	9	10
Research paper	7	9	8	9	7	8	8
Project display	9	10	9	8	10	8	9
Creativity of idea	8	7	8	10	6	8	7

13. An opera company is planning a cross-country tour. It plans to perform *Carmen* and *La Traviata* in Atlanta in May. The person in charge of logistics wants to make airplane reservations for the two troupes. *Carmen* has 2 stars, 25 other adults, 5 children, and 5 staff members. *La Traviata* has 3 stars, 15 other adults, and 4 staff members. There are 3 airlines from which to choose. Redwing charges round-trip fares to Atlanta of $630 for first class, $420 for coach, and $250 for youth. Southeastern charges $650 for first class, $350 for coach, and $275 for youth. Air Atlanta charges $700 for first class, $370 for coach, and $150 for youth. Assume that stars travel first class, other adults and staff travel coach, and children travel for the youth fare.

a. Find the total cost for each opera troupe on each airline.

b. If each airline will give a 30% discount to the *Carmen* troupe because troupe members will stay over Saturday night, what is the total cost on each airline?

c. Suppose instead that each airline will give a discount to both troupes just to get their business. Redwing gives a 30% discount, Southeastern 20%, and Air Atlanta 25%. Use matrix multiplication to find a new 3 × 3 cost matrix. (See the hint for Exercise 11.)

14. On the Sunday before the 1998 NCAA basketball championship, a national survey was taken of people's preferences for the winner of the game, along with their income. The following information was collected:

for individuals with incomes greater than $40,000 per year,
   435 people preferred the University of Kentucky,
   105 people preferred the University of Utah, and
   115 people had no preference;
for individuals with incomes less than $40,000 per year,
   125 people preferred the University of Kentucky,
   205 people preferred the University of Utah, and
   231 people had no preference.

A survey was done in a restaurant in New York on the night of the game to determine the incomes of the people eating there. The following information was collected:

   302 individuals had incomes greater than $40,000 per year, and
   276 individuals had incomes less than $40,000 per year.

Based on the survey from Sunday, use matrix operations to estimate the number of Kentucky fans, the number of Utah fans, and the number of fans with no preference in the restaurant. Use probabilities based on the Sunday survey and the data collected on the night of the game. Before you attempt to find answers, the information from each survey should be converted to ratios and displayed in matrices.

15. A company that produces and markets stuffed animals has three manufacturing plants: one on the East Coast, one on the West Coast, and one in the central part of the country. Among other items, each plant manufactures stuffed pandas, kangaroos, and rabbits. Personnel are needed to cut fabric, sew appropriate parts together, and provide finish work for each animal. Matrix A gives the time (in hours) of each type of labor required to make each type of stuffed animal, matrix B gives the daily production capacity at each plant, matrix C provides hourly wages of the different workers at each plant, and matrix D contains the total orders received by the company in October and November.

$$
A = \begin{array}{c} Panda \\ Kang. \\ Rabbit \end{array} \begin{bmatrix} Cut & Sew & Finish \\ 0.5 & 0.8 & 0.6 \\ 0.8 & 1.0 & 0.4 \\ 0.4 & 0.5 & 0.5 \end{bmatrix}
\qquad
B = \begin{array}{c} East \\ Central \\ West \end{array} \begin{bmatrix} Panda & Kang. & Rabbit \\ 25 & 15 & 12 \\ 10 & 20 & 15 \\ 20 & 15 & 15 \end{bmatrix}
$$

$$
C = \begin{array}{c} East \\ Central \\ West \end{array} \begin{bmatrix} Cut & Sew & Finish \\ 7.50 & 9.00 & 8.40 \\ 7.00 & 8.00 & 7.60 \\ 8.40 & 10.50 & 10.00 \end{bmatrix}
\qquad
D = \begin{array}{c} Panda \\ Kang. \\ Rabbit \end{array} \begin{bmatrix} Oct. & Nov. \\ 1000 & 1100 \\ 600 & 850 \\ 800 & 725 \end{bmatrix}
$$

Use the matrices on page 483 to compute the following quantities:

a. the number of hours of each type of labor needed each month (October and November) to fill all orders;

b. the production cost per item at each plant;

c. the cost of filling all October orders at the East Coast plant;

d. the number of daily hours of each type of labor needed at each plant if production levels are at capacity;

e. the daily amount each plant will pay its personnel when producing at capacity.

SECTION 7.4 | Computer Graphics

The applications of computer graphics are commonplace today. We see them everyday in newspapers, in magazines, on television, and in video games. This section focuses on the use of matrices and matrix operations in the implementation of computer graphics. It differs from previous sections in that it places greater emphasis on abstraction and generalizations. For individuals who enjoy computer programming, this section can be the basis for many worthwhile projects and interesting investigations.

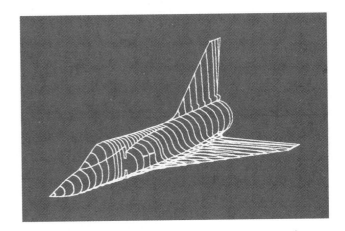

Rotations in Two Dimensions

Suppose we rotate a set of points in the plane through an angle θ. (An angle θ is considered positive if the direction of rotation is counterclockwise.) If a point P has coordinates (x, y), what are the coordinates (x', y') of the new point P' after rotation by a positive θ?

In Figure 6, the point P is located in the first quadrant along $\overrightarrow{OP}$ at a distance r from the origin and at an angle α to the positive x-axis.

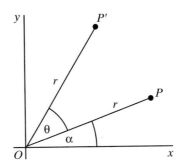

Figure 6   Rotation of a point in the plane

We can express the coordinates of P in terms of r and α as follows:

$$x = r \cos \alpha$$

$$y = r \sin \alpha.$$

The point P' is also a distance r from the origin, but on $\overrightarrow{OP'}$ at an angle θ to $\overrightarrow{OP}$. Since $\overrightarrow{OP'}$ is at an angle $\theta + \alpha$ to the x-axis, the coordinates of P' can be expressed as follows:

$$x' = r\cos(\theta + \alpha)$$

$$y' = r\sin(\theta + \alpha).$$

We can use the trigonometric identities for the sine and cosine of the sum of two angles to write

$$x' = r\cos\theta\cos\alpha - r\sin\theta\sin\alpha$$

$$y' = r\sin\theta\cos\alpha + r\cos\theta\sin\alpha,$$

which can be rearranged as

$$x' = (r\cos\alpha)\cos\theta - (r\sin\alpha)\sin\theta$$

$$y' = (r\cos\alpha)\sin\theta + (r\sin\alpha)\cos\theta.$$

Substituting for the original x- and y-values, $x = r\cos\alpha$ and $y = r\sin\alpha$, yields the following equations for transforming (x, y) to (x', y') through a rotation by θ:

$$x' = x\cos\theta - y\sin\theta \qquad \text{(1)}$$

$$y' = x\sin\theta + y\cos\theta. \qquad \text{(2)}$$

Equations (1) and (2) for rotating a point in the plane can be conveniently represented with matrix multiplication as

$$\begin{bmatrix} x' \\ y' \end{bmatrix} = \begin{bmatrix} \cos\theta & -\sin\theta \\ \sin\theta & \cos\theta \end{bmatrix} \begin{bmatrix} x \\ y \end{bmatrix}.$$

The efficiency of matrix representation is apparent if we wish to rotate a set of n points in the plane, such as the vertices of a many-sided polygon. Instead of using n pairs of $(x, y) \rightarrow (x', y')$ transformation equations, the following single matrix equation will accomplish the rotation of all n points:

$$\begin{bmatrix} x_1' & x_2' & \cdots & x_n' \\ y_1' & y_2' & \cdots & y_n' \end{bmatrix} = \begin{bmatrix} \cos\theta & -\sin\theta \\ \sin\theta & \cos\theta \end{bmatrix} \begin{bmatrix} x_1 & x_2 & \cdots & x_n \\ y_1 & y_2 & \cdots & y_n \end{bmatrix}.$$

If we let R represent the rotation matrix

$$R = \begin{bmatrix} \cos\theta & -\sin\theta \\ \sin\theta & \cos\theta \end{bmatrix},$$

then notice that

$$R^2 = \begin{bmatrix} \cos^2\theta - \sin^2\theta & -2\sin\theta\cos\theta \\ 2\sin\theta\cos\theta & \cos^2\theta - \sin^2\theta \end{bmatrix}.$$

Using the double angle identities, we can write

$$R^2 = \begin{bmatrix} \cos 2\theta & -\sin 2\theta \\ \sin 2\theta & \cos 2\theta \end{bmatrix}.$$

Comparing the entries in R and R^2 reveals that R differs from R^2 only in that θ has been replaced by 2θ. Since the argument of the sine and cosine functions in R is the angle of rotation, R^2 must produce a rotation through an angle 2θ. This result indicates that successive rotations by θ can be implemented through successive multiplications by matrix R.

Also notice that if we let

$$R_1 = \begin{bmatrix} \cos\theta_1 & -\sin\theta_1 \\ \sin\theta_1 & \cos\theta_1 \end{bmatrix}$$

and

$$R_2 = \begin{bmatrix} \cos\theta_2 & -\sin\theta_2 \\ \sin\theta_2 & \cos\theta_2 \end{bmatrix},$$

then

$$R_1 R_2 = \begin{bmatrix} \cos\theta_1\cos\theta_2 - \sin\theta_1\sin\theta_2 & -\cos\theta_1\sin\theta_2 - \sin\theta_1\cos\theta_2 \\ \sin\theta_1\cos\theta_2 + \cos\theta_1\sin\theta_2 & \cos\theta_1\cos\theta_2 - \sin\theta_1\sin\theta_2 \end{bmatrix}$$

$$= \begin{bmatrix} \cos(\theta_1 + \theta_2) & -\sin(\theta_1 + \theta_2) \\ \sin(\theta_1 + \theta_2) & \cos(\theta_1 + \theta_2) \end{bmatrix}.$$

Comparing the entries of $R_1 R_2$ with the general form of R shows that the angle θ in R has been replaced by $(\theta_1 + \theta_2)$; therefore, the matrix $R_1 R_2$ above produces a rotation by an angle $(\theta_1 + \theta_2)$. The matrix R_1 represents a rotation through an angle θ_1, and R_2 represents a rotation through an angle θ_2. The final form of $R_1 R_2$ indicates that successive rotations by θ_1 and θ_2, or a total rotation of $(\theta_1 + \theta_2)$, can be implemented through successive matrix multiplications by R_1 and R_2.

Homogeneous Coordinates

We have seen that rotation of coordinates can be accomplished through multiplication by the matrix R. A much simpler transformation of coordinates in the plane is represented by the following equations:

$$x' = x + a$$

$$y' = y + b.$$

This transformation, called a translation through (a, b), corresponds to shifting a point a units in the x-direction and b units in the y-direction. A translation can be implemented with matrices using the equation

$$\begin{bmatrix} x \\ y \end{bmatrix} + \begin{bmatrix} a \\ b \end{bmatrix} = \begin{bmatrix} x + a \\ y + b \end{bmatrix}.$$

The matrix equation for translating n points through (a, b) is

$$\begin{bmatrix} x_1 & x_2 & \cdots & x_n \\ y_1 & y_2 & \cdots & y_n \end{bmatrix} + \begin{bmatrix} a & a & \cdots & a \\ b & b & \cdots & b \end{bmatrix} = \begin{bmatrix} x_1 + a & x_2 + a & \cdots & x_n + a \\ y_1 + b & y_2 + b & \cdots & y_n + b \end{bmatrix}.$$

We have now implemented translations using matrix addition. We can also translate coordinates by using matrix multiplication. However, left-multiplying the matrix

$$\begin{bmatrix} x \\ y \end{bmatrix}$$

by a matrix will not accomplish our task because the matrix equation

$$A \begin{bmatrix} x \\ y \end{bmatrix} = \begin{bmatrix} x + a \\ y + b \end{bmatrix}$$

has no solution for A. Mathematicians have found that expanding the dimensions of the matrices leads to a matrix multiplication for translation of coordinates. The rationale for implementing translations with matrix multiplication will become clearer later in this section. If the matrix representation of the coordinates (x, y) is changed to

$$\begin{bmatrix} x \\ y \\ 1 \end{bmatrix},$$

then translation of (x, y) by (a, b) corresponds to the multiplication

$$\begin{bmatrix} 1 & 0 & a \\ 0 & 1 & b \\ 0 & 0 & 1 \end{bmatrix} \begin{bmatrix} x \\ y \\ 1 \end{bmatrix} = \begin{bmatrix} x + a \\ y + b \\ 1 \end{bmatrix}.$$

Likewise, translation of n points corresponds to

$$\begin{bmatrix} 1 & 0 & a \\ 0 & 1 & b \\ 0 & 0 & 1 \end{bmatrix} \begin{bmatrix} x_1 & x_2 & \cdots & x_n \\ y_1 & y_2 & \cdots & y_n \\ 1 & 1 & \cdots & 1 \end{bmatrix} = \begin{bmatrix} x_1 + a & x_2 + a & \cdots & x_n + a \\ y_1 + b & y_2 + b & \cdots & y_n + b \\ 1 & 1 & \cdots & 1 \end{bmatrix}.$$

homogeneous coordinates   Adding a third row to the matrix representation for the point (x, y) provides a new representation called *homogeneous coordinates*. In general, homogeneous coordinates are not required to contain a 1 in the bottom row of the matrix. If h is the number in the bottom row, then the point (x, y) is represented by the homogeneous coordinates

$$\begin{bmatrix} hx \\ hy \\ h \end{bmatrix}.$$

For example, the homogeneous coordinates

$$\begin{bmatrix} 16 \\ 10 \\ 4 \end{bmatrix}$$

represent the point $(4, \frac{5}{2})$ in the plane.

Homogeneous coordinates for a specific point are not unique. For example, the point $(5, 3)$ can be represented in homogeneous coordinates by

$$\begin{bmatrix} 5 \\ 3 \\ 1 \end{bmatrix},$$

or, equivalently, by

$$\begin{bmatrix} 10 \\ 6 \\ 2 \end{bmatrix}.$$

The homogeneous coordinates

$$\begin{bmatrix} 7 \\ 4 \\ 2 \end{bmatrix}$$

represent the point $(\frac{7}{2}, 2)$ as do the homogeneous coordinates

$$\begin{bmatrix} \frac{7}{4} \\ 1 \\ \frac{1}{2} \end{bmatrix}.$$

In addition to allowing translations to be done with matrix multiplication, homogeneous coordinates can be used for scaling operations. Scaling refers to changing the size of an object according to a fixed factor. If a set of n points corresponds to the vertices of a polygon, then the polygon is scaled by a factor of 2 by multiplying the x- and y-coordinates of the n points by 2, which doubles the length of each side of the polygon. With matrices, this can be accomplished by either multiplying the matrix of nonhomogeneous coordinates by a scalar 2 or by changing the bottom entry in the homogeneous coordinates to $\frac{1}{2}$. The second option is accomplished by the following matrix multiplication.

$$\begin{bmatrix} 1 & 0 & 0 \\ 0 & 1 & 0 \\ 0 & 0 & \frac{1}{2} \end{bmatrix} \begin{bmatrix} x_1 & x_2 & \cdots & x_n \\ y_1 & y_2 & \cdots & y_n \\ 1 & 1 & \cdots & 1 \end{bmatrix} = \begin{bmatrix} x_1 & x_2 & \cdots & x_n \\ y_1 & y_2 & \cdots & y_n \\ \frac{1}{2} & \frac{1}{2} & \cdots & \frac{1}{2} \end{bmatrix}$$

The homogeneous coordinates

$$\begin{bmatrix} x \\ y \\ \frac{1}{2} \end{bmatrix}$$

correspond to the point $(2x, 2y)$; therefore, the multiplication above corresponds to transforming all points (x_i, y_i) to $(2x_i, 2y_i)$. The distances between all pairs of points are doubled.

EXAMPLE 1   What are the new coordinates of the points $(1, 2)$ and $(3, 7)$ after they have been rescaled by a factor of $\frac{1}{2}$?

Solution   To halve the size of a figure, left-multiply the matrix of homogeneous coordinates by

$$\begin{bmatrix} 1 & 0 & 0 \\ 0 & 1 & 0 \\ 0 & 0 & 2 \end{bmatrix}$$

The effect of multiplication by this matrix on the points $(1, 2)$ and $(3, 7)$ is shown below.

$$\begin{bmatrix} 1 & 0 & 0 \\ 0 & 1 & 0 \\ 0 & 0 & 2 \end{bmatrix} \begin{bmatrix} 1 & 3 \\ 2 & 7 \\ 1 & 1 \end{bmatrix} = \begin{bmatrix} 1 & 3 \\ 2 & 7 \\ 2 & 2 \end{bmatrix}$$

The right side of the equation gives homogeneous coordinates for the points $\left(\frac{1}{2}, 1\right)$ and $\left(\frac{3}{2}, \frac{7}{2}\right)$ in the plane, so the coordinates of the original points have been halved.

Now that we have examined translations and scaling with homogeneous coordinates, how are rotations accomplished with homogeneous coordinates? If the point (x, y), represented by

How are rotations accomplished with homogeneous coordinates?

$$\begin{bmatrix} x \\ y \\ 1 \end{bmatrix},$$

is rotated by θ to the point (x', y'), represented by

$$\begin{bmatrix} x' \\ y' \\ 1 \end{bmatrix},$$

then the transformation equations relating (x, y) and (x', y') are the same as the ones we saw earlier. Inserting the 2×2 rotation matrix in a 3×3 matrix and multiplying by the homogeneous coordinates for (x, y) yields

$$\begin{bmatrix} \cos\theta & -\sin\theta & 0 \\ \sin\theta & \cos\theta & 0 \\ 0 & 0 & 1 \end{bmatrix} \begin{bmatrix} x \\ y \\ 1 \end{bmatrix} = \begin{bmatrix} x\cos\theta - y\sin\theta \\ x\sin\theta + y\cos\theta \\ 1 \end{bmatrix}$$

$$= \begin{bmatrix} x' \\ y' \\ 1 \end{bmatrix},$$

where (x', y') corresponds to the rotation of (x, y) through an angle θ.

With homogeneous coordinates, matrix multiplication can be used to implement the translation, scaling, and rotation of points in the plane. Earlier, we observed that successive rotations correspond to successive matrix multiplications. Similarly, translation, scaling, and rotation can be performed in sequence through successive multiplication by the appropriate matrices.

EXAMPLE 2 Transform the point $(2, 5)$ by scaling by a factor of 2, and then translating through $(-1, 3)$.

Solution The matrices that accomplish these transformations are shown below. Notice that the matrix for the second transformation, the translation, is placed to the left of the matrix for the first transformation, the scaling.

$$\begin{bmatrix} 1 & 0 & -1 \\ 0 & 1 & 3 \\ 0 & 0 & 1 \end{bmatrix} \begin{bmatrix} 1 & 0 & 0 \\ 0 & 1 & 0 \\ 0 & 0 & \frac{1}{2} \end{bmatrix} \begin{bmatrix} 2 \\ 5 \\ 1 \end{bmatrix}$$

Multiplying the two transformation matrices yields

$$\begin{bmatrix} 1 & 0 & -\frac{1}{2} \\ 0 & 1 & \frac{3}{2} \\ 0 & 0 & \frac{1}{2} \end{bmatrix} \begin{bmatrix} 2 \\ 5 \\ 1 \end{bmatrix}.$$

Completing the multiplication results in the matrix

$$\begin{bmatrix} \frac{3}{2} \\ \frac{13}{2} \\ \frac{1}{2} \end{bmatrix},$$

which corresponds to the point in the plane with coordinates $(3, 13)$. Is this the result we should expect? Scaling $(2, 5)$ by a factor of 2 transforms the point to $(4, 10)$. A subsequent translation through $(-1, 3)$ gives us the point $(3, 13)$—the result obtained by matrix multiplication.

If we were to reverse the order of the transformations by translating first, then we would get

$$
\begin{bmatrix} 1 & 0 & 0 \\ 0 & 1 & 0 \\ 0 & 0 & \frac{1}{2} \end{bmatrix}
\begin{bmatrix} 1 & 0 & -1 \\ 0 & 1 & 3 \\ 0 & 0 & 1 \end{bmatrix}
\begin{bmatrix} 2 \\ 5 \\ 1 \end{bmatrix}
=
\begin{bmatrix} 1 & 0 & -1 \\ 0 & 1 & 3 \\ 0 & 0 & \frac{1}{2} \end{bmatrix}
\begin{bmatrix} 2 \\ 5 \\ 1 \end{bmatrix}
$$

$$
=
\begin{bmatrix} 1 \\ 8 \\ \frac{1}{2} \end{bmatrix}.
$$

The transformed homogeneous coordinates represent the point in the plane with coordinates $(2, 16)$, which corresponds to translating $(2, 5)$ through $(-1, 3)$ (yielding the point $(1, 8)$) and then scaling by 2 (resulting in the point $(2, 16)$). Notice that changing the order of the transformations leads to different results. You should expect that changing the order of the transformation matrices would change the final results, since matrix multiplication is not commutative. ■

The concepts illustrated in Example 2 can be expanded by examining a more general question: How do we transform a point (x, y) by first scaling by a factor of k, and then translating through (a, b)? To accomplish these transformations, we use the matrix multiplication shown below.

How do we transform a point (x, y) by first scaling by a factor of k, and then translating through (a, b)?

$$
\begin{bmatrix} 1 & 0 & a \\ 0 & 1 & b \\ 0 & 0 & 1 \end{bmatrix}
\begin{bmatrix} 1 & 0 & 0 \\ 0 & 1 & 0 \\ 0 & 0 & \frac{1}{k} \end{bmatrix}
\begin{bmatrix} x \\ y \\ 1 \end{bmatrix}
$$

This multiplication simplifies to

$$
\begin{bmatrix} 1 & 0 & \frac{a}{k} \\ 0 & 1 & \frac{b}{k} \\ 0 & 0 & \frac{1}{k} \end{bmatrix}
\begin{bmatrix} x \\ y \\ 1 \end{bmatrix},
$$

which results in

$$
\begin{bmatrix} x + \frac{a}{k} \\ y + \frac{b}{k} \\ \frac{1}{k} \end{bmatrix}.
$$

The final matrix corresponds to the point in the plane with coordinates $(kx + a, ky + b)$. This is the result we expected—a scaling by k and a translation through (a, b).

EXAMPLE 3 Combine the following transformations: rotate by an angle θ, scale by a factor k, and then translate through (a, b).

Solution The matrix multiplication that accomplishes the transformations in the order stated is shown below.

$$\begin{bmatrix} 1 & 0 & a \\ 0 & 1 & b \\ 0 & 0 & 1 \end{bmatrix} \begin{bmatrix} 1 & 0 & 0 \\ 0 & 1 & 0 \\ 0 & 0 & \frac{1}{k} \end{bmatrix} \begin{bmatrix} \cos\theta & -\sin\theta & 0 \\ \sin\theta & \cos\theta & 0 \\ 0 & 0 & 1 \end{bmatrix} \begin{bmatrix} x \\ y \\ 1 \end{bmatrix}$$

$$= \begin{bmatrix} 1 & 0 & a \\ 0 & 1 & b \\ 0 & 0 & 1 \end{bmatrix} \begin{bmatrix} \cos\theta & -\sin\theta & 0 \\ \sin\theta & \cos\theta & 0 \\ 0 & 0 & \frac{1}{k} \end{bmatrix} \begin{bmatrix} x \\ y \\ 1 \end{bmatrix}$$

$$= \begin{bmatrix} \cos\theta & -\sin\theta & \frac{a}{k} \\ \sin\theta & \cos\theta & \frac{b}{k} \\ 0 & 0 & \frac{1}{k} \end{bmatrix} \begin{bmatrix} x \\ y \\ 1 \end{bmatrix}$$

$$= \begin{bmatrix} x\cos\theta - y\sin\theta + \frac{a}{k} \\ x\sin\theta + y\cos\theta + \frac{b}{k} \\ \frac{1}{k} \end{bmatrix}$$

$$= \begin{bmatrix} x' + \frac{a}{k} \\ y' + \frac{b}{k} \\ \frac{1}{k} \end{bmatrix}$$

The point (x', y') corresponds to the rotation of (x, y) through an angle θ. The final transformed homogeneous coordinates correspond to the point $(kx' + a, ky' + b)$. The result is the point rotated through an angle θ, scaled by a factor k, and translated through (a, b). The results above demonstrate that the single matrix

$$\begin{bmatrix} \cos\theta & -\sin\theta & \frac{a}{k} \\ \sin\theta & \cos\theta & \frac{b}{k} \\ 0 & 0 & \frac{1}{k} \end{bmatrix}$$

performs a rotation by θ, a scaling by k, and a translation through (a, b) when it left-multiplies a matrix of points given in homogeneous coordinates. ■

Your intuition may suggest that rotation and scaling are independent and can be performed in either order. This can be demonstrated mathematically by noting that the transformation matrices for rotation and for scaling are commutative with respect to multiplication, as shown below.

$$\begin{bmatrix} \cos\theta & -\sin\theta & 0 \\ \sin\theta & \cos\theta & 0 \\ 0 & 0 & 1 \end{bmatrix} \begin{bmatrix} 1 & 0 & 0 \\ 0 & 1 & 0 \\ 0 & 0 & \frac{1}{k} \end{bmatrix} = \begin{bmatrix} \cos\theta & -\sin\theta & 0 \\ \sin\theta & \cos\theta & 0 \\ 0 & 0 & \frac{1}{k} \end{bmatrix}$$

$$\begin{bmatrix} 1 & 0 & 0 \\ 0 & 1 & 0 \\ 0 & 0 & \frac{1}{k} \end{bmatrix} \begin{bmatrix} \cos\theta & -\sin\theta & 0 \\ \sin\theta & \cos\theta & 0 \\ 0 & 0 & 1 \end{bmatrix} = \begin{bmatrix} \cos\theta & -\sin\theta & 0 \\ \sin\theta & \cos\theta & 0 \\ 0 & 0 & \frac{1}{k} \end{bmatrix}$$

Homogeneous coordinates provide a unified, compact notation for transformations. Translation, scaling, and rotation can all be represented with matrix multiplication. To accomplish this, the matrix associated with the last transformation is positioned farthest to the left.

Exercise Set 7.4

Determine the matrix or sequence of matrices that can be multiplied to perform the following transformations. Use these matrices to find the new coordinates under each transformation of the vertices $A(2, 1)$, $B(1.5, 2)$, and $C(3, 4)$ of $\triangle ABC$.

1. Reflect over the x-axis.

2. Reflect over the y-axis.

3. Triple in size, rotate $75°$, and then translate through $(-3, -2)$.

4. Translate through $(-2, -1)$, rotate $-45°$, and then scale by a factor of $\frac{1}{2}$.

SECTION 7.5 | The Leontief Input-Output Model and the Inverse of a Matrix

We have seen that matrices are useful for organizing and manipulating data and that the arithmetic of matrices makes sense in light of their applications. Another example of the use of matrices is the Leontief Input-Output Model of an economy. In the April 1965 issue of *Scientific American*, the economist Wassily Leontief explained his input-output system using the 1958 American economy. He divided the economy into 81 sectors grouped into 6 families, viewing the economy as an 81×81 matrix. For his work, Leontief won the 1973 Nobel Prize for economics, and his model is now used worldwide. We will use a simplified version of his model to demonstrate an important application of matrices.

Leontief divided the economy into 81 sectors (transportation, manufacturing, steel, utilities, etc.), each of which relies on input resources taken from the output of other sectors. For example, the steel industry uses outputs of the utilities, heavy manufacturing, and transportation sectors, and even some steel, as inputs in its production. Therefore, not all of the steel manufactured is available to meet consumer demand. To develop the details of the Leontief model, we will examine a simplified economy that has only three sectors: agriculture, manufacturing, and transportation. The model hinges on the fact that some of the output from each sector is used in the production process so that not all output is available to meet consumer demand.

Suppose the following technology information has been gathered by a research team:

- Production of a unit of output of agriculture requires inputs consisting of $\frac{1}{10}$ unit of agriculture, $\frac{1}{5}$ unit of manufacturing, and $\frac{1}{5}$ unit of transportation. (A unit here refers to the value of a unit of output from the agriculture sector. Generally, a unit is assigned some fixed monetary equivalence.)

- Production of a unit of output of manufacturing requires inputs consisting of $\frac{1}{15}$ unit of agriculture, $\frac{1}{4}$ unit of manufacturing, and $\frac{1}{5}$ unit of transportation. (A unit here refers to the value of a unit of output from the manufacturing sector.)

- Production of a unit of output of transportation requires inputs consisting of no agriculture, $\frac{1}{4}$ unit of manufacturing, and $\frac{1}{6}$ unit of transportation. (A unit here refers to the value of a unit of output from the transportation sector.)

state diagram

The diagram in Figure 7, called a *state diagram*, illustrates the flow of resources from one sector to another. For example, the arrow from manufacturing to transportation is assigned the number $\frac{1}{4}$, indicating that the production of one unit of output of transportation requires the input of $\frac{1}{4}$ unit of manufacturing. The arrow from agriculture to agriculture is assigned the number $\frac{1}{10}$, indicating that each unit of output of agriculture requires the input of $\frac{1}{10}$ unit of agriculture. In general, an arrow from sector i to sector j is assigned a number indicating the units of input required from sector i to produce one unit of output from sector j.

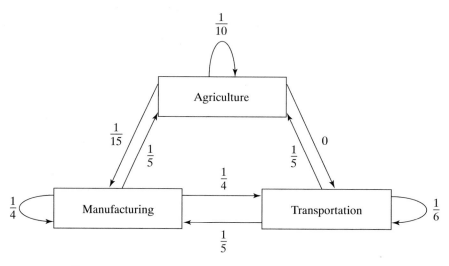

Figure 7 State diagram for a three-sector economy

technology matrix

The matrix T reflecting this information, called the *technology matrix*, is shown below.

$$
\begin{array}{c}
 \\
T = \begin{array}{c} \textit{Agriculture} \\ \textit{Manufacturing} \\ \textit{Transportation} \end{array}
\end{array}
\begin{array}{c}
\textit{Agr. Manu. Tran.} \\
\begin{bmatrix} \frac{1}{10} & \frac{1}{15} & 0 \\[4pt] \frac{1}{5} & \frac{1}{4} & \frac{1}{4} \\[4pt] \frac{1}{5} & \frac{1}{5} & \frac{1}{6} \end{bmatrix}
\end{array}
$$

For the sake of uniformity, the units of output for each sector are usually measured in dollars rather than physical units such as tons. Assuming a unit equals one dollar, we observe from matrix T that to produce one dollar's worth of manufactured goods requires about 7 cents worth (approximately $\frac{1}{15}$ dollar) of agriculture, 25 cents worth of manufacturing, and 20 cents worth of transportation. In general, a column of the technology matrix gives the fraction of one dollar's worth of input from each sector needed to produce one dollar's worth of output in the sector represented by that column.

It is conventional, but not essential, for the column headings to represent output and for the row labels to represent input, as in matrix T. The interpretations of rows and columns could be exchanged, but if we did so, the matrix operations that we are about to describe would need to be modified to achieve correct results.

Suppose we know that the economy produces 100 million dollars worth of agriculture, 120 million dollars worth of manufacturing, and 120 million dollars worth of transportation. This information is displayed in the *production matrix P* shown below. Note that in the production matrix each entry represents millions of dollars.

production matrix

$$
P = \begin{array}{c} \textit{Agriculture} \\ \textit{Manufacturing} \\ \textit{Transportation} \end{array}
\begin{array}{c}
\textit{Production} \\
\begin{bmatrix} 100 \\ 120 \\ 120 \end{bmatrix}
\end{array}
$$

Since the output of each sector requires input produced in the other sectors, the total production is not available for the demands of consumers. Some of this production is consumed internally by the economy. How much of each sector's output must be used to achieve the production levels given in matrix P? By determining which resources are used in production, we will be able also to determine what amount remains for consumers. We will answer this question for each of the three sectors.

First, how many units of agriculture are used in the production of all three sectors? Each unit of agricultural output requires 0.1 unit of agriculture as input; therefore, the 100 units of agriculture specified in P require $(\frac{1}{10})(100)$ units of agriculture as input. Likewise, 120 units of manufacturing output require $(\frac{1}{15})(120)$ units of agricultural input. The 120 units of transportation output require $(0)(120)$ units of agricultural input. The total input, in millions of dollars, of agriculture necessary to meet the production levels given in P is

$$\left(\frac{1}{10}\right)(100) + \left(\frac{1}{15}\right)(120) + (0)(120) = 18.$$

How many units of manufacturing are used to produce at a level given by P? By the same reasoning as above, the total input, in millions of dollars, of manufacturing required is

$$\left(\frac{1}{5}\right)(100) + \left(\frac{1}{4}\right)(120) + \left(\frac{1}{4}\right)(120) = 80.$$

Finally, how many units, in millions of dollars, of transportation are needed to achieve the production given by matrix P? Using the information in the technology matrix, we find that we require

$$\left(\frac{1}{5}\right)(100) + \left(\frac{1}{5}\right)(120) + \left(\frac{1}{6}\right)(120) = 64.$$

In the calculations above, notice that the second factors in the products are the numbers found in P. The first factors in the expression for agriculture come from the agriculture row in the technology matrix. Likewise, the first factors in the expression for manufacturing come from the manufacturing row, and the first factors for transportation come from the transportation row. Clearly, the concept of matrix multiplication is at work in these calculations. In fact, the matrix product TP has entries corresponding to the calculations above. Specifically,

$$TP = \begin{bmatrix} \frac{1}{10} & \frac{1}{15} & 0 \\ \frac{1}{5} & \frac{1}{4} & \frac{1}{4} \\ \frac{1}{5} & \frac{1}{5} & \frac{1}{6} \end{bmatrix} \begin{bmatrix} 100 \\ 120 \\ 120 \end{bmatrix} = \begin{bmatrix} 18 \\ 80 \\ 64 \end{bmatrix}.$$

In general, if T is a technology matrix and P is a production matrix, then TP is a matrix that represents the amount of the output consumed by the system internally. The matrix product TP is called the *internal consumption matrix* and shows the amount of each sector's output needed as input by the other sectors to meet production goals.

How much output is left to meet the demands of consumers after the input requirements of each sector are met? This can be represented by the *demand matrix* D, which is the difference between production and internal consumption. In symbols, D is defined as

$$D = P - TP.$$

Using the numbers provided above gives

$$D = P - TP$$

$$= \begin{bmatrix} 100 \\ 120 \\ 120 \end{bmatrix} - \begin{bmatrix} 18 \\ 80 \\ 64 \end{bmatrix} = \begin{bmatrix} 82 \\ 40 \\ 56 \end{bmatrix}.$$

The amounts left for distribution to consumers are 82 units of agriculture, 40 units of manufacturing, and 56 units of transportation, where a unit represents one million dollars.

Viewing this problem from a different perspective, we see that to have 82 units of agriculture available for consumer demand requires the production of 100 units of agriculture. Furthermore, to have 40 units of manufacturing available for demand, 120 units of manufacturing must be produced. Finally, to meet a demand of 56 units of transportation requires the production of 120 units of transportation.

We have demonstrated a method for determining the quantity of resources available for consumer demand given a certain level of production. Generally, however, professional analysts will estimate society's demand for specific goods and services. The question usually investigated is the following: Given a certain demand by society, how much should each sector of the economy produce to meet this demand? To produce less than the estimated demand may cause shortages and hardships, while producing more than the estimated demand leads to waste and inefficiency.

From the discussion above, we have $D = P - TP$. Since $P = IP$, where I is an identity matrix of the same dimension as T, this equation can be rewritten as

$$D = (I - T)P.$$

How can we solve a
matrix equation for an
unknown matrix?

Because matrix multiplication is not commutative, P must be factored out to the right, since it is to the right of T. In this equation, we know the entries in the matrices D and $(I - T)$, and we wish to determine the matrix P. How can we solve a matrix equation for an unknown matrix?

The analogous situation with numbers is the equation

$$ax = b,$$

in which the values of a and b are known. To determine the value of x, we multiply both sides of the equation by the *multiplicative inverse* of a, namely $\frac{1}{a}$. This strategy transforms the equation as follows:

$$\frac{1}{a} \cdot a \cdot x = \frac{1}{a} \cdot b$$

$$\left(\frac{1}{a} \cdot a\right) \cdot x = \frac{1}{a} \cdot b$$

$$1 \cdot x = \frac{b}{a}$$

$$x = \frac{b}{a}.$$

multiplicative inverse

The strategy for solving a matrix equation is similar. To isolate P on one side of the equation $D = (I - T)P$, we need to multiply both sides of the equation by the *multiplicative inverse* of $(I - T)$. The inverse of $(I - T)$ is a matrix that when multiplied by the matrix $(I - T)$ yields an identity matrix. After left-multiplying both sides of the equation $D = (I - T)P$ by the inverse of $(I - T)$, the right side of the equation is the product of an identity matrix and P, or simply P. This process isolates P so that economic production can be determined for a given amount of consumer demand.

Before going further with the problem of determining P, we need a method for finding the inverse of a matrix.

Finding the Inverse of a Matrix

The inverse of a square matrix R is a matrix S such that the product of R and S is an identity matrix. For example, if

$$R = \begin{bmatrix} 1 & -1 \\ 2 & 0 \end{bmatrix} \quad \text{and} \quad S = \begin{bmatrix} 0 & \frac{1}{2} \\ -1 & \frac{1}{2} \end{bmatrix},$$

then

$$RS = I$$

since

$$\begin{bmatrix} 1 & -1 \\ 2 & 0 \end{bmatrix} \begin{bmatrix} 0 & \frac{1}{2} \\ -1 & \frac{1}{2} \end{bmatrix} = \begin{bmatrix} 1 & 0 \\ 0 & 1 \end{bmatrix}.$$

also true that $SR = I$. Since $RS = SR = I$, the matrices R and S are inverses of ach other. The inverse of R is symbolized by R^{-1}, so

$$R^{-1} = S$$

and

$$S^{-1} = R.$$

Given the matrix A shown below, what is the inverse of A?

$$A = \begin{bmatrix} a_{11} & a_{12} & a_{13} \\ a_{21} & a_{22} & a_{23} \\ a_{31} & a_{32} & a_{33} \end{bmatrix}$$

We wish to find a matrix such that when it is multiplied by A, the product is an identity matrix. In other words, what is the matrix A^{-1} such that $A^{-1}A = I$?

In approaching this problem, think of the entries in A as the coefficients in a system of linear equations:

$$a_{11}x_1 + a_{12}x_2 + a_{13}x_3 = c_1$$
$$a_{21}x_1 + a_{22}x_2 + a_{23}x_3 = c_2.$$
$$a_{31}x_1 + a_{32}x_2 + a_{33}x_3 = c_3$$

A common procedure for finding the solution of a system involves transforming the system using the following operations:

- Interchange the positions of two equations.

- Multiply an equation by a nonzero constant.

- Replace an equation with the sum of the equation and a multiple of another equation.

Each of these operations produces a system with a solution identical to that of the original system. Eventually, a system results whose solution can be determined by inspection. A similar strategy can be applied to the problem of finding the inverse of a matrix.

The system of equations

$$a_{11}x_1 + a_{12}x_2 + a_{13}x_3 = c_1$$
$$a_{21}x_1 + a_{22}x_2 + a_{23}x_3 = c_2$$
$$a_{31}x_1 + a_{32}x_2 + a_{33}x_3 = c_3$$

can be expressed as the matrix equation

$$\begin{bmatrix} a_{11} & a_{12} & a_{13} \\ a_{21} & a_{22} & a_{23} \\ a_{31} & a_{32} & a_{33} \end{bmatrix} \begin{bmatrix} x_1 \\ x_2 \\ x_3 \end{bmatrix} = \begin{bmatrix} c_1 \\ c_2 \\ c_3 \end{bmatrix}.$$

If we let

$$X = \begin{bmatrix} x_1 \\ x_2 \\ x_3 \end{bmatrix} \quad \text{and} \quad C = \begin{bmatrix} c_1 \\ c_2 \\ c_3 \end{bmatrix}$$

then the matrix equation above is equivalent to

$$AX = C.$$

There are matrix operations analogous to the three transformations for systems of equations. These can be used to find the inverse of a matrix and are called *elementary row operations* (EROs). The three EROs are as follows:

elementary row operations

- Interchange two rows.

- Multiply one row by a nonzero constant.

- Replace a row with the sum of the row and a multiple of another row.

Suppose we perform EROs on the coefficient matrix A to produce $I_{3 \times 3}$. If we perform the same EROs on $I_{3 \times 3}$, the result will be A^{-1}. In other words, if we can transform A into I, then the same operations will transform I into A^{-1}.

EXAMPLE 1   Use EROs to determine the inverse of matrix A given by

$$A = \begin{bmatrix} 2 & 1 & 0 \\ 1 & 0 & 2 \\ 0 & 1 & -1 \end{bmatrix}.$$

Solution   The goal is to transform A using EROs until we get a 3×3 identity matrix. The inverse of A results from performing the same EROs on $I_{3 \times 3}$, so we write $I_{3 \times 3}$ alongside A. This form is called an *augmented matrix* and is

augmented matrix

$$\left[\begin{array}{ccc|ccc} 2 & 1 & 0 & 1 & 0 & 0 \\ 1 & 0 & 2 & 0 & 1 & 0 \\ 0 & 1 & -1 & 0 & 0 & 1 \end{array} \right].$$

We begin the sequence of EROs by interchanging rows 1 and 2 so that a 1 is in the upper left corner, which results in the new augmented matrix

$$\left[\begin{array}{ccc|ccc} 1 & 0 & 2 & 0 & 1 & 0 \\ 2 & 1 & 0 & 1 & 0 & 0 \\ 0 & 1 & -1 & 0 & 0 & 1 \end{array}\right].$$

Multiply the first row by -2 and add the result to row 2. Replace row 2 with the sum, which yields

$$\left[\begin{array}{ccc|ccc} 1 & 0 & 2 & 0 & 1 & 0 \\ 0 & 1 & -4 & 1 & -2 & 0 \\ 0 & 1 & -1 & 0 & 0 & 1 \end{array}\right].$$

We are finished transforming the first column of A since it is identical to the first column of $I_{3\times3}$. Multiply the second row by -1 and add the result to the third row, which finishes the second column and results in

$$\left[\begin{array}{ccc|ccc} 1 & 0 & 2 & 0 & 1 & 0 \\ 0 & 1 & -4 & 1 & -2 & 0 \\ 0 & 0 & 3 & -1 & 2 & 1 \end{array}\right].$$

To transform the third column, multiply the third row by $\frac{1}{3}$, yielding

$$\left[\begin{array}{ccc|ccc} 1 & 0 & 2 & 0 & 1 & 0 \\ 0 & 1 & -4 & 1 & -2 & 0 \\ 0 & 0 & 1 & -\frac{1}{3} & \frac{2}{3} & \frac{1}{3} \end{array}\right].$$

Multiply row 3 by 4 and add the result to row 2, and multiply row 3 by -2 and add the result to row 1, yielding the matrix

$$\left[\begin{array}{ccc|ccc} 1 & 0 & 0 & \frac{2}{3} & -\frac{1}{3} & -\frac{2}{3} \\ 0 & 1 & 0 & -\frac{1}{3} & \frac{2}{3} & \frac{4}{3} \\ 0 & 0 & 1 & -\frac{1}{3} & \frac{2}{3} & \frac{1}{3} \end{array}\right].$$

The left half of the augmented matrix is an identity matrix, so the process is now complete, and the inverse of A is the right half of the augmented matrix, or

$$A^{-1} = \left[\begin{array}{ccc} \frac{2}{3} & -\frac{1}{3} & -\frac{2}{3} \\ -\frac{1}{3} & \frac{2}{3} & \frac{4}{3} \\ -\frac{1}{3} & \frac{2}{3} & \frac{1}{3} \end{array}\right].$$

Calculating the product of A^{-1} and A verifies that $A^{-1}A = I$. ▪

1. Use EROs to find A^{-1} for matrix A.

$$A = \begin{bmatrix} 1 & 2 & 3 \\ 1 & 1 & 0 \\ 3 & 1 & -1 \end{bmatrix}$$

2. Use EROs to find B^{-1} for matrix B.

$$B = \begin{bmatrix} 1 & 1 & 1 \\ 3 & 5 & 2 \\ 0 & 0 & 1 \end{bmatrix}$$

Do All Matrices Have Inverses?

If B is the inverse of A, then A is the inverse of B, which means that

$$AB = BA = I.$$

Inverses can be multiplied in either of the two orders, and the result will be an identity matrix. An implication of this property is that only square matrices have inverses.

invertible

Are all square matrices invertible?

coefficient matrix

If a matrix has an inverse, it is said to be *invertible*. Are all square matrices invertible? If not, what are the general conditions for a square matrix A to have an inverse? Our method for finding an inverse involves using EROs to reduce matrix A to an identity matrix. If this is possible, then the same sequence of EROs will transform the identity matrix into A^{-1}. Recall that EROs were introduced as analogies of operations used to solve a system of linear equations. Suppose A is the *coefficient matrix* for a system of equations so that the matrix equation $AX = C$ represents the system. If A is invertible, then the system can be solved with EROs. Therefore, the existence of A^{-1} is equivalent to the existence of a solution for the matrix equation $AX = C$.

The equations in the 2×2 system

$$2x_1 + x_2 = 3$$
$$4x_1 + 2x_2 = 8$$

inconsistent system

describe parallel lines because the coefficients of x_1 and x_2 in the second equation are twice the coefficients in the first equation, but the constant terms are not in a 2-to-1 ratio. This system has no solution. A system of linear equations that has no solution is called an *inconsistent system*.

The equations in the 2×2 system

$$3x_1 + x_2 = 5$$

$$9x_1 + 3x_2 = 15$$

dependent system

both describe the same line because the coefficients and the constant term of the second equation are 3 times the corresponding numbers in the first equation. The system has an infinite number of solutions. A system of linear equations that has an infinite number of solutions is called a *dependent system*.

consistent system

If a linear system does not have a unique solution, then the matrix equation $AX = C$ that represents the system cannot be transformed to $X = A^{-1}C$. Therefore, the coefficient matrix of an inconsistent or dependent system does not have an inverse. On the other hand, if a linear system represented by $AX = C$ does have a unique solution, the system is called a *consistent system* and A is invertible.

linear combination

To find the inverse of a matrix, elementary row operations are used to transform A. If at any point in the process two rows of A are found to be equal, then A is not invertible. This means that if any row of a matrix can be expressed as a sum of multiples of the other rows, then that matrix does not have an inverse. Adding multiples of rows is called *linear combination*. The property of a matrix that determines if it is invertible can be stated as follows: *A square matrix is invertible if and only if no row can be expressed as a linear combination of the other rows.*

The system

$$2x_1 + x_2 = 3,$$

$$4x_1 + 2x_2 = 8$$

has a coefficient matrix

$$\begin{bmatrix} 2 & 1 \\ 4 & 2 \end{bmatrix}.$$

This matrix is not invertible because the second row is twice the first row.

In the matrix

$$\begin{bmatrix} 5 & -3 & 0 \\ 1 & 2 & -2 \\ 9 & -8 & 2 \end{bmatrix},$$

note that the third row is equal to twice the first row minus the second row, so this matrix is not invertible.

Leontief's Model Revisited

Recall that Leontief's Input-Output Model uses the relationship between consumer demand (D), the economy's production (P), and the input-output interaction between sectors represented by the technology matrix (T) that can be expressed as

$$D = P - TP$$

or

$$D = (I - T)P.$$

Each entry in the technology matrix T represents the output of one sector utilized as input by another sector. More specifically, the entry in row i, column j of T is the amount of goods and services needed as input from sector i for the production of one unit of the output resources of sector j. Although the technology matrix changes over time, the entries can be determined through research for a certain time period. The technology matrix from the beginning of this section is rewritten below with approximate decimal values.

$$
T = \begin{array}{c} \\ Agriculture \\ Manufacturing \\ Transportation \end{array}
\begin{array}{c} \begin{array}{ccc} Agri. & Manu. & Trans. \end{array} \\
\left[\begin{array}{ccc} 0.1 & 0.067 & 0 \\ 0.2 & 0.25 & 0.25 \\ 0.2 & 0.2 & 0.167 \end{array} \right] \end{array}
$$

The entries of the demand matrix D are often estimated through market research. We wish to determine a matrix P such that $P - TP$ will yield the given entries of D.

To solve for the matrix P, left-multiply both sides of the equation $D = (I - T)P$ by the inverse of $(I - T)$ to get

$$(I - T)^{-1} D = (I - T)^{-1}(I - T)P,$$

which simplifies to

$$(I - T)^{-1}D = IP$$

or

$$(I - T)^{-1}D = P.$$

Once again, it is important to maintain the correct order of multiplication on both sides of the equation.

Suppose the consumer demand is for 100 units of agriculture, 120 units of manufacturing, and 90 units of transportation, where each unit represents one million dollars worth of a sector's goods. This is displayed in a demand matrix as

$$
D = \begin{array}{c} \\ Agriculture \\ Manufacturing \\ Transportation \end{array}
\begin{array}{c} \begin{array}{c} Demand \end{array} \\
\left[\begin{array}{c} 100 \\ 120 \\ 90 \end{array} \right] \end{array} .
$$

Given this level of consumer demand, we wish to determine the level of production necessary to satisfy this demand as well as the amount of internal consumption given by the product TP.

The first step in calculating the matrix P is to determine $(I - T)^{-1}$. We start with

$$I - T \approx \begin{bmatrix} 1 & 0 & 0 \\ 0 & 1 & 0 \\ 0 & 0 & 1 \end{bmatrix} - \begin{bmatrix} 0.1 & 0.067 & 0 \\ 0.2 & 0.25 & 0.25 \\ 0.2 & 0.2 & 0.167 \end{bmatrix}$$

$$\approx \begin{bmatrix} 0.9 & -0.067 & 0 \\ -0.2 & 0.75 & -0.25 \\ -0.2 & -0.2 & 0.833 \end{bmatrix}.$$

Using EROs or the inverse operation on a calculator, we find that

$$(I - T)^{-1} \approx \begin{bmatrix} 1.14 & 0.11 & 0.03 \\ 0.43 & 1.49 & 0.45 \\ 0.38 & 0.38 & 1.32 \end{bmatrix}.$$

Now we can calculate P as follows:

$$P = (I - T)^{-1}D$$

$$= \begin{bmatrix} 1.14 & 0.11 & 0.03 \\ 0.43 & 1.49 & 0.45 \\ 0.38 & 0.38 & 1.32 \end{bmatrix} \begin{bmatrix} 100 \\ 120 \\ 90 \end{bmatrix}$$

$$\approx \begin{matrix} Agriculture \\ Manufacturing \\ Transportation \end{matrix} \begin{bmatrix} 130.5 \\ 262.2 \\ 202.3 \end{bmatrix}.$$

The matrix P was computed using more decimal places but is presented rounded to tenths. The economy must produce approximately 130.5 million dollars worth of agriculture, 262.2 million dollars worth of manufacturing, and 202.3 million dollars worth of transportation to satisfy consumer demand after the requirements of internal consumption are met. The internal consumption is given by $TP = P - D$, which is

$$TP = P - D = \begin{matrix} Agriculture \\ Manufacturing \\ Transportation \end{matrix} \begin{bmatrix} 30.5 \\ 142.2 \\ 112.3 \end{bmatrix}.$$

1. Consider a four-sector economic system consisting of petroleum, textiles, transportation, and chemicals. The production of one unit of petroleum requires 0.2 unit of transportation, 0.4 unit of chemicals, and 0.1 unit of itself. The production of one unit of textiles requires 0.4 unit of petroleum, 0.1 unit of textiles, 0.15 unit of transportation, and 0.3 unit of chemicals. The production of one unit of transportation requires 0.6 unit of petroleum, 0.1 unit of itself, and 0.25 unit of chemicals. Finally, the production of one unit of chemicals requires 0.2 unit of petroleum, 0.1 unit of textiles, 0.3 unit of transportation, and 0.2 unit of chemicals.

   a. Write a technology matrix to represent this information.

   b. On which sector is petroleum most dependent? least dependent?

   c. If the textiles sector has an output of $4 million, what is the input in dollars from petroleum?

   d. Suppose the production matrix is

   $$P = \begin{array}{r} \textit{Petroleum} \\ \textit{Textiles} \\ \textit{Transportation} \\ \textit{Chemicals} \end{array} \begin{bmatrix} 800 \\ 200 \\ 700 \\ 750 \end{bmatrix}.$$

   What is the internal consumption matrix? How much petroleum is leftover for external use?

   e. Suppose the demand matrix, in millions of dollars, is

   $$D = \begin{array}{r} \textit{Petroleum} \\ \textit{Textiles} \\ \textit{Transportation} \\ \textit{Chemicals} \end{array} \begin{bmatrix} 25 \\ 14 \\ 30 \\ 42 \end{bmatrix}.$$

   How much in each sector must be produced?

2. Suppose the demand matrix given in Exercise 1, Part e, is doubled. What is the new production matrix? How does the new production matrix compare with the original production matrix?

3. An economy with the four sectors manufacturing, petroleum, transportation, and hydroelectric power has the following technology matrix:

$$T = \begin{array}{c} \\ \textit{Manufacturing} \\ \textit{Petroleum} \\ \textit{Transportation} \\ \textit{Hydroelectric} \\ \textit{Power} \end{array} \begin{array}{cccc} \textit{Manu.} & \textit{Petr.} & \textit{Tran.} & \textit{HP} \\ \left[\begin{array}{cccc} 0.15 & 0.18 & 0.3 & 0.1 \\ 0.22 & 0.12 & 0.37 & 0 \\ 0.09 & 0.3 & 0.11 & 0 \\ 0.27 & 0.05 & 0.07 & 0.1 \end{array}\right] \end{array}$$

Find the production matrix if all the entries of the demand matrix are 200.

4. Consider a three-sector system consisting of steel, coal, and transportation. The production of one unit of coal requires 0.23 unit of transportation, 0.19 unit of itself, and 0.2 unit of steel. The production of one unit of steel requires 0.2 unit of itself, 0.3 unit of coal, and 0.15 unit of transportation. Finally, producing one unit of transportation requires 0.1 unit of itself, 0.15 unit of coal, and 0.35 unit of steel.

    a. Write a technology matrix to represent this information.

    b. On which sector does coal rely most? rely least?

    c. Which sector depends most on steel?

    d. Suppose the production matrix is

$$P = \begin{array}{c} \textit{Steel} \\ \textit{Coal} \\ \textit{Transportation} \end{array} \left[\begin{array}{c} 18 \\ 23 \\ 15 \end{array}\right].$$

        What is the surplus available beyond internal consumption?

    e. Suppose the demand matrix is

$$D = \begin{array}{c} \textit{Steel} \\ \textit{Coal} \\ \textit{Transportation} \end{array} \left[\begin{array}{c} 24 \\ 19 \\ 12 \end{array}\right].$$

        Find the production matrix.

SECTION 7.6 Additional Applications of the Inverse of a Matrix

This section includes an example and exercises to illustrate other applications of the inverse of a matrix.

EXAMPLE 1 McDougal's Restaurant sponsors special funding for three projects: scholarships for employees, special public service projects, and beautification of the exteriors of the restaurants. Each of the three locations of McDougal's in the Durham area, Hillsborough, Northgate, and Boulevard, made requests for funds, with the relative amounts requested by each location for the three projects distributed as shown in Figure 8. Headquarters decided to allocate $100,000 for these projects to the Durham area. The money was to be distributed with 43% to scholarships, 28% to public service projects, and the remaining 29% to beautification. How much will each of the three locations receive?

Figure 8 Distribution of funding requests

Project	Hillsborough (H)	Northgate (N)	Boulevard (B)
Scholarships	50%	30%	40%
Public service	20%	30%	40%
Beautification	30%	40%	20%

Solution Assign the variables x, y, and z to the amounts that each of the three locations will receive. A 3×3 system of equations can be written and solved using matrices. Make the following variable assignments:

$$x = \text{the amount of money for Hillsborough,}$$

$$y = \text{the amount of money for Northgate,}$$

$$z = \text{the amount of money for Boulevard.}$$

Information about the money allocated for the three projects can now be written as the following system of linear equations:

Scholarships:	$0.5x + 0.3y + 0.4z = \$43{,}000$
Public Service:	$0.2x + 0.3y + 0.4z = \$28{,}000$
Beautification:	$0.3x + 0.4y + 0.2z = \$29{,}000.$

A matrix equation for this system is shown below.

$$
\begin{array}{c}
\text{Scholarships} \\
\text{Public Service} \\
\text{Beautification}
\end{array}
\begin{bmatrix}
0.5 & 0.3 & 0.4 \\
0.2 & 0.3 & 0.4 \\
0.3 & 0.4 & 0.2
\end{bmatrix}
\times
\begin{array}{c} H \\ N \\ B \end{array}
\begin{bmatrix} x \\ y \\ z \end{bmatrix}
=
\begin{array}{c}
\text{Scholarships} \\
\text{Public Service} \\
\text{Beautification}
\end{array}
\begin{bmatrix} 43{,}000 \\ 28{,}000 \\ 29{,}000 \end{bmatrix}
$$

This matrix equation is of the form $AX = B$, so we can solve for X by finding A^{-1} and left-multiplying on both sides, giving $X = A^{-1}B$. Using a computer or calculator to find the inverse of the coefficient matrix gives

$$\begin{bmatrix} x \\ y \\ z \end{bmatrix} \approx \begin{bmatrix} 3.333 & -3.333 & 0 \\ -2.667 & 0.667 & 4 \\ 0.333 & 3.667 & -3 \end{bmatrix} \begin{bmatrix} 43{,}000 \\ 28{,}000 \\ 29{,}000 \end{bmatrix}.$$

After performing this multiplication, we find that

$$\begin{bmatrix} x \\ y \\ z \end{bmatrix} \approx \begin{bmatrix} 50{,}000 \\ 20{,}000 \\ 30{,}000 \end{bmatrix}.$$

The $100,000 should be distributed so that $50,000 goes to the Hillsborough location, $20,000 goes to the Northgate location, and $30,000 goes to the Boulevard location. ▨

For the preceding example, we did not actually need to know the entries in A^{-1}. All we were really interested in was the result of the multiplication $A^{-1}B$. Recall that we can find A^{-1} by performing EROs on the augmented matrix $[A \,|\, I]$. If we perform the same EROs on B, the result will be $A^{-1}B$. Therefore, we can actually find the product $A^{-1}B$ by performing the same EROs on $[A \,|\, B]$ that are performed on $[A \,|\, I]$ in finding A^{-1}. In the example above, this is accomplished by first forming the augmented matrix

$$\begin{bmatrix} 0.5 & 0.3 & 0.4 & | & 43{,}000 \\ 0.2 & 0.3 & 0.4 & | & 28{,}000 \\ 0.3 & 0.4 & 0.2 & | & 29{,}000 \end{bmatrix},$$

and then using EROs to transform the left three columns into an identity matrix. The numbers in the right-hand column will be the solution to the system of equations found in Example 1.

Exercise Set 7.6

1. Explain why a matrix that contains a column of all zeros cannot have an inverse.

2. Solve the following systems of equations.

   a. $2x - 5y + 7z = 4$
      $3x + y - 12z = -8$
      $5x + 2y - 4z = 3$

   b. $2x - 5y + 7z = 4$
      $3x + y - 12z = -8$
      $5x - 4y - 5z = -4$

   c. $2x - 5y + 7z = 4$
      $3x + y - 12z = -8$
      $7x - 9y + 2z = 1$

3. Find the inverses of the following matrices.

a.
$$\begin{bmatrix} 2 & -7 & 5 \\ 1 & -3 & -10 \\ 3 & 4 & -5 \end{bmatrix}$$

b.
$$\begin{bmatrix} 1 & 2 & 4 & 8 \\ 1 & 3 & 9 & 27 \\ 1 & 4 & 16 & 64 \\ 1 & 5 & 25 & 125 \end{bmatrix}$$

4. In Example 1 on page 509, suppose that the Hillsborough location of McDougal's changes its request and asks for 43% for scholarships, 28% for public service, and 29% for beautification, using the same allocation scheme as headquarters has. The other stores keep their original requests. How much money will each store receive? Could another location change its request to obtain more money?

5. In Exercise 10 in Section 7.3 on page 481, we found that the matrix below rotates a point (x, y) in the plane through an angle θ about the origin. Find the inverse of this matrix. What transformation does the inverse perform?

$$\begin{bmatrix} \cos\theta & -\sin\theta \\ \sin\theta & \cos\theta \end{bmatrix}$$

6. A total of $30,000 is available in a school for student groups to spend. The Student Council, the Beta Club, and the 4-H Club were asked to submit proposals describing how each would spend its portion of the $30,000. The principal accepted the proposals shown in Figure 9 for how each group would spend its allocation of the money.

Figure 9 Proposed allocation of funds

Student Group	Activities	Community Service	Club Expenses
Student Council	40%	40%	20%
Beta Club	20%	50%	30%
4-H Club	20%	30%	50%

a. The principal wants to allocate 30% of the funds to activities, 50% to community service, and 20% to club expenses. How much money would each club receive under these conditions? What problem results from this allocation?

b. Given the principal's allocation scheme in Part a, how much would each club receive if the Beta Club changed its proposal so that it would spend 30% on activities, 50% on community services, and 20% on club expenses?

c. The principal actually decides to allocate 25% of the funds to activities, 40% to community service, and 35% to club expenses. On the basis of the original proposals from each group, how much money does the principal allocate to each group?

The Leslie Matrix Model

Population growth is a significant phenomenon for which many mathematical models have been developed. A frequently used model is the exponential function

$$P(t) = P_0 e^{kt}$$

in which $P(t)$ is the size of a population growing without limits over time. Constrained growth of a population can be modeled by the recursive function

$$P_n = P_{n-1} + k \cdot P_{n-1}(M - P_{n-1}), \qquad n = 1, 2, 3, \ldots,$$

in which M is the maximum sustainable population and P_0 is the initial population. Both of these models are macromodels, meaning that the models consider the population as a whole. In this section, a micromodel is developed that allows us to investigate questions about different age groups within an entire population.

Modeling Age-Specific Population Growth

The future of social security, the future of veterans' benefits, and the changing school population in different regions of the country are current issues in public policy. The principal question arising in each of these discussions is how many people will be of a certain age after a specified period of time. The total population can be modeled with the equations shown above, but these macromodels provide little help in answering age-specific growth questions. We would like to be able to examine the growth and decline of future populations within various age groups. The model developed in this section will enable us to make these age-specific projections.

A fundamental assumption we will use in our model is that the proportion of males in the population is the same as the proportion of females, an assumption justified for most species. Thus, we need to consider only the number of females in the population. Consider a female population of small woodland mammals. Figure 10 gives this population for three-month age groups. The total population in each age group is assumed to be twice the female population. The life span of this mammal is assumed to be $15-18$ months, so none advance beyond the final column in Figure 10. Our primary task with these data is to derive a mathematical model that will allow us to predict the number of animals in each age group after some number of years. To proceed, we first need to know something about the birthrate and death rate for each age group, rates that vary with age for most animal populations.

Figure 10 **Female population of small woodland mammals**

Age (months)	0–3	3–6	6–9	9–12	12–15	15–18
Number of females	14	8	12	4	0	0

The birthrate depends on a variety of factors, such as the probability that an animal will become pregnant, the number of pregnancies that can occur in any age group, and the average number of newborns in a litter. For animals that bear only one young and require a long gestation period, the birthrate is low. However, for insects, fish, and other species that bear thousands of young at a time, the birthrate is high.

Generally, the birthrate is given as a proportion of the total population. For example, if an animal population of 100 has a birthrate of 0.4, then it is understood that the 100 animals will produce 40 newborns, or 0.4 of the population. If we consider only the female population, the 50 females of the 100 animals (half the total) will be reproducing at a rate of 0.8 of their population, since the 50 females will produce 40 young. Only 20 of these 40 young are expected to be female, however, so the birthrate of females will be 0.4 of the female population. In our model, the birthrate will represent the average number of daughters born to each female in the population during a specified time interval. Under the assumption of equal female and male proportions, this definition of birthrate is equivalent to viewing birthrate as a proportion of the total population.

For the species with the age distribution given in Figure 10, the birthrates and death rates by age are listed in Figure 11. To investigate the mammal population in each age group over time, we begin by considering the following question: After three months, how many females will there be in each age group?

Figure 11 **Birthrate and death rate for each age group**

Age (months)	0–3	3–6	6–9	9–12	12–15	15–18
Birthrate	0	0.3	0.8	0.7	0.4	0
Death rate	0.4	0.1	0.1	0.2	0.4	1

The populations of age groups 3–6, 6–9, 9–12, 12–15, and 15–18 are easily found. The calculations can be simplified by introducing a quantity called the survival rate (SR), which is equal to one minus the death rate (DR). In symbols, the relationship is

$$SR = 1 - DR.$$

Whereas the death rate indicates the proportion of a population group that dies during a three-month interval, the survival rate is the proportion of a population group that survives during this three-month period. The survival rate of the 0–3 age group is $1 - 0.4$, or 0.6, and the number of females from the original 14 that advance to the 3–6 age group after three months is

$$(0.6)(14) = 8.4.$$

The survival rate of the 3–6 age group is $1 - 0.1$, or 0.9, and of the original 8 mammals in this group, the number that advance to the 6–9 age group after three months is

$$(0.9)(8) = 7.2.$$

The calculations for the number of females moving up to the next age group after three months are summarized in Figure 12.

Figure 12   Movement of females through the age groups

Age	DR	SR	Number	Number moving up
0–3	0.4	0.6	14	(0.6)(14) or 8.4 move up to the 3–6 age group
3–6	0.1	0.9	8	(0.9)(8) or 7.2 move up to the 6–9 age group
6–9	0.1	0.9	12	(0.9)(12) or 10.8 move up to the 9–12 age group
9–12	0.2	0.8	4	(0.8)(4) or 3.2 move up to the 12–15 age group
12–15	0.4	0.6	0	(0.6)(0) or 0 move up to the 15–18 age group
15–18	1.0	0	0	no animal advances beyond the 15–18 age group

The numbers in Figure 12 may seem strange because the populations are not rounded to integers but contain fractional parts. These fractions of animals should remain in our analysis. The birthrates and survival rates used in the model are probabilistic quantities. They represent the probable rates for a given time, perhaps found by averaging data on a species for a number of years. We do not expect these rates to be exact at any one time, but we expect that over the long run they accurately reflect age-specific population growth. The fractional parts can make a significant difference in calculations over time, so they must be retained. To obtain an estimate of the population at a definite time, or a "snapshot" of the process, one would of course round off numbers to the nearest integer.

A question we have left unanswered is how many females enter the 0–3 age group? This is the sum of all the births in each of the other age groups. To find the births in each age group, multiply the birthrate by the population in that age group. Using the data from Figure 10, this sum is

$$14(0) + 8(0.3) + 12(0.8) + 4(0.7) + 0(0.4) + 0(0) = 14.8.$$

Consolidating this number with the information in Figure 12 gives us the female population in each age group after three months, as presented in Figure 13. Notice that the population has grown from 38 to about 44 animals after three months.

Figure 13   Female population after three months

Age (months)	0–3	3–6	6–9	9–12	12–15	15–18
Number of females	14.8	8.4	7.2	10.8	3.2	0

After another three-month period, how many females will be in each group? Sixty percent of the 14.8 females in the 0–3 age group, or 8.88 females, will survive to move into the 3–6 age group. Of the 8.4 females in the 3–6 age group, ninety percent (7.56) will survive to move into the 6–9 age bracket. Ninety percent of the 7.2 females in the 6–9 age bracket, or 6.48 females, will survive to move into the 9–12 age group, while eighty percent (8.64) of the 10.8 females in the 9–12 age group will survive. Of the 3.2 females in the 12–15 age group, sixty percent (1.92) will move into the 15–18 age group. The 0–3 age group will be populated by newborns from each group. The total moving into the 0–3 age group will be

$$14.8(0) + 8.4(0.3) + 7.2(0.8) + 10.8(0.7) + 3.2(0.4) + 0(0) = 17.12.$$

Figure 14 shows the number of animals in the female population after six months. The total population of females is now about 51, which implies, by the equal-proportions assumption, that the total animal population has reached about 102.

Figure 14   Female population after six months

Age (months)	0–3	3–6	6–9	9–12	12–15	15–18
Number of females	17.12	8.88	7.56	6.48	8.64	1.92

CLASS PRACTICE   **1.**   Find the total population of females after 9 months and after 12 months.

The Leslie Matrix

In the previous example, we saw an increase in the female population from 38 to about 44 during the initial three-month interval, followed by an increase to about 51 after six months. Further calculations reveal that the population remains approximately 51 or 52 for the next two intervals. During subsequent intervals, will the population remain stable at approximately 51, begin to grow again, or start to die out? We can answer this question by performing the same sequence of operations to calculate the population every three months. However, the matrix data structure provides an easier way to determine future age-specific population distributions.

The form of the preceding calculations provides clues for developing a matrix model of age-specific population growth. The sums of products that were used in the calculations can each be represented as a row of one matrix multiplied by a column of another matrix. We will develop a matrix representation for the data that determine the age distribution of a population from one interval to the next, namely the birthrates and survival rates. This matrix is called the *Leslie matrix* and will be symbolized by L. The survival rate for the kth age group is denoted by S_k. Similarly, the birthrate for the initial age group (0–3 months) is denoted by B_1, the next (3–6 months) by B_2, and, in general, the birthrate of the kth age group by B_k.

The number of females in each age group at a given time can be represented by a column vector called an *age distribution vector*. In our example, we have z

Leslie matrix

age distribution vector

$$X_0 = \begin{bmatrix} 14 \\ 8 \\ 12 \\ 4 \\ 0 \\ 0 \end{bmatrix} \quad \text{and} \quad X_1 = \begin{bmatrix} 14.8 \\ 8.4 \\ 7.2 \\ 10.8 \\ 3.2 \\ 0 \end{bmatrix},$$

where the subscript of X signifies the number of three-month intervals that have elapsed. X_0 was given, and X_1 was calculated based on X_0, the birthrates and the survival rates.

The first entry in X_1 was obtained by multiplying the values in X_0 by the birthrates for the n different age groups. If the first row of L is

$$[B_1 \quad B_2 \quad B_3 \quad B_4 \quad \cdots \quad B_n],$$

then multiplying the first row of L by the column vector X_0 will yield the first entry in X_1. The second entry in X_1 is the product of the survival rate for the first age group and the number of females in the first age group from the previous cycle, which is the first entry in X_0. If the second row of L is

$$[S_1 \quad 0 \quad 0 \quad 0 \quad \cdots \quad 0],$$

then multiplying this by X_0 will provide the second entry in X_1. The third entry in X_1 is the product of the survival rate for the second age group and the number of females in the second age group from the previous cycle, which is the second entry in X_0. If the third row of L is

$$[0 \quad S_2 \quad 0 \quad 0 \quad \cdots \quad 0],$$

then the result of multiplying this row by X_0 is the third entry in X_1.

This pattern is continued for the remaining rows of L; therefore, the Leslie matrix L is defined as

$$L = \begin{bmatrix} B_1 & B_2 & B_3 & B_4 & \cdots & B_{n-1} & B_n \\ S_1 & 0 & 0 & 0 & \cdots & 0 & 0 \\ 0 & S_2 & 0 & 0 & \cdots & 0 & 0 \\ 0 & 0 & S_3 & 0 & \cdots & 0 & 0 \\ \vdots & \vdots & \vdots & \vdots & & \vdots & \vdots \\ 0 & 0 & 0 & 0 & \cdots & S_{n-1} & 0 \end{bmatrix}.$$

For the woodland mammal example, we have

$$L = \begin{bmatrix} 0 & 0.3 & 0.8 & 0.7 & 0.4 & 0 \\ 0.6 & 0 & 0 & 0 & 0 & 0 \\ 0 & 0.9 & 0 & 0 & 0 & 0 \\ 0 & 0 & 0.9 & 0 & 0 & 0 \\ 0 & 0 & 0 & 0.8 & 0 & 0 \\ 0 & 0 & 0 & 0 & 0.6 & 0 \end{bmatrix}.$$

Why have we chosen the definition of the Leslie matrix explained above? To see the rationale for the definition, examine the product LX_0 shown below.

$$LX_0 = \begin{bmatrix} 0 & 0.3 & 0.8 & 0.7 & 0.4 & 0 \\ 0.6 & 0 & 0 & 0 & 0 & 0 \\ 0 & 0.9 & 0 & 0 & 0 & 0 \\ 0 & 0 & 0.9 & 0 & 0 & 0 \\ 0 & 0 & 0 & 0.8 & 0 & 0 \\ 0 & 0 & 0 & 0 & 0.6 & 0 \end{bmatrix} \begin{bmatrix} 14 \\ 8 \\ 12 \\ 4 \\ 0 \\ 0 \end{bmatrix}$$

$$= \begin{bmatrix} 0(14) + 0.3(8) + 0.8(12) + 0.7(4) + 0.4(0) + 0(0) \\ 0.6(14) + 0(8) + 0(12) + 0(4) + 0(0) + 0(0) \\ 0(14) + 0.9(8) + 0(12) + 0(4) + 0(0) + 0(0) \\ 0(14) + 0(8) + 0.9(12) + 0(4) + 0(0) + 0(0) \\ 0(14) + 0(8) + 0(12) + 0.8(4) + 0(0) + 0(0) \\ 0(14) + 0(8) + 0(12) + 0(4) + 0.6(0) + 0(0) \end{bmatrix}$$

$$= \begin{bmatrix} 14.8 \\ 8.4 \\ 7.2 \\ 10.8 \\ 3.2 \\ 0 \end{bmatrix}$$

The last matrix above is just X_1. Thus, the structure of the Leslie matrix leads to the matrix equation

$$LX_0 = X_1.$$

CLASS PRACTICE

1. Verify that $LX_k = X_{k+1}$ for $k = 1, 2,$ and 3 by comparing these products with the values shown in the text and in the preceding Class Practice exercises.

2. Verify that $L^k X_0 = X_k$ for $k = 1, 2, 3,$ and 4 by comparing these products with the X_k vectors evaluated previously. L^k is defined as the matrix L raised to the kth power, or in other words, L taken as a factor k times.

The Leslie matrix has been defined so that left-multiplication by it will generate the age distribution vectors in the sequence $X_0, X_1, ..., X_{k-1}, X_k$, as shown in the first Class Practice Problem on page 517. Multiplication by L determines the next age distribution vector from the current vector, which is represented by the recursive equation

$$X_k = LX_{k-1}.$$

The equation $X_k = LX_{k-1}$ is an iterative model; however, we need not go through all the $k-1$ preceding vectors just to find the kth vector. Since we always left-multiply X_{k-1} by L to find X_k, the associative property of matrix multiplication implies that we can simply multiply L by itself k times (which is L^k) and then right-multiply the result by the initial age distribution vector X_0. This property was demonstrated in the second Class Practice Problem on page 517 and is shown in general in the calculations below.

We have observed that

$$X_1 = LX_0 \quad \text{and} \quad X_2 = LX_1.$$

By substituting the first expression for X_1 into the second equation, we find that

$$X_2 = L(LX_0) = L^2 X_0.$$

Similarly substituting the expression for X_2 into the equation

$$X_3 = LX_2$$

yields

$$X_3 = L(L^2 X_0) = L^3 X_0.$$

In general, we have the following sequence of equivalent expressions:

$$X_k = LX_{k-1} = L(LX_{k-2}) = L(L(L...(LX_0)))$$

$$= \overbrace{(L(L(L\cdots L)\cdots))}^{k \text{ factors}} X_0$$

$$= L^k X_0.$$

Therefore, an explicit equation for the kth age distribution vector in the Leslie model is

$$X_k = L^k X_0.$$

An interesting pattern emerges in the long-term behavior of a process modeled with the Leslie matrix. The total female population of the woodland mammals is shown in Figure 15 for the first five cycles (a total elapsed time of 15 months).

Figure 15 Total female population

Cycle	0	1	2	3	4	5
Total female population	38	44.4	50.6	52.14	51.33	53.76

No obvious pattern is apparent from examining the total female population during the first five cycles; however, a pattern does emerge further along in the process. The total female population during cycles 10–13 increases by a gradually larger percent as shown in Figure 16.

Figure 16 Total female population and percent growth

Cycle	10	11	12	13
Total female population	62.7874	64.5985	66.5059	68.5738
Percent growth		2.88	2.95	3.11

Does the percentage growth of the population during each cycle continue to increase? Moving even further along in the process sheds light on this question. The results of calculations for cycles 20–26 are shown below.

Figure 17 Total female population and percent growth

Cycle	20	21	22	23	24	25	26
Total female population	84.5867	87.1639	89.8168	92.5485	95.3648	98.2676	101.258
Percent growth		3.05	3.04	3.04	3.04	3.04	3.04

The growth rate appears to be stabilizing at about 3.04% for each three-month cycle. After 30 cycles, the total female population is 114.158, and after 31 cycles it is 117.632, which is an increase of 3.04%. Additional calculations confirm that the growth rate of the population converges to about 3.04%. This is called the *long-term growth rate* of the total population. The growth rate of the total population after a small number of cycles, called the *short-term growth rate*, is variable and does not reveal a pattern. After a large number of cycles, however, a stable long-term growth rate appears.

Examination of the age distribution vector X_k for large values of k also leads to some interesting results. The age distributions of the population after 20 and 21 cycles are

$$X_{20} \approx \begin{bmatrix} 27.46 \\ 15.99 \\ 13.97 \\ 12.20 \\ 9.47 \\ 5.51 \end{bmatrix} \quad \text{and} \quad X_{21} \approx \begin{bmatrix} 28.29 \\ 16.47 \\ 14.39 \\ 12.57 \\ 9.76 \\ 5.68 \end{bmatrix}.$$

The matrices X_{20} and X_{21} were computed using all possible digits, but values are presented rounded to hundredths. Using all available decimal digits, the total female population after 20 cycles is 84.59 and after 21 cycles it is 87.16. If the entries in an age distribution vector are each divided by the current total female population, then the resulting entries are the proportions of the total population found in each age group at that time. If each entry in X_{20} is divided by the total female population 84.59 and each entry in X_{21} is divided by the total female population 87.16, then the proportions found are as shown in the matrices that follow.

$$\left(\frac{1}{84.59}\right)X_{20} \approx \begin{bmatrix} 0.3246 \\ 0.1890 \\ 0.1651 \\ 0.1442 \\ 0.1119 \\ 0.0652 \end{bmatrix} \quad \text{and} \quad \left(\frac{1}{87.16}\right)X_{21} \approx \begin{bmatrix} 0.3246 \\ 0.1890 \\ 0.1651 \\ 0.1442 \\ 0.1120 \\ 0.0652 \end{bmatrix}.$$

The distribution of the proportions in each age group appears to stabilize. Examining the data after cycles 30 and 31, we find that the differences in the proportions in each age group from X_{30} to X_{31} are less than 10^{-6}. In the long run, therefore, the woodland mammal population tends to the following approximate distribution: 32% are 0–3 months old, 19% are 3–6 months old, 17% are 6–9 months old, 14% are 9–12 months old, 11% are 12–15 months old, and 7% are 15–18 months old.

A final property of the Leslie matrix model concerns the growth rates of the different age groups. Since the long-term proportions in each age group are fixed and the growth rate of the total population eventually remains at 3.04%, the growth rate of each age group must also eventually converge to 3.04%. This implies that for a large enough value of k, the age distribution vector X_k is equal to the scalar 1.0304 times the previous age distribution vector X_{k-1},

$$X_k = 1.0304 X_{k-1}.$$

Amazingly, after the proportions of the population in each age group approach fixed values, we can simply multiply X_{k-1} by a scalar to find X_k, and we no longer need to use the Leslie matrix L.

In the calculations above, we have seen evidence of the following general characteristics of the long-term behavior of the Leslie matrix model:

- The growth rate of the total population converges to a constant percentage growth rate, called the long-term growth rate.

- The proportions of the population in each age group eventually approach fixed values, and the growth rate of each age group stabilizes at the same value (the long-term growth rate of the total population).

- For large values of k, successive age distribution vectors are related by $X_k = (1 + r)X_{k-1}$, where r is the long-term growth rate.

1. Use the data from Figure 12 to determine how many mammals will be in the 6–9 month age group in 5 years and in 10 years.

2. Based on the data in Figure 12, how long will it take for the total number of mammals to exceed 500? Note: Instead of adding the entries in an age distribution vector by hand to find the total population, matrix software and some calculators can compute the total population if the age distribution vector is left-multiplied by the row vector $[1 \quad 1 \quad \cdots \quad 1]$.

3. For each initial age distribution vector given below, use the Leslie matrix from the woodland mammal example to determine the length of time before the total population reaches 500.

   a.
   $$X_0 = \begin{bmatrix} 20 \\ 10 \\ 8 \\ 0 \\ 0 \\ 0 \end{bmatrix}$$
   b.
   $$X_0 = \begin{bmatrix} 38 \\ 0 \\ 0 \\ 0 \\ 0 \\ 0 \end{bmatrix}$$
   c.
   $$X_0 = \begin{bmatrix} 6 \\ 6 \\ 6 \\ 6 \\ 6 \\ 8 \end{bmatrix}$$

4. For each of the initial age distributions given in Exercise 3, determine the long-term growth rate of the total population. Use the same Leslie matrix. How does the initial age distribution appear to be related to the long-term growth rate?

5. Using the Leslie matrix from the woodland mammal example, compare the long-term proportions of the population in each age group for the initial age distribution vectors below.

   $$X_0 = \begin{bmatrix} 20 \\ 15 \\ 10 \\ 0 \\ 0 \\ 0 \end{bmatrix} \qquad X_0 = \begin{bmatrix} 22 \\ 16 \\ 12 \\ 9 \\ 8 \\ 10 \end{bmatrix}$$

   Write a conjecture based on your results.

6. Suppose an animal population has the characteristics described in the table below.

Figure 18   Data for Exercise 6, Parts a and b

Age (years)	0–5	5–10	10–15	15–20	20–25	25–30	30–35
Birthrate	0	0	1.2	0.8	0.7	0.2	0
Death rate	0.5	0.2	0.1	0.1	0.3	0.5	1

a.   What is the expected life span of this animal?

b.   Construct the Leslie matrix for this animal.

c.   For the initial female population given in the table shown below, find the female age distribution and the total female population after 300 years.

Figure 19   Data for Exercise 6, Parts c–e

Age (years)	0–5	5–10	10–15	15–20	20–25	25–30	30–35
Number of females	30	30	26	28	32	15	10

d.   Determine the long-term growth rate for this population.

e.   If the maximum sustainable population for this animal in its native habitat is 700, when will the maximum population be reached?

SECTION 7.8 Markov Chains

This section begins with an example that illustrates many of the essential features of a class of matrix models called Markov chains.

EXAMPLE 1   The Taxi Problem

A taxi company has divided the city into three regions—Northside, Downtown, and Southside. By keeping track of pickups and deliveries, the company has found that of the fares picked up in Northside, 50% stay in that region, 20% are taken Downtown, and 30% go to Southside. Of the fares picked up Downtown, only 10% go to Northside, 40% stay Downtown, and 50% go to Southside. Of the fares picked up in Southside, 30% go to each of Northside and Downtown, while 40% stay in Southside.

We would like to know what the distribution of taxis will be over time as they pick up and drop off successive fares. This is a difficult analysis that we will approach first through a simpler question in this example. Examples later in this section will gradually work toward the analysis of the long-term distribution of taxis throughout the city. The question we will examine for now is the following: If a taxi starts off Downtown, what is the probability that it will be Downtown after letting off its third fare?

state diagram

transition

Solution   The information in this example can be represented with a *state diagram*, which includes (a) three states D, N, and S corresponding to the three regions of the city; and (b) the probabilities of moving from one region to another. The state diagram for the Taxi Problem is shown in Figure 20. In general, the movement from one state to another is called a *transition*. In this example, a transition corresponds to a customer being picked up in a region and dropped off in a region. Transitions to all three regions are feasible from every region. Therefore, each state in the state diagram is connected to all other states, including itself.

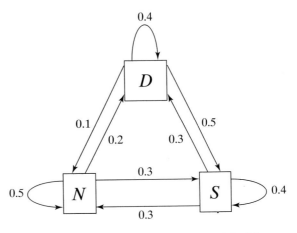

Figure 20   State diagram for the Taxi Problem

A taxi that starts off Downtown and ends up Downtown after three fares can follow several different paths through the state diagram. It could go Southside after one fare, then go Northside after two fares, and end up Downtown after three fares; or, it could go Northside, Downtown, and Downtown on its three fares. This reasoning shows that we must consider all possible combinations of fares such that the third fare ends up Downtown. The diagram in Figure 21, called a *tree diagram*, shows all possible paths starting Downtown and picking up three fares, with the third fare ending up Downtown. With a taxi picking up a fare Downtown, there is a probability of 0.1 that the taxi will go to Northside, a probability of 0.4 of dropping off the fare Downtown, and a probability of 0.5 of ending in Southside. These probabilities are indicated in the tree diagram on the lines, called *branches*, that represent the possible destinations of the first fare. The branches in the first set start at D and end at N, D, and S. The branches representing the second fare likewise have the associated probabilities shown in the tree diagram. Each path through the tree from the starting D to the ending D represents a possible sequence of three fares in which the first fare starts Downtown and the third fare ends Downtown. Every such path has a probability that is determined by multiplying the probabilities for each branch in the path. Our objective is to determine the probabilities associated with following each of the possible paths through the tree diagram. We will denote the probability of being in some region R after n fares as $P(R_n)$.

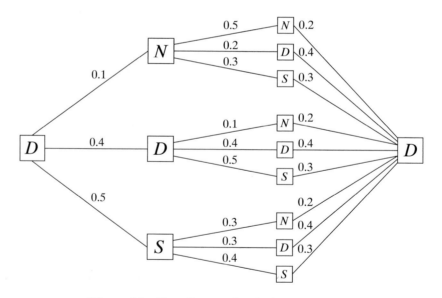

Figure 21   Tree diagram for the Taxi Problem

After one fare, the probabilities of being Northside (N), Downtown (D), and Southside (S) are

$$P(N_1) = 0.1, \qquad P(D_1) = 0.4, \qquad P(S_1) = 0.5.$$

After two fares, the probability of being Northside is

$$P(N_2) = P(N_1)P(NN) + P(D_1)P(DN) + P(S_1)P(SN),$$

where $P(NN)$ is the probability of going from Northside to Northside, $P(DN)$ is the probability of going from Downtown to Northside, and $P(SN)$ is the probability of going from Southside to Northside. Similarly, the probabilities of being Downtown and Southside after two fares are

$$P(D_2) = P(N_1)P(ND) + P(D_1)P(DD) + P(S_1)P(SD)$$

and

$$P(S_2) = P(N_1)P(NS) + P(D_1)P(DS) + P(S_1)P(SS).$$

Each term is the product of two probabilities. For example, $P(N_1)P(NS)$ is the probability of going to Northside on the first fare times the probability of going from Northside to Southside on the second fare. Substituting the appropriate values from the tree diagram in the sums gives the following:

$$P(N_2) = (0.1)(0.5) + (0.4)(0.1) + (0.5)(0.3)$$
$$= 0.24,$$
$$P(D_2) = (0.1)(0.2) + (0.4)(0.4) + (0.5)(0.3)$$
$$= 0.33,$$
$$P(S_2) = (0.1)(0.3) + (0.4)(0.5) + (0.5)(0.4)$$
$$= 0.43.$$

Notice that $P(N_2) + P(D_2) + P(S_2) = 1$, meaning that the taxi must be in one of the three regions after two fares, as expected. The probability of being Downtown after three fares is

$$P(D_3) = P(N_2)P(ND) + P(D_2)P(DD) + P(S_2)P(SD)$$
$$= (0.24)(0.2) + (0.33)(0.4) + (0.43)(0.3)$$
$$= 0.309.$$

A taxi starting Downtown has a probability of 0.309 of being Downtown after three fares. ■

The preceding example describes mathematical modeling techniques for analyzing the behavior of a system that includes a finite number of states, the three regions of the city, and probabilities of transitions from each state to every other possible state. The probabilities are constant and independent of previous behavior. We assume that a transition, picking up and dropping off a fare, occurs each time the system is observed, and that observations occur at regular intervals. Systems with these characteristics are called *Markov chains* or *Markov processes*.

Markov chains

Suppose we wish to take the Taxi Problem further and determine the probability of a taxi being Downtown after five fares if it starts Downtown. At this point, finding the probability of being Downtown after five fares appears to be a tedious endeavor. Let us return to the calculations we used to determine the probability of being Downtown after three fares. From our previous experiences with matrices, we might guess that the calculations above can be simplified considerably with the use of matrix multiplication. The probabilities of being in each region after two fares, represented by $P(N_2)$, $P(D_2)$, and $P(S_2)$, were calculated by finding sums of products. The first product in each sum has a factor of 0.1, the second product in each sum has a factor of 0.4, and the third product in each sum has a factor of 0.5. These observations lead us to deduce that the probabilities of being in each region after two fares are given by the row vector resulting from the matrix multiplication shown below:

$$\begin{bmatrix} 0.1 & 0.4 & 0.5 \end{bmatrix} \begin{bmatrix} 0.5 & 0.2 & 0.3 \\ 0.1 & 0.4 & 0.5 \\ 0.3 & 0.3 & 0.4 \end{bmatrix} = \begin{bmatrix} 0.24 & 0.33 & 0.43 \end{bmatrix}.$$

The matrix T defined by

$$T = \begin{array}{c} \\ N \\ D \\ S \end{array} \begin{array}{c} \begin{array}{ccc} N & D & S \end{array} \\ \begin{bmatrix} 0.5 & 0.2 & 0.3 \\ 0.1 & 0.4 & 0.5 \\ 0.3 & 0.3 & 0.4 \end{bmatrix} \end{array}$$

transition matrix

is crucial to our calculations. T is called a *transition matrix* because it contains entries such that T_{ij} is the probability of a transition from region i to region j. For example, T_{13} is the probability that a fare that originates in Northside goes to Southside. The probability of a fare that originates in Southside going to Downtown is T_{32}. Continuing with our analysis of the previous calculations, observe that the probability of being Downtown after three fares is equal to the product of the row vector after two fares and the column vector, which is the second column of T as shown below:

$$\begin{bmatrix} 0.24 & 0.33 & 0.43 \end{bmatrix} \begin{bmatrix} 0.2 \\ 0.4 \\ 0.3 \end{bmatrix} = 0.309.$$

Substituting the matrix product that produced the matrix [0.24   0.33   0.43] in the equation above yields

$$\begin{bmatrix} 0.1 & 0.4 & 0.5 \end{bmatrix} \begin{bmatrix} 0.5 & 0.2 & 0.3 \\ 0.1 & 0.4 & 0.5 \\ 0.3 & 0.3 & 0.4 \end{bmatrix} \begin{bmatrix} 0.2 \\ 0.4 \\ 0.3 \end{bmatrix} = 0.309.$$

The rightmost matrix in the product above is the second column of T. Using the entire matrix T yields

$$\begin{bmatrix} 0.1 & 0.4 & 0.5 \end{bmatrix} \begin{bmatrix} 0.5 & 0.2 & 0.3 \\ 0.1 & 0.4 & 0.5 \\ 0.3 & 0.3 & 0.4 \end{bmatrix} \begin{bmatrix} 0.5 & 0.2 & 0.3 \\ 0.1 & 0.4 & 0.5 \\ 0.3 & 0.3 & 0.4 \end{bmatrix} = \begin{bmatrix} 0.282 & 0.309 & 0.409 \end{bmatrix}.$$

The matrix [0.1 0.4 0.5] is the second row of T. It can be found by left-multiplying T by the vector [0 1 0]. Therefore, the calculations through three fares can be written as

$$[0\ \ 1\ \ 0]\begin{bmatrix} 0.5 & 0.2 & 0.3 \\ 0.1 & 0.4 & 0.5 \\ 0.3 & 0.3 & 0.4 \end{bmatrix}\begin{bmatrix} 0.5 & 0.2 & 0.3 \\ 0.1 & 0.4 & 0.5 \\ 0.3 & 0.3 & 0.4 \end{bmatrix}\begin{bmatrix} 0.5 & 0.2 & 0.3 \\ 0.1 & 0.4 & 0.5 \\ 0.3 & 0.3 & 0.4 \end{bmatrix} = [0.282\ \ 0.309\ \ 0.409],$$

which is equivalent to

$$[0\ \ 1\ \ 0]T^3 = [0.282\ \ 0.309\ \ 0.409].$$

The probability of being Downtown after three fares when starting out Downtown is 0.309, the middle entry in the row vector shown above. What is the meaning of the other entries in this row vector? The result of multiplying T^3 by [0 1 0] is the second row of T^3, specifically [0.282 0.309 0.409]. Rather than focus only on the second row of T^3, let us examine the following more general question: What meaning should we attach to all of the entries of T^3?

Before answering this question, consider the entries in T^2 shown below.

$$T^2 = \begin{bmatrix} 0.5 & 0.2 & 0.3 \\ 0.1 & 0.4 & 0.5 \\ 0.3 & 0.3 & 0.4 \end{bmatrix}\begin{bmatrix} 0.5 & 0.2 & 0.3 \\ 0.1 & 0.4 & 0.5 \\ 0.3 & 0.3 & 0.4 \end{bmatrix}$$

$$= \begin{array}{c} N \\ D \\ S \end{array}\begin{array}{ccc} N & D & S \end{array}\begin{bmatrix} 0.36 & 0.27 & 0.37 \\ 0.24 & 0.33 & 0.43 \\ 0.3 & 0.3 & 0.4 \end{bmatrix}$$

The entry in row 1, column 2 of T^2 is derived by multiplying row 1 of T by column 2 of T as follows:

$$(0.5)(0.2) + (0.2)(0.4) + (0.3)(0.3) = 0.27.$$

Row 1 of T contains the probabilities for the transitions from N to each of the regions, whereas column 2 of T contains the probabilities for the transitions from each of the regions to D. In symbols, the product of row 1 and column 2 is

$$P(NN)P(ND) + P(ND)P(DD) + P(NS)P(SD),$$

which is the probability of going from region N to region D after two transitions. In the matrix T^2, the entry in row 1, column two has row label N and column label D, and it represents the probability of going from N to D in two transitions. Using similar reasoning, it can be shown that the entry in row 1, column 1 of T^2 gives the probability of starting in N and ending in N after two transitions. In general, the entry in row i, column j of T^2 gives the probability of starting in region i and ending in region j after two transitions.

What do the entries in T^3 represent?

$$T^3 = \begin{array}{c} \\ N \\ D \\ S \end{array} \begin{array}{ccc} N & D & S \\ \left[\begin{array}{ccc} 0.318 & 0.291 & 0.391 \\ 0.282 & 0.309 & 0.409 \\ 0.3 & 0.3 & 0.4 \end{array} \right] \end{array}$$

The row and column labels were useful in interpreting T^2, and they are useful with T^3 as well. The probability of being in region D three transitions after starting in region D, which we found to be 0.309, is the entry with row label D and column label D. By reasoning similar to that used with T^2, we can deduce that the entry with row label D, column label S is the probability of ending in region S three transitions after starting in region D. In general, the entry in row i, column j of T^3 gives the probability of starting in region i and ending in region j after three transitions.

We now return to the question on page 526: If a taxi starts off Downtown, what is the probability that it will be Downtown after letting off its fifth fare? The answer to this question is found in the entry in row 2, column 2 of T^5, which is 0.30081, as shown in the matrix below that was found by using a calculator.

$$T^5 = \begin{array}{c} \\ N \\ D \\ S \end{array} \begin{array}{ccc} N & D & S \\ \left[\begin{array}{ccc} 0.30162 & 0.29919 & 0.39919 \\ 0.29838 & 0.30081 & 0.40081 \\ 0.3 & 0.3 & 0.4 \end{array} \right] \end{array}$$

A taxi that starts off Northside must be either Northside, Downtown, or Southside after letting off its fifth fare. Therefore, the sum of these probabilities must be 1. Notice that the sum of the elements of the first row [0.30162 0.29919 0.39919] is 1. A very important check of your work in Markov chains is to add the appropriate probabilities in each row to see if you get 1. To maintain the accuracy of the entries of these matrices, all digits created in the process of multiplication must be kept.

EXAMPLE 2   In the Taxi Problem, where should a taxi start to have the best chance of being Northside after three fares?

Solution   The probability of being in each region for each possible starting region is given by matrix T^3 shown below.

$$T^3 = \begin{array}{c} \\ N \\ D \\ S \end{array} \begin{array}{ccc} N & D & S \\ \left[\begin{array}{ccc} 0.318 & 0.291 & 0.391 \\ 0.282 & 0.309 & 0.409 \\ 0.3 & 0.3 & 0.4 \end{array} \right] \end{array}$$

The probability of ending up in Northside for each starting place is found in column 1, the N column, of T^3. The Northside entry is the largest in this column, so starting in Northside offers the best chance of being Northside after three fares, with a probability of 0.318. ■

On the basis of the work in the first two examples, the following general observations can be made about a transition matrix T for a Markov chain.

- A transition matrix is square. This is true because the number of rows and the number of columns are both the same as the number of states.

- All entries are between 0 and 1 inclusive. This follows from the fact that the entries correspond to transition probabilities from one state to another.

- The sum of the entries in any row must be 1. The sum of the entries in an entire row is the sum of the transition probabilities from one state to all other states. Since a transition is sure to take place, this sum must be 1.

- The ij entry in the matrix T^n gives the probability of being in state j after n transitions, if i is the initial state.

- The entries in the transition matrix are constant. A Markov chain model depends upon the assumption that the transition matrix does not change throughout the process. This implies that to determine the state of the system after any transition, it is necessary to know only the immediately preceding state of the system. Knowledge of the prior behavior of the system is not needed, provided that the immediately preceding state is known.

EXAMPLE 3 A bag contains 3 red and 4 green jelly beans. Suppose you take out 3 beans, one at a time, and eat them. What is the probability that the third bean chosen is green?

Solution This is a probabilistic situation in which all the previous behavior of the system must be examined; therefore, it is not modeled as a Markov chain. The set of beans chosen in all previous selections will affect the probability of choosing each color in the present selection. The outcomes that have a green jelly bean chosen third are *RRG, RGG, GRG,* and *GGG,* where R stands for choosing a red jelly bean and G stands for choosing a green jelly bean. The probabilities are listed below.

$$P(RRG) = \left(\frac{3}{7}\right)\left(\frac{2}{6}\right)\left(\frac{4}{5}\right)$$

$$P(RGG) = \left(\frac{3}{7}\right)\left(\frac{4}{6}\right)\left(\frac{3}{5}\right)$$

$$P(GRG) = \left(\frac{4}{7}\right)\left(\frac{3}{6}\right)\left(\frac{3}{5}\right)$$

$$P(GGG) = \left(\frac{4}{7}\right)\left(\frac{3}{6}\right)\left(\frac{2}{5}\right)$$

The solution to this problem is the sum of the 4 products above, which is

$$\frac{24 + 36 + 36 + 24}{7 \cdot 6 \cdot 5} = \frac{120}{210} = \frac{4}{7}.$$

The probability that the third jelly bean drawn from the bag is green is $\frac{4}{7}$. ▨

Example 3 is a simple example of a *stochastic process*. In a generalized stochastic process, the transition probabilities are not necessarily constant or independent of the previous behavior of the system. Notice that in Example 3, the probability of drawing a red jelly bean will vary depending on the number of beans of each color that have been chosen previously. A Markov chain is a special case of a stochastic process in which the transition probabilities are constant and independent of the previous behavior of the system. In a Markov chain, the next state is determined solely by the unchanging transition probabilities and the current state of the system. The route that is followed to arrive at the current state does not affect the transition matrix; only the current state of the system is relevant.

EXAMPLE 4   In the Taxi Problem, the cab company initially places 25% of the cabs Northside, 40% of the cabs Downtown, and 35% of the cabs Southside. What will be the distribution of cabs after each has made three pickups?

Solution   Represent the initial distribution of cabs with the row vector

$$\begin{array}{ccc} N & D & S \end{array}$$
$$X_0 = \begin{bmatrix} 0.25 & 0.40 & 0.35 \end{bmatrix}.$$

To find the percentage of cabs Northside after one pickup, multiply the percentage of cabs in each region by the probability of going from that region to Northside and add the resulting products:

$$(0.25)(0.5) + (0.40)(0.1) + (0.35)(0.3) = 0.27.$$

This sum is the product of X_0 and the first column of the transition matrix T. Likewise, the percentage of cabs Downtown after one pickup is equal to the product of X_0 and the second column of T. The percentage of cabs Southside after one pickup is equal to the product of X_0 and the third column of T. The distribution X_1 of cabs after one pickup is therefore given by

$$X_1 = X_0 T$$

$$= \begin{bmatrix} 0.25 & 0.40 & 0.35 \end{bmatrix} \begin{bmatrix} 0.5 & 0.2 & 0.3 \\ 0.1 & 0.4 & 0.5 \\ 0.3 & 0.3 & 0.4 \end{bmatrix}$$

$$\begin{array}{ccc} N & D & S \end{array}$$
$$= \begin{bmatrix} 0.27 & 0.315 & 0.415 \end{bmatrix}.$$

Continuing this line of reasoning, we see that X_2, the distribution of cabs after two fares, is given by

$$X_2 = X_1 T = X_0 T^2.$$

The distribution of cabs after three pickups is

$$X_3 = X_2 T = X_0 T^3,$$

which is

$$[0.25 \quad 0.40 \quad 0.35] \begin{bmatrix} 0.318 & 0.291 & 0.391 \\ 0.282 & 0.309 & 0.409 \\ 0.3 & 0.3 & 0.4 \end{bmatrix} = \begin{matrix} N & D & S \end{matrix} \\ [0.2973 \quad 0.30135 \quad 0.40135].$$

After three fares, about 30% of the taxis are Northside, about 30% are Downtown, and about 40% are Southside. ∎

state vector

The vector X_k is called the *state vector* for a Markov chain after k transitions with an initial distribution X_0. The jth entry of X_k is the probability of being in state j after k transitions. The matrix model we have used to calculate state vectors is an iterative model, where the kth state vector is calculated from the previous state vector by multiplying by the transition matrix. We can calculate X_k using the iterative equation

$$X_k = X_{k-1}T.$$

In the previous examples, we have also seen that the kth state vector can be calculated using the explicit equation

$$X_k = X_0 T^k.$$

EXAMPLE 5   For the Taxi Problem with the initial state vector given in Example 4, what is the long-term distribution of cabs?

Solution   We found X_3 in Example 4. After five fares, we have

$$\begin{matrix} & N & D & S \end{matrix}$$
$$X_5 = [0.25 \quad 0.40 \quad 0.35]\, T^5$$

$$\begin{matrix} & N & D & S \end{matrix}$$
$$= [0.2997570 \quad 0.3001215 \quad 0.4001215].$$

Moving along even further, the state vector after 10 fares is

$$\begin{matrix} & N & D & S \end{matrix}$$
$$X_{10} = X_0 T^{10} = [0.299999 \quad 0.300000 \quad 0.400000].$$

After 15 pickups, we find the entries in $X_{15} = X_0 T^{15}$ to be very close to

$$\begin{matrix} N & D & S \end{matrix}$$
$$[0.3 \quad 0.3 \quad 0.4].$$

It seems that the state vector X_k converges to the vector [0.3 0.3 0.4]. Once the state vector X_k converges to [0.3 0.3 0.4], further transitions will not change this distribution.

The system in the Taxi Problem stabilizes. That is, after a certain number of transitions, the distribution of taxis is not changed by additional transitions. Stability means that for sufficiently large values of k, multiplication by the transition matrix does not change the distribution so that eventually $X_k T = X_k$ or $X_{k+1} = X_k$. When

stable state

stable state vector

this occurs, the system has reached a *stable state*. A *stable state vector* of a Markov process is a vector X such that $XT = X$, where T is the transition matrix.

In the Taxi Problem, our work shows that X_k converges to [0.3 0.3 0.4]. This vector is in fact a stable state vector, because

$$\begin{bmatrix} 0.3 & 0.3 & 0.4 \end{bmatrix} \begin{bmatrix} 0.5 & 0.2 & 0.3 \\ 0.1 & 0.4 & 0.5 \\ 0.3 & 0.3 & 0.4 \end{bmatrix} = \begin{bmatrix} 0.3 & 0.3 & 0.4 \end{bmatrix}.$$

If the distribution of taxis reaches 30% Northside, 30% Downtown, and 40% Southside, subsequent transitions will not change this distribution. ▨

CLASS PRACTICE

1. Let V be a transition matrix for a Markov chain.

$$V = \begin{bmatrix} 0.2 & 0.4 & 0.4 \\ 0.4 & 0.2 & 0.4 \\ 0 & 0.4 & 0.6 \end{bmatrix}$$

   a. What is the probability of moving from State 1 to State 3? from State 3 to State 1?

   b. If the system is in State 2, what is the probability of staying there on the next transition?

2. Find X_{15}, X_{16}, and X_{17}, if $X_0 = [1 \ \ 0 \ \ 0]$, if $X_0 = [0.5 \ \ 0.1 \ \ 0.4]$, and if $X_0 = [0 \ \ 0.3 \ \ 0.7]$.

3. Based on your answer to Problem 2, does the long-term behavior of the system depend on the initial distribution?

4. Based on your answer to Problem 2, what is the stable state vector for the Markov chain with transition matrix V? That is, for what distribution vector X is it true that $XV = X$?

5. Calculate the entries in V^{15}, V^{16}, and V^{17}.

6. Compare the entries in V^k for large values of k with the entries in the stable state vector. What do you notice?

In the Class Practice, you saw that $\begin{bmatrix} \frac{1}{6} & \frac{1}{2} & \frac{1}{3} \end{bmatrix}$ is a stable state vector because

$$\begin{bmatrix} \frac{1}{6} & \frac{1}{3} & \frac{1}{2} \end{bmatrix} \begin{bmatrix} 0.2 & 0.4 & 0.4 \\ 0.4 & 0.2 & 0.4 \\ 0 & 0.4 & 0.6 \end{bmatrix} = \begin{bmatrix} \frac{1}{6} & \frac{1}{3} & \frac{1}{2} \end{bmatrix}.$$

You also saw that

$$V^k = \begin{bmatrix} \frac{1}{6} & \frac{1}{3} & \frac{1}{2} \\[6pt] \frac{1}{6} & \frac{1}{3} & \frac{1}{2} \\[6pt] \frac{1}{6} & \frac{1}{3} & \frac{1}{2} \end{bmatrix}$$

for large values of k. A similar result occurs in the Taxi Problem. Since $XT = X$, $X = [0.3 \ 0.3 \ 0.4]$ is a stable state vector. The calculation of T^5, T^{10}, and T^{15}, shown below, reveals that T^k converges to a matrix in which each row is identical.

$$T^5 \approx \begin{array}{c} \\ N \\ D \\ S \end{array} \begin{array}{ccc} N & D & S \\ \left[\begin{array}{ccc} 0.30162 & 0.29919 & 0.39919 \\ 0.29838 & 0.30081 & 0.40081 \\ 0.3 & 0.3 & 0.4 \end{array}\right] \end{array}$$

$$T^{10} \approx \begin{array}{c} \\ N \\ D \\ S \end{array} \begin{array}{ccc} N & D & S \\ \left[\begin{array}{ccc} 0.300004 & 0.299998 & 0.399998 \\ 0.299996 & 0.300002 & 0.400002 \\ 0.3 & 0.3 & 0.4 \end{array}\right] \end{array}$$

$$T^{15} \approx \begin{array}{c} \\ N \\ D \\ S \end{array} \begin{array}{ccc} N & D & S \\ \left[\begin{array}{ccc} 0.3 & 0.3 & 0.4 \\ 0.3 & 0.3 & 0.4 \\ 0.3 & 0.3 & 0.4 \end{array}\right] \end{array}$$

We see that for large values of k, T^k converges to

$$\begin{array}{c} \\ N \\ D \\ S \end{array} \begin{array}{ccc} N & D & S \\ \left[\begin{array}{ccc} 0.3 & 0.3 & 0.4 \\ 0.3 & 0.3 & 0.4 \\ 0.3 & 0.3 & 0.4 \end{array}\right] \end{array}.$$

stable state matrix

This is called a *stable state matrix*, because $T^k = T^{k+1}$. Notice that each row of the stable state matrix, T^k, has the same entries as the stable state vector. As a matter of fact, whenever powers of a transition matrix converge to a matrix in which all rows are identical, the row is the unique stable state vector associated with the transition matrix.

Since the entries in T^k are the probabilities of moving from state to state after k transitions, the vector $X_0 T^k$ will represent the distribution of taxis k transitions after the initial distribution X_0. How does the long-term distribution, $X_0 T^k = X_k$, depend on the initial distribution X_0?

We will use $X_0 = [n \ d \ s]$ to represent an arbitrary initial distribution of taxis. Since every taxi is in one of the three states, it must be true that $n + d + s = 1$. We know that

$$T^k \approx \begin{array}{c} \\ N \\ D \\ S \end{array} \begin{array}{ccc} N & D & S \\ \left[\begin{array}{ccc} 0.3 & 0.3 & 0.4 \\ 0.3 & 0.3 & 0.4 \\ 0.3 & 0.3 & 0.4 \end{array}\right] \end{array}$$

for large values of k. Therefore,

$$X_0 T^k = [n \ d \ s] \begin{bmatrix} 0.3 & 0.3 & 0.4 \\ 0.3 & 0.3 & 0.4 \\ 0.3 & 0.3 & 0.4 \end{bmatrix}$$

$$= [0.3n + 0.3d + 0.3s \quad 0.3n + 0.3d + 0.3s \quad 0.4n + 0.4d + 0.4s]$$

$$= [0.3(n + d + s) \quad 0.3(n + d + s) \quad 0.4(n + d + s)]$$

$$= [0.3 \quad 0.3 \quad 0.4].$$

In this example any initial distribution $X_0 = [n \ d \ s]$ will eventually yield $X_k = [0.3 \ 0.3 \ 0.4]$ after a large number of transitions. We can conclude that the distribution of taxis converges to $[0.3 \ 0.3 \ 0.4]$ regardless of the initial distribution.

How do we know if a Markov chain has a stable state vector? A sufficient condition, which is stated without proof, for a Markov chain to have a stable state vector is that some power of the transition matrix has only nonzero entries. A Markov chain satisfying this criterion is called *regular*. For example, a Markov chain with transition matrix

regular Markov chain

$$B = \begin{bmatrix} 0.2 & 0.8 \\ 1 & 0 \end{bmatrix}$$

is regular since

$$B^2 = \begin{bmatrix} 0.84 & 0.16 \\ 0.2 & 0.8 \end{bmatrix}.$$

For regular Markov chains, we can find the stable state vector by raising the transition matrix to a large power. As the power increases, the rows of the transition matrix each approach the stable state vector.

Absorbing Markov Chains

After completing the Taxi Problem you may wonder if all Markov chains have a stable state vector. In the Class Practice that follows, you will explore the long-term behavior of another transition matrix.

1. Consider the following transition matrix.

$$A = \begin{bmatrix} 0.3 & 0.1 & 0.4 & 0 & 0.2 \\ 0 & 1 & 0 & 0 & 0 \\ 0.2 & 0.3 & 0.1 & 0.1 & 0.3 \\ 0 & 0 & 0 & 1 & 0 \\ 0.2 & 0.2 & 0.1 & 0.5 & 0 \end{bmatrix}$$

a. Find A^{10}, A^{45}, and A^{55}.

b. For large values of k, to what matrix does A^k converge?

c. If the initial distribution vector is $X_0 = [1\ 0\ 0\ 0\ 0]$, what is X_1, the distribution after one transition? What is the long-term distribution X_k for large values of k?

d. If the initial distribution vector is $Y_0 = [0.3\ 0\ 0.2\ 0.1\ 0.4]$, what is Y_1? What is the long-term distribution vector Y_k for large values of k?

e. If the initial distribution vector is $Z_0 = [0\ 0.3\ 0\ 0.7\ 0]$, what is Z_1? What is the long-term distribution vector Z_k for large values of k?

The entries in rows 2 and 4 of transition matrix A in the Class Practice imply that once the system enters state 2 or state 4, it can never leave that state. Zeros in rows 2 and 4 indicate a probability of zero that the system will change to another state once it is in state 2 or state 4. These states are called *absorbing states*. The row corresponding to an absorbing state contains a 1 on the diagonal of the matrix and 0s for all the other entries. An *absorbing Markov chain* is a Markov chain that has the following properties:

absorbing states

absorbing Markov chain

- it has at least one absorbing state;

- it is possible to move from any nonabsorbing state to an absorbing state in a finite number of transitions.

In the Class Practice you determined that the transition matrix A converges to the stable state matrix

$$A^k = \begin{bmatrix} 0 & 0.587097 & 0 & 0.412903 & 0 \\ 0 & 1 & 0 & 0 & 0 \\ 0 & 0.589247 & 0 & 0.410753 & 0 \\ 0 & 0 & 0 & 1 & 0 \\ 0 & 0.376344 & 0 & 0.623656 & 0 \end{bmatrix}.$$

Unlike the rows of the stable state matrix in the Taxi Problem, the rows of A^k are not identical. The only nonzero entries in A^k are found in the absorbing-state columns. This means that the system modeled by an absorbing Markov chain will eventually end up in one of the absorbing states, as can be observed in matrix A^k above.

We also observe from the Class Practice that the long-term distribution of the system depends on the initial distribution. It is also true that an absorbing Markov chain can have many stable state vectors. For example, in the Class Practice we saw $Z_0 A = Z_0$, so Z_0 is a stable state vector. Another example of a stable state vector for matrix A is $[0 \quad 0.25 \quad 0 \quad 0.75 \quad 0]$ since

$$
\begin{bmatrix} 0 & 0.25 & 0 & 0.75 & 0 \end{bmatrix}
\begin{bmatrix}
0.3 & 0.1 & 0.4 & 0 & 0.2 \\
0 & 1 & 0 & 0 & 0 \\
0.2 & 0.3 & 0.1 & 0.1 & 0.3 \\
0 & 0 & 0 & 1 & 0 \\
0.2 & 0.2 & 0.1 & 0.5 & 0
\end{bmatrix}
= \begin{bmatrix} 0 & 0.25 & 0 & 0.75 & 0 \end{bmatrix}.
$$

In fact, any vector that has nonzero entries in at least one absorbing state and zeros elsewhere is a stable state vector. Summarizing, absorbing Markov chains converge to a stable state matrix in which the rows will differ. If there is more than one absorbing state in an absorbing Markov chain, the long-term distribution is dependent on the initial distribution.

Exercise Set 7.8

1. A rat is placed in the maze shown in Figure 22. During a fixed time interval, the rat randomly chooses one of the doors available to it (depending upon which room it is in) and moves through that door to the next room; it does not remain in the room it occupies.

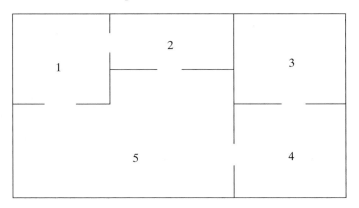

Figure 22 Maze for Exercise 1

Each movement of the rat is taken as a transition in a Markov chain in which a state is identified by the room the rat is in. The first row of the transition matrix is

$$
\begin{array}{ccccc} 1 & 2 & 3 & 4 & 5 \end{array}
$$
$$
1 \begin{bmatrix} 0 & \tfrac{1}{2} & 0 & 0 & \tfrac{1}{2} \end{bmatrix}.
$$

a. Construct the entire transition matrix for this process.

b. If the rat starts in room 1, what is the probability that it is in room 3 after two transitions? after three transitions?

c. Determine the stable state vector.

d. After a large number of transitions, what is the probability that the rat is in room 4?

e. In the long run, what percentage of the time will the rat spend in rooms 2 or 3?

2. Wade, Donald, and Andrea are playing Frisbee®. Wade always throws to Donald, Donald always throws to Andrea, but Andrea is equally likely to throw to Wade or Donald.

   a. Represent this information as a transition matrix of a Markov chain.

   b. Notice that this transition matrix has zero entries in several places. Compare the values of the transition matrix if raised to the second, fourth, sixth, and tenth powers. Are the zero entries still there? Explain why or why not.

3. The snack bar at school sells three items that students especially like: onion rings, french fries, and chocolate chip cookies. The manager noticed that what each student ordered depended on what he or she ordered on the previous visit. A survey found that 50% of those who ordered onion rings on their last snack break ordered them again, while 35% switched to french fries, and 15% switched to chocolate chip cookies. Of those who ordered french fries on their last visit, 40% did so the next time, but 30% switched to onion rings, and another 30% switched to chocolate chip cookies. Of the students who ordered chocolate chip cookies on their last visit, 20% switched to onion rings, and 55% switched to french fries.

   a. Set up the transition matrix for this Markov process.

   b. On Monday, 30 students buy french fries, 40 buy onion rings, and 25 buy chocolate chip cookies. If these same students come in on Tuesday and each buys one of these items, how many orders of french fries should the manager expect to sell?

   c. Suppose the students in Part b continue buying from the snack bar every day for two weeks. How many orders of onion rings, french fries, and cookies should the manager expect to sell on the third Monday?

   d. If these same students come all year, how many orders of onion rings, french fries, and cookies should the manager expect to sell each day?

   e. In the long run, what percent of the orders will be onion rings? french fries? chocolate chip cookies?

4. The manager of the snack bar in Exercise 3 decided to add soft custard ice cream cones to the menu. Many students tried the new item, but few of them liked it. (The machine didn't work properly, and the ice cream came out lumpy.) A two-week survey gave the results in the transition matrix below.

$$
\begin{array}{c}
\quad\;\; Rings\;\; Fries\;\; Cookies\;\; Cones \\
\begin{array}{c} Rings \\ Fries \\ Cookies \\ Cones \end{array}
\begin{bmatrix}
0.4 & 0.3 & 0.1 & 0.2 \\
0.25 & 0.35 & 0.2 & 0.2 \\
0.15 & 0.4 & 0.2 & 0.25 \\
0.3 & 0.35 & 0.3 & 0.05
\end{bmatrix}
\end{array}
$$

    a. Does this system reach a stable state?

    b. If everyone really dislikes the ice cream, why does the stable state matrix show that many students still buy it?

    c. Why is a Markov chain not a good model for this system? Are people likely to forget that they did not like something only two days after they have eaten it? (Recall the principal assumption of a Markov process.) Why would this model work if the customers liked the ice cream?

5. A mouse is in the maze shown in Figure 23. Doors are shown by openings between rooms. Arrows indicate one-way doors and the direction of passage through these doors.

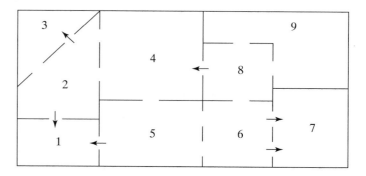

Figure 23   Maze for Exercise 5

The mouse does not have to change rooms at each transition but can stay in a room. Notice that some of the rooms are impossible to leave once they are entered. During each transition, the mouse has an equal chance of leaving a room by a particular door or staying in the room. For example, during a single transition, a mouse in room 2 has a $\frac{1}{6}$ chance of moving to room 1, a $\frac{1}{6}$ chance of staying in room 2, a $\frac{1}{3}$ chance of moving to room 3, and a $\frac{1}{3}$ chance of moving to room 4.

    a. Find the transition matrix which describes the movement of the mouse.

    b. If the mouse starts in room 4, what is the probability that it will eventually be trapped in room 1?

c. From which room, besides 7, does the mouse have the best chance of being trapped in room 7? From which room, besides 1, does the mouse have the worst chance of being trapped in room 7?

6. A research article (Marshall, A.W., and H. Goldhamer, "An Application of Markov Processes to the Study of the Epidemiology of Mental Diseases," *American Statistical Association Journal*, March 1955, pp. 99–129) on the application of Markov chains to mental illness suggests that we consider a person to be in one of four states:

> State I—severely insane and hospitalized
> State II—dead, with death occurring while unhospitalized
> State III—sane
> State IV—insane and unhospitalized.

For this model, assume that States I and II are absorbing states. Suppose that of the people in State III, after one year, 1.994% will be in State II, 98% will be in State III, and 0.006% will be in State IV. Also suppose that of the people in State IV, after one year, 2% will be in State I, 3% will be in State II, and 95% will be in State IV.

a. Set up the transition matrix that describes this model.

b. To what matrix do powers of the transition matrix converge?

c. Determine the probability that a person who is currently in State III eventually will be severely insane and hospitalized.

7. The election process can be viewed as a Markov chain—each new election is a transition and the states of the process are Democrat and Republican. Design a transition matrix based on the probability of electing Democratic and Republican presidents in the twentieth century. Begin your analysis with the transition to Theodore Roosevelt in 1904. In the transition matrix, count each election as a transition. For example, Ronald Reagan's election and reelection constitute two transitions, and Gerald Ford was not elected. According to your matrix, what is the probability a Democrat will be elected in the 2000 presidential election? a Republican? What are the probabilities for the 2008 presidential election? What is the main weakness with this model of presidential elections?

8. A dreaded strain of flu is studied by research biologists. Statistics are taken each week in an effort to describe the probabilities after exposure of staying well, getting ill, becoming immune, and dying from the flu. A person becomes immune to this flu by having a mild case of it. Any well person once exposed to this flu has a 20% chance of getting the illness. Once a person becomes ill there is a 55% chance of remaining ill for more than a week, a 40% chance of being permanently immune after a mild illness, and a 5% chance of dying from the illness.

a. Construct a transition matrix to represent the information given.

b. Compute the probability of becoming immune after 10 weeks for groups that begin as follows:
  i. 100% well and exposed;
  ii. 50% well and 50% sick;
  iii. 80% well and 20% immune;
  iv. 100% sick.

c. Does this Markov process have a stable state matrix?

9. An ant walks from one corner to another of square *ABCD*. Assume that between successive observations of the process, the ant has moved from one corner to another corner. The transition probabilities between corners are shown in Figure 24. Construct a transition matrix for this situation.

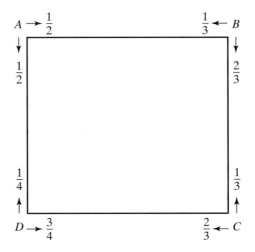

Figure 24 Ant walking around a square

a. Investigate the behavior of successive powers of the transition matrix for this Markov chain. For example, consider T^7, T^8, T^{16}, and T^{17}. What pattern do you observe?

b. If the ant starts at corner *A* and makes a large, even number of transitions, where will the ant end up?

Chapter 7 Review Exercises

1. Suppose in a recent survey of men it was found that the occupation of a father can help predict the occupation of the son. The results showed that if the fathers are professionals, 63% of their sons were professionals and 20% of their sons were in service-oriented jobs. For fathers with service-oriented jobs, 31% of their sons were professionals and 45% of their sons also held service-oriented jobs. For fathers who worked in manufacturing, 18% of their sons were in manufacturing and 41% went into service-oriented jobs.

   a. Based on the survey results, if a man's great grandfather was in a service-oriented job, what is the probability that the man will be in a service-oriented job?

   b. What assumption(s) did you make in answering Part a?

   c. If the survey results showed that 41% were professionals, 32% were in manufacturing, and 27% were in service-oriented jobs, and if the transition probabilities remain constant over a number of generations, what percentage of the population will be professionals in 10 generations?

2. a. Under what conditions can we perform matrix addition?

   b. Under what conditions can we perform matrix multiplication?

3. Show that $\begin{bmatrix} 3 & 1 \\ 5 & 2 \end{bmatrix}$ and $\begin{bmatrix} 2 & -1 \\ -5 & 3 \end{bmatrix}$ are multiplicative inverses.

4. The technology matrix of an economy with sectors oil (O), agriculture (A), electricity (E), machinery (M), and chemical (C) is shown below.

$$
\begin{array}{c}
 \\ O \\ A \\ E \\ M \\ C
\end{array}
\begin{array}{ccccc}
O & A & E & M & C \\
\begin{bmatrix} 0.2 & 0.1 & 0.1 & 0.1 & 0.1 \\ 0.1 & 0.05 & 0.05 & 0.05 & 0.2 \\ 0.3 & 0.02 & 0.05 & 0.1 & 0.05 \\ 0.05 & 0.2 & 0.2 & 0.05 & 0.1 \\ 0.1 & 0.1 & 0.05 & 0.2 & 0.02 \end{bmatrix}
\end{array}
$$

   Find the production matrix if the demand matrix in millions of dollars is

$$
\begin{array}{c}
\textit{Demand} \\
\begin{array}{c} O \\ A \\ E \\ M \\ C \end{array}
\begin{bmatrix} 22 \\ 31 \\ 12 \\ 9 \\ 24 \end{bmatrix}
\end{array}.
$$

5. Suppose an animal population has the characteristics described in Figure 25.

Figure 25   Data for Exercise 5

Age (years)	0–3	3–6	6–9	9–12	12–15	15–18	18–21
Birthrate	0	0.5	1.3	0.9	0.6	0.1	0
Death rate	0.5	0.2	0.1	0.1	0.3	0.5	1

a. What is the expected life span of this animal?

b. Construct the Leslie matrix for this animal.

c. For the initial female population given in Figure 26, find the female age distribution and the total female population after 15 years. How did you obtain your answer?

Figure 26   Data for Exercise 5, Part c

Age (years)	0–3	3–6	6–9	9–12	12–15	15–18	18–21
Population	30	25	29	32	32	15	10

6. Suppose 1017 tickets were sold to a local high school football game. The prices of the tickets were: students $2, adults $5, and senior citizens $3. The revenue generated from the sale of the tickets was $3140, and there were 224 more student tickets sold than adult tickets. How many tickets of each type were sold?

Appendix A
Complex Numbers

A complex number is a number of the form $a + bi$, where $i = \sqrt{-1}$ and a and b are real numbers. The first place you may have encountered such numbers was in algebra, when you used the quadratic formula to find solutions to a quadratic equation. The first term, a, is called the ***real part*** and the second term, bi, is called the ***imaginary part*** of the complex number.

EXAMPLE 1 Find the solutions to $x^2 - 4x + 13 = 0$.

Solution Using the quadratic formula, we get $x = \frac{4 \pm \sqrt{16 - 52}}{2} = \frac{4 \pm \sqrt{-36}}{2} = \frac{4 \pm 6\sqrt{-1}}{2} = 2 \pm 3\sqrt{-1}$. Mathematicians in the 17th Century decided to define $i = \sqrt{-1}$. This allows us to write the solutions to the quadratic as $2 \pm 3i$. ∎

The systems of numbers of the form $a + bi$, where a and b are real numbers and $i = \sqrt{-1}$, turns out to have most of the same properties as the real numbers. The only significant loss is that of ordering. (For example, which is greater, $1 + 2i$ or $2 + i$?) The operations of addition and multiplication are defined logically as

$$(a + bi) + (c + di) = (a + c) + (b + d)i$$

and
$$(a + bi)(c + di) = ac + adi + bci + bdi^2$$
$$= ac + (ad + bc)i - bd$$
$$= (ac - bd) + (ad + bc)i$$

Notice that both addition and multiplication are performed as if the numbers $(a + bi)$ and $(c + di)$ were binomials. For addition you add *like* terms and for multiplication you use the distributive property. Notice also that the term bdi^2 simplifies to $-bd$ since $i^2 = -1$.

It is not difficult to show that the complex numbers with the operations defined above are closed under addition and multiplication (that is, if you add or multiply two complex numbers the result is a complex number) and have identities for both operations. Also, additive inverses exist for complex numbers and multiplicative inverses exist for all nonzero numbers; both operations are commutative and associative; and multiplication still distributes over addition.

EXAMPLE 2   Add the complex numbers $2 + 3i$ and $3 - 7i$. Find the product of these numbers.

Solution   Addition is simply $(2 + 3i) + (3 - 7i) = (2 + 3) + (3 - 7)i = 5 - 4i$. The product of these two numbers is

$$(2 + 3i)(3 - 7i) = (2 \cdot 3) + 2(-7)i + 3(3)i + (-7)(3)i^2$$
$$= 6 - 14i + 9i - 21i^2 = 27 - 5i.$$

Subtraction of complex numbers is completely analogous to subtraction of real numbers, as is division. However, with division, simplification of answers may be preferred.

EXAMPLE 3   Determine the quotient of $3 + 4i$ and $4 - 9i$.

Solution   We could write the quotients as $\frac{3+4i}{4-9i}$ and this is indeed correct, but we can write the quotient in a simpler form. In order to do this, we use the same technique you may have used in algebra to remove radicals from the denominators of fractions. If we had $3 + \sqrt{5}$ in the denominator of a fraction, we could multiply both the numerator and the denominator of the fraction by the conjugate $3 - \sqrt{5}$. Since the i in the denominator of our complex quotient is actually a radical, we will do the same thing to simplify the quotient.

$$\frac{3 + 4i}{4 - 9i} = \frac{(3 + 4i)(4 + 9i)}{(4 - 9i)(4 + 9i)} = \frac{-24 + 43i}{97} = -\frac{24}{97} + \frac{43}{97}i$$

We can graph all the real numbers on a number line, but complex numbers have two parts, the real part and the imaginary part, so we cannot graph them on a line. A set of perpendicular lines, or axes, are used to define a plane, and complex numbers are graphed in this plane. The horizontal axis is used to locate the real part of the number and the vertical axis is used to locate the imaginary part. Along the horizontal axis the units are simply real numbers; but along the vertical or imaginary side, the units are in terms of i.

EXAMPLE 4   Graph the complex numbers $3 + 2i$ and $-2 - i$.

Solution   The real part of $3 + 2i$ is 3 and the imaginary part is $2i$. The coordinates for $3 + 2i$ are $(3, 2)$. Similarly, the coordinates for $-2 - i$ are $(-2, -1)$.

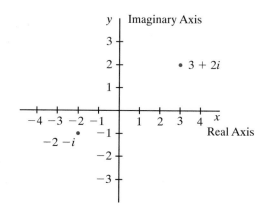

Appendix B
Derivation of Linear Least Squares Parameters

Presented here is an algebraic derivation of the mathematical formulas for the slope and the y-intercept of the linear least squares line for a set of data points.

Given a set of data points (x_1, y_1), (x_2, y_2), (x_3, y_3), (x_4, y_4), ..., (x_n, y_n), with $n \geq 2$, let $y = mx + b$ be a line through the data.

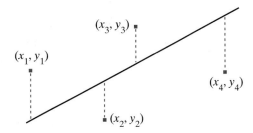

The residuals are: $(y_1 - (mx_1 + b))$, $(y_2 - (mx_2 + b))$, $(y_3 - (mx_3 + b))$, $(y_4 - (mx_4 + b))$, ..., $(y_n - (mx_n + b))$.

To find the linear least squares line, we want the squares of these residuals to be a minimum. Thus we need to square these quantities. Doing this gives the following expressions:

$$(y_1 - (mx_1 + b))^2 = y_1^2 - 2mx_1y_1 - 2by_1 + m^2x_1^2 + 2bmx_1 + b^2$$
$$(y_2 - (mx_2 + b))^2 = y_2^2 - 2mx_2y_2 - 2by_2 + m^2x_2^2 + 2bmx_2 + b^2$$
$$(y_3 - (mx_3 + b))^2 = y_3^2 - 2mx_3y_3 - 2by_3 + m^2x_3^2 + 2bmx_3 + b^2$$
$$(y_4 - (mx_4 + b))^2 = y_4^2 - 2mx_4y_4 - 2by_4 + m^2x_4^2 + 2bmx_4 + b^2$$
$$\vdots$$
$$(y_n - (mx_n + b))^2 = y_n^2 - 2mx_ny_n - 2by_n + m^2x_n^2 + 2bmx_n + b^2$$

If we now add all the squared residuals together we have the following sum S.

$$S = (y_1^2 + y_2^2 + y_3^2 + y_4^2 + \ldots + y_n^2) - 2m(x_1y_1 + x_2y_2 + x_3y_3 + x_4y_4 + \ldots + x_ny_n)$$
$$- 2b(y_1 + y_2 + y_3 + y_4 + \ldots + y_n) + m^2(x_1^2 + x_2^2 + x_3^2 + x_4^2 + \ldots + x_n^2)$$
$$+ 2bm(x_1 + x_2 + x_3 + x_4 + \ldots + x_n) + (b^2 + b^2 + b^2 + b^2 + \ldots + b^2).$$

This simplifies to: $S = \sum_{i=1}^{n} y_i^2 - 2m \sum_{i=1}^{n} x_iy_i - 2b \sum_{i=1}^{n} y_i + m^2 \sum_{i=1}^{n} x_i^2 + 2bm \sum_{i=1}^{n} x_i + nb^2.$

The sum, S, is quadratic in both m and b. That is, m and b are the only variables; all of the summations are constants. We can find the minimum of this expression by completing the square while first treating the expression as a function of only m and then as a function of only b. The first step then, is to treat b as a constant and complete the square for m.

First look at the quadratic in m. We have:

$$S = \left(\sum_{i=1}^{n} x_i^2\right) \cdot m^2 + \left(2b\sum_{i=1}^{n} x_i - 2\sum_{i=1}^{n} x_i y_i\right) \cdot m + \left(\sum_{i=1}^{n} y_i^2 - 2b\sum_{i=1}^{n} y_i + nb^2\right).$$

Recall that in the quadratic $y = Ax^2 + Bx + C$, the minimum value occurs when $x = \frac{-B}{2A}$. Thus, the minimum of S in the above expression occurs when

$$m = \frac{-\left(2b\sum_{i=1}^{n} x_i - 2\sum_{i=1}^{n} x_i y_i\right)}{2\left(\sum_{i=1}^{n} x_i^2\right)} = \frac{-b\sum_{i=1}^{n} x_i + \sum_{i=1}^{n} x_i y_i}{\sum_{i=1}^{n} x_i^2}.$$

Now look at S as a quadratic in b with m as a constant. We have:

$$n \cdot b^2 + \left(2m\sum_{i=1}^{n} x_i - 2\sum_{i=1}^{n} y_i\right) \cdot b + \left(\sum_{i=1}^{n} y_i^2 - 2m\sum_{i=1}^{n} x_i y_i + m^2 \sum_{i=1}^{n} x_i^2\right).$$

The minimum occurs when $b = \dfrac{-\left(2m\sum_{i=1}^{n} x_i - 2\sum_{i=1}^{n} y_i\right)}{2n} = \dfrac{\left(\sum_{i=1}^{n} y_i - m\sum_{i=1}^{n} x_i\right)}{n} = \bar{y} - m\bar{x}.$

If you now solve the system of equations involving m and b, you will have the slope and the y-intercept of the least squares line. Since all of the sums are just constants, let's call them k_1, k_2, k_3, k_4, k_5, as follows:

$$k_1 = \sum_{i=1}^{n} x_i; \quad k_2 = \sum_{i=1}^{n} y_i; \quad k_3 = \sum_{i=1}^{n} x_i y_i; \quad k_4 = \sum_{i=1}^{n} x_i^2; \quad k_5 = \sum_{i=1}^{n} y_i^2.$$

Using these k_i's, we can rewrite the two equations involving m and b as

$$m = \frac{-bk_1 + k_3}{k_4} \quad \text{and} \quad b = \frac{-mk_1 + k_2}{n}.$$

Substituting the second equation into the first, we obtain $m = \dfrac{-\left(\frac{-mk_1 + k_2}{n}\right)k_1 + k_3}{k_4}$,

which we can simplify to $m = \frac{mk_1^2}{nk_4} + \frac{-k_2 k_1}{nk_4} + \frac{k_3}{k_4}$.

Solving for m produces $m = \left(\frac{-k_2 k_1}{nk_4} + \frac{k_3}{k_4}\right) \div \left(1 - \frac{k_1^2}{nk_4}\right)$, and with a little more algebra we get

$$m = \frac{nk_3 - k_2 k_1}{nk_4 - k_1^2} = \frac{n \cdot \sum_{i=1}^{n} x_i y_i - \left(\sum_{i=1}^{n} y_i\right)\left(\sum_{i=1}^{n} x_i\right)}{n \cdot \sum_{i=1}^{n} x_i^2 - \left(\sum_{i=1}^{n} x_i\right)^2}.$$

Now we can use this expression for m in the equation $b = \bar{y} - m\bar{x}$ to get b.

Answers to Selected Exercises

CHAPTER 1 Data Analysis

EXERCISE SET 1.2, *page 9*

1. The slope $-\frac{200}{39}$ describes the change in the number of tickets students can expect to sell with each one-dollar increase in the price of the ticket. For example, if the price is increased by 1 dollar, the expected number of tickets sold will decrease by $\frac{200}{39}$ tickets; or if the price is increased by \$39, then the expected number of tickets sold will decrease by 200 tickets.

If the students charge too much for the tickets, no tickets will be sold. The *P*-intercept represents the lowest price that is "too much." In other words, the *P*-intercept is the lowest price for the tickets that, according to the model, will result in no tickets being sold.

The *T*-intercept represents the number of families that would be interested in receiving free season tickets.

3. Figure 2 on page 4 changes to the following:

Maximum price ($)	50	75	90	95	115	135	150	175
Expected ticket sales	154	85	48	90	127	85	64	159

(Since numbers have been rounded to the nearest integer, the "numbers" total 812 responses rather than 811.)

The results in Figure 4 change to:

Price ($)	50	75	90	95	115	135	150	175
Expected ticket sales	812	658	573	525	435	308	223	159

The equation of the (*Price, Tickets*) line is something close to: *Tickets* $= -5.44 \times Price + 1060$, which gives the revenue function $R(P) = P(-5.44P + 1060)$. This revenue function has a maximum near $P = \$97.50$.

EXERCISE SET 1.3, *pages 17–19*

1. b. The average daily cost of gas is negatively associated with the average temperature, and the relationship appears linear. The average daily cost of electricity seems to be independent of the average temperature.

c. Sketches and equations will vary. Sample equations are provided. For the (*temperature, gas cost*) graph, the line is approximately $G = -\frac{3}{80}T + 3$.

For the (*temperature, electricity cost*) graph, the line is approximately $E = 0.8$.

The average daily cost of gas in February should be $G(19) \approx \$2.29$, and the average daily cost of electricity in February should be $E(19) \approx \$0.80$.

2. a. The shape is curved. The relationship is positive, with no gaps, clusters, or outliers. The association is strong; we would feel comfortable making predictions.

b. Answers will vary. One answer would be to describe the relationship as linear, and the association as somewhat weak, so we would not be very confident in our predictions. The relationship is negative. The first point is an outlier. Another answer would be to describe the relationship as curved with no outlier.

c. There does not seem to be a relationship between the variables. The two highest points could be considered outliers.

EXERCISE SET 1.5, *pages 33–34*

1. Summary points: (3.8, 15.5), (4.4, 13.5), (5.1, 8.5)

   Slope: $\frac{15.5 - 8.5}{3.8 - 5.1} = -5.3846$

   $y - 15.5 = -5.3846\,(x - 3.8)$

   $y = -5.3846x + 35.9615$

   Point on line corresponding to middle summary point is

   $y = -5.3846\,(4.4) + 35.9615 = 12.269$.

   Must slide up $\frac{1}{3}(13.5 - 12.269) = 0.4103$.

   $y = -5.3846x + 36.3718$

   If average March temperature is 3.5°, this model predicts that the first blossoms will appear on April 17.

3. The (*average temperature, average daily cost of gas*) data set has the median-median line:

   gas = $-0.037 \times$ *temperature* + 2.971.

   The (*average temperature, average daily cost of electricity*) data set has the median-median line:

   electricity = $0.00214 \times$ *temperature* + 0.693.

   a. The median-median line for gas predicts an average daily cost of $2.27, and the median-median line for electricity predicts an average daily cost of approximately $0.73. The predicted cost for gas here is close to that found in Section 1.3, but the predicted cost for electricity is lower using the median-median line than judging by sight.

   b. For the gas line, the slope gives the expected decrease in the average daily cost of gas for each one-degree increase in average daily temperature, and the "gas-intercept" gives the expected average daily cost of gas ($2.97) when the temperature is 0° F. The "temperature-intercept" doesn't have much practical meaning, since it indicates the temperature at which the model predicts that the average daily cost of gas would be $0.00.

   Since the average daily cost of electricity seems unrelated to the average daily temperature, the slope of the electricity line has little meaning. Since that is the case, the "electricity-intercept" is a reasonable estimate of the average daily cost of electricity at any temperature between 23° F and 70° F.

EXERCISE SET 1.6, *pages 37–39*

1. a. (Figure 41) The line is a reasonable model although some residuals are relatively large; no pattern.

   b. (Figure 42) The fit here is very similar to the fit for Figure 42. There is no pattern, although some residuals are relatively large.

   c. (Figure 43) Since the residuals are noticeably positive at both ends and negative in the middle, a linear model is not a good model for the data. The data should be modeled by a function that is concave up.

EXERCISE SET 1.7, *pages 46–48*

1. The sum of squared residuals can be zero only if all of the residuals are zero. This happens only when each point of the data falls on the line.

3. a. Predictions will vary.

   b. The resulting least squares line is approximately $y = 0.983x + 0.083$.

   c. The effects of the two variant points do not offset each other, since the least squares line is not the line $y = x$. The point (9, 8) has greater impact on the slope, which is now less than 1. Variant points have greater impact on the slope if they are farther from the middle of the data.

   d. (2, 2), (5, 5), (8, 8)

   e. They have no effect; the medians do not change.

5. a. $y = 0.239x + 37.269$ where x is the number of chirps per minute and y is the temperature in degrees Fahrenheit. The slope suggests that for each additional chirp per minute, the temperature will be about 0.239 degree warmer. The y-intercept suggests that when there are zero chirps, the temperature is 37.3°. Since there cannot be a negative number of chirps, the model suggests that 37.3° is the warmest temperature at which there are no chirps.

   b. Counting the chirps in 15 seconds and adding 37 gives a reasonable estimate of the temperature.

   c. It is more likely that temperature is the independent variable, since temperature does not depend on how fast crickets are chirping. The new least squares line is $c = 4.148t - 153.696$. The slope suggests that for each increase of 1°F in temperature, there will be an increase of about 4 chirps.

7. a. (A, B):

median-median line	$y = 0.556x + 2.470$
least squares line	$y = 0.500x + 3.000$

(A, C):

median-median line	$y = 0.500x + 3.213$
least squares line	$y = 0.500x + 3.001$

(A, D):

median-median line	$y = 0.370x + 3.840$
least squares line	$y = 0.500x + 3.002$

(E, F):

least squares line	$y = 0.500x + 3.002$

b. Answers will vary. They are similar for (A, C), but there are substantial differences for (A, B) and (A, D).

c.–d. All least squares lines are very close. Without scatter plots or tables of data, it could be mistakenly concluded that all data sets are similar.

e. A linear model is appropriate for set (A, B). Set (A, C) has obvious curvature, so a line is not an appropriate model. Data set (A, D) appears linear with the exception of one outlier.

This point should be examined before fitting a line, since it could be an error in the data. All points but one in set (E, F) line up vertically; there does not appear to be any relationship among these variables.

EXERCISE SET 1.8, *pages 51–52*

1. For the radioactive contamination data, the *standard deviation of the residuals* from the least squares line is 13.035. Using twice the standard deviation to find error bounds, we obtain the lines below as reasonable upper and lower error bounds for the data.

$$y = 9.27x + 140.75 \text{ and } y = 9.27x + 88.71$$

For the Leaning Tower of Pisa data, the *standard deviation of the residuals* is 0.4003. Using twice the standard deviation to find error bounds, we obtain the lines below as reasonable upper and lower error bounds for the data.

$$y = 0.9319x + 1124.1391 \text{ and}$$
$$y = 0.9319x + 1122.5379$$

3. a. Least squares line: $k = 0.125 r - 41.430$, where k is the number of manatees killed and r is the number of powerboat registrations in thousands. The slope of this line suggests that for every ten thousand additional powerboat registrations, 1.25 more manatees will be killed in boating accidents. The x-intercept 331.81 suggests that 0 manatees would be killed if the number of registrations were less than 332 thousand. The residual plot shows that the line is a reasonably good model for the data.

b. Least squares line: $r = 21.602 y + 427.086$, where r is the number of powerboat registrations in thousands and y is the number of years since 1977. The slope of this line estimates that each year, there are roughly 21.6 (thousand) more powerboats registered than in the previous year. The y-intercept estimates that there were 427 (thousand) powerboat registrations in 1977. This number is reasonably close to the value of 447 thousand in the data. The "v" shape in the residual plot implies that the line is not a good model for this data.

c. Least squares line: $k = 2.629y + 12.343$, where y is the number of years since 1977 and k is the number of manatees killed. Based on this model, a reasonable estimate for the number of manatees accidentally killed in the year 2000 is 73. The standard deviation of the residuals is 5.259, therefore, reasonable error bounds are $k = 2.629y + 12.343 \pm 10.518$. We anticipate that the number of manatees killed will lie on the interval from 62 to 83.

CHAPTER 1 Review Exercises

page 53

1. a. Data do appear linear.

b. $y = 0.183x + 1.261$

c. The time needed to complete the wave increases approximately 0.18 second for every additional student in the wave.

d. The vertical intercept represents the time required by 0 students to complete the wave, which we expect should be 0 seconds; 1.26 seconds is close and probably represents human error in timekeeping.

e. $3.4 - y\,(12) \approx -0.05$

f. $y\,(300) \approx 58.06$ seconds; There should be very low confidence because you are extrapolating far beyond the domain of the data used to find the model.

g. $y = 0.183x + 0.452$
$y = 0.183x + 2.069$

These error bands are based on two standard deviations of the residuals.

h. $(7, 2.9); (15, 4.2); (22, 5.3)$.

2. The median-median line will provide a better fit when the data has outliers.

CHAPTER 2 Functions

EXERCISE SET 2.1, *pages 58–60*

1. In each case, the independent variable is graphed on the horizontal axis; the dependent variable is graphed on the vertical axis.

a.

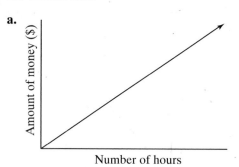

Number of hours

As the number of hours worked increases, the amount of money earned increases.

c.

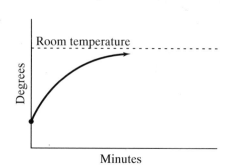

The temperature rises until it reaches room temperature.

e.

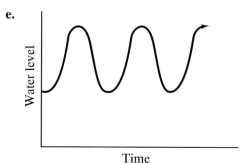

As time passes, the tides rise and fall.

g.

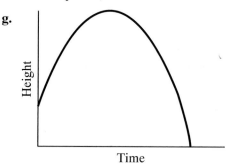

After the ball is hit, it rises and then falls until it hits the ground.

i.

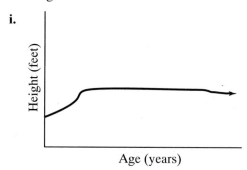

As a person ages, his height increases, maintains a certain level, and finally decreases slightly.

k.

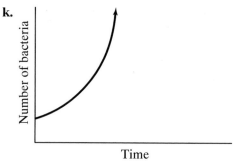

Assuming an unlimited supply of space, food, and other resources, the population can increase without bound.

3. a. $f(4) = 11$

b. $f(x) = 4 = 3x - 1$; $5 = 3x$; $x = \dfrac{5}{3}$

c. $f(2x) = 6x - 1$; $2f(x) = 2(3x - 1) = 6x - 2$;
$f(x + 2) = 3x + 5$; $f(x) + 2 = 3x + 1$

5. a. $x^2 + 3x + 2$

c. $0.25x^2 + 0.5x$

e. $x^2 + |x|$

6. a. $y = f(x - 4)$

b. $y = 3g(x)$

EXERCISE SET 2.2, *page 63*

1.

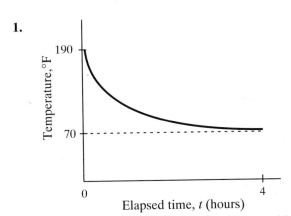

Answers will vary. The function illustrated has a range of $70 < y \le 190$, making the assumption that room temperature is 70Υ and therefore the temperature of the hot chocolate gets close to 70Υ.

3. a. $-4 \le x \le -2$, $-1 < x \le 4$

b. $-4 \le y \le 1$, $2 < y \le 3$

c. f is increasing for $-4 < x < -2$ and $-1 < x < 0$.
f is decreasing for $0 < x < 4$.

d. f is never concave up.
f is concave down for $-1 < x < 1$.

EXERCISE SET 2.3, *pages 69–72*

1. Check graphs using graphing calculator. Suggested viewing window: $-2 \le x \le 2$ and $-7 \le y \le 7$

If n is even, then the graph of $y = x^n$ is concave up, decreases to zero at $x = 0$, has a turning point at $x = 0$, and increases for $x > 0$. If n is odd, then the graph of $y = x^n$ is always increasing. For $x < 0$ it is concave down and for $x > 0$ it is concave up.

3. $x, x^2, x^3, \sqrt{x}, \dfrac{1}{x}$, constant

5. For $0 < x < 1$, $x > x^2 > x^3$. They all coincide at $x = 0$ and $x = 1$.

Squaring or cubing a number between 0 and 1 yields a result smaller than the original number. Cubing a number yields a smaller result than squaring the number.

7. a. $f(x) = |x|$, $f(x) = c$

b. $f(x) = \dfrac{1}{x}$, $f(x) = \sin x$, $f(x) = x$

9. If f is odd, $f(-x) = -f(x)$.

If f is even, $f(-x) = f(x)$.

These together imply $-f(x) = f(x)$, which implies $f(x) = 0$. A function cannot be both even and odd unless the function is $f(x) = 0$.

11. The graph of f is symmetric about the line $y = -x + 2$.

The graph of f is symmetric about the line $y = x$.

13. Three intersection points: $(-0.7667, 0.5877)$, $(2, 4)$, $(4, 16)$

15. a.

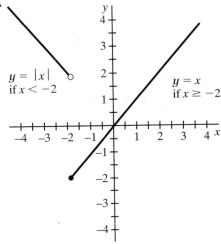

c.

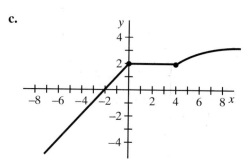

17. $f(x) = \dfrac{3}{2}x + 4$

EXERCISE SET 2.4, *pages 77–78*

1. a. $x \le 3$

   c. $-2 \le x \le 0$ or $x \ge 1$

   e. $x \in \Re, x \ne 2, x \ne 3$

   g. $x \in \Re, x \ne 1$

   i. $x \ge -3$

3. $Df: x > 2$ and $Dg: x \le -1$ or $x > 2$.

   Therefore, the functions are not identical.

EXERCISE SET 2.5, *pages 87–92*

1. a. $D = \{t: 0 \le t \le 7.355\}$

   $R = \{h: 0 \le h \le 218.9\}$

   b. Maximum height: 218.9 feet

   c. When $2 \le t \le 5.3125$

   d. Graph $h(t)$ and the horizontal line $h = 175$. Use "intersect" option on your graphing calculator.

   e. Approximately 7.355 seconds

3. a. Minimum value of distance occurs at $x \approx 0.24$ so the submarine should be located at approximately (0.24, 0.0576).

   b. $d < 4$ for $-1.325 < x < 1.568$

5. a. $R(x) = (15 + x)(50,000 - 2500x)$

   b. $0 \le x \le 20$ (x must be a multiple of 0.01.)

   c. $17.50 per ticket

7. a. $n = 1$ $P = 10,100$ $n = 36$ $P = 332.14$

     $n = 12$ $P = 888.49$ $n = 48$ $P = 263.34$

   b. $n \ge 1$, n an integer

     $100 < P \le 10,100$

   c. The outstanding balance would never be paid off.

11. Don't cut the wire; make a circle with all the wire.

13. The cable should be laid so that approximately 474 meters is underground.

15.
$$C(x) = \begin{cases} 5, & \text{for } 0 < x \le 5 \\ 5 + \lfloor x - 5 \rfloor(0.10), & \text{for } 5 < x \le 50 \end{cases}$$

   $D = (0; 50]$

   $R = \{\$5.00, \$5.10, \$5.20, \ldots, \$9.50\}$

17. $0 < r \le 0.009$; annual interest rate no more than 10.8%

19. 20 months

EXERCISE SET 2.8, *pages 101–103*

1. a. The graph of the absolute value function will be shifted vertically upward 5 units. Check graph using graphing calculator. Suggested window:
$-10 \le x \le 10$ and $0 \le y \le 10$

   c. The graph of $y = \dfrac{1}{x}$ is shifted 2 units to the left. Vertical asymptote is $x = -2$. Check graph using graphing calculator. Suggested window:
$-5 \le x \le 5$ and $-5 \le y \le 5$

   e. The graph of $f(x) = x^3$ is shifted 4 units to the right. Check graph using graphing calculator. Suggested window:
$-5 \le x \le 10$ and $-10 \le y \le 10$

   g. The graph of the sine function is shifted π units to the left. Check graph using graphing calculator. Suggested window:
$-2\pi \le x \le 2\pi$ with scale of $\dfrac{\pi}{2}$ and $-2 \le y \le 2$

   i. The graph of $f(x) = x^2$ is compressed vertically by a factor of 5. Check graph on graphing calculator. Suggested window:
$-5 \le x \le 5$ and $-5 \le y \le 10$

k. The graph of $f(x) = 2^x$ is stretched horizontally by a factor of 3. Check graph on graphing calculator. Suggested window: $-5 \le x \le 10$ and $-5 \le y \le 10$

4. a.

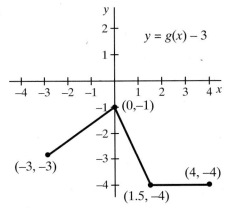

$y = g(x) - 3$

domain: $-3 \le x \le 4$

range: $-4 \le y \le -1$

c.

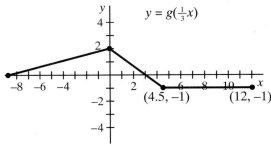

$y = g(\tfrac{1}{3}x)$

domain: $-9 \le x \le 12$

range: $-1 \le y \le 2$

5. a.

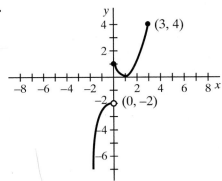

c.

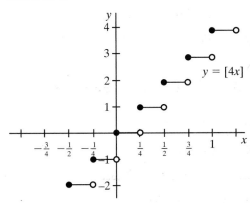

7. a. $y = \sqrt{x+2}$

   b. $y = \sin\dfrac{1}{2}x$

   c. $y = \dfrac{1}{x} + 2$

   d. $y = \dfrac{1}{2}|x|$

9. All new y-values are c times their previous value. If $c > 1$, the graph is stretched vertically away from the x-axis. If $0 < c < 1$, the graph is compressed vertically toward the x-axis.

11. a. The graph has been compressed horizontally by a factor of 4.

$y = [4x]$

EXERCISE SET 2.9, *pages 109–111*

1. a. This graph is flipped over the x-axis and vertically stretched by a factor of 2 compared to the toolkit function $T(x) = \sqrt{x}$. Check graph using graphing calculator. Suggested window: $-10 \le x \le 10$ and $-10 \le y \le 10$

   c. The graph is vertically stretched by a factor of 3, shifted 1 unit to the left, and shifted down 2 units compared to the toolkit function $T(x) = x^3$. Check graph using graphing calculator. Suggested window: $-5 \le x \le 5$ and $-10 \le y \le 10$

e. Complete the square to rewrite the function as $f(x) = 4(x + 3.5)^2 + 4$. This graph is vertically stretched by a factor of 4, shifted 3.5 units to the left, and shifted 4 units upward compared to the toolkit function $T(x) = x^2$. Check graph on graphing calculator. Suggested window: $-10 \leq x \leq 10$ and $-10 \leq y \leq 15$

g. This graph is vertically stretched by a factor of 2, flipped over the x-axis, and shifted to the left π units compared to the toolkit function $T(x) = \sin x$. Check graph on graphing calculator. Suggested window: $-2\pi \leq x \leq 2\pi$ with $\frac{\pi}{2}$ scale and $-3 \leq y \leq 3$

i. This graph is vertically compressed by a factor of 5 and flipped over the x-axis compared to the graph of the toolkit function $T(x) = x^2$. Check graph on graphing calculator. Suggested window: $-10 \leq x \leq 10$ and $-10 \leq y \leq 10$

k. This graph is flipped over the x-axis, then shifted 3.5 units to the right, and then up 0.25 unit compared to the graph of the toolkit function $T(x) = x^2$. Note: $g(x) = -(x - 3.5)^2 + 0.25$. Check graph on graphing calculator. Suggested window: $-5 \leq x \leq 8$ and $-5 \leq y \leq 5$

m. This graph is vertically stretched by a factor of 4 and shifted 3 units to the right compared to the graph of the toolkit function $T(x) = \frac{1}{x}$. Check graph on graphing calculator. Suggested window: $-10 \leq x \leq 10$ and $-10 \leq y \leq 10$

o. This graph is vertically stretched by a factor of 2, then flipped over the x-axis, shifted 4 units to the right, and shifted up 3 units compared to the graph of the toolkit function $T(x) = \sqrt{x}$. Check graph on graphing calculator. Suggested window: $-5 \leq x \leq 10$ and $-5 \leq y \leq 5$

q. This graph is flipped over the y-axis and then shifted 1 unit to the right compared to the graph of the toolkit function $T(x) = 2^x$. Check graph on graphing calculator. Suggested window: $-5 \leq x \leq 10$ and $-2 \leq y \leq 8$

s. This graph is flipped over the x-axis, then flipped over the y-axis, and then shifted 2 units to the right compared to the graph of the toolkit function $T(x) = \sqrt{x}$. Check graph on graphing calculator. Suggested window: $-10 \leq x \leq 5$ and $-5 \leq y \leq 5$

2. a.

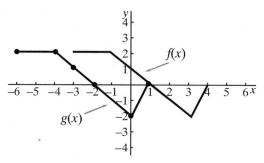

Domain of g: $-6 \leq x \leq 1$

Range of g: $-2 \leq y \leq 2$

c.

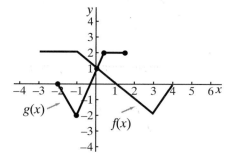

Domain of g: $-2 \leq x \leq \frac{3}{2}$

Range of g: $-2 \leq y \leq 2$

e.

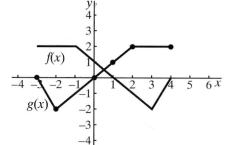

Domain of g: $-3 \leq x \leq 4$

Range of g: $-2 \leq y \leq 2$

g.

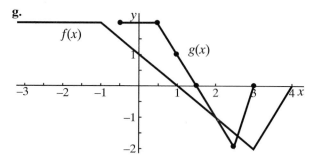

Domain of g: $-\frac{1}{2} \leq x \leq 3$

Range of g: $-2 \leq y \leq 2$

3. a. The graph is shifted up 11 units compared to $T(x) = 2^x$. Change to $10 \le y \le 30$ for a better window.

c. The graph of $T(x) = \sqrt{x}$ is flipped across the y-axis and shifted left 50 units to get this graph. A better window would be $-60 \le x \le -40$; $-1 \le y \le 5$.

5. i. a. Quadratic function

b. The toolkit quadratic function is reflected across the x-axis, stretched vertically, and shifted right by 2 units and up by 144 units.

c. $y = -16(x - 2)^2 + 144$, where $x =$ time in seconds and $y =$ height in feet.

iii. a. Cubic function

b. The toolkit cubic function is compressed vertically.

c. $y = 0.0045x^3$, where $x =$ diameter in inches and $y =$ volume in hundreds of board feet.

EXERCISE SET 2.10, *pages 115–116*

1. a. $f \circ g = -\dfrac{1}{x} - 1$; $\{x: x \ne 0, -1\}$

$g \circ f = 1 - \dfrac{1}{x}$; $\{x: x \ne 0, 1\}$

c. $f \circ g = \sqrt{\dfrac{1 - 5x^2}{x^2}}$;

$\left\{x: -\dfrac{\sqrt{5}}{5} \le x < 0 \text{ or } 0 < x \le \dfrac{\sqrt{5}}{5}\right\}$

$g \circ f = \dfrac{1}{x-5}$; $\{x: x > 5\}$

e. $f \circ g = |\sqrt{x - 1} + 1| = \sqrt{x - 1} + 1$; $\{x: x \ge 1\}$

$g \circ f = \sqrt{|x + 1|} - 1$;

$\{x: x \ge 0 \text{ or } x \le -2\}$

g. $f \circ g = \dfrac{\sqrt{x + 2}}{\sqrt{x + 2} - 1}$; $\{x: x \ge -2, x \ne -1\}$

$g \circ f = \sqrt{\dfrac{3x - 2}{x - 1}}$; $\{x: x \le \tfrac{2}{3} \text{ or } x > 1\}$

i. $f \circ g = \sqrt{\sin x}$; $\{x: 2k\pi \le x \le (2k + 1)\pi,$ where k is an integer$\}$

$g \circ f = \sin \sqrt{x}$; $\{x: x \ge 0\}$

3. $g(g(1)) = \dfrac{5}{2}$

$g(g(g(1))) = \dfrac{11}{4}$

5. a. $f(102) = 92$

c. $f(97) = 91$

e. $f(301) = 291$

The range $= \{91, 92, 93 \ldots\}$.

EXERCISE SET 2.12, *pages 126–127*

1. $D_f: \{x : x > 0\}$; $R_f = \{y : y \in \Re\}$

$D_{1/f}: \{x : x > 0, x \ne 1\}$; $R_{1/f} = \{y : y \ne 0\}$

a. $f(x) = \dfrac{1}{x}$, $g(x) = \sin x$. Suggested viewing window: $-2\pi \le x \le 2\pi$ with $\dfrac{\pi}{2}$ scale and $-5 \le y \le 5$

c. $f(x) = (x - 2)^3$, $g(x) = |x|$ Suggested viewing window: $-5 \le x \le 5$ and $-10 \le y \le 10$

e. $f(x) = \dfrac{1}{x}$, $g(x) = 2^x - 1$. Suggested viewing window: $-5 \le x \le 5$ and $-5 \le y \le 5$

g. $f(x) = \sqrt{x}$, $g(x) = |x - 4|$. Suggested viewing window: $-10 \le x \le 10$ and $-5 \le y \le 5$

i. $f(x) = \sqrt{x}$, $g(x) = \sin x$. Suggested viewing window: $0 \le x \le 9.4$ and $-0.5 \le y \le 5.8$.

k. $f(x) = \dfrac{1}{x}$, $g(x) = x^2 - 3x + 2$. Suggested viewing window: $-4 \le x \le 5$ and $-7 \le y \le 7$

m. $f(x) = (x - 2)^2 - 1$, $g(x) = |x|$. Suggested viewing window: $-10 \le x \le 10$ and $-10 \le y \le 10$

3. Where $f(x) = 0$ or $f(x) = \pm 1$, $(f(x))^2 = 0$ and $(f(x))^2 = 1$ respectively.

Where $|f(x)| > 1$, $(f(x))^2 > f(x)$, and as $f(x) \to \pm \infty$, $(f(x))^2 \to \infty$.

Where $0 < f(x) < 1$, $(f(x))^2 < f(x)$ and $(f(x))^2 < 1$.

Where $-1 < f(x) < 0$, $(f(x))^2 < |f(x)|$, and $(f(x))^2 < 1$.

5. c. As a transformation of $T(x) = |x|$, $h(x)$ has a horizontal shift of 3 units to the left. As a composition, $h(x) = |T(x)|$ if $T(x) = x + 3$.

  d. If $T(x) = \sqrt{x}$, shift left 1 unit.

  If $T(x) = x + 1$, $s(x) = \sqrt{T(x)}$.

7. a. $p(q(3)) = 10$

  $p(q(3.3)) = 10$

  $p(q(3.5)) = 10$

  $p(q(3.9)) = 10$

  b. $q(p(3)) = 10$

  $q(p(3.3)) = 11$

  $q(p(3.5)) = 13$

  $q(p(3.9)) = 16$

9.

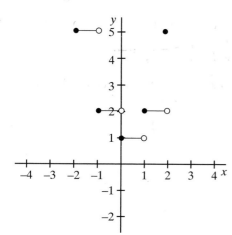

EXERCISE SET 2.13, *pages 136–137*

1. a. $f^{-1}(x) = (x - 2)^{\frac{1}{3}} + 1$ or $\sqrt[3]{x - 2} + 1$

  c. $f^{-1}(x) = 5 - x^2$, $x \geq 0$

  e. $f^{-1}(x) = \sqrt{9 - x^2}$, $x \geq 0$

  g. $f^{-1}(x) = 3 - \sqrt{9 - x}$

3. Domain of f Domain of f^{-1}

	Domain of f	Domain of f^{-1}
a.	$\Re$	$\Re$
c.	$x \leq 5$	$x \geq 0$
e.	$0 \leq x \leq 3$	$0 \leq x \leq 3$
g.	$x \leq 3$	$x \leq 9$

5. a. $h^{-1}(x) = \sqrt[3]{1 - x^3}$

  c. $h^{-1}(x) = \frac{3x - 1}{2x - 3}$

7. a. If domain of $f(x)$ is restricted to $x \geq 1$, $f^{-1}(x) = 1 + \sqrt{1 + x}$.

  b. If domain of $f(x)$ is restricted to $x \geq 1$, then $f^{-1}(x) = x$.

  c. If the domain of $f(x)$ is restricted to $0 \leq x \leq 2$, then $f^{-1}(x) = \sqrt{4 - x^2}$, $-2 \leq x \leq 0$.

9. a. $g^{-1}(x) = \begin{cases} x + 2, & x \leq 0 \\ 0.5x + 2, & x > 0 \end{cases}$

  b. $g^{-1}(x) = \begin{cases} 1 - \sqrt{3 - x}, & x \leq 3 \\ 2x - 5, & 3 < x < 6 \\ (x - 6)^2 + 7, & x \geq 6 \end{cases}$

EXERCISE SET 2.14, *pages 142–145*

1. a. A quadratic function would be a good choice.

  b. Ordered pairs of the form $(x, \sqrt{y})$ look linear.

  c. $y = (3.4171x + 0.0977)^2$

  d. Although the curve appears to fit the data well when it is superimposed on the data, there is a pattern in the residual plot that might imply a different model would be more appropriate.

3. If we assume the data can be modeled by a square root function, then (x, y^2) would be a reasonable re-expression.

5. a. $y = \dfrac{1}{0.00285x + 0.01567}$

  b. $y(42) \approx 73$. We are reasonably confident in the result, since we are interpolating.

  c. The residual plot shows that the error is greater for short waterfalls.

7. $y = (0.14238x - 0.00058)^{\frac{2}{3}}$

9. a. The quadratic least squares model is $y = 11.390x^2 + 1.096x - 0.043$.

  c. The residual plot for this fit is more random than that obtained in Exercise 1. This quadratic function seems to be the better fit.

EXERCISE SET 2.16, *pages 153–155*

1. Using $y(t) = -9t^2 + 6.25$ and $x(t) = 58.\overline{6}t$, the cap clears the fence when $t = 0.8\overline{3}$.

3.

t	$x(t)$	$y(t)$
-2	-4	-6
-1	-2	-3
3	6	9
2	4	6
1	2	3
0	0	0

The data (x, y) are linear; slope $= \frac{3}{2}$.

The graph is in fact linear. The reason is that both x and y change by a constant amount for a given increase in t.

5. **a.** $y = \frac{3}{2}x - \frac{7}{2}$

   b. $y = 2x^2$

   c. $y = \pm\sqrt{x}$

   d. $y = x, x \geq 0$

7. **a.** At $t = 4.7$, Coco can get within 1.5 feet of the spider.

   b. 5496 ft

   c. It changes direction every time the function $y(t) = 1.5 \sin t + 6$ hits a maximum or minimum for $0 \leq t \leq 10$. This happens for $t = \frac{\pi}{2}, \frac{3\pi}{2}, \frac{5\pi}{2}$.

CHAPTER 2 REVIEW EXERCISES,
pages 156–157

1. $f(s) = \begin{cases} 10s & \text{if } s \leq 10 \\ 9s & \text{if } 10 < s \leq 20 \\ 7.5s & \text{if } s > 20 \end{cases}$

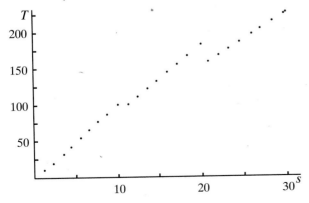

2. **a.** $x \in \Re, x^2 - 5 \neq 0$, so $x \neq \sqrt{5}$

   b. We need a nonnegative radicand. Therefore, $x^2 + 5x + 6 \geq 0$ or $(x + 3)(x + 2) \geq 0$. Using a number line analysis, we find $x \geq -2$ or $x \leq -3$.

   c. $x \in \Re, x - 2 \neq 0$, so $x \neq 2$.

3. Since our domain of f was restricted to $x \geq 1$, the range of f^{-1} is $y \geq 1$.
   Therefore, $f^{-1}(x) = 1 + \sqrt{\frac{x + 12}{2}}$

4. **a.** $f(x) = 1 - |x|$

   b. $f(x) = \frac{1}{x - 2} - 1$

   c. $f(x) = |x^2 - 1|$

   d. $f(x) = 2 \sin \frac{x}{2}$

   e. $f(x) = \sqrt{|x| - 2}$

5. Let $C = $ cost in dollars and $x = $ speed in mph.
   $$C = \$10\left(\frac{240}{x}\right) + \left(0.5 + \frac{x}{40}\right)240$$

   Graph this function on your calculator and find minimum at $x = 20$ mph.

6. **a.** $D: -2 \leq x \leq 2; R: -1 \leq y \leq 1$

   b. i. $g(x) = |f(x)| + 2$

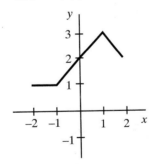

   ii. $g(x) = 2f(\frac{1}{2}x)$

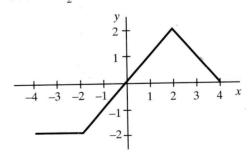

iii. $g(x) = f(-x)$

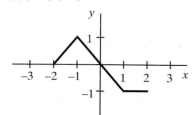

iv. $g(x) = f(|x|)$

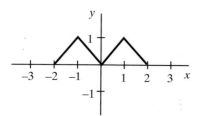

v. $g(x) = f(2x - 3)$

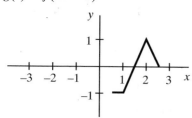

7. The graph is decreasing, concave up, and has a horizontal asymptote at room temperature.

8. a. The graph is a circle of radius 5 centered at $(1, -2)$.

b. The graph is reasonable because the x-values cycle between -4 and 6, while the y-values cycle between -7 and 3.

c. $x(t) = 5 \sin t - 1$

$y(t) = 5 \sin\left(x + \dfrac{\pi}{2}\right) - 1$

9. A possible graph follows.

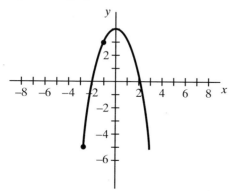

10. $f \circ g(x) = \dfrac{1}{(\sqrt{x+2})^2 - 4} = \dfrac{1}{x - 2}$

Domain: $x \geq -2, x \neq 2$

11. Let $h(x) = \dfrac{1}{x}$ and $g(x) = x^2 - 9$. The $f(x) = h(g(x))$. First graph $y = x^2 - 9$, then take the reciprocal of all the y-values.

12. Knowing something about the phenomena often gives you an idea of the type of function that may be used to model the data. Then, you can make an educated guess at the family of functions that models the data. Use the inverse of the hypothesized function to linearize the data, if your guess is correct. Using the inverse on y-values has the effect of composing a function with its inverse, or doing and undoing the original operation that produced the curvature.

CHAPTER 3 Exponential and Logarithmetic Functions

EXERCISE SET 3.1, *pages 163–166*

1. a. Using the recursive system $D_0 = 400$, $D_n = D_{n-1} - 0.67D_{n-1} + 400$, it takes only 1 iteration to reach 532, which is within the therapeutic range. This would be after she takes the second dose.

b. About 597

3. a. r represents the percent filtered out of the body in four hours.

c. $\dfrac{b}{r}$

d. 400 mg

5. a. Using the recursive system $C_0 = 2.5$, $C_n = C_{n-1} - 0.15C_{n-1}$, it will take 6 iterations or 6 days for the chlorine level to go below 1 ppm.

b. Using the recursive system $C_0 = 2.5$, $C_n = C_{n-1} - 0.15C_{n-1} + 0.5$, we see that the level will eventually stabilize at 3.33 ppm.

c. The level will stabilize at 0.667 ppm.

d. 0.27 ppm must be added each day.

7. b. The level ranges from 293.33 to 733.33.

c. Dosages between 450 mg and 480 mg will work.

EXERCISE SET 3.2, *pages 171–172*

1. a. 28 payments of $200 plus a final payment of $182.43

b. 31 payments of $200 plus a final payment of $113.97

c. 26 payments of $200 plus a final payment of $154.69

d. 22 payments of $250 plus a final payment of $106.74

3. a. 5 annual payments of $4985.33 for a total of $24,926.65

b. 5 annual payments of $6172.31 for a total of $30,861.55

c. 5 annual payments of $6647.10 for a total of $33,235.50

5. The 8% loan over 15 years will have 15 annual payments of $5841.48 for a total of $87,622.20. The 7% loan will have 20 annual payments of $4719.65 for a total of $94,393. Using the total repaid as the criteria for picking the "better" loan, the shorter loan is better.

EXERCISE SET 3.3, *pages 178–179*

1. Recursive system: $P_0 = 1,000,000$, $P_n = P_{n-1} + 0.14P_{n-1}$

Explicit function: $P(t) = 1,000,000(1.114)^t$

Human growth is discrete, but with such a large population, the whole population growth is nearly continuous.

3. Recursive system: $P_0 = 230$, $P_n = P_{n-1} - 0.05P_{n-1}$

Explicit function: $P(t) = 230(0.95)^t$

The population changes discretely with each death, but the changes occur more often than once a year.

5. Assuming that the growth is geometric (or exponential), the growth rate is $\frac{7422 - 6712}{6712} \approx 0.106$. The number of cases diagnosed by the end of the year 2005 will be about 18,346.

7. $15,104.81

9. $A(t) = A_0(0.5)^{\frac{t}{25}}$. After 300 days, only 0.0244% remains.

11. a. $P(t) = 1.2(1.013)^t$ (P in billions)

b. Between 53 and 54 years

c. $P(t) = 1.2(2)^{\frac{t}{54}}$, if we assume a 54-year doubling time.

d. $P_0 = 1.2$, $P_n = P_{n-1} + 0.013P_{n-1}$

e. The doubling time changes very little. It is closer to 53.

13. $P_0 = a$, $P_n = P_{n-1} + k$

EXERCISE SET 3.5, *pages 185–187*

1. a. 1257.79   261,002.52 $3.0926 \cdot 10^{24}$

b. 8.02526   19.88159   19.99999

c. 1.99804   2   2

d. 0.66602   0.666667   0.666666

e. 0.33333   0.333333   0.333333

f. -22.6667 $-1.62645 \cdot 10^{17}$ $-4.93562 \cdot 10^{175}$

3. a. $B_0 = 1000, B_n = B_{n-1} + \left(\frac{0.10}{12}\right)B_{n-1} + 150$

b. $7615.46

c. $B(4) = A_0\left(1 + \frac{0.10}{12}\right)^4 + \dfrac{150 - 150\left(1 + \frac{0.10}{12}\right)^4}{1 - \left(1 + \frac{0.10}{12}\right)}$

d. $B(n) = A_0\left(1 + \frac{0.10}{12}\right)^n + \dfrac{150 - 150\left(1 + \frac{0.10}{12}\right)^n}{1 - \left(1 + \frac{0.10}{12}\right)}$

5. 3336

9. $B(n) = B(1 + i)^n - P\left(\frac{(1 + i)^n - 1}{i}\right)$

11. a. $405,391.08

b. 8.98% (very close to 9%)

EXERCISE SET 3.7, *pages 193–196*

3. $290.06; $290.32; $290.43; $290.45; $290.46

5. $113,086,417,610 in the year 2000

9. $1,695,887.95

13. $3901.44

EXERCISE SET 3.8, *pages 203–206*

1. a. $y = e^{(2x)}$ is a horizontal compression of $y = e^x$, written as $y = (e^2)^x \approx 7.4^x$. Both graphs contain $(0,1)$ and horizontal asymptote $y = 0$.

   c. Written as $y = (-4)2^x$, the graph of $y = 2^x$ is flipped over the x-axis and stretched vertically by a factor of 4. Written as $y = -1(2^2)(2^x) = -1(2^{x+2})$, the graph of $y = 2^x$ is shifted 2 units to the left and then flipped over the x-axis.

5. a. The graph contains the points $(0, 14.91)$ and $(1, 12.454)$.

   b. 14.91 mg

   c. 7.26 mg

   d. Slightly more than 6 hours

7. a. Use your calculator to graph $y = 1000\left(1 + \frac{0.0514}{4}\right)^{4t}$; about 7.94 years.

   b. Between 7.88 and 7.89 years

EXERCISE SET 3.9, *pages 211–213*

1. a. $4^3 = 64$

   c. $10^4 = 10,000$

3. a. 4 since $3^4 = 81$

   c. $\frac{3}{5}$ since $\log_2 \sqrt[5]{2^3} = \log_2 2^{\frac{3}{5}}$

5. a. $f^{-1}(x) = -\frac{1}{5}\ln x; x > 0$

   c. $f^{-1}(x) = \frac{1}{2}e^x; \Re$

7. a. It is one-to-one.

   b. It is a function.

   c. No; for $f(x) = x^2$, $f(-3) = f(3)$, but $-3 \neq 3$.

9. a. $x = \frac{1}{3}$

   c. $x = 36$

   e. $x = -6$ or $x = 1$

EXERCISE SET 3.10, *pages 218–219*

1. If $b > 1$, the function is increasing. For $0 < b < 1$, the function is decreasing.

4. a. $\log\left(\frac{1}{x}\right) = \log(x^{-1}) = -\log x$

   c. $\frac{1}{\ln x} = f(g(x))$, where $g(x) = \ln x$ and $f(x) = \frac{1}{x}$.

   e. $\ln(x^2 - 5) = f(g(x))$, where $g(x) = x^2 - 5$ and $f(x) = \ln x$.

   g. $\log(10^{-x^2 + 2}) = -x^2 + 2$

EXERCISE SET 3.11, *pages 226–228*

1. $g(x) = \left(\frac{1}{2}\right)^x = \left(10^{\log(\frac{1}{2})}\right)^x \approx 10^{-0.301x}$

3. $66,400.92

7. $\frac{\ln 2}{\ln 1.03} \approx 23.45$ years; $10 \cdot 2^3 = 80$ million

9. About 19.5 minutes

11. 8.7%

15. $\log_b a = \frac{\log_c a}{\log_c b}$

19. a. $t = \frac{\ln 2}{r}$

   c. $t = \frac{e^{S/A} - C}{B}$

   e. $y = a \cdot e^{bx} + 10$

EXERCISE SET 3.12, *pages 232–234*

1. 74 dB

3. a. 5.77

   b. $178 \cdot 10^{17}$

7. Distilled water: 1.0×10^{-7}; rain and snow: 2.5×10^{-6}; 25 times greater

EXERCISE SET 3.13, *pages 246–250*

1. The residuals are strongly patterned; the residuals for the quadratic model are "U" shaped.

3. We would fit a line to $(x, \log y)$. The line is $\log y = 0.00639x + 0.6047$ and the model is $y = 10^{0.6047} \cdot 10^{0.00639x}$. Both this and the model in the section can be rewritten as $y = 4.02(1.0148^x)$.

5. $\ln y = -1.9870 \ln x + 9.1824$

$$y = 9724.8x^{-1.9870}$$

7. Log-log re-expression linearizes the data. Least squares line using skid length as independent variable and speed as dependent variable is: $\ln y = 0.5002 \ln x + 1.6484$ and $y = 5.20x^{0.5002}$.

9. a. The data is reasonably linear when re-expressed as $\left(\frac{1}{x}, y\right)$.

b. $y = 720.564\left(\frac{1}{x}\right) + 38.118$

c. $y = 720.564\left(\frac{1}{25}\right) + 38.118 \approx 67°$

11. The least squares line through the ordered pair $(\log x, \log y)$ is $\log y = 1.4999 \log x - 2.9515$. Exponentiating both sides, we have

$$y = 10^{1.4999} \cdot 10^{-2.9515}$$

$$y = 0.001118x^{1.4999}$$

$$y = 0.001x^{1.5}.$$

EXERCISE SET 3.14, *pages 253–254*

1. Two standard deviations of residuals for linearized data is (0.24554).

$$P_{\text{upper}}(t) = 5.145e^{0.0147t}$$

$$P_{\text{lower}}(t) = 3.148e^{0.0147t}$$

Population per square mile in 1995 is between $P_{\text{lower}}(205) = 64.09$ and $P_{\text{upper}}(205) = 104.74$.

Population per square mile in 1823 is between $P_{\text{lower}}(33) = 5.11$ and $P_{\text{upper}}(33) = 8.36$.

Population per square mile in 1998 is between $P_{\text{lower}}(208) = 66.98$ and $P_{\text{upper}}(208) = 109.45$.

CHAPTER 3 REVIEW EXERCISES, *pages 264–265*

1. a. $A_0 = 150$

$$A_n = A_{n-1}\left(1 + \frac{.06}{12}\right) + 150$$

b. 58 months (4 years, 10 months)

Note that A_{57} means that you have made 57 deposits after the initial one, for a total of 58 deposits over 58 months.

c. $A = \dfrac{150(1.005)^{t+1} - 150}{0.005}, t \geq 0$

2. Monthly: $1 + r = \left(1 + \frac{0.06}{12}\right)^{12}; r = 0.06168$ or 6.17%

Continuously: $e^{0.06} \approx 1.06184; r \approx 6.18\%$

3. $A \cdot \left(1 + \frac{x}{365}\right)^{9 \cdot 365} = 2A$

$$1 + \frac{x}{365} = 1.000211026$$

$$\frac{x}{365} = 0.000211026$$

$$x = 0.0770244793$$

7.70% annual interest

4. a.

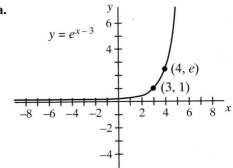

$y = e^{x-3}$ with points $(4, e)$ and $(3, 1)$

b.

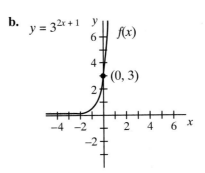

$y = 3^{2x+1}$, $f(x)$ with point $(0, 3)$

c. $y = \ln(x^2 - 4)$

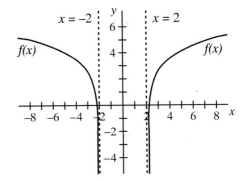

with vertical asymptotes $x = -2$ and $x = 2$

d.

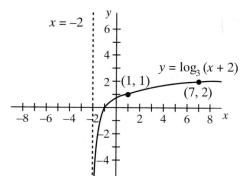

$x = -2$

$y = \log_3 (x + 2)$

$(1, 1)$

$(7, 2)$

e. $y = \log(2x)$

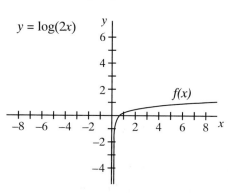

$f(x)$

f. $y = \dfrac{1}{e^x - 1}$

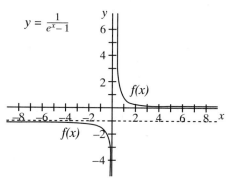

$f(x)$

$f(x)$

5. a. -3

b. 17

c. x^2

6. a. $4^x = 12$

$\log 4^x = \log 12$

$x \log 4 = \log 12$

$x = \dfrac{\log 12}{\log 4}$

$x \approx 1.79$

b. $\log x + \log(x - 3) = 1$

$\log(x^2 - 3x) = 1$

$x^2 - 3x = 10$

$x^2 - 3x - 10 = 0$

$(x - 5)(x + 2) = 0$

$x = 5 \text{ or } x = -2$

5 is the only solution because -2 is not in the domain of $\log x$ or $\log(x - 3)$.

c. $6000 = 4000e^{0.06t}$

$1.5 = e^{0.06t}$

$\ln 1.5 = \ln e^{0.06t}$

$\ln 1.5 = 0.06t$

$t = \dfrac{\ln 1.5}{0.06}$

$t \approx 6.76$

d. $\ln(\ln x) = 2$

$e^{\ln(\ln x)} = e^2$

$\ln x = e^2$

$e^{\ln x} = e^{e^2}$

$x = e^{e^2}$

$x \approx 1618.18$

7. $\log_b \dfrac{1}{x} = \log_b 1 - \log_b x$

$= 0 - \log_b x$

$= 0 - 2$

$= -2$

8. a. $f(x) = e^{2x + 1} - 2$

$x = e^{2y + 1} - 2$

$x + 2 = e^{2y + 1}$

$\ln(x + 2) = 2y + 1$

$2y = \ln(x + 2) - 1$

$y = \dfrac{\ln(x + 2) - 1}{2}$

$f^{-1}(x) = \dfrac{1}{2} \ln(x + 2) - \dfrac{1}{2}$

Domain: $x > -2$

b. $f(x) = 2 \log(x - 4)$

$x = 2 \log(y - 4)$

$\frac{x}{2} = \log(y - 4)$

$10^{\frac{x}{2}} = y - 4$

$y = 10^{\frac{x}{2}} + 4$

$f^{-1}(x) = 10^{\frac{1}{2}x} + 4$

Domain: $\Re$

9.

$25 = 22.5e^{b \cdot 1} + 22.9$

$\frac{25}{22.5} = e^{b}$

$\ln\left(\frac{25}{22.5}\right) = e^{b}$

$b = 0.1053$

$45 = 22.5e^{0.1053t}$

$2 = e^{0.153t}$

$\ln 2 = 0.1053t$

$t = \frac{\ln 2}{0.1053} \approx 6.6$

6.6 years

10. a. Re-express temperature values by taking $\ln(\text{temp} - 30)$ and fit a least squares line through linearized data.

$\ln(y - 30) = -0.04075x + 3.80041$

$y - 30 = e^{-0.04075x} \cdot e^{3.80041}$

$y = 44.7195e^{-0.04075x} + 30$

b. $35 = 44.7195e^{-0.04075x} + 30$

$5 = 44.7195e^{-0.04075x}$

$\ln\left(\frac{5}{44.7195}\right) = -0.04075x$

$x \approx 53.8$ minutes

11. a. $60 = 80(1 - e^{-0.08t})$

$0.75 = 1 - e^{-0.08t}$

$-0.25 = -e^{-0.08t}$

$\ln(0.25) = -0.08t$

$t = \frac{\ln(0.25)}{-0.08}$

$t \approx 17.3$ weeks

b. Upper limit to typing speed is 80 words per minute. As values of t get very large, values of w get close to 80.

CHAPTER 4 Modeling

EXERCISE SET 4.1, *pages 271–272*

1. a. All darts hit the board, and the darts are randomly thrown.

   b. $\frac{16\pi - 4\pi}{18^2} \approx 0.116$

   c. 2.27 inches

2. a. 0.7071

   b. $x = 7$

3. 0.64

EXERCISE SET 4.2, *page 274*

1. The long-run ratio is $\frac{A_n}{B_n} \approx 4$.

2. a. The long-run ratios are $\frac{Z_n}{Y_n} = \frac{10}{3}$, $\frac{Z_n}{Y_n} = \frac{5}{6}$ in both cases.

   b. If $X_0 = 5$, $Y_0 = 25$, $Z_0 = 10$, then the long-run ratios are $\frac{Z_n}{Y_n} = \frac{10}{3}$, $\frac{Z_n}{Y_n} = \frac{5}{6}$.

3. In the long run, 24% of the buffalo are calves, 13% are yearlings, and 63% are adults. The initial amount does not affect long term behavior.

EXERCISE SET 4.3, *page 278*

1. 20 shots

2. Yes, he could now have 30 hits out of 98 times at bat or 33 hits out of 108 times at bat.

EXERCISE SET 4.4, *page 287*

1. Using the distance method, D is the best player.

2. Using the distance method, the New Air Flow is most like the Old Air Flow.

CHAPTER 5 Circular Functions and Trigonometry

EXERCISE SET 5.1, *pages 310–312*

1. a. -1

 b. -1

 c. 0

 d. 1

 e. 0

 f. 0

3. Period $= \pi$

 Asymptotes at $k\pi$, for k an integer

 x-intercepts at $\frac{\pi}{2} + k\pi$, for k an integer.

 Check graph on calculator. Graph $y = \frac{1}{\tan x}$ with viewing window of $-2\pi \le x \le 2\pi$ and $-4 \le y \le 4$.

5. Shift $y = \sec x$ to the right $\frac{\pi}{2}$ units to obtain

 $y = \csc x = \sec\left(x - \frac{\pi}{2}\right)$.

7. a. $x = \frac{\pi}{2}$

 b. $0 < x < \pi$

 c. $x = 0$ or $x = \pi$ or $x = 2\pi$

 d. $x = 0$ or $x = 2\pi$

9. On the interval $-\frac{\pi}{4} < x < \frac{\pi}{4}$, the graphs of $y = \sin x$ and $y = x$ look almost identical.

EXERCISE SET 5.2, *pages 317–320*

1. $d = \pm 1.3$ when $t = k$, where k is an integer.

3. $y = 1.3 \sin\left(\pi\left(t - \frac{1}{2}\right)\right)$.

5. a.

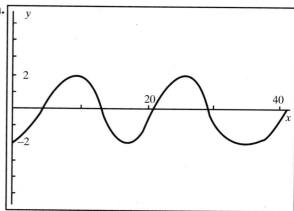

 b.

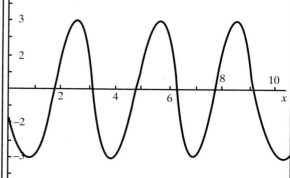

7. a. $y = 4 \sin x$

 b. $y = 0.1 \sin\left(6\left(x - \frac{\pi}{4}\right)\right)$

 c. $y = \frac{1}{3} \sin\left(2(x + 1)\right)$

 d. $y = \sin\left(\pi(x - \pi)\right)$

 e. $y = 2 \sin\left(8\pi(x + 2)\right)$

 f. $y = \frac{1}{2} \sin\left(20{,}000\pi x\right)$

9. a. $\frac{1}{2\pi}$

 b. $\frac{3}{\pi}$

 c. $\frac{1}{\pi}$

 d. $\frac{1}{2}$

 e. 4

 f. $10{,}000$

11. Period $= 1$ second

 Frequency $= 1$ cycle per second

 $y = 1.3 \cos\left(2\pi t\right) + 23.1$

13. $v(t) = -0.685 \cos\left(\frac{20\pi}{17.1}t\right) + 1.015$

$h(t) = -2.26\sin\left(\frac{10\pi}{17.1}t\right)$ or $h(t) = 2.26\sin\left(\frac{10\pi}{17.1}t\right)$

EXERCISE SET 5.4, *pages 326–328*

3. Ant stops in fourth quadrant.

$(x, y) = (\cos 5, \sin 5) = (0.2837, -0.9589)$

5. a. $(x, y) = (0.3624, 0.9320)$

   b. $(x, y) = (-0.3624, 0.9320)$

     distance $= (\pi - 1.2)$ miles

   c. $[(\pi - 1.2) + 2k\pi]$ miles, where k is a positive integer

   d. $\sin d = 0.9798$

   e. $(x, y) = (0.9798, -0.2)$

   f. Along the circle, he is $\left(\frac{\pi}{2} - 0.4\right)$ miles from start.

   g. Maggie's coordinates are $(\cos 0.4, \sin 0.4) = (0.9211, 0.3894)$. Michael's coordinates are $(0.3894, 0.9211)$.

EXERCISE SET 5.5, *pages 335–336*

1. If x represents the length of an arc on the unit circle, then moving the same arc length clockwise or counterclockwise represents x or $-x$. In both cases, the x-coordinates of the terminal point have the same value. Therefore, $\cos(-x) = \cos x$, and the cosine function is an even function.

3. $\cos\frac{\pi}{6} = \frac{\sqrt{3}}{2}$; $\sin\frac{\pi}{6} = \frac{1}{2}$

5. a. $\frac{\sqrt{2}}{2}$

   b. $-\frac{1}{2}$

   c. $-\frac{\sqrt{3}}{2}$

   d. $-\frac{1}{\sqrt{3}}$

   e. $\frac{\sqrt{2}}{2}$

   f. undefined

   g. $\frac{1}{\sqrt{3}}$

   h. -1

   i. $-\sqrt{2}$

7. a. $\frac{9}{25}$

   b. $\frac{4\sqrt{34}}{25}$

   c. $\frac{9}{4\sqrt{34}}$

   d. $-\frac{9}{4\sqrt{34}}$

9. $t = 0.5548 + 2k\pi$ or $t = -0.5548 + 2k\pi$, where k is an integer

11. $\pi + 0.9908 + 2\pi k$ where k is an integer

13. a. $-\sqrt{1 - w^2}$

   b. $\frac{-\sqrt{1 - w^2}}{w}$

   c. $-\sqrt{1 - w^2}$

   d. $\frac{-1}{w}$

   e. w

   f. $\frac{w}{\sqrt{1 - w^2}}$

EXERCISE SET 5.6, *pages 346–349*

1.

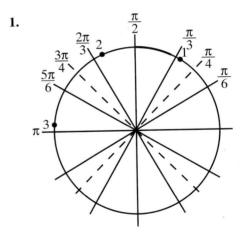

3. a. $\frac{3\pi}{2}$

   b. $\frac{3\pi}{4}$

   c. $\frac{4\pi}{3}$

   d. $\frac{2\pi}{5}$

   e. $\frac{5\pi}{9}$

   f. $-\frac{5\pi}{36}$

   g. $\frac{\pi}{180}$

5. $b = 51.2$

$A = 34.4°$

$B = 55.6°$

7. 3.9 meters; 20.9°

9. 12.7°

11. 54.74°

13. In radian mode, $\cos 1 = 0.5403$. In degree mode, $\cos 1 = 0.9998$.

15. a. $(0.5402, 0.8416)$

b. 1 radian or 57.3°

17. 25π or 78.54 feet

19. 1.6 radians

21. a. $f(0) = 2$

$f(\pi) = 0$

$f(2\pi) = -2$

b. The domain includes all real numbers. The range is $-2 \le f(t) \le 2$.

c. The period is 4π. As the person walks around the circle, she eventually retraces her steps.

e. $f(t) = 2\cos\left(\frac{1}{2}t\right)$

23. $f(t) = r \cdot \cos\left(\frac{t}{r}\right)$ and $g(t) = r \cdot \sin\left(\frac{t}{r}\right)$

EXERCISE SET 5.7, *pages 355–357*

1. Answers may vary. Several are: $-17.2496, -7.8248, 1.6, 11.0248, 20.4496, 29.8743$.

3. a. $x = 0.7227 + 2k\pi$ or $-0.7227 + 2k\pi$

b. $x = 1.4115 + 2k\pi$ or $3.7301 + 2k\pi$

c. No solution

d. $x = 0.1007 + k\pi$ or $1.4701 + k\pi$

e. $x = 0.8453 + k\pi$ or $0.2018 + k\pi$

f. $-2.7677 + 2k\pi < x < 1.1969 + 2k\pi$

g. $-1.4596 + k\pi < x < -0.4292 + 2k\pi$

5. a. 0.2 meter

b. $\frac{1}{4}$ second

c. $t = \frac{3}{8}$

d. $t = \frac{5}{8}$

7. $t = 0.4493 + 2.8k$

9. a. $x = \frac{\pi}{6} + k\pi$ or $\frac{5\pi}{6} + k\pi$

b. $x = \frac{\pi}{3} + 2k\pi$ or $-\frac{\pi}{3} + 2k\pi$ or $\pi + 2k\pi$

c. $x = \frac{k\pi}{2}$ or $0.1699 + k\pi$ or $1.4009 + k\pi$

d. $x = -0.6235 + k\pi$ or $-1.2256 + k\pi$

e. $x = \frac{\pi}{2} + k\pi$ or $1.3181 + 2k\pi$ or $-1.3181 + 2k\pi$

f. $x = 1.1765$ or -1.1765

g. $\frac{\pi}{4} + 2k\pi < x < \frac{5\pi}{4} + 2k\pi$

h. $x = k\pi$

i. $x = 0.9477$ or -0.9477

j. $x = \pi k$, where k is an integer

11. a. $y = \cos\left(\frac{\pi}{6.2}t\right) + 4$

b. 3.1 hours or 186 minutes before and after noon.

c. 4 hours and 8 minutes before and after noon. or 7:52 A.M. to 4:08 P.M.

d. The ship must leave port by 3:53 P.M.

EXERCISE SET 5.9, *pages 363–365*

1. a. $\frac{-\sqrt{2} - \sqrt{30}}{8}$

b. $\frac{7}{8}$

c. $-\frac{\sqrt{15}}{8}$

d. $\frac{\sqrt{30} - \sqrt{2}}{8}$

2. a. $0, \frac{2\pi}{3}, \frac{4\pi}{3}, 2\pi$

b. $0, \frac{\pi}{6}, \frac{\pi}{2}, \frac{5\pi}{6}, \pi, \frac{7\pi}{6}, \frac{3\pi}{2}, \frac{11\pi}{6}, 2\pi$

c. $0, \frac{\pi}{4}, \pi, \frac{5\pi}{4}, 2\pi$

d. $0, \frac{\pi}{2}, \pi, \frac{3\pi}{2}, 2\pi$

e. $0, \frac{\pi}{3}, \frac{2\pi}{3}, \pi, \frac{4\pi}{3}, \frac{5\pi}{3}, 2\pi$

f. No solution

g. $\frac{\pi}{2}$

3. a. $2\pi k$, where k is an integer

b. $x = \frac{3\pi}{4} + \pi k$ or $0.5404 + \pi k$, where k is an integer

c. $x = 0.7227 + 2\pi k,\ 2.4189 + 2\pi k,\ 5.5605 + 2\pi k,$ or $3.08642 + 2\pi k,$ where k is an integer

d. $0.1148 + \frac{2}{7}\pi k$ or $0.3340 + \frac{2}{7}\pi k,$ where k is an integer

e. $x = \pi k$ or $\frac{\pi}{6} + 2\pi k,$ or $\frac{5\pi}{6} + 2\pi k,$ where k is an integer

5. a. $\dfrac{\sqrt{3}}{3}$

   b. $\dfrac{\sqrt{6}}{3}$

9. a. QP: $d = \sqrt{(\cos A - \cos B)^2 + (\sin A - \sin B)^2}$

      SR: $d = \sqrt{(\cos(A - B) - 1)^2 + (\sin(A - B) - 0)^2}$

   b. $\cos(A - B) = \cos A \cos B + \sin A \sin B$

EXERCISE SET 5.10, *pages 370–372*

1. a. $\sin^{-1}(0.95) = 1.2532$

   This value is the angle measure in radians between $-\frac{\pi}{2}$ and $\frac{\pi}{2}$ whose sine is 0.95.

   b. $\cos^{-1}(-0.67) = 2.305$

   This value is the angle measure in radians between 0 and π whose cosine is -0.67.

2. a. $\sin(\sin^{-1}(0.9)) = 0.9$

   b. $\sin^{-1}(\sin(2)) = 1.1416$

5. a. Domain: $x \le -1$ or $x \ge 1$

   Range: $-\frac{\pi}{2} \le y < 0$ or $0 < y \le \frac{\pi}{2}$

   b. $-\frac{\pi}{2}\csc^{-1}x = \sin^{-1}\left(\frac{1}{x}\right)$ only if you choose the range for $\csc^{-1}x$ to be $-\frac{\pi}{2} \le y < 0$ or $0 < y \le \frac{\pi}{2}$.

7. a. Check graph on calculator. Suggested viewing window: $-1.2 \le x \le 1.2$ and $-2 \le y \le 2$

   b. Check graph on calculator. Suggested viewing window: $-10 \le x \le 10$ and $-4 \le y \le 4$

   c. Check graph on calculator. Suggested viewing window: $-2 \le x \le 2$ and $0 \le y \le 5$

   d. Check graph on calculator. Suggested viewing window: $-3 \le x \le 2$ and $-2 \le y \le 2$

8. a. $f^{-1}(x) = \frac{1}{2}\sin x$

   Domain: $-\frac{\pi}{2} \le x \le \frac{\pi}{2}$

   Range: $-\frac{1}{2} \le y \le \frac{1}{2}$

9. Domain of $y = \cos^{-1}x + \sin^{-1}x$ is $-1 \le x \le 1$.

EXERCISE SET 5.12, *pages 381–382*

1. a. $\dfrac{\sqrt{15}}{4}$

   b. $-\sqrt{3}$

   c. $\dfrac{4\sqrt{2}}{9}$

5. a. D: $-1 \le x \le 1$; R: $-1 \le y \le 1$

   Check graph on calculator. Suggested viewing window: $-2 \le x \le 2$ and $-2 \le y \le 2$

   b. D: all real numbers; R: $0 \le y \le \pi$

   Check graph on calculator. Suggested viewing window: $-10 \le x \le 10$ and $-2 \le y \le 5$

   c. D: all real numbers; R: all real numbers

   Check graph on calculator. Suggested viewing window $-10 \le x \le 10$ and $-10 \le y \le 10$

   d. D: $\{x\colon x \ne (2k + 1)\frac{\pi}{2},$ where k is an integer$\}$; R: $-\frac{\pi}{2} < y < \frac{\pi}{2}$

   Check graph on calculator. Suggested viewing window $-5 \le x \le 5$ and $-5 \le y \le 5$; use dot mode.

7. For x-values $2\pi k \le x \le \pi + 2\pi k$, $\cos^{-1}(\cos(x)) = x - 2\pi k$, for k an integer. For x-values $\pi + 2\pi k \le x \le 2\pi + 2\pi k$, $\cos^{-1}(\cos(x)) = -x + 2\pi(k + 1)$, for k an integer. If x is a value for which $\pi + 2\pi k \le x \le 2\pi + 2\pi k$, its corresponding cosine value will occur at $2\pi - x$.

9. Maximum when $x \approx 36.9$ meters from the taller building.

11. a. $x = \dfrac{\sqrt{5}}{5}$

   b. $x = \dfrac{3}{5}$

   c. No solution

EXERCISE SET 5.13, *pages 393–395*

1. a. $b = 7.8201$

   $C = 51.5976°$

   $A = 78.4024°$

b. $B = 90°$

   $c = 10.3923$

   $C = 60°$

c. There are two triangles.
   $C = 58.9973°$ or $C = 121.0027°$

   If $C = 121.0027°$ then $A = 28.9973°$ and $a = 6.7868$.

   If $C = 58.9973°$, then $A = 91.0027°$ and $a = 13.9979$.

d. $C = 120°$

   $A = 21.7868°$

   $B = 38.2132°$

3. 146.2965 meters

5. 20.8323 feet

7. 11.5370°

9. 3760 feet

11. 72.1399 m and 46.1752 m

13. 33.9049 feet

15. a. 26.9288°

b. 18.5365°

17. a. 69.1150 miles

b. 55.9152 miles

c. 52.9452 miles

d. 64.2720°

CHAPTER 5 REVIEW EXERCISES,
pages 398–399

1. a. $\dfrac{3\pi}{2}$

b. $\dfrac{3\pi}{2} + 2\pi k$, where k is any integer

2. $t = -\dfrac{16}{3}, -\dfrac{8}{3}, -\dfrac{4}{3}, \dfrac{4}{3}, \dfrac{8}{3}, \dfrac{16}{3}$

3. $y = 3 \sin\left(\dfrac{\pi x}{2}\right) + 1$

4. Think of the angles as the two nonright angles in a right triangle.

   The sum of the two angles is $\dfrac{\pi}{2}$. The angles must first be restricted to a domain between $-\dfrac{\pi}{2}$ and $\dfrac{\pi}{2}$. 5.70 is equivalent to $-.583$, and 7.2708 is equivalent to 0.988. The sum of the absolute values of these angles is $\dfrac{\pi}{2}$, so both students are correct.

5. 0.8203 radian

6. 38.7 miles

7. a. $-\dfrac{2\sqrt{2}}{3}$

b. $\dfrac{1}{3}$

c. $-\dfrac{4\sqrt{2}}{9}$

d. $\dfrac{7}{9}$

e. $\dfrac{-4 - \sqrt{2}}{6}$

8. a. $-0.98278 + k\pi$

b. $\dfrac{\pi}{3} + k\pi$ or $\dfrac{2\pi}{3} + k\pi$

c. $\dfrac{\sqrt{10}}{10}$

d. $0.6155 + 2k\pi,\ 2.526 + 2k\pi,$
   $-0.6155 + 2k\pi,\ -2.526 + 2k\pi$

9. Check graphs on calculator.

   a. Suggested viewing window: $-3 \leq x \leq 3$ and $-3 \leq y \leq 3$

   b. Suggested viewing window: $-2\pi \leq x \leq 2\pi$ and $-3 \leq y \leq 3$

   c. Suggested viewing window: $-2\pi \leq x \leq 2\pi$ and $-5 \leq y \leq 5$

   d. Suggested viewing window: $-\pi \leq x \leq \pi$ and $-2 \leq y \leq 2$

   e. Suggested viewing window: $-7 \leq x \leq 7$ and $-3 \leq y \leq 3$

10. a. $f^{-1}(x) = \cos\left(\dfrac{x}{2}\right) - 1$

   $D_{f^{-1}}: 0 \leq x \leq 2\pi$

   $R_{f^{-1}}: -2 \leq y \leq 0$

568   ANSWERS TO SELECTED EXERCISES

b. $f^{-1}(x) = \sin^{-1}(x - 1) + \frac{\pi}{4}$

$D_{f^{-1}}: 0 \le x \le 2$

$R_{f^{-1}}: -\frac{\pi}{4} \le y \le \frac{3\pi}{4}$

11. a. $\angle C = 68°$

$b = 3.75$

$c = 9.27$

b. There are two triangles.

$c = 4.35$ or $c = 18.63$

If $c = 4.35$, $C = 13.47°$ and $B = 126.53°$.

If $c = 18.63$, $C = 86.31°$ and $B = 53.69°$.

12. The sine definition is the y-value for a point on the unit circle, but also one side of a right triangle (hypotenuse 1) with that angle.

13. If the ratio of the two legs of a triangle $\left(\frac{\text{opposite}}{\text{adjacent}}\right)$ is 0.7, what is the angle?

14. Between about 7:08 A.M. and 4:52 P.M.

CHAPTER 6 Combinations of Functions

EXERCISE SET 6.1, *page 403*

2. Answers will vary. One reasonable model for the last two cycles is $f(x) = 0.6 \sin(0.82x + 2.65)$.

4. The size of the residuals indicates a reasonably good fit.

EXERCISE SET 6.3, *pages 416–420*

1. a. Use a window of $-12 \le x \le 12$ and $-9 \le y \le 9$. h has a vertical asymptote at $x = -2$ and an oblique asymptote $y = x - 1$.

c. Use a window of $-6 \le x \le 6$ and $-2 \le y \le 6$. The domain of h is $x > -4$. There is a vertical asymptote at $x = -4$.

3. a. Domain: $\{x: x \ne 5\}$

b. $k(x) > 0$ when $x > 5$. $k(x) < 0$ when $x < 5$. $k(x)$ is never 0.

c. As $x \to \infty$, $k(x) \to \infty$. As $x \to -\infty$, $k(x) \to 0$.

d. The graph of k is asymptotic to the x-axis as $x \to -\infty$. When $x > 5$, $k(x)$ values are very large and positive, but this part of the graph is not shown because of the viewing window.

5. a. Domain: $(x: x \ne 0\}$; there is a vertical asymptote at $x = 0$, and the x-intercepts are at $x = \frac{\pi}{4} + k \cdot \frac{\pi}{2}$. The graph oscillates within the envelopes formed by $y = \pm\frac{1}{x}$.

c. Domain: $\{x: x \ne 0\}$; asymptote at $x = 0$

7. The product of two even functions is even.

9. a. Use a window of $0 \le x \le 10$ and $-15 \le y \le 25$. Domain: $x > 1$; zero: $x = 2$; $k(x) > 0$ when $x > 2$. As $x \to \infty$, $k(x) \to \infty$. As $x \to -\infty$, $k(x)$ is undefined.

c. Use a window of $0 \le x \le 6$ and $-1 \le y \le 1$. Domain: $\Re$; zeros: $x = \frac{\pi}{3}k$, k an integer; $k(x) > 0$ when $0 + \frac{2\pi}{3}k < x < \frac{\pi}{3} + \frac{2\pi}{3}k$. As $x \to \infty$, $k(x) \to 0$. As $x \to -\infty$, $k(x)$ oscillates more and more wildly. Use a window of $-6 \le x \le 0$ and $-100 \le y \le 100$.

e. Look at two viewing windows: $0 \le x \le 5$ and $0 \le y \le 1000$, and $-5 \le x \le 0$ and $-1 \le y \le 1$. Domain: $\Re$; zero: $x = 1$, $x = -4$; $k(x) > 0$ when $x > 1$. As $x \to \infty$, $k(x) \to \infty$. As $x \to -\infty$, $k(x) \to -\infty$.

EXERCISE SET 6.6, *pages 429–430*

1. a. $y = (x + 3)(x - 9)$

c. $y = (x^2 + 3)(x + 3)(x - 3)$

2. a. $y = 2x^2 + 25x - 42$

c. $y = 3x^3 - 17x^2 - 8x + 112$

3. a. Graph using window $-20 \le x \le 5$ and $-150 \le y \le 50$; $(-\infty, -14) \cup (1.5, +\infty)$

c. Graph using window $-5 \le x \le 10$ and $-25 \le y \le 125$; $[-\frac{7}{3}, \infty]$.

5. a. 1

c. -4

EXERCISE SET 6.8, *pages 439–441*

1. $p(x) = k(x - a)(x - b)(x - c)(x - d)$, for some number k. This is a degree 4 polynomial.

3. One possible answer: $y = (x + 1)(x - 3)(x)$

5. Possible answers: $y = -1(x + 2)^2(x - 5)$ or $y = \frac{2}{5}(x + 2)(x - 5)^2$

7. $p(x) = -1(x - 2)(x + 3)(x - 1)(x + 4)$
$= -x^4 - 4x^3 + 7x^2 + 22x - 24$

9. a. $p(x) = -\frac{1}{32}(x + 4)(x + 1)^2(x - 4)$

  c. $p(x) = \frac{1}{36}(x + 3)(x + 1)^2(x - 2)^2(x - 4)$

11. $p(x) = -\frac{1}{3}x^3 + \frac{1}{2}x^2 + \frac{11}{6}x + 4$

EXERCISE SET 6.9, *pages 453–454*

1. Check graphs on calculator.

  a. Intercepts: $(1, 0)$, $\left(0, -\frac{1}{4}\right)$

  Asymptotes: $x = 2$, $y = 0$

  Window: $-5 \leq x \leq 5$ and $-10 \leq y \leq 10$

  c. Intercept: $(0, 0)$

  Asymptotes: $x = -2$, $x = 3$, $y = 0$

  Window: $-5 \leq x \leq 5$ and $-10 \leq y \leq 10$

  e. Intercepts: $(2, 0)$, $\left(0, \frac{2}{9}\right)$

  Asymptotes: $x = -3$, $x = 3$, $y = 0$

  Window: $-4.7 \leq x \leq 4.7$ and $-3.1 \leq y \leq 3.2$

  g. Intercepts: $(4, 0)$, $\left(0, \frac{4}{3}\right)$

  Asymptotes: $x = 3$, $y = 1$

  Hole at $\left(-2, \frac{6}{5}\right)$

  Window: $-4.7 \leq x \leq 4.7$ and $-2.1 \leq y \leq 4.2$

  i. Intercepts: $(-5, 0)$, $(4, 0)$, $(5, 0)$, $(0, -50)$

  Asymptotes: $x = -1$, $x = 2$, $y = 1$

  Window: $-6 \leq x \leq 6$ and $-60 \leq y \leq 60$

  k. Intercept: $(0, -2)$

  Vertical asymptote: $x = 2$

  Oblique asymptote: $y = x - 1$

  Window: $-2.7 \leq x \leq 6.7$ and $-4.2 \leq y \leq 8.4$

3. a. $f(x) = \frac{x^2 + 1}{x^2 - 2x + 1}$ has a horizontal asymptote of $y = 1$, and crosses at $x = 0$.

  b. $y = \frac{x^3 - 4x}{(x - 1)^2(x + 1)}$ has a horizontal asymptote of $y = 1$, and crosses at $x = \frac{3 + \sqrt{13}}{2}$.

5. $k(x) = \frac{-x^2 + 5x - 13}{x - 2}$

7. a. There are no x-intercepts. There is a y-intercept at 1.

  c. There are x-intercepts at $x = \frac{\pi}{2} \cdot n$ and a vertical asymptote at $x = \pi$.

EXERCISE SET 6.10, *pages 459–460*

1. a. The scatter plot is curved, increasing, and concave down, so the relationship could be square root, cube root, etc.

  b. Radius = (volume)$^{0.335}$ (0.553)

  c. $r = 6.801$

  d. $\frac{h}{2r} = \frac{h}{d} \approx 1$

3. 6.9 by 6.9 by 5.25

CHAPTER 6 REVIEW EXERCISES, *pages 464–465*

1. We can think of this as $f(x) = \sin 4x$ "wrapped around" $g(x) = \frac{1}{x - 1}$.

2. a. $D: \{x: x \neq 4\}$

  b. If $x > 4$, $k(x) > 0$. If $x < 4$, $k(x) < 0$.

  There is no x for which $k(x) = 0$.

  c. As $x \to +\infty$, $h(x) \to +\infty$.

  As $x \to -\infty$, $k(x) \to 0^-$.

3. $D = \mathfrak{R}$

4. a. $D: x > -3$

  b. $x = -2$ is the only zero.

c. $k(x) > 0$ if $x > -2$.

d. As $x \to \infty$, $k(x) \to 0^+$.

  As $x \to -\infty$, $k(x)$ is undefined.

5. $f(t) = \sin(20\pi t) + \sin(30\pi t) = 2\sin(25\pi t)\cos(5\pi t)$

6. One possible answer:
$$f(x) = \frac{1}{12}(x + 3)(x - 1)(x - 4)(x + 2)$$

7. $f(x) = -\frac{1}{6}(x + 3)^2(x - 4)$

  $g(x) = \frac{1}{8}(x + 3)(x - 4)^2$

8. One possible solution is $y = (x + 1)(x - 1)^2$. Another possible solution is $y = (x + 1)(x - 4)^4$.

9. a. x-intercept: 4; y-intercept: -1; vertical asymptote: $x = -2$; horizontal asymptote: $y = 0$

  b. This is just $y = \frac{1}{x}$ with a hole at $x = -3$. No intercepts; vertical asymptote at $x = 0$; horizontal asymptote at $y = 0$.

10.a. A rational function has a hole in the graph if both the numerator and the denominator have the same linear factor; the degree of this factor in the numerator is greater than or equal to the degree in the denominator.

  b. A rational function has a horizontal asymptote at $y = 2$ if the numerator and denominator are of the same degree and the coefficient of the highest degree term in the numerator is twice that of the highest degree term in the denominator.

  c. A rational function has a slant asymptote if the degree of the numerator is one more than the degree of the denominator.

CHAPTER 7 Matrices

EXERCISE SET 7.2, *pages 472–473*

1. a.
	t	f	n	r
H	5280	1680	2320	1890
P	1940	2810	1490	2070

b.
	t	f	n	r
H	6340	2220	1790	1980
P	2050	3100	1720	2710

c.
	t	f	n	r
H	11,620	3990	4110	3870
P	3990	5910	3210	4780

EXERCISE SET 7.3, *pages 479–484*

1. *MO*: 2×1, *MP*: 2×2, *PM*: 2×2, *MR*: 2×2, *RM*: 2×2, *NQ*: 3×1, *NU*: 3×4, *PO*: 2×1, *PR*: 2×2, *RP*: 2×2, *RO*: 2×1, *SM*: 4×2, *SO*: 4×1, *SP*: 4×2, *SR*: 4×2, *US*: 3×2, *UT*: 3×1

3. Yes

5. No, not necessarily

7.
1994	1995	1996
30,175	31,800	34,050

9. a.
t	ca	co	d
2810	1656	358	228

  b. 14 Budget, 16 Economy, 15 Executive, 9 President

  c. 36 employees

11.
	c	s	m
N	48,448	13,683	779
D	18,430	11,814	160
S	30,991	16,453	148

  Total
N	62,910
D	30,404
S	47,592

13. a.

	C	LT
R	15,110	9870
S	13,175	8600
AA	13,250	9130

b.

	Cost
R	10,577
S	9222.5
AA	9275

c.

	1st	C	Y
R	441	294	175
S	520	280	220
AA	525	277.5	112.5

15. a.

	Oct.	Nov.
Cut	1300	1520
Sew	1800	2092.5
Finish	1240	1362.5

b.

	East	Cent.	West
Panda	15.99	14.46	18.60
Kangaroo	18.36	16.64	21.22
Rabbit	11.70	10.60	13.61

c. $36,366

d.

	Cut	Sew	Finish
East	29.3	41	27
Central	27	35.5	21.5
West	28	38.5	25.5

e. East Coast will pay $815.55; Central will pay $636.40; West will pay $894.45.

EXERCISE SET 7.4, *page 494*

1. The transformed points have coordinates $(2, -1)$, $(1.5, -2)$, and $(3, -4)$.

3. The transformed points have coordinates $(-4.34, 4.57)$, $(-7.63, 3.90)$, and $(-12.26, 9.80)$.

EXERCISE SET 7.5, *pages 506–508*

1. a.

$T =$

	Petr.	Text.	Tran.	Chem.
Petroleum	0.1	0.4	0.6	0.2
Textiles	0	0.1	0	0.1
Transportation	0.2	0.15	0.1	0.3
Chemicals	0.4	0.3	0.25	0.2

b. Petroleum is most dependent on chemicals, least dependent on textiles.

c. $1,600,000

d.

Petroleum	730
Textiles	95
Transportation	485
Chemicals	705

e.

Petroleum	195.6
Textiles	39.3
Transportation	154.4
Chemicals	213.3

3.

Manufacturing	579.25
Petroleum	572.31
Transportation	476.21
Hydroelectric Power	464.83

EXERCISE SET 7.6, *pages 510–511*

2. a. $x = 1, y = 1, z = 1$

3. a.

$$\begin{bmatrix} \frac{11}{70} & -\frac{3}{70} & \frac{17}{70} \\ -\frac{1}{14} & -\frac{1}{14} & \frac{1}{14} \\ \frac{13}{350} & -\frac{29}{350} & \frac{1}{350} \end{bmatrix}$$

b.

$$\begin{bmatrix} 10 & -20 & 15 & -4 \\ -\frac{47}{6} & 19 & -\frac{31}{2} & \frac{13}{3} \\ 2 & -\frac{11}{2} & 5 & -\frac{3}{2} \\ -\frac{1}{6} & \frac{1}{2} & -\frac{1}{2} & \frac{1}{6} \end{bmatrix}$$

5.

$$\begin{bmatrix} \cos\theta & \sin\theta \\ -\sin\theta & \cos\theta \end{bmatrix}$$

This matrix rotates points through an angle θ in the clockwise direction.

EXERCISE SET 7.7, *pages 521–522*

1. 27.9304 after 5 years; 50.866 after 10 years

3. a. Between 56 and 57 quarters

   b. Between 66 and 67 quarters

   c. Between 69 and 70 quarters

5. First initial distribution:

	Proportion
0–3	0.324597
3–6	0.189007
6–9	0.165082
9–12	0.144186
12–15	0.111942
15–18	0.065200

Second initial distribution:

	Proportion
0–3	0.324598
3–6	0.189007
6–9	0.165083
9–12	0.144187
12–15	0.111943
15–18	0.065200

The long-term proportions of the population in each age group seem to be independent of the initial age distribution.

EXERCISE SET 7.8, *pages 536–540*

1. a.

$$\begin{array}{c} \\ 1 \\ 2 \\ 3 \\ 4 \\ 5 \end{array} \begin{array}{ccccc} 1 & 2 & 3 & 4 & 5 \\ \begin{bmatrix} 0 & \frac{1}{2} & 0 & 0 & \frac{1}{2} \\ \frac{1}{2} & 0 & 0 & 0 & \frac{1}{2} \\ 0 & 0 & 0 & 1 & 0 \\ 0 & 0 & \frac{1}{2} & 0 & \frac{1}{2} \\ \frac{1}{3} & \frac{1}{3} & 0 & \frac{1}{3} & 0 \end{bmatrix} \end{array}$$

b. $0; \frac{1}{12}$

c.

$$\begin{array}{ccccc} 1 & 2 & 3 & 4 & 5 \end{array}$$
$$\begin{bmatrix} 0.2 & 0.2 & 0.1 & 0.2 & 0.3 \end{bmatrix}$$

d. 0.2

e. 30% of the time

3. a.

$$\begin{array}{c} \\ OR \\ FF \\ CC \end{array} \begin{array}{ccc} OR & FF & CC \\ \begin{bmatrix} 0.5 & 0.35 & 0.15 \\ 0.3 & 0.4 & 0.3 \\ 0.2 & 0.55 & 0.25 \end{bmatrix} \end{array}$$

b. 40 orders of french fries

c. 33 orders of onion rings, 40 orders of fries, and 22 cookies

d. 33 orders of onion rings, 40 orders of fries, and 22 cookies each day

e. 34.5% onion rings, 41.8% french fries, and 23.6% cookies

7.

$$T = \begin{array}{c} Democrat \\ Republican \end{array} \begin{array}{cc} \textit{Dem.} & \textit{Rep.} \\ \begin{bmatrix} \frac{7}{11} & \frac{4}{11} \\ \frac{5}{13} & \frac{8}{13} \end{bmatrix} \end{array}$$

In the election in 2000, there is a $\frac{7}{11}$ chance of a Democrat being elected and a $\frac{4}{11}$ chance of a Republican. In the election in 2008, the probability is about 52% of a Democrat being elected and about 48% of a Republican being elected.

9. b. If the ant starts at corner A and makes a large, even number of moves, there is about a 28% chance it will be in corner A and a 72% chance it will be in corner B.

CHAPTER 7 REVIEW EXERCISES,
pages 541–542

1. a. If a man's great grandfather was in the service-oriented job, the probability that the man is in a service-oriented job is 0.333544.

b. We assume that the percents do not change over long periods of time.

c. In 10 generations, about 48% of the population will be professionals.

2. a. The matrices must have the same dimensions.

b. We can calculate the product AB if the number of columns in A is equal to the number of rows in B.

3. Their product is the identity matrix.

4.

$$\begin{array}{c} Oil \\ Agriculture \\ Electricity \\ Machinery \\ Chemical \end{array} \begin{bmatrix} 47.87 \\ 50.44 \\ 34.72 \\ 34.48 \\ 43.33 \end{bmatrix}$$

5. a. 21 years

b.
$$\begin{bmatrix} 0 & 0.5 & 1.3 & 0.9 & 0.6 & 0.1 & 0 \\ 0.5 & 0 & 0 & 0 & 0 & 0 & 0 \\ 0 & 0.8 & 0 & 0 & 0 & 0 & 0 \\ 0 & 0 & 0.9 & 0 & 0 & 0 & 0 \\ 0 & 0 & 0 & 0.9 & 0 & 0 & 0 \\ 0 & 0 & 0 & 0 & 0.7 & 0 & 0 \\ 0 & 0 & 0 & 0 & 0 & 0.5 & 0 \end{bmatrix}$$

c. Left-multiply the column vector of initial age distribution by L^5 to obtain approximately:

$$\begin{bmatrix} 97 \\ 46 \\ 29 \\ 28 \\ 32 \\ 7 \\ 6 \end{bmatrix}.$$

Total is 245.

6. 537 student tickets; 313 adult tickets; 167 senior citizen tickets

Index

A

Absolute value 41, 105
Absolute value function 68–69
Absorbing Markov chain 535
Absorbing state 535
Age distribution vector 515
Age-specific population 467, 512
Age-Specific Population Data 467
AIDS Research Project 261
Aircraft Data 19
Ambiguous case 390
Amortization 167
Amplitude 303, 424
Angle 337–349
Angle of elevation 383
Annuity 183
Annuity Problem 183–184
Anscombe's Data 48
Arccosine 332
Areas of Triangles Data 247
Argument 58, 98, 99, 105, 351
Associative property (matrices) 479
Asymptote 67, 97
Asymptotic 66, 443
Augmented matrix 502

B

Bakery Problem 86
Ball off the Roof Data 110
Balloon payment 168
Ball Toss Data Collection Problem
    147–148
Base, change of 222–225
Baseline 409
Base of an exponential function 197, 222
Beats 424–425
Beats Investigation 424–425
Binary search 169–170
Binary Search Problem 169–170
Birthrate 512
Bison Problem 274
Body Proportions Data Collection
    Problem 20–21
Bouncing Ball Data Collection
    Problem 22
Box Volume Example 79
Braking Distance Data 462
Branches 524

C

Camera Lens Problem 89
Cane Toad Problem 179
Carbon-14 dating 225
Car Sales Problem 468
Change of base 222
Chernobyl disaster 227
Cherry Blossom Data 31
Cherry Blossom Problem 30–32, 49
Chicago Bulls Data 43, 44
Chicago Bulls Problem 43–45
Circles Data Collection 22
Circular functions 325
Closed form 160
Closed interval 76
Coco, the Wonder Cat Problem 155
Coefficient matrix 503
College Costs Research Project 260
Columns 468, 477
Column vector 478
Common logarithm 208
Common ratio 182
Commutative property
    (matrices) 479
Complementary angles 383
Completing the square 108, 134
Composition 132
Composition of functions 112–116,
    119, 133, 376
Composition of Functions
    Investigation 117
Compound interest 190–196
Computer graphics 485–494
Computer Graphics Problem 485
Computer Problem 279–288
Computer Rating Data 279
Concave down 62
Concave up 62, 65
Cone 93
Conical Container Investigation 93–94
Consistent (system of equations) 504

Constant function 64, 85
Continuous 57, 62, 65, 175
Continuous compounding 193
Continuous function 62
Continuous graph 57, 65
Cooling Data 39, 243
Cooling Data Collection Problem
    256–257
Cosecant function 309
Cosine 323–326
Cosine function 303
Cotangent function 309
CO_2 Concentration Data 422
CO_2 Concentration Problem 421–423
Cricket Chirp Data 47, 248
Cubic function 65
Curie, Marie 222
Curvature 13, 62, 65, 138
Cutting Board Problem 474–475
Cycle 67, 99, 304

D

Data analysis 1–53, 10, 16, 56, 142,
    235–250, 263
Data collection investigations 20–24
Death rate 512
Decay rate 174
Decibels 231
Decomposing functions 120
Decrease without bound 124, 413, 428,
    435
Decreasing 62
Degree mode 340
Degree of a polynomial 429, 431, 433,
    447, 448, 451
Demand matrix 498, 505
Dependent (system of equations) 504
Dependent variable 5, 56, 61, 129
Dice Data Collection Problem 256
Dimension (matrix) 468, 472
Discontinuous 57, 66, 72
Discontinuous graph 57
Discrete mathematics 469
Discrete points 175
Displacement 313

Distance 285
Distribution of Funding
    Requests Problem 509
Dog Food Data 144
Domain 61, 99, 113, 129
    restrictions 73, 133, 176, 368
Double Ferris Wheel Problem 373–375
Doubling time 224
Doubling Time Problem 176–177

E

Effective annual yield 194
Elementary row operations (ERO)
    500-501
Elevator Problem 297
Eliminate the parameter 152
Embedded right triangle 343
Empirical model 290
Entries 468
Envelopes 411, 424
Equilibrium 62, 63, 163, 313
Error bounds 49–53, 251–253
Euler, Leonhard 192
Even function 70, 304, 411
Even multiplicity 436
Event space 269
Explicative model 290
Explicit expressions 160
Explicit function 174
Exponential decay 198
Exponential equations 220–228
Exponential function 66, 159–265, 193
    graphing 197–206
Extrapolate 15

F

Factors 73, 429
Factor Theorem 434, 441
Falling Object Data 39
Fitting a linear model 16
Floor function 72
Focal length 89
Focal Length Data 111, 262
Focal Length Problem 262–263
Free-Fall Data 139
Free Throw Problem 275–280
Frequency 319, 424
Function 56–157
    argument 58

basic transformation 97–103
    characteristics 61–63
    composition 112–116
    definition 58
    domain 73–78
    mathematical models 56–60, 79–92
    notation 7, 95
    sums and products 404–420
    toolkit 64–72
    value 58
Future value 190

G

Garbage Disposal Problem 188–189
General linear function 65
Geometric growth 173–179
    models 173–179
    summing 182–187
Geometric probability 271
Geometric series 182
George Washington Gale Ferris 301
Global behavior of a function 75,433
Global view 83
Graph 7
Greatest integer function 72, 103
Great Lakes Problem 298
Great Pyramid of Cheops 393
Growth rate 174

H

Half-life 179, 224
Half-open interval 76
Hanging Picture Problem 396–397
Hat Data Collection 153
Hat Problem 149–150
Hawks and Doves Problem 321–322
Heading 339
Heating Bill Data 17
Hertz (Hz) 424
Hobby Shop Problem 470–471
Homogeneous coordinates 487–494
Horizontal
    asymptote 199, 214
    axis 5, 35, 57
    component 150
    compression 100, 105
    stretch 106, 108
    translation 98
Hunger threshold 180

I

Identity 325, 358
Identity function 65, 132, 376
Identity matrix 478, 499, 500
Inclined Plane Data 142
Inconsistent 503
Increasing 57, 65
Increase without bound 124, 192 307,
    413, 428, 435
Increment 151
Independent variable 5, 35, 56, 61, 129
Index Card Data Collection Problem 23
Inequality 74, 85, 106
Infinity symbol (∞) 76
Initial distribution 531
Initial value 160
Inner function 112
Input 58, 131, 495
Intensity and Loudness Data 231
Intercept 97
Interest rate 167
Internal consumption matrix 498, 506
Interpolate 15
Interval 74
    closed 75
    estimate 50
    half-open 76
    notation 76
    open 76
Inverse cosecant 369
Inverse cosine 332, 367, 368
Inverse cotangent 369
Inverse function 128, 132, 207,
    368–372, 376–382
Inverses 128–137
    to straighten curves 138–145
Inverse secant 369
Inverse sine 367
Inverse tangent 368
Invertible (matrix) 503
Iterations 163
Iterative equations (process) 160, 170,273

J

Joan's Knee Problem 162

K

Kepler, Johann 249
Kepler Data 249
Kepler's Third Law 249

L

Lattice points 276
Law of cosines 385
Law of sines 387
Laws of exponents 210
Laws of logarithms 210
Leaning Tower of Pisa Data 13
Leaning Tower of Pisa Problem 12–13, 33
Learning curves 200
Least squares line 40–48
    versus median-median line 42
Least squares principle 41–42
Left-multiply 479
Leontief, Wassily 495
Leontief Input-Output Model 495–508
Leslie matrix model 512–522
Levels of Brightness Data 234
Light Intensity Data 246
Light Intensity Data Collection
    Problem 258
Limiting magnitude 234
Limiting value 191, 192
Linear combination 504
Linear factor 427, 429, 437
Linear function 65
Linear Irrigation Problem 299
Linearize data 138, 238
Linear model 8, 43
    fitting a 16
Loan Amortization Problem 167
Loan Payment Problem 184–185
Loans 167–172
Loan Repayment Problem 169–170
Local behavior
    of a function 433
    of a graph 74
Local view 83
Logarithmic equations 220–228
Logarithmic functions 207–265
Logarithmic graphs characteristics 215
Logarithmic scales 229–232
Logarithms 207–213
Logistic growth 206
Log-log plot 242
Log-log re-expression 242
Long division 434, 444
Long-term distribution 531
Long-term growth rate 519

M

Macromodels 512
Magnitude 234
Main diagonal 478
Manatee Data 52
Mantid Data 180
Mantid Problem 159, 180–181
Mapping 58
Markov chains 523–540
Markov processes 525
Mathematical model 4–9, 15, 56, 245,
    268, 295–296, 461
Mathematics Enrollment Data 229
Matrices 468–542
Matrix 468
Matrix addition 470–473
Matrix equation 498
Matrix multiplication 476, 478
Matrix subtraction 472
Mauna Loa Observatory 421
Maximum 8, 67, 80, 81, 302, 436
Measuring Circles Data Collection
    Problem 22
Median-fit line 25
Median-median line 25–34, 44
    versus least squares line 42
Median x-value 26
Median y-value 26
Medical Costs Research Project 260
Megan Bisk Problem 233
Micromodels 512
Midge Data 298
Midge Problem 298
Mile Run Data 34
Minimum 67, 302, 436
Model 2–9
Mortgage Payments Problem 84–85
Mount Everest 459
Multiplicative inverse 499
Multiplicity 434

N

Natural logarithm 207
Negative association 12
Negative residuals 35
Nonlinear data 138
Nonabsorbing state 535

O

Oblique asymptote 446
Oblique triangles 384
Odd function 70, 309, 358
Odd multiplicity 435–436
One-to-one function 131, 209, 368
Open Box Problem 79–81
Open interval 76
Ordered pairs 5, 10, 56, 61
Oscillate 62, 67
Outer function 112
Outlier 14, 25, 36, 42, 44, 45
Output 58, 131, 495
Overhead Projector Data Collection
    Problem 21

P

Paper-and-pencil graphing 64, 97
Parabola 7, 65, 86, 108
Parameter 150
Parametric equations 150, 373
Parametric mode 151, 373
Pendulum Problem 312
Pendulum Swing Data Collection
    Problem 23, 146–147
Penny Data Collection Problem 21–22
Period 67, 305, 424
Periodic (motion) 63, 67, 313
Phase portrait 321
Phase shift 315
pH 230, 457
pH Data 460
pH Problem 457–458
Piecewise-defined function 68
Pogson, Norman 234
Point of intersection 85
Polynomial function 426–441, 429
Polynomial Investigation 431–432
Population Density Data 235
Population Research Project 259–260
Positive association 13
Positive linear relationship 26
Positive residuals 35
Postage Rate Data 59
Power function 69, 235
Predator-Prey Problem 321–322
Prediction 49–53
Present value 194

Pressure Data Collection Problem 147, 257–258
Principal 167
Principle square root 65–66
Probability 268, 269
Probability of transition 526
Production matrix 496
Products of functions 411
Products of Functions Investigation 405
Pythagorean Theorem 47, 82, 378

Q

Quadratic factor 435
Quadratic formula 434, 458
Quadratic function 8, 65, 86

R

Radian 337–349
Radian mode 67, 326, 341
Radioactive Chain Problem 273–274
Radioactive Contamination Data 25, 40
Radioactive Contamination Problem 25–30
Range 61, 99, 113, 129
Rational functions 442–460
Rebound Height Data Collection Problem 255–256
Reciprocal function 66, 123
Recursive system 161
Re-express data 138, 238, 402
Reflect about x-axis 104
Reflect about y-axis 104
Reflect about y = x 129, 207
Refrigerator Cooling Data 402
Regression line 42
Regular Markov chain 534
Relation 57
Relationship between variables 5, 56, 64
Residual 35
Residual plot 36, 49, 141, 237, 403
Resistant line 25, 42
Richter scale 231
Richter Scale Problem 232–233
Right-multiply 479
Right triangle trigonometry 377, 383
Road Map Data Collection Problem 24
Rotate in the plane 485, 487, 493
Rotation matrix 486, 493

Rotations in two dimensions 485–487
Row 468, 477
Row vector 478
Rule of 69 227

S

Sample space 269
SAT Data 10–11
Scalar 471
Scalar multiplication 470–473
Scaling 489, 493
Scaling operations 489
Scatter plot 5, 6, 12
    to analyze data 10–19
Secant 348
Secant function 309
Seismic Reflection Problem 81–84
Semi-log plot 237
Semi-log re-expression 237
Shift 98, 107
Short-term growth rate 519
Sign-Line Analysis 74, 114, 445
Simplifying assumptions 5
Sine 323–326
Sine function 67
Sinusoid 303
Sinusoidal function 110
Slant asymptote 446
Slope 27, 29, 42
Solving triangles 383–395
Sound 231
Sound waves 424
Speed and Skid Length Data 247
Spring Data Collection Problem 147
Square matrix 468
Square root function 65–66
Stable distribution 520
Stable state matrix 533
Stable state vector 532
Standard deviation 50
State diagram 495, 523
State vector 531
Step function 72
Stochastic process 530
Stopping Distance Data 111
St. Paul's Cathedral Bells Data 254
Strong relationship 16
Subscripts 160

Summary point 26
Sums of functions 408–420, 456
Sums of Functions Investigation 404
Sunspot Problem 357
Surface Area of a Can Problem 455
Survival rate 513
Swimming Pool Problem 164
Swing Problem 320
Symmetric about a point 70, 309
Symmetric about the y-axis 70, 304

T

Table of Transformations 101
Table of values 80, 85
Tangent 348
Tangent function 309
Tape Counter Data 289
Tape Counter Problem 288–294
Tape Erasure Problem 268–274
Taxi Problem 523
Tax Rate Data 59
Technology matrix 496, 505
Ticket Data 4
Ticket Problem 4–9
Tidal River Problem 357
Tim and Tom Problem 186
Toolkit functions 64–72, 95, 104
Traffic Flow Problem 461–463
Transformation 69, 95
    combinations of 104–111
    equations 486
Transformed data 140
Transforming Graphs Investigations 95–96
Transition matrix 468, 523, 526, 529
Translation 98, 487, 493
Tree diagram 524
Tree Height Data 51
Tree Volume Data 111, 238
Triangle
    solving 383–395
    (30-60-90) 336
Trigonometric equations 352–357
Trigonometric function 303
Trigonometric identities 358–359, 362–365, 486
Trigonometry
    of angles 337–349
    curves of 302–312